SORPTION AND FILTRATION METHODS FOR GAS AND WATER PURIFICATION

NATO ADVANCED STUDY INSTITUTES SERIES

Proceedings of the Advanced Study Institute Programme, which aims at the dissemination of advanced knowledge and the formation of contacts among scientists from different countries.

The series is published by an international board of publishers in conjunction with NATO Scientific Affairs Division

A	Life Sciences	Plenum Publishing Corporation
B	Physics	London and New York
C	Mathematical and Physical Sciences	D. Reidel Publishing Company Dordrecht and Boston
D	Behavioural and Social Sciences	Sijthoff International Publishing Company Leyden
E	Applied Sciences	Noordhoff International Publishing Leyden

Series E: Applied Sciences
Volume 13 — Sorption and Filtration Methods
for Gas and Water Purification

SORPTION AND FILTRATION METHODS FOR GAS AND WATER PURIFICATION

edited by

M. BONNEVIE-SVENDSEN

NOORDHOFF - LEYDEN - 1975

Proceedings of the NATO Advanced Study Institute
on Sorption and Filtration Methods for Gas and Water Purification
Fauske, Norway
June 23-29, 1974

ISBN-13: 978-94-010-1905-7   e-ISBN-13: 978-94-010-1903-3
DOI: 10.1007/978-94-010-1903-3

PREFACE

This book contains the papers presented at the NATO Advanced Study Institute on "Scientific Aspects of Sorption and Filtration Methods for Gas and Water Purification". The Study Institute was held at Fauske Hotel, Fauske, a small town in the northern part of Norway, 23rd-29th June, 1974. The members of the Scientific Advisory Committee were:

T. Halmø   The Engineering Research Foundation at
                  the Norwegian Institute of Technology,
                  Trondheim, Norway

W.H. Hardwick   AERE, Harwell, Didcot, Berks., U.K.

B. Ottar   Norwegian Institute of Air Research,
                  Kjeller, Norway

J.A. Wilhelm   Karlsruhe Nuclear Research Center,
                  Karlsruhe, Germany

R. Berg   Institutt for Atomenergi, Kjeller, Norway

The members of the Organizing Committee were:

E. Andersen
M. Bonnevie-Svendsen
G. Jarrett

all from Institutt for Atomenergi, Kjeller, Norway.

The Advanced Study Institute was financially sponsored by the NATO Scientific Affairs Division.

The aim of the Study Institute was to bring together scientists concerned with fundamental aspects of sorption, solid state physics and reaction kinetics and workers who are occupied with the development of filter systems for controlled and efficient removal of impurities and poisons from air, off-gases, potable water and industrial effluents. The papers presented covered both theoretical and practical aspects of sorption and membrane filtration. The emphasis was on factors which may effect the filter efficiency, on evaluation, optimalization, controlled development of "tailor made" systems, mathematical models and

economical consideration. The publication of these lectures
was made possible through the kind cooperation of the lecturers.

   The number of participants was sixty. Their names are
given at the end of the book.

   We wish to express our gratitude to all those who have
contributed to the realization of the Study Institute. Special
thanks are due to our friends in Fauske, who, under the excellent
leadership of Mr. Storm Halvorsen, extended their hospitality
to us and helped with practical arrangements, accomodation and
social events, including fishing on the fjord under the golden
midnightsun.

Kjeller, September 1974   The Organizing Committee

TABLE OF CONTENTS

THE SCOPE OF THE NATO ADVANCED STUDY INSTITUTE

Moj Bonnevie-Svendsen

Institutt for Atomenergi
Kjeller, Norway.

In recent years the increasing pollution problems and growing concern for environmental protection has imposed more stringent demands on off-gas and effluent purification, on methods for control and supply of clean air and drinking-water. In this connection much effort has been directed towards the improvements of technical installations, of mechanical lay-out and performance. A considerable number of conferences has dealt with these topics.

There is also an increasing trend to attack the problem from a more scientific level and to utilize fundamental data from thermo-dynamics and solid state physics for studies of reaction mechanisms and for the controlled development of more efficient filter materials and purification systems. The present Study Institute is concerned with this approach. Both fundamental and practical aspects of gas and water purification are treated. The emphasis is on the evaluation of factors which may affect the efficiency and performance of filter systems, on optimalization, modelling and controlled development of "tailor-made" systems.

The topic is limited to the filtration of molecules from a homogeneous phase (gas or liquid) by means of

- sorption on activated carbon, molecular sieves and other inorganic surface-active solid materials

- diffusion through semipermeable membranes, reversed osmosis and ultrafiltration.

These methods are by no means new. Removal of impurities by sorption on carbon has been applied for thousands of years, and

diffusion through semipermeable membranes is a well known basis
for most biological functions. But systematic research and process development based on a balanced combination of scientific
aspects and empirical data is of relatively recent date, and is
still impeded by the gap between the triviality of sewage treatment and the advanced theoretical and experimental background
which is needed to study the complicated reaction mechanisms and
solid state physics involved in these processes.

The incitament for a scientific approach has been much stronger in the field of respiratory protection against toxic gases and
vapours. Consequently, much of the early fundamental research
and development work on sorption of gases by activated carbon has
been carried out at chemical defence research establishments.

Even in the nuclear field stringent safety precautions have
incited systematic studies of gas and water purification, development of filters for retention of gasborne radioactivities (such as
Xe, Kr, I) and of procedures for removal of fission products and
activation products from effluents and waste solutions. Like the
chemical defence establishments, the nuclear research institutes
are well equippped for fundamental research. The established nuclear techniques can also be used to control industrial and commu-
nal gas- and water filtration by means of tracer techniques and
labelled molecules.

Recently, the stringent limits imposed upon industrial re-
leases to the environment and the concern for supplies of air and
water have also given rise to a more systematic research about
purification methods for industrial off-gases and effluents, for
air and potable water.

Although the actual processes, technical installations and
lay-out used for the above mentioned purposes may have little in
common, the problems related to sorbent properties, reaction mechanisms, interactions between sorbent and sorbate are often much the
same. Questions of common interest may include

- the choice of methods, sorption versus alternative procedures (here only represented by membrane filtration)

- the choice between sorbents, activated carbon versus inorganic sorbents

- the relevance of criteria and the evaluation of sorbent
  qualities based on producers' specifications

- the prediction of performance, process steering and modelling

- the need for pretreatments and conditioning of the sorbate

- control methods, analyses of sorbents, of their capacity, selectivity and performance

- the effect of BET-surface, pore size distribution, pore structures and surface chemistry on the efficiency, selectivity and stability of filter materials

- the relationship between static isothermes and dynamic sorption

- the influence of activation and regeneration on sorbent properties

- interactions between impregnation and sorbate

- means to improve sorbent properties and eventually to produce "tailor-made" sorbents for special purposes.

The overlapping of problems and interests may also be briefly illustrated by a more specific example from our own work.

Aging and poisoning of iodine filters in nuclear installations are recognized problems, which have given rise to the establishment of strict control routines and to various research programmes. Reduced efficiency in standby filters could be particularly troublesome and involve some control problems. At Institutt for Atomenergi it was found that studies of the control of such effects could with advantage be combined with similar efforts to control aging of protective filters for toxic gases. Thus a cooperation with the neighbouring Defence Research Institute was initiated, which for one thing has resulted in a simple X-ray test for a rapid control of aging effects.

The present meeting has brought together experts from scientific and applied research institutes, from the carbon industry, from chemical defence and nuclear research establishments, from industrial firms and institutions concerned with gas and water purification. The following lectures should furnish a basis for discussion between these representatives from the different fields dealing with sorption, gas and water filtration.

A main session is devoted to fundamental surface characteristics, reaction mechanisms and kinetics. The following sessions deal with test and control methods, with activation, regeneration and impregnation of sorbents. In the last sessions industrial offgas and water purification, removal of radioactivity from offgases and waste solutions, and respiratory protection against toxic gases are treated.

The main part of the meeting is alotted to sorption. Physical adsorption, chemical sorption, ion exchange reactions, catalytic reactions, the formation of surface compounds and cage compounds are treated. Unimpregnated and impregnated activated carbon as well as inorganic sorbents are included. Since carbon is by far the most used sorbent for gas and water purification, it is naturally even a main topic of this ASI, but contrary to the established carbon conference it is considered in relation to other sorbents and methods.

Alternative methods are mainly represented by membrane filtration. It might have been desirable to include further methods, but this was not possible due to time limitations.

It is hoped that the present ASI will help to bridge the gap between the different approaches to sorption, gas and water purification and will promote research and cooperation in this field.

THE USE OF ADSORPTIVE FILTERS IN AIR PURIFICATION

F.A.P. Maggs

Chemical Defence Establishment, Porton Down,
Salisbury, Wiltshire, England

The most useful introductory lecture to an audience of varied interests takes the form of an enunciation of principles and qualitative aspects of the subjects under subsequent discussion. In this way nearly everyone will understand and appreciate more fully the later detailed proceedings, even when these may not be entirely apposite to their own topic. This, then, is the course I intend to follow in introducing the subject of air purification by use of adsorptive filters.

At this early stage of our proceedings it seems useful to outline those aspects of the subject which will be discussed, after first noting that the filtration of aerosol or particulate matter will not be dealt with. We shall limit our consideration to filters dealing with the production of respirable air from air contaminated with either toxic or merely obnoxious gases or vapours, in field concentrations of less than 1%. Indeed, even for high volatility liquids, a concentration of 0.1% is difficult to maintain for longer than a few minutes in the open air. In confined spaces higher concentrations can be set up, but under such circumstances the respirator would only be used in escaping from the room, and an air-fed type of breathing apparatus would be a necessity for more prolonged activity.

At the other extreme, many practical situations refer to the removal of low concentrations (i.e. in the p.p.m. range) of very toxic or very malodorous compounds (e.g. H_2S, HCN, biological decomposition products, radio-active products). Generally, large volumes of contaminated air need purification.

The majority of filters act by a mechanism of physical adsorption and this imposes a further limitation on their use in that the surface interaction must be sufficient to retain the contaminant in the adsorbent. As a general guide, substances of boiling points below ca. 50°C are not satisfactorily removed by physical adsorption. In some cases, however, this limitation may be by-passed by the incorporation of substances in the charcoal with which the adsorbate will interact chemically. The mode of action of these impregnants may be catalytic (changing the adsorbate to a non-toxic or an involatile substance), or chemisorptive (binding the adsorbate strongly), or a straight-forward chemical interaction, the charcoal merely acting as a carrier.

Dynamic Adsorption

The detailed mode of interaction of an air-contaminant with the filter material is by no means straightforward. However, the following simplified picture of the situation will provide a basis for a better understanding of the principles underlying the more rigorous treatments.

The adsorption isotherm relates the amount of vapour adsorbed in equilibrium with various vapour pressures or concentrations. The typical type I isotherm of Brunauer[1] is very commonly encountered (Fig.1). If, for the moment, we assume that the filter will adsorb vapour to an extent given by the isotherm, it is readily seen[2] that if the contaminant concentration lies in the region AB (Fig.1) then the penetration time $T = wK_1/F$. At higher concentrations $T = wK_2/FC$.

$\llcorner$ T = penetration time, min; w = weight of adsorbent, g.

F = volume flow rate, l/min; c = concentration of vapour, g/l.

K_1 = slope of isotherm along AB.

K_2 = weight adsorbed along BC, g/g. $\lrcorner$

Many instances of the operation of these two approximate relationships are known, in which the penetration time varies either (i) with the inverse flow rate only, and is independent of concentration over the low range; or (ii) with both flow and concentration.

We may now introduce some refinement to this simple picture (following Hinshelwood et al.[3]). It is assumed that at any point in the bed the <u>rate</u> of adsorption is proportional to the concentration of vapour at that point and thus that the rate of decrease of concentration in the gas phase is also proportional to the concentration at any point, i.e. $dc/dt = -K_3c$, where K_3 is a rate constant. On integrating and evaluating the integration constant we obtain:

$$\ln c/c_o = -K_3t, \quad c_o \text{ being the influent concentration}$$

This expression shows that the concentration falls exponentially with time (or distance) through the bed, a behaviour which is also qualitatively reasonable.

In the particular case of a bed of finite length d, the value of the concentration at distance d (i.e. the effluent concentration) is c'; the time of passage through the bed will be t', and will equal the residence time; (so that $t' = d/L = V/F\varepsilon$, where L is the air velocity, V is the volume of the bed, and ε its porosity.) Then

$$\ln c'/c_o = -K_3t' = -K_3d/L$$

The rate constant K_3 is thus an umbrella constant of considerable importance, describing the efficiency of adsorption in the system. Its numerical value will be related to intergranular diffusion, intra-granular diffusion, grain size, adsorptive capacity, and the form of the adsorption isotherm.[4] We have observed, for instance, that its value, determined under given conditions, is over an order of magnitude greater for CDE charcoal cloth than for a good respirator charcoal - a property which is reflected in the much higher capacity of the charcoal cloth at penetration.

As the flow of vapour laden air continues, the influent end of the bed will steadily reach equilibrium with the vapour concentration (according to its position on the adsorption isotherm), and the wave front will move through the bed. At the same time the effluent concentration will rise as the effective length of the bed is progressively shortened. At an arbitrary value of the effluent concentration the "penetration" of the bed is said to occur, and the air flow is stopped.

Hinshelwood took his analysis further to give further quantitative values to the various components, and others have developed the analysis in different directions. Thus the assumption of irreversible adsorption is clearly untenable in many systems, but as long as the concentrations are low, and the adsorbate is strongly adsorbed, little error will be involved. The shape of the complete wave front curve will be dependent on the shape of the adsorption isotherm (our treatment has assumed a linear isotherm), and if the concentration of vapour is high (past point B on Figure 1) this will clearly be reflected in the wave front. However, the following equation was developed by Hinshelwood, making several reasonable assumptions:

$$T = \frac{N_o V}{c_o F} - \frac{1}{K_3 c_o} \ln c_o/c'$$

N_o is the adsorptive capacity per unit volume of bed in equilibrium with the concentration c_o. The relevance of the rate constant K_3 and of the choice of the value of c' with respect to the penetration time are made apparent in this equation. This equation can be shown to be formally similar to the well-known Mecklenburg equation:

$$T = \frac{N_o}{c_o F} (V - V_o) \quad \text{(where } V_o \text{ is the critical bed volume)}$$

This equation relates the penetration time to two sections of the bed: that which has reached equilibrium with the influent concentration, and that part - the so-called critical bed - which is not fully utilised at penetration. The simple equations given earlier in this section assume that the beds are big enough for the critical part to be neglected.

Effect of moisture

The most severe practical limitations, however, relate to the presence of ubiquitous water vapour. A moment's thought will show that in most air filters the adsorbents are already spent before use by exposure to air containing moisture, and that the filter can only operate if the adsorbed water can be displaced by the contaminant. Whether this is possible or not is of course determined by the surface free energy changes involved; whether it is probable is determined by rates of displacement, flow rates, heats of adsorption and recondensation of bulk water at the effluent end of the bed. In our experience, the last factor - difficult to predict and control - can upset the most ingenious of calculations. In considering the problem of adsorbed water, the solid/water and solid/vapour interactions are of course very important. A carbon surface, for instance, gives energy changes in a favourable direction for displacement, and for this reason, amongst others, charcoal retains its popularity as a respirator adsorbent. Even so, adsorbed water in most cases reduces the penetration time, unless a secondary interaction between water and the adsorbate intervenes.

It is appropriate to mention here that modification of the
nature of the surface - by, for instance, heating a charcoal in
hydrogen, or by oxidising it - can be achieved, so that the
surface free energy changes vis-à-vis water adsorption can be
altered; in effect, the shape of the water adsorption isotherm
can be changed by such pretreatments.

Adsorbents

In this introductory talk I have so far said little about
the properties of the adsorbent. Whilst this topic will
undoubtedly arise later during this Conference, some general
observations will not be out of place here.

In any gas/solid interaction, the interfacial area will be
of prime importance in a dynamic system. An extensive accessible
internal surface area of the solid, or extreme subdivision of the
solid can achieve this. Generally, fine powders are impractical,
if only because of the high air flow resistance offered by such
an assembly, and nearly all filters are packed with granular
porous adsorbents.

Reference will be made in later papers to methods of determining the often extensive specific surface areas (values up to
$1200 \ m^2g^{-1}$ are quoted), and in this context the theories associated with Freundlich, Langmuir, B.E.T., and Dubinen[5] will
undoubtedly be discussed. I shall take this opportunity of
reminding you of aspects which may not be immediately apparent
to everyone.

The Freundlich isotherm is frankly empirical: a curve such
as that of Figure 1 will probably follow a power law, and a
representational equation is sometimes all that is required.
The Langmuir equation, in effect, implies that the surface will
be covered with a monomolecular layer at high pressures, and its
application gives us a value for the coverage which is a little
more precise than mere inspection of the isotherm would yield.
One may also venture on extrapolation from measurements in the
lower pressure range.

The B.E.T. method[1] started off by showing that isotherms of
several vapours on a given solid gave roughly the same surface
area values if the monolayer completion was postulated to occur
near the shoulder of type 2 isotherms. The subsequent theory
enabled this empirical point to be calculated reproducibly.

Dubinen, au fond, reminded us of two important points.
Firstly, if we plot isotherms obtained at different temperatures
for a given adsorbent/adsorbate system, in the form of weight
adsorbed versus T ln p - a form stemming from Polyani's theory -

then a single common curve is obtained.[6] Secondly, the difference in the adsorption forces between various adsorbates (with
respect to a given surface) is likely to be reflected (approximately) in the difference between the mutual interactions of
the adsorbate molecules. That is, latent heats, electrical
polarisabilities etc., are a good guide to the behaviour of
various adsorbates. The subsequently developed Dubinen equations
represent an _empirical_ attempt to add precision and to find a
convenient linear form of the isotherm.

Empiricism has, of course, its virtues; in particular, it
is likely to describe the phenomena with some accuracy and in
this respect has considerable value. Here I merely remind you
that the values of the derived surface areas should be viewed
with at least a little scepticism: the main attribute of these
isotherm equations lies in their ability to select in a
reproducible manner a particular point on isotherms at which
monolayer _might_ be said to be complete. The possibility of the
occurrence of phase changes in the adsorbed film (cf. trough
films[7]) would, of course, preclude the application of a single
equation over the whole of the isotherm. In pores of molecular
dimensions, one may, in any case, doubt an interpretation in
terms of area.

Some adsorbents do, in fact, behave as molecular sieves;
the zeolites provide a well-established example to be discussed
later in this Conference. Sieving is also well demonstrated for
charcoals by heat of wetting (H.O.W.) measurements. The H.O.W.
of a non-porous carbon black is found to be approximately
independent of the organic wetting liquid: the heat liberation
per unit area is the same for many organic liquids. For a
charcoal, the H.O.W. remains constant with increasing molecular
size of the wetting liquid up to a certain size; beyond this,
the H.O.W. falls with further increase in molecular size.
Further, the critical size increases with increasing activation
of the charcoal. At CDE this approach is used to assess rapidly
the activity of a charcoal by measuring the H.O.W. in benzene
and in a silicone fluid (large molecular size). For charcoals,
elaborate calorimetry and sample evacuation are not required
(except for great accuracy), and equipment for the simple and
rapid determination of heat of wetting has been devised.[8]
The possibility of molecular sieve effects must thus be borne
in mind when selecting an adsorbent for a given use.

Concluding Remarks

For atmospheric purification in particular, the water/
surface interaction may be decisive, as indicated earlier.
During the Conference we shall also be discussing chemical inter-

actions at the gas/solid interface and here the adsorbent per se may be involved, or it may merely form a substrate enabling an impregnant to offer a large interfacial area for reaction.

In all cases, you will observe our earlier considerations of the dynamic interaction will apply. The contaminant molecule must diffuse through the air to hit the external surface, and be able to migrate to the much more extensive internal surface; note that the rate of removal from the external surface by diffusion should exceed the rate of impingement. In some cases the products of the interaction, if gaseous, must be removed; in other cases, the pore structure must be sufficiently open to allow further contaminant molecules to pass molecules already chemisorbed in the pores.

A word of caution may be appropriate, in relation to our discussions during this week. Dynamic adsorption is a fascinating subject allowing full play to experiment, theory and imagination. But in practice, it must be remembered that (i) parts of the system do not reach equilibrium conditions; (ii) heat is evolved, so that the temperature at the effluent end of the bed is always higher than that at the influent end - I have even seen the bed catch alight!; (iii) water is often displaced, and may even condense as liquid near the effluent end - wave fronts of heat and water! Elaborate theoretical treatments of adsorbents, diffusion and rates of reaction, often founder in practice.

Future developments

Prognostication is always hazardous - especially in print - and all I shall indulge in is a tentative extrapolation.

In the next few years a radical improvement in the equilibrium adsorptive capacity of adsorbents is improbable, although modification of their physical form may introduce greater rates of dynamic adsorption (as with CDE charcoal cloth), to give greater efficiency. Interest may well develop in modifying the nature of the surface of adsorbents, to introduce specificity towards certain contaminants, and to reduce the usually deleterious effects of adsorbed water. Similarly, a renewed study of both metallic and organic impregnants (and of methods of incorporating these in the adsorbent) may lead to improved filtration of several obnoxious gases of industrial origin. We may, I think, look forward to a steadily increasing concern in the problems of aerial pollution and in the solution provided by adsorptive filters.

12

REFERENCES

1. Brunauer, S., Adsorption of Gases and Vapours, O.U.P. 1944.

2. Maggs, F.A.P., Design and Use of Respirators, p.24, Pergamon Press, 1962.

3. Hinshelwood, C.N. et al., J.Chem.Soc., p.918, 1946.

4. Maggs, F.A.P., Filtr. and Separ., p.413, 10, 1973.

5. cf. Gregg, S.J., and Sing, K.S.W., Adsorption, Surface Area and Porosity; Academic Press, 1967.

6. Maggs, F.A.P., Carbon, p.113, 10, 1972.

7. Gregg, S.J., J.Chem.Soc., p.696, 1942;
    Maggs, F.A.P., Trans.Far.Soc., p.123, 44, 1948.

8. Maggs, F.A.P., Schwabe, P.H., J.Sc.Inst., p.60, 37, 1960.

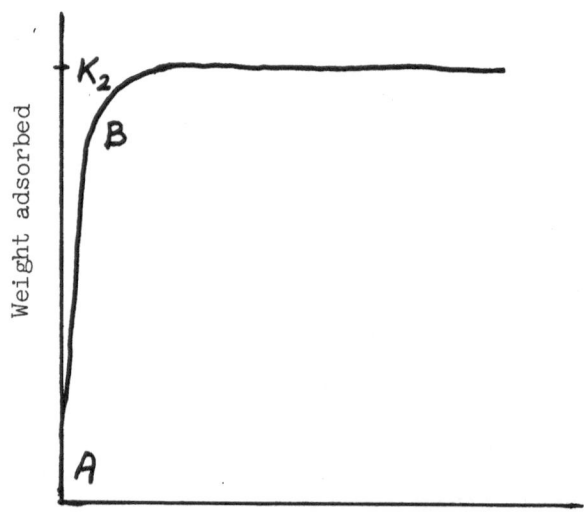

Figure 1. Adsorption isotherm, type 1

PURIFICATION OF WATER AND EFFLUENTS BY ADSORPTION*

Walter J. Weber, Jr.

Professor of Environmental and Water Resources
Engineering and Chairman, Water Resources Program,
College of Engineering, The University of Michigan,
Ann Arbor, Michigan, U.S.A.

1. INTRODUCTION

Sorption reactions play important roles in a number of
physical, biological, and chemical processes and operations in
the environmental field. In specific process applications,
purification of gases by sorption is an effective technique for
air pollution control, sorption of dissolved impurities from
solution is widely employed for water purification, and sorption
is now viewed as a potentially superior method for effluent
treatment and water reclamation.

Applications of sorption for industrial chemical processing
are well known and reasonably understood. Water purification
applications and newer applications for effluent treatment and
water pollution control, conversely, are generally not as well
understood. Although the process has been demonstrated as widely
effective for removing dissolved organic substances from water
supplies and effluents, it should not be viewed as a catholicon,
in either case, nor should its application for water and effluent
treatment be made in rote fashion. This lecture will discuss
broadly these application areas, highlighting advantages over
other purification processes, and defining factors and considerations involved in the use of sorption systems. Subsequent lectures
will deal with specific theoretic, design, and operational aspects
of these systems.

---

* Lecture 3. Session I NATO Advanced Study Institute, Scientific
Aspects of Sorption and Filtration Methods for Gas and Water
Purification, Fauske, Norway, 23-29 June 1974

2. SORPTION SYSTEMS

Of particular significance for application of sorption for large-scale treatment of water and effluents is the manner in which to contact, most effectively, the sorbent with the solution to be treated. Rates of sorption from solution on granular sorbents have been found to depend, in part, upon the particle size of the sorbent. It is desirable to employ sorbent media of as small a diameter as conditions of efficient operation allow, so that high rates of sorption obtain. The term "efficient operation" is a key one here, for the size of particle chosen will dictate to some extent the type of reactor system in which contact of the sorbent with the water or effluent will be accomplished. For example, powdered carbon must be used in either a batch or stirred-tank flow reactor; head loss through a bed or column reactor would be prohibitive in most cases. The type of reactor, on the other hand, will dictate to some extent the efficiency of contact, and therefore the efficiency of the sorption reaction.[1]

In batch-type contact processes, a quantity of sorbent is mixed continuously with a specific volume of water until the contaminants have been decreased to a desired level. The sorbent is then removed and either discarded or regenerated for use with another volume of solution. If finely divided sorbent, such as powdered carbon, is used in this type of system, separation of the spent sorbent from the water may present difficulties. Conversely, the use of large particles, which may be removed more readily when exhausted, requires longer periods of contact between solution and sorbent, necessitating larger basins or tanks in which to retain the water during treatment.

Continuous-flow operations have an advantage over batch-type operations because rates of sorption depend upon the concentration of solute in the solution being treated. Further, plug-flow (PF) or column, reactors have this same type of advantage over completely-mixed flow (CMF) reactors.[1] For column operation, the sorbent is continuously in contact with a fresh solution. Consequently, the concentration in the solution in contact with a given layer of sorbent in a column changes very slowly. For batch treatment, the concentration of solute in contact with a specific quantity of sorbent decreases much more rapidly as sorption proceeds, thereby decreasing the effectiveness of the sorbent for removing the solute.

Rates of exhaustion of adsorbents in water treatment are usually not sufficient to justify the use of moving-bed adsorbers for column or bed-type systems. Thus a fixed-bed adsorber is generally preferred for water treatment.

Conversely, to provide sufficient removal of the higher organic loads normally associated with effluents, and to utilize the sorbent most effectively, an approach to countercurrent contact is commonly required for effluent treatment. This can be achieved by having the effluent flow through a number of contactors or stages in series in one direction while the sorbent moves in the opposite direction. In powdered-sorbent contact systems this is the procedure used. With granular sorbents the procedure is generally simplified to avoid unnecessary handling of the carbon. For most granular-media contact systems the head contactor in a series of adsorption columns is removed from service when the sorbent it contains is exhausted (or nearly so) and, after being refilled with fresh sorbent, is placed at the end of the series. Each contactor is thus advanced one position in the series by piping and valving arrangements which permit shifting of inflow and outflow points of the series accordingly. As the number of stages increases, the piping and valving arrangement becomes more complex and costly. A compromise between the advantage of employing multiple stages to more effectively utilize the sorbent and the cost of each additional stage must be achieved.

Upflow expanded operation of fixed beds of granular sorbent permits the use of small particle sizes for faster adsorption rates, without the associated problems of excessive head-loss, air-binding, and fouling with particulate matter common to packed-bed operation with fine media. In expanded-bed operation, the water flows upward through a column of relatively fine granular sorbent at a velocity sufficient to suspend this media. Packed-bed adsorption techniques have conventionally been used for water treatment. Expanded-bed technology is relatively new. The advantages of expanded-bed adsorbers over packed-bed adsorbers have been demonstrated and discussed.[2] These advantages are most clearly recognized in effluent treatment, for readily apparent reasons.

For fixed-bed (either packed or expanded) sorption operations, the water or effluent to be treated is passed through a stationary bed. Non steady-state conditions prevail in that the sorbent continues to remove increasing amounts of impurities from solution, over the entire period of useful operation.

Figure 1 is a plot of the sorption pattern which normally obtains for a fixed-bed non-steady-state sorber. The impurity is sorbed most rapidly and effectively by the first few layers of fresh sorbent during the initial stages of operation. These first layers are in contact with the solution at its highest concentration level, C_0. The small amounts of solute which escape adsorption in the first few layers are then removed from solution in subsequent strata, and essentially no solute escapes from the sorber initially (C=0). The primary sorption zone is concentrated near the influent

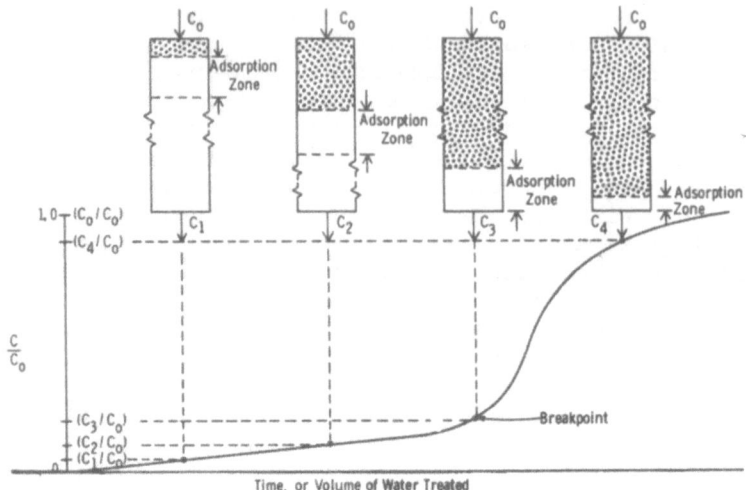

Fig. 1. Schematic representation of the movement of the adsorption zone and the resulting breakthrough curve. (after Weber)[1]

end of the column, the first layers of carbon become practically saturated with solute and less effective for further sorption. Thus, the primary sorption zone moves through the column to regions of fresher sorbent. The wave-like movement of this zone, accompanied by a movement of the C_0 concentration front, occurs at a rate much slower than the linear velocity of the water or effluent. As the primary sorption zone moves through the bed, more and more solute tends to escape in the column effluent, as indicated in the sequence of schematic drawings in Figure 1. The plot of C/C_0 vs time (for a constant flow rate), or volume treated, depicts the increase in the ratio of effluent to influent concentrations as the zone moves through the column. The breakpoint on this curve represents that point in operation where -- for all practical purposes -- the column is in equilibrium with the influent water, and beyond which little additional removal of solute occurs. At this point, it is desirable to reactivate or replace the carbon.

The method chosen for operation of a fixed-bed sorber is dependent on the shape of the curve given by plotting C/C_0 vs time or volume. As noted previously, this curve is referred to as a breakthrough curve. For most sorption operations in water and effluent treatment, breakthrough curves exhibit characteristic non-symetric "S" shapes, but with varying degrees of steepness and positions of break-point. Factors which affect the actual shape of the curve include the shape of the adsorption isotherm, solute concentration, pH, rate-limiting mechanism for adsorption and nature of the equilibrium conditions, particle size, depth of the column of sorbent and the velocity of flow. As a general rule, the time to break-point is decreased by: 1) increased particle size of the sorbent; 2) increased concentration of solute in the influent; 3) increased pH of the water; 4) increased flow rate; and, 5) decreased bed depth. These factors are discussed in detail in Lecture 2, Session II. If the total bed depth is smaller than the length of the primary sorption zone required for effective removal of solute from solution, then the concentration of solute in the effluent will rise sharply from the time the effluent is first discharged from the sorber. Thus, for each type of sorption operation there exists a critical minimum bed depth.

Quantitative prediction of the performance of fixed-bed sorbers involves prediction of the shape and position of the breakthrough curve, representing the movement of the sorption front through an adsorber. This requires application of appropriate mathematical modelling techniques for operation on information developed from component and condition analysis. This subject will be discussed in detail in Lecture 3, Session III.

3. WATER PURIFICATION

Evidence of the use of charcoal for improving the quality of water for drinking can be traced through writings pre-dating the evolution of Christianity. Not unlike early practices prescribed by Hippocrates for boiling and straining water for protection of health, the practice of purification by treatment with charcoal pre-dated, by over 30 centuries, any real understanding of either the manner or effect of such practice. Not until the last quarter of the 18th Century was sorption clearly recognized as a phase-partitioning phenomenon; first by Scheele for gases, and shortly thereafter by Lowitz for solutions.

Rudimentary as it was, charcoal functioned in basically the same manner for water purification then as does activated carbon today; namely, by sorption of certain types of impurities. Charcoal is simply the residue obtained by partial burning of carbonaceous animal or vegetable matter in an oxygen deficient atmosphere; the end product of pyrolysis, or "destructive distillation" of organic matter. Despite relatively common use of charcoal for household and municipal water treatment for over a hundred years following the observations of Lowitz, it was not until near the end of the 19th Century, that von Ostrejko determined that the sorptive properties of ordinary charcoals could be markedly enhanced by partial oxidation in the presence of steam or carbon dioxide at temperatures of 800 to 1000 °C. The materials so prepared by von Ostrejko were the immediate forerunners of the activated carbons used today for a multitude of separation and purification purposes. A variety of sorbents have been tested for potential applications for water and effluent purification, but activated carbon is by far the most successful and widely used; the remainder of this lecture will therefore focus on carbon as the adsorbent of choice.

As early materials for adsorption have evolved and become more sophisticated and effective, so the problems addressed by this form of water treatment have advanced in complexity and magnitude. Indeed, the very technology that has permitted an understanding of sorption processes, and therefore the development of better techniques, has thrust upon us an increased urgency for more extensive and refined application of these processes in water treatment. Early use of charcoal was largely motivated by the occasional presence of undesirable tastes and odors in drinking waters. This problem exists today, but it has intensified in at least two of its dimensions. First, increased populations and increased per capita use of water force us to turn more frequently to less desirable water sources to provide adequate amounts of supply. Second, pollution of available water sources has increased by virtue both of increased use and of the evolution of a broader and more complicated spectrum of pollutants; consequences of population expansion and industrialization. The problems of taste and odor resulting from natural substances, such as those produced by algae and the decay of vegetation, have existed for centuries. These are, and will continue to be, of concern for those in whom responsibility for providing adequate, safe, and palatable resources of water is vested. Within the relatively brief span of the last 50 years or so, however, we have been confronted increasingly with the problem of pollution of our water supplies by synthetic by-products of an advanced society. Phenols and related industrial compounds have been commonplace pollutants in water supplies for several decades. Chemicals that 10 years or so ago were considered "exotic" now appear with greater frequency in many surface and ground-water supplies. Our society is geared to increased production and application of environmentally persistent chemicals to meet domestic, agricultural, and industrial needs. It is unrealistic to think that these materials can be totally excluded from all sources which provide for our water supplies.

To give but one example, Table 1 presents a list of 34 different organic chemicals identified in the lower Mississippi River, which serves as a drinking water supply for over 50% of the population of the state of Louisiana.[3] Some of these materials are known to have sever physiologic impacts as toxicants and carcinogens, even in low concentrations; for others we can only "guess the worst."

The point to be underscored is that although sorption processes, and more specifically sorption processes using activated carbon, have been, and will continue to be, useful for treatment of waters for reduction of "conventional" tastes and odors, their significance for water treatment may be much broader and their universal application more imminent than we realize. Fortunately, a large percentage of the compounds with which we need be concerned are amenable to removal by sorption, and activated carbon has a demonstrated ability to sorb many different types of organic substances. This is doubly important because we often do not have clear-cut identification of all organic contaminants in a water source. Some substances may present potential long-term hazards in concentrations which defy even the most sophisticated analytical procedures.

Table 1. Refractory industrial wastes found in water supplies

Acetone	Dichloroethyl ether
Benzene*	Dinitrotoluene*
2-Benzothiozole	Ethylbenzene*
Borneol (bornyl alcohol)	Ethylene dichloride*
Bromobenzene	2-Ethylhexanol
Bromochlorobenzene	Guaiacol (methoxy phenol)
Bromophenylphenyl ether	Isoborneol
Butylbenzene	Isocyanic acid*
Camphor	Isopropylbenzene
Chlorobenzene	Methylbiphenyl
Chloroethyl ether	Methylchloride*
Chloroform	Nitrobenzene*
Chloromethyl ethyl ether	Styrene*
Chloronitrobenzene	Tetrachloroethylene
Chloropyridine	Trichloroethane
Dibromobenzene	Toluene*
Dichlorobenzene	Veratrole (1,2-dimethoxy benzene)

*In trace amounts, these compounds also have been found to impart taste and odor to drinking water supplies.

(after Miller)[3]

It is unfortunate that at present relatively little is known concerning the nature of organic contaminants in water supplies. Much of the organic matter in many water supplies derives from natural sources, and is likely quite harmless. There is growing concern, however, that rivers receiving large volumes of domestic and industrial effluents might possibly contain substances that are toxic, carcinogenic, mutagenic or teratogenic. Upper concentration limits for certain chemical substances in drinking water have been recommended by the World Health Organization and a number of other agencies, but these lists are seldom sufficiently comprehensive. Such limits might be regarded as maximum safe concentrations, and it is desirable to keep the concentration of all toxic substances in water as low as possible.

It is difficult to specify a limit for organic compounds in water in terms of any one of the general parameters commonly used for analysis; most are so non-specific that they are of little use for this purpose. Possibly a limit on Total Organic Carbon (TOC) concentration, say less than 1 mg/l might constitute a reasonable criterion for drinking water quality. The need for more research on the nature of organic compounds in water must be stressed, as must the need for more quantitative information on the chronic physiologic effects of such materials at trace levels of concentration. In the absence of more definitive information in these regards, the best philosophy to apply is probably that if water contains unknown substances that could conceivably constitute a potential risk to health, then everything within reason should be done to reduce their concentration as much as possible. This philosophy will doubtless lead to increased use of activated carbon in water treatment. Indeed, the time when sorption is a routine part of water treatment practice probably lies in the relatively near future.

A common question with respect to the application of activated carbon treatment of water is whether powdered carbon or granular carbon should be used. The former generally involves a low capital cost, high operating cost system; the latter a higher capital cost, lower operating cost system. Aside from the question of the relative efficiencies of the different reactor systems employed with these different carbon types, it must be recognized that, historically, the principal use of carbon in water treatment has been for removal of tastes and odors, whereas interest has extended now to include removal of organic materials of suspect health implications. The removal of such materials requires a continuous process, whereas taste and odor control can frequently be accomplished on an intermittent basis. There is a distinct advantage in using granular carbon beds to deal with material requiring continuous removal and powdered carbon to deal with emergencies.

Once the type of carbon and type of contacting or reactor system has been chosen, the location of the sorption process in the overall treatment scheme must be determined. It is highly desirable that this be determined in the perspective of the particular treatment objectives, raw water character, and operating constraints of each treatment plant. In essence, although certain operating parameters and design factors can be extrapolated from one raw water and its associated treatment facility to another, it is highly desirable to tailor a sorption system to a particular application to realize maximum benefits from that system. Indeed, this is a reasonable maxim for each unit process involved in a water treatment scheme. For sorption, such factors as flow rate, contact time, concentration fluctuations, pH control, and regeneration frequency require reasonably specific process design consideration.

As a general rule, when powdered carbon is used it should be applied prior to sedimentation or clarification. This has particular advantages if the clarifier is of the floc blanket type, in that considerably increased contact time is provided. Regardless of the type of clarification involved, addition of powdered carbon prior to this point results in a much lower load on the filters than if addition follows clarification.

Granular carbon beds can be used either as a final polishing and safeguard treatment, or just prior to final disinfection. If chlorine is used as a disinfectant, sorption treatment after chlorination provides removal of excess chlorine and of objectionable chlorinated compounds formed during the disinfection process.

The growth of bacteria on granular activated carbon beds and their effect both on carbon performance and on the bacteriological quality of the treated water is a subject of some concern in water purification applications. The nature and extent of biological growth that may develop depends in large measure on the treatment processes preceeding sorption. Prechlorination, ozonation or lime softening results in low bacteria levels in the water entering the sorption system. The bacteria level in the effluent from the sorber may frequently be much higher, however, and cases have been cited where the count was comparable to that in the raw water. It appears that bacteria surviving the early stages of treatment can multiply in contact with organic nutrient accumulated on carbon during sorption treatment. Because such bacteria grow in an environment of declining chlorine concentration, they tend to develop a measure of chlorine resistance, and can therefore often survive final disinfection with chlorine. Such regrowth phenomena can frequently be remedied by sterilizing the carbon beds with either high concentrations of chlorine in the backwash water, the use of a caustic backwash, or by occasional wet steam sterilization of the carbon beds.

4. EFFLUENT PURIFICATION

A number of physicochemical separation and conversion processes have been studied over the past two decades for potential applications to municipal and industrial effluent purification. Among these have been adsorption, coagulation, chemical oxidation, solvent extraction, ion exchange, distillation, freezing, reverse osmosis, ultrafiltration, electrodialysis, electrochemical degradation, flotation, and foam separation.[4-6] The process combination of coagulation and precipitation for removal of insoluble impurities followed by sorption on activated carbon for removal of soluble organic impurities has emerged as the treatment sequence of greatest promise in terms of both technologic and economic feasibility. In this lecture I will attempt to summarize the salient features of this physicochemical treatment system in the perspective of its application to water pollution control.

Development of physicochemical processes for higher levels of treatment centered initially on "tertiary" systems designed to follow "primary" sedimentation and "secondary" biological treatment.[5,6] There are, however, several fundamental shortcomings to this approach. First, the implementation of tertiary systems depends upon prior implementation of primary and secondary systems. Second, the addition of tertiary processes to primary and secondary processes incurs capital and operating expenses of such magnitude as to discourage this development in many instances. Third, the effective operation of a tertiary process is dependent to a large extent on the consistent and efficient operation of a biological secondary process, which is normally subject to problems arising from transients in waste composition and flow (often requiring at least partial diversion) and from the occasional presence of toxic materials.

The concept of applying coagulation-sorption processes directly to raw wastes rather than to secondary effluents therefore has derived partially from considerations regarding the effectiveness and reliability of treatment and partially from the relative economics of "direct" versus "tertiary" treatment systems.[7] Direct treatment subsequently has been demonstrated to be an attractive technical and economic alternative to biological treatment.[8-13]

With pretreatment of raw waste by chemical clarification -- which results in significant removal of both total and soluble organic matter, phosphates and suspended solids -- activated carbon treatment commonly produces a clear effluent of low organic content, suitable to meet requirements for pollution control and for many reuse applications.

To give some illustration of this, Table 2 summarizes overall treatment results obtained at six different pilot-plant installations of physicochemical treatment by coagulation and sorption. More than twenty municipalities in the United States are currently designing, constructing or operating physicochemical facilities for effluent purification.[14] The number of industrial treatment facilities using physicochemical processes is an order of magnitude larger. With urgency impressed by recent more stringent legislation requiring higher effluent standards and receiving water quality, these numbers will increase sharply within the next decade.

Sorption is fundamental to physicochemical treatment for removal of soluble organic impurities. As noted previously, application of sorption for purification of water has been

Table 2. Operating results of pilot physicochemical treatment plants

Plant	Organic Removal, %	Effluent Concentration
Ewing-Lawrence (New Jersey)	95-98	TOC[1] = 3-5
Blue Plains (Washington, D.C.)	95-98	TOC = 6
Lebanon (Ohio)		
a. powdered carbon	95	TOC = 11
b. granular carbon	97	TOC = 6
New Rochelle (New York)	95	COD[2] = 8
Rocky River (Ohio)	93	BOD[3] = 8
Salt Lake City (Utah)		
powdered carbon	91	BOD = 13
Owosso (Michigan)	94	BOD = 8

(1) TOC - total organic carbon
(2) COD - chemical oxygen demand
(3) BOD - biochemical oxygen demand

practiced for years. Impurities removed in such applications are usually present in low concentration, and the need to provide removal often has been intermittent or occasional. Conversely, relatively large amounts of organic impurities are present in effluents, and treatment must be continuous. More efficient utilization of the capacity of activated carbon is consequently required in effluent treatment than is commonly achieved in water treatment. Further, the fact that large amounts of carbon are used requires an efficient scheme of regeneration and reuse of this material. These requirements suggest the use of granular carbon continuous contacting systems; granular carbon because of the relative ease of handling and regenerating this material relative to powdered carbon.

The most common type of continuous system is one in which the effluent is passed through fixed beds of carbon. In such systems hydraulic application rates generally range from 2 gpm/ft^2 to 8 gpm/ft^2 (81-326 l/min/m^2). In this flow range essentially equivalent sorption efficiency is obtained for equivalent contact times. At flow rates below 2 gpm/ft^2 (81 l/min/m^2) sorption efficiency is reduced, while at flow rates above 8 gpm/ft^2 (326 l/min/m^2) excessive pressure drop takes place in packed beds. Contact times employed are in the range of 30 minutes to 60 minutes on an empty bed basis. In general, increases in contact time up to 30 minutes yield proportionate increases in organic removal. Beyond 30 minutes the rate of increase falls off with increases in contact time, and at about 60 minutes the effects of additional contact time become negligible. Carbon beds operated at the lower end of the flow range are generally designed for gravity flow. Systems designed for higher flow rates must employ pressure vessels if packed beds are used. A pressure vessel is more expensive to construct than a gravity flow vessel, but commonly requires less land area, and provides greater ability to handle fluctuations in flow.

Provision must be made to regularly backwash packed-bed carbon systems because they collect suspended solids and tend to develop attached biologic growths in this application. Backwashing alone generally relieves clogging due to suspended solids, but does not completely remove attached biologic growth. It is advisable to include a surface wash and air scour to be assured of removal of gelatinous biologic growth.

This attached growth can lead to development of anaerobic conditions in packed beds. Aeration of the feed is partially effective in preventing anaerobic conditions, but ·this also accelerates biologic growth to the extent that excessive backwash is required; air-binding can also result. Effective control of biological growth can be accomplished in most instances by regular chlorination of the influent to the adsorbers, and/or by chlorination during regular backwash operations.

Packed beds of granular carbon are well suited for treatment of effluents containing little or no suspended solids, and under such circumstances normally operate effectively for extended periods without clogging or excessive pressure loss. However, the suspended solids invariably present in many municipal and industrial effluents, and the potential for biologic growth on the surfaces of the carbon can present problems for the use of packed beds. Because solids and biologic activity usually cause progressive clogging and high head loss in packed beds, increased interest has developed in the potential of expanded-bed adsorbers, which have certain inherent operating advantages over packed-bed adsorbers for treating solutions containing suspended solids. By passing wastewater upward through a bed of carbon at velocities sufficient to expand the bed, problems of fouling, plugging and increasing pressure drop are minimized. Effective operation over longer periods of time results, as has been demonstrated in comparative laboratory studies and in field investigations in both "tertiary" and direct physicochemical applications.[2,10,11] Another advantage of the expanded bed, as noted earlier, is the relatively small dependence of pressure drop on particle size. It is possible to use carbon of smaller particle size in an expanded bed than is practical in a packed bed, thus taking advantage of somewhat higher adsorption rates which obtain for smaller particles.

Perhaps the most significant potential benefit provided by expanded-bed adsorption systems for effluent treatment is the apparent extention of the operational capacity of activated carbon

observed by Weber et al,[10,11] who found that apparent sorption
capacities in excess of 100 weight-percent as organic matter and
150 weight-percent as chemical oxygen demand (COD) could be
obtained in expanded-beds of activated carbon in which biologic
growth was allowed to fully develop.[11] Because expanded beds
require little maintenance, extended periods of undisturbed
operation facilitate the development and continuous growth of
bacteria on the carbon surfaces. This biologic activity, primarily
anaerobic, degrades some of the organic matter which adsorbs on
the carbon, functioning to provide in-situ partial regeneration
by renewing a portion of the carbon surface for continued sorption.
The direct feeding of settled but uncoagulated (not chemically
coagulated) raw sewage to expanded-bed activated carbon systems
in which biologic activity was encouraged, followed by chemical
coagulation of the carbon-treated wastewater, has been investigated
and described.[15] The results of these studies indicated no real
advantage to this sequence of treatments, and considerable disadvantage due to fouling of the carbon by the solids present in the
uncoagulated feed.

Table 3 gives carbon capacities obtained in field operations
at several physicochemical pilot plants. In that the wastes,
effluent criteria, number of contact stages, etc. varied from
plant to plant, it is not surprising that some spread in the
results is observed. For general planning purposes a COD capacity
of 50 weight-percent is reasonable if no biologic extension of
carbon capacity is taken into account. This is approximately
equivalent to a requirement of 500 pounds of activated carbon per
million gallons (60 grams per cubic meter) of sewage treated.
However, the results obtained by Weber et al[11] with biologically-
extended adsorption systems suggest that it may be possible to
achieve higher effective capacities, reducing the carbon exhaustion rate to less than 200-250 pounds per million gallons (24-30 grams
per cubic meter).

Table 3. Carbon capacities obtained in physicochemical pilot plants

Plant	Capacities, weight-percent	
	TOC	COD
Blue Plains (Washington)	15	41
Ewing-Lawrence (New Jersey)*	50	150
New Rochelle (New York)	20-24	60
Lebanon (Ohio)	22	50
Owosso (Michigan)	--	65
Salt Lake City (Utah)	--	36

*Biologically-extended expanded-bed operation[11]

Even for the highest capacities observed, the initial costs of carbon are such as to make regeneration and reuse of this material highly desirable. Technically and economically feasible regeneration of granular activated carbon can be accomplished by controlled heating in a multiple-hearth or rotary-kiln furnace in the presence of steam. During each regeneration cycle some carbon is lost by burning and attrition, and some by alteration of surface properties. The overall loss, expressed as percent by weight of virgin carbon required to restore the total original capacity of the batch, ranges from 5 to 10 percent. For planning purposes, carbon make-up requirements can be considered to range from 25 to 50 lbs. per million gallons (3-6 grams per cubic meter) of waste-water treated, again not taking account of in-situ biological regeneration.

At present, regeneration systems for powdered carbon are being developed and tested at the pilot stage. A successful process for regeneration of the powdered form would represent a significant step toward making an effluent treatment system utilizing this lower cost material a technical and economic reality. The key factor will be maintaining carbon loss at a sufficiently low level during regeneration.

A suggested flow sheet for physicochemical treatment of waste-waters is given in Figure 2. In this scheme, coagulant is added to the effluent, and flocculation takes place in a chamber which

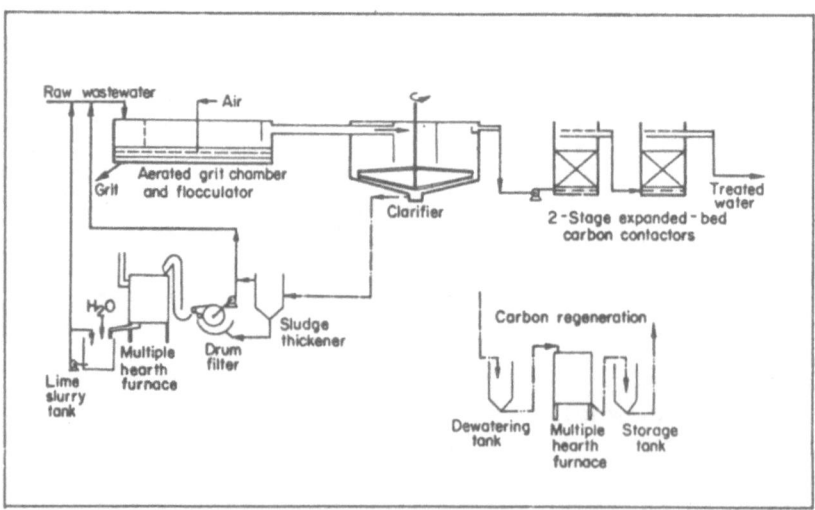

Fig. 2. Typical flowsheet for treatment of effluent by chemical clarification and adsorption.

provides moderate agitation for an average detention time of 15 minutes. Clarification takes place in a sedimentation basin with an average detention time of two hours. The particular flow sheet presented here is a single-stage coagulation system.

The clarified effluent is then passed through activated carbon adsorption units for removal of dissolved organics. The preferred mode of operation is an expanded bed, which permits the use of simple open-top concrete contacting basins and relatively trouble-free operation. The use of open tanks with overflow weirs at the surface of the contacting basin provides a means for additional aeration of the wastewater during treatment, thus helping to control anaerobic conditions in subsequent reactors. Two-stage contacting of the activated carbon is outlined in the treatment sequence given in Figure 2. However, a larger number of stages can be utilized if desired for a particular application. A typical plant layout for a design capacity of 10 million gallons (37,850 cubic meters) per day might be based on five parallel adsorption units of two stages each. When the granular carbon in the first stage of one unit is spent, that unit can be taken off stream while the spent carbon is removed and regenerated in a furnace provided for this purpose. During the time this unit is off-stream for regeneration, the other four units can run at 25% higher feed rate each. Upon completion of the regeneration, the carbon is returned to the adsorber, which then becomes the second stage of that unit; the former second stage with partially spent carbon becoming the first stage. Feed is then evenly divided to the five units until another carbon bed is spent.

The water resulting from the clarification and activated carbon treatment will enhance the quality of most surface waters, and with disinfection is suitable for many reuse applications. A final filtration may be desirable to insure a crystal clear effluent for some uses. This post-filtration would remove any suspended matter passing through, or biologically generated in, the carbon columns.

Physicochemical treatment consisting of coagulation and adsorption holds significant promise as a means of economically meeting today's higher effluent standards and water reuse requirements.

The system has a high degree of stability and reliability. It is more resistant to shock loads and toxic waste constituents than biological treatment systems. Biological systems are notoriously sensitive to changes in environmental conditions. If a toxic material gains even temporary entrance to a biological plant, or a hydraulic peak occurs, not only will the efficiency of the plant drop off, but recovery may take from several days to several weeks. In a physicochemical plant serious upsets are unlikely. Further, it can be expected that an immediate recovery of the plant will take place once the source of the upset is eliminated. This inherent stability of performance is also reflected in greater design and operational flexibility. Entire sections of a physicochemical plant can be cut in or out of the process stream as required, and a temporary overload can be absorbed with little effect. The major advantages of a physico-chemical system over a biological system are summarized below:

1. less land area required (1/2 to 1/4 that for a biological system);
2. lower sensitivity to diurnal variations;
3. not affected by toxic substances;
4. potential for significant heavy metal removal;
5. superior removal of phosphates;
6. greater flexibility in design and operation; and,
7. superior removal of organic waste constituents.

REFERENCES

1. Weber, W. J., Jr., Physicochemical Processes for Water Quality Control, Wiley-Interscience, New York, N.Y., 1972.
2. Weber, W. J., Jr., Hopkins, C. B. and Bloom, R., Jr., A comparison of expanded-bed and packed-bed adsorption systems, Report No. TWRC-2 U.S. Dept. of the Interior, Federal Water Pollution Control Administration (formerly, now EPA), Cincinnati, Ohio, 1968.
3. Miller, S. S., Are you drinking bio-refractories too, Envir. Sci. and Tech., (1), 1, 14, 1973.
4. Morris, J. C. and Weber, W. J., Jr., Preliminary appraisal of advanced waste treatment processes, SEC TR W62-24, U.S. Dept. of Health, Education and Welfare, Public Health Service, R.A. Taft Sanitary Engineering Center, Cincinnati, Ohio, 1962.
5. AWTR-1, Summary report - the advanced waste treatment research program, SEC TR W62-9, U.S. Dept. of Health, Education and Welfare, Public Health Service, R.A. Taft Sanitary Engineering Center, Cincinnati, Ohio, 1962.
6. AWTR-14, Summary report - the advanced waste treatment research program, 999-WP-24, U.S. Dept. of Health, Education and Welfare, Public Health Service, R.A. Taft Sanitary Engineering Center, Cincinnati, Ohio, 1965.
7. Weber, W. J., Jr. and Kim, J. G., Preliminary evaluation of the treatment of raw sewage by coagulation and adsorption, Technical Memorandum, TM-2-65, San. and Water Resources Eng. Div., The University of Michigan, Ann Arbor, Michigan, 1965.
8. Rizzo, J. L. and Schade, R. E., Secondary treatment with granular activated carbon, Water and Sewage Works, 116, 307, 1969.
9. Hager, D. G. and Reilly, D. B., Clarification-adsorption in

the treatment of municipal wastewaters, Jour. Water Pollution Control Fed., (42), 5, 794, 1970.

10. Weber, W. J., Jr., Hopkins, C. B. and Bloom, R., Jr., Physicochemical treatment of wastewater, Jour. Water Pollution Control Fed., (42), 1, 83, 1970.

11. Weber, W. J., Jr., Friedman, L. D. and Bloom, R., Jr., Biologically-extended physicochemical treatment, Proceedings, Sixth Conference on Water Pollution Research, Jerusalem, 18-24 June, 1972.

12. Bishop, D. F., O'Farrell, T. P. and Stamberg, J. D., Physical-chemical treatment of municipal wastewater, Jour. Water Pollution Control Fed., (44), 3, 361, 1972.

13. Atkins, P. F., Scherger, D. A. and Barnes, R. A., Ammonia removal in a physical-chemical wastewater treatment process, Proceedings, 27 Annual Industrial Waste Conference, Purdue University, Lafayette, Indiana, 1972.

14. U.S. Environmental Protection Agency, Technology Transfer, p. 3, March 1, 1973.

15. Hopkins, C. B., Weber, W. J., Jr. and Bloom, R., Jr., Granular carbon treatment of raw sewage, Report No. ORD-17050DAL05/70, Water Pollution Control Research Series, U.S. Dept. of the Interior, Federal Water Quality Administration (formerly, now EPA), Cincinnati, Ohio, 1970.

THE PRINCIPLES OF PRESSURE-DRIVEN MEMBRANE FILTRATION PROCESSES
(REVERSE OSMOSIS AND ULTRAFILTRATION)

W.H. Hardwick

Process Technology Division, AERE Harwell,
Didcot, Oxfordshire, England.

ABSTRACT. The relevance of reverse osmosis (RO) and ultrafiltration (UF) to the problems of water supply and purification is indicated. The development of osmotic pressure is explained and the two equations which relate the fluxes of water (F_1) and salt (F_2) through an RO membrane to pressure and concentration are derived. These equations are

$$F_1 = A(\Delta P - \Delta \pi) \tag{1}$$

where ΔP is the applied pressure and $\Delta \pi$ the osmotic pressure difference across the membrane, and where A is a constant and

$$F_2 = B \, \Delta C \tag{2}$$

where ΔC is the difference in salt concentration and B is a constant across the membrane. For RO, operating pressures of 25-100 atmospheres are required depending on the feed concentration.

Since the flow of water through the membrane is normally much greater than the salt flow, a relatively concentrated layer of solution tends to form on the membrane surface. This concentration polarisation is undesirable since it tends to increase salt flow, increase osmotic pressure, and can lead to saturation and precipitation. Ways of minimising the effects of concentration polarisation are described.

The membranes used in industrial plants are briefly described, together with the basic design requirements for such plants.

The process of ultrafiltration (UF) differs from RO in the

use of membranes, that are permeable to most inorganic solutions
and osmotic pressures are not developed. However, the passage of
larger organic molecules and colloids is restricted and the process may be used to concentrate such species at modest pressures
of only a few atmospheres. Fractional selectivity is also possible.

SYMBOLS

A Solvent permeation constant   R Gas constant
B Solute permeation constant   T Absolute temperature
P Applied pressure   V Volume of solvent

For component i:

a_i Activity n_i Molar concentration

C_i Concent. mass per unit vol. v_i Partial molar volume

F_i Total flux μ_i Chemical potential of
                                           component i
M_i Mobility

p_i Vapour pressure π Osmotic pressure

N_i Number of molecules λ Membrane thickness

Subscripts:

1 Solvent   i For component i
2 Solute   m Within membrane
f Feed   p Permeate

1. INTRODUCTION

The purification of drinking water, or more strictly the purification of water for drinking, is a very wide ranging subject
indeed. But the supply of high quality water for drinking is now
such an accepted fact in the so called "developed" countries that
the supply of water must be one of the most effective and reliable
of modern technologies. We are also conditioned to think of this
water being supplied at very modest cost indeed. Industrial growth
has required plentiful supplies of water, firstly for power, and
later for process use and so the industrial settlement has had
access to large sources of water for domestic supply. As industrial development has intensified and the demand for water has increased, there has at the same time been a worsening of quality
because of industrial pollution. This has meant that additional
supplies are sought, and these are often expensive to develop, and
the cost of water both for industrial and potable use has increased
very much in recent years. In the UK, for example, water was traditionally thought of as costing about $2p/m^3$, but the cost of
supply to the new town of Milton Keynes was 10 p/m^3 in 1970, and

in the south east of England, the price has risen much above this.

There has also been a growing concern for the environment in which we live and the pollution of rivers and the drowning of pleasant valleys are now accepted as anti-social. There are then both economic and social reasons for recycling water after purification both to prevent pollution and to augment supplies, and it is here that membrane processes can be particularly effective. It is certainly possible to recycle sewage effluent using conventional treatments and produce water that the citizens of Windhoek, South Africa, drink without apparent harmful effects. This is not so surprising when one considers that a river like the River Thames in England may be used as a supply of potable water for several towns which in turn abstract water for drinking and return their treated sewage effluent to the river which serves the next town downstream. There is dilution in the river and a few miles of residence for the river to effect some magical purification but, in fact, there is a high proportion of recycle of water that has already served as domestic supply. What then can reverse osmosis/ultrafiltration do with advantage over existing practice?

The first and most important feature of reverse osmosis is desalination. Water supply throughout the world is having more and more to resort to sources which are increasingly saline while several small countries in the Middle East are greatly dependent upon the distillation of sea water at the rate of many tens of millions of gallons per day. The fraction of water used for irrigation which returns to its source is often of increased salinity through leaching salts from the soil and the water that is discharged from a sewage treatment plant has increased in salinity - principally in chlorides - by about 300 ppm. The Colorado River in the United States of America has become so saline as a result of both over- and mis-use that it is proposed to build a 100 mgd membrane desalting plant to restore some of the river quality before it enters Mexico. The removal of salts is a very important part of water purification and the levels which should be attained are illustrated in Table 1, the World Health Organisation recommended values for potable water.

The second property of reverse osmosis is its ability to reject large organic species and organisms giving a product which is virtually sterile. This is also achieved by ultrafiltration, but if desalination is necessary then the ultrafiltration function is automatically included.

2. DESCRIPTION OF THE PROCESSES

Reverse osmosis and ultrafiltration are established industrial processes which are now probably treating some $20 \cdot 10^6$ gallons

(10^4 m^3) per day of saline water and effluents throughout the world. They are both pressure driven filtration processes where the filter is a membrane and they differ only in that in reverse osmosis the species to be filtered are ions and small molecules in true solution while ultrafiltration is used to separate large molecules and colloidal species where osmotic effects are very small. Thus the pressures to drive reverse osmosis are larger because in addition to the pressure imposed to obtain useful water flow through the membrane, osmotic pressures must be overcome. With ultrafiltration, using more permeable membranes, pressures of a few atmospheres suffice.

The four lectures that will be given on reverse osmosis and ultrafiltration attempt to cover the underlying principles with special reference to the properties and behaviour of membranes, and also to review the technology and applications. I shall try to restrict my remarks on membranes to all that is necessary to make this introductory lecture intelligible and Dr. Sammon will deal with the nature, properties, and behaviour of membranes in depth in his contributions.

3. MEMBRANES

The phenomenon of osmosis can be demonstrated by interposing a semipermeable membrane between a solution and a pure solvent. The hydrostatic head which must be applied to the solution to prevent ingress of the solvent is a measure of the osmotic pressure.(Fig. 1).

Semipermeable membranes have been made and used in laboratory experiments for many years - the copper-ferro-cyanide membrane formed by chemical reaction in the pores of an unglazed ceramic pot is one example - but membranes that were relevant to the economic desalting of sea water were not prepared until about 1960 when Loeb and Sourirajan, helped by a publication by Mlle Dobry in 1936, produced their assymetric cellulose acetate membranes. These were cast at a thickness of 0.025 cm from a quaternary mixture of cellulose acetate - $Mg(ClO_4)_2$ - H_2O - acetone 22.2 - 1.1 - 10.0 - 66.7% by weight with the temperature of components and casting kept between -5°C and -10°C. After three minutes for evaporation, the glass casting plate with membrane attached was immersed in ice water for an hour and then transferred to a hot bath for five minutes at a temperature between 65°C - 85°C. Such membranes in laboratory tests could give water containing only 500 ppm from a brine feed containing 52,500 ppm of brine (cf. sea water 35,000 ppm) at an applied pressure of 1500 psi (10 MN/m^2); but at the rate of only 6 US gallons per sq ft per day. They were not truly semipermeable membranes as they passed some salt, but they appeared to have very nearly the properties that were required for use in commercial desalination equipment.

4. OSMOSIS

The pressure difference between solute and solution, the osmotic pressure, derives from the difference in chemical potential, and in turn the activity, since

$$\mu_1 = \mu_1^o + RT \ln a_1 \tag{1}$$

where R = gas constant, a_1 = activity of component i, and μ_i^o called the standard chemical potential of i, is dependent on temperature and pressure only and not on concentration (Reid.[1]). For the solvent in the treatment of solutions μ_i^o is taken to be the chemical potential of pure solvent μ_1^* and hence the activity of pure solvent is unity. For osmotic equilibrium the chemical potential of the solvent on both sides of the membrane should be the same (the solute is not transferred) and by considering the reduction in potential when solute is added to the solvent on one side of the membrane and the pressure adjustment necessary to restore this equilibrium we find that

$$- RT \ln a_1(p'') = (p' - p'')v_1 = \pi v_1 \tag{2}$$

where π is the osmotic pressure. The activity, a, is often derived from vapour pressure measurements by the equation

$$a_1 = \frac{P_1}{P_1^*} \tag{3}$$

which allows

$$\pi v_1 = RT \ln \frac{P_1^*}{P_1} \tag{4}$$

Despite the approximations used in deriving this equation, results gained are usually very good. Assuming solutions are very dilute, Raoult's Law relating activity of solvent to mole fraction can be applied to derive the van't Hoff equation

$$\pi V = N_2 RT \tag{5}$$

or

$$\pi = n_2 RT \tag{6}$$

where $V = N_1 v_1 \simeq$ volume of solvent or $\approx$ volume of solution and n_2 is the molar concentration of solute.

5. OSMOTIC PRESSURES

The van't Hoff equation allows us to calculate approximate values of osmotic pressure from concentrations but the relationship expressed in equation 4 gives much more accurate values as Table 2[1] shows. To return to measured experimental values which are relevant to practical reverse osmosis operations, Table 3 gives values for 0.6 M solutions (data ex Sourirajan[3]) and assuming complete dissociation of salts the osmotic pressure contribution per ion or molecule is fairly constant with sulphates giving lower values than chlorides and nitrates. The pressures are quite considerable - some 600 psi (4 MN/m²) for uni-divalent salts at these concentrations and it is at once apparent that reverse osmosis must be concerned with pressures of several MN/m² if solutions of this strength are to be processed.

In addition to overcoming the resistance due to osmotic pressure, there is a resistance due to viscous forces opposing solvent flow through the membrane. The effectiveness of the applied pressure in producing water flow through a membrane may be deduced using thermodynamic arguments. The flow is a function of the chemical potential gradient which in an isothermal osmotic pressure process is a function both of concentration and pressure gradients. The difference in chemical potential across a membrane is shown to be (Merten)[2]

$$\Delta\mu_1 = v_1 (\Delta P - \Delta\pi) \tag{7}$$

Now the water flux

$$F_1 = - C_{1m} M_{1m} \frac{d\mu_1}{dy} = - C_{1m} M_{1m} \frac{\Delta\mu_1}{\lambda} \tag{8}$$

where C_{1m} is the concentration of water within the membrane material, M_{1m} is its mobility, and both have been assumed constant over the small μ_1 range of interest, and where λ is the membrane thickness.

This leads to

$$F_1 = - C_{1m} M_{1m} v_1 \frac{\Delta P - \Delta\pi}{\lambda} \tag{9}$$

The mobility can be related to a Fick's Law Diffusion coefficient

..or water in membrane material through the relationship

$$D_{1m} = M_{1m} RT \tag{10}$$

where m is in the membrane. So

$$F_1 = \frac{-C_{1m} D_{1m} v_1}{RT \lambda} (\Delta P - \Delta \pi) = A(\Delta P - \Delta \pi) \tag{11}$$

A = membrane constant

Similar arguments plus the assumption that at reasonable concentration differences across the membrane the pressure gradient term is not important in salt diffusion, and assuming a constant distribution coefficient K_2 leads to

$$F_2 = - D_{2m} \frac{\Delta C_{2m}}{\lambda} = - D_{2m} K_2 \frac{\Delta C_2}{\lambda} = B \Delta C_2 \tag{12}$$

where D_{2m} is the diffusion coefficient for the solute in the membrane material. From these two equations (11) and (12) the main principles of operation of the reverse osmosis process are evident.

6. OPERATING PRESSURE

Since

$$F_1 = A(\Delta P - \Delta \pi) \tag{11}$$

the water flux is proportional to the applied pressure reduced by the osmotic pressure difference across the membrane.

The salt flux is independent of pressure but dependent upon the difference in salt concentration. For solutions containing a few hundred parts per million where the osmotic pressure is only a few N/m^2, increasing the pressure on the solution against a membrane will induce an increasing water flux which will be approximately proportional to the applied pressure. The salt flux is however independent of pressure and so the increasing water flow will contain relatively less salt leading to the situation expressed in Figure 2 where the salt rejection increases with pressure but the absolute transport of salt is constant.

The percentage salt rejection, R, is:

$$\frac{C_f - C_p}{C_f} \cdot 100 \tag{13}$$

38

and

$$\frac{C_f - C_p}{C_f} = 1 - \frac{1}{DF} \tag{14}$$

where DF, the desalination factor, $= C_f/C_p$, C_f is the feed concentration, C_p is the permeate concentration.

Thus high operating pressure at first sight appears desirable both to obtain more and purer product by reverse osmosis from a given membrane area but other factors such as membrane compaction, power requirements, suggest reducing pressure to the minimum and an optimisation must be sought.

7. CONCENTRATION POLARISATION

It is evident that in an unstirred system, the removal of solvent by forcing it through a membrane will leave a more concentrated solution in the vicinity of the membrane. The concentration gradient can only be removed by diffusion back into the bulk which is normally slow compared with the permeation rate. This effect is called concentration polarisation. It means that:

1. The increased solute concentration at the membrane surface will lead to increased solute flow through the membrane.

2. $\Delta\pi$ will increase and the effective driving force ($\Delta P - \Delta\pi$) reduce.

3. It might also lead to saturation at the membrane surface and precipitation of relatively insoluble species.

We will return to this subject later.

8. MEMBRANES FOR REVERSE OSMOSIS PLANTS

8.1 Desirable Properties

The properties we seek in a reverse osmosis membrane are:

1. high water flux
2. high salt rejection
3. good chemical stability
4. good physical stability
5. cheapness, especially for water treatment.

The cellulose acetate membrane, which has been improved considerably since Loeb and Sourirajan, satisfies (1, 2 and 5). It

is less satisfactory with regard to 4 and less so with regard to
3. We think at Harwell that a membrane which has a water flux of
1 m³/m²/d (20 g/ft²/d) at 600 psi (4 MN/m²) and a desalination
factor of 20 to be a good working membrane for plants desalting
brackish water and we are well able to achieve this using blends
of cellulose acetate which have different acetyl values. Such
membranes also possess adequate physical stability and they are
cheap to produce. However, cellulose acetate is hydrolysed by
acids and bases and is only really useful over the pH range 2 - 8
which fortunately embraces most natural waters; even so the pH
of feeds is often adjusted to about 5 to reduce hydrolysis to a
minimum.

8.2 Limitations of Cellulose Acetate

A great deal of effort has been expended on trying to overcome the limitations of cellulose acetate or to provide better
membranes derived from other polymers. While there are several
alternatives to cellulose acetate for ultrafiltration, only two
compete with cellulose acetate in reverse osmosis, and one is
partly cellulose acetate. This is the composite form where a
porous substrate of one polymer carries an extremely thin desalting film of another polymer, and cellulose triacetate has been
used very successfully in this latter role. The other alternative
to cellulose acetate, which is commercially successful, is the
DuPont hollow fibre membrane which is made from a nylon - an aromatic polyamide. These fibres are some 40 microns inside diameter
and 80 microns outside diameter, and are asymmetric in that they
carry an external "skin" that effects the desalting duty. The
fibres do not require support as they are able to withstand the
applied pressure but a thin film or sheet membrane must be mounted
on a porous support against pressures of up to 1500 psi (10 MN/m²)
or more. The ways of achieving this are dealt with later.

8.3 Dynamic Membranes

Oak Ridge National Laboratory in the USA has pursued this
idea for many years but has not yet succeeded in producing a viable
commercial version. The membrane is formed on a porous support by
a process such as hydrolysing a salt of zirconium to give a hydrous
oxide while organic additives such as polyacrylic acid may be incorporated. Such membranes undoubtedly can show very high fluxes,
hundreds of gallons per ft² per day and high rejections of monovalent ions, but these properties are only realised or maintained
under conditions rarely obtained in normal water treatment
practice.

9. BASIC REVERSE OSMOSIS PLANT REQUIREMENTS

Fig. 3 shows the flow diagram for an RO plant.

9.1 Membrane Area

Assuming a membrane of adequate flux and salt rejection on a porous pressure-resistant support, we may consider what it is necessary to do to achieve a successful yield of desalted water.

We must provide the membrane area to give the product flow at the design pressure and maintain this over the life of the membrane. Thus for a plant to produce 1 mgd, (4546 cu m/d) 100,000 sq ft (9290 m^2) are needed if the membrane flux is 10 $g/ft^2/d$. If a cellulose acetate membrane is used and the membrane is 250 μ thick, the weight of cellulose acetate is only some 2,000 kg. The overall plant desalination factor is also specified and will be lower than the membrane desalination factor as measured on the feed because the salt concentration rises through the plant and the pressure falls due to frictional losses.

9.2 Pump and Flow Regime

The pressure is generated by a pump which also drives the feed across the membrane surface. The distance the feed has to travel and rate of flow will depend on whether laminar or turbulent flow regimes are chosen.

9.3 Seals and Controls

The membrane surface must be sealed against the pressure and the plant must have controls to ensure that the design conditions are maintained during operation.

9.4 Temperature

The temperature is an important design consideration because the viscosity change with temperature affects the rate of flow - very roughly the product flow rate increases by 3% per 1°C rise. While this seems advantageous, membrane compaction rates also tend to increase with temperature and the consequent loss of performance may well be unacceptable.

9.5 Pretreatment

The feed may require pretreatment to remove substances that might foul the membrane surface; it may need an adjustment of pH, or the addition of polyphosphate to control calcium scaling.

The practical realisation of reverse osmosis operation consists essentially then of a high pressure pump feeding to an area of membrane surface capable of withstanding the applied pressure. The equipment is very simple in concept and also in practice. There are complications due to the interaction of the parameters such as temperature and membrane compaction, etc. which affect design and these will be treated in detail later.

10. ULTRAFILTRATION (UF)

So far we have dealt only with reverse osmosis. Ultrafiltration is in some ways even simpler than reverse osmosis being essentially free of the complication of the high pressures necessary for reverse osmosis. Again a pump and adequate membrane surface are required with suitable controls but, since the process is usually employed for the separation or concentration of organic species and foodstuffs, there is need to ensure that equipment can easily be sterilised. In reverse osmosis the membrane characteristics can be varied to give the maximum flux for the desalination factor required but the selectivity of the membrane for rejection of divalent as opposed to monovalent ions is not changed. In ultrafiltration the membrane is acting much more as a sieve and can be selected to give a particularly molecular weight cut-off point. Figure 4 illustrates the characteristics of a number of cellulosic membranes which are commercially available (from one supplier). In practice such membranes do not necessarily perform as might be expected from this graph.

The molecular weight is not the sole criterion of rejection but shape and charge are also involved. Also separation characteristics, as measured with single species, may be different in mixed solutions where the rejection of a simpler species is increased by the presence of highly rejected large molecules at the membrane surface.

Again, although ultrafiltration membranes have a more open structure than reverse osmosis membranes and are much more permeable to water, the higher water fluxes that should be obtained are often not realised because of concentration polarisation effects which become increasingly important as water transfer rates increase and also with the increased viscosities that are characteristic of many feeds treated by ultrafiltration.

It should be made quite clear that where the feed must be purified by removal of both salts and organic or colloidal material a reverse osmosis membrane must be used and this of course is more effective in removing large molecules than simple ions. The latter function however may be impaired because of fouling of the membrane surface with slimes of rejected organic colloidal matter and the operation of plants under these conditions requires special techniques.

REFERENCES

1.   Reid, C.E., Desalination by Reverse Osmosis, Merten, U. Ed.
      M.I.T. Press, Cambridge, Mass., U.S.A., 1966, Pages 5 - 7.

2.   Merten, U., Desalination by Reverse Osmosis, M.I.T. Press,
      Cambridge, Mass., U.S.A., 1966, Pages 20 - 21.

3.   Sourirajan, S., Reverse Osmosis, Logos Press Limited, London,
      1970.

TABLE 1

W.H.O. INTERNATIONAL STANDARDS FOR DRINKING WATER

1. Bacteriological Requirements

    Coliform micro organisms

        90% of samples - less than 1 count/100 ml
          all samples - less than 10 counts/100 ml

2. Chemical and Physical Requirements

    Toxic substances - maximum allowable concentration

Lead | (Pb) | 0.1 mg/l | Arsenic | (As) | 0.2 mg/l
Selenium | (Se) | 0.05 mg/l | Cyanide | (CN) | 0.01 mg/l
Chromium | (Cr^6) | 0.05 mg/l

    Other substances - permissible concentration

Fluoride (F) 1.5 mg/l — Total solids 500 mg/l
Nitrate (NO_3) 50 mg/l — Colour (Hazen units) 5 units
Iron (Fe) 0.3 mg/l — Turbidity 5 units
Manganese (Mn) 0.1 mg/l — Copper (Cu) 1.0 mg/l
Zinc (Zn) 5.0 mg/l — Magnesium (Mg) 50 mg/l
Calcium (Ca) 75 mg/l — Sulphate (SO_4) 200 mg/l
Phenol 0.001 mg/l — Chloride (Cl) 200 mg/l
Magnesium and Sodium Sulphate 500 mg/l

3. Also Biological and Radiological Requirements.

Table 2
Osmotic Pressure of Aqueous Sucrose Solutions at 30°C

Molality	Osmotic pressure/atm.		
	van't Hoff's Eq.	Eq. 4	Experimental
0.991	20.3	26.8	27.2
1.646	30.3	47.3	47.5
2.366	39.0	72.6	72.5
3.263	47.8	107.6	105.9
4.108	54.2	143.3	144.0
5.332	61.5	199.0	204.3

TABLE 3

Osmotic Pressures of 0.6M Aqueous Solutions at 25°C

Salt	Osmotic Pressure psi*	Osmotic Pressure psi per ion
$LiNO_3$	414	207
$NaNO_3$	374	187
$NaCl$	398	199
KCl	387	193
NH_4Cl	386	193
KNO_3	345	172
$CaCl_2$	607	202
$BaCl_2$	566	189
$Ca(NO_3)_2$	537	179
$Mg(NO_3)_2$	625	208
Na_2SO_4	439	146
K_2SO_4	440	147
$MgSO_4$	223	112
$CuSO_4$	200	100

*psi = 6.895×10^3 N/m^2

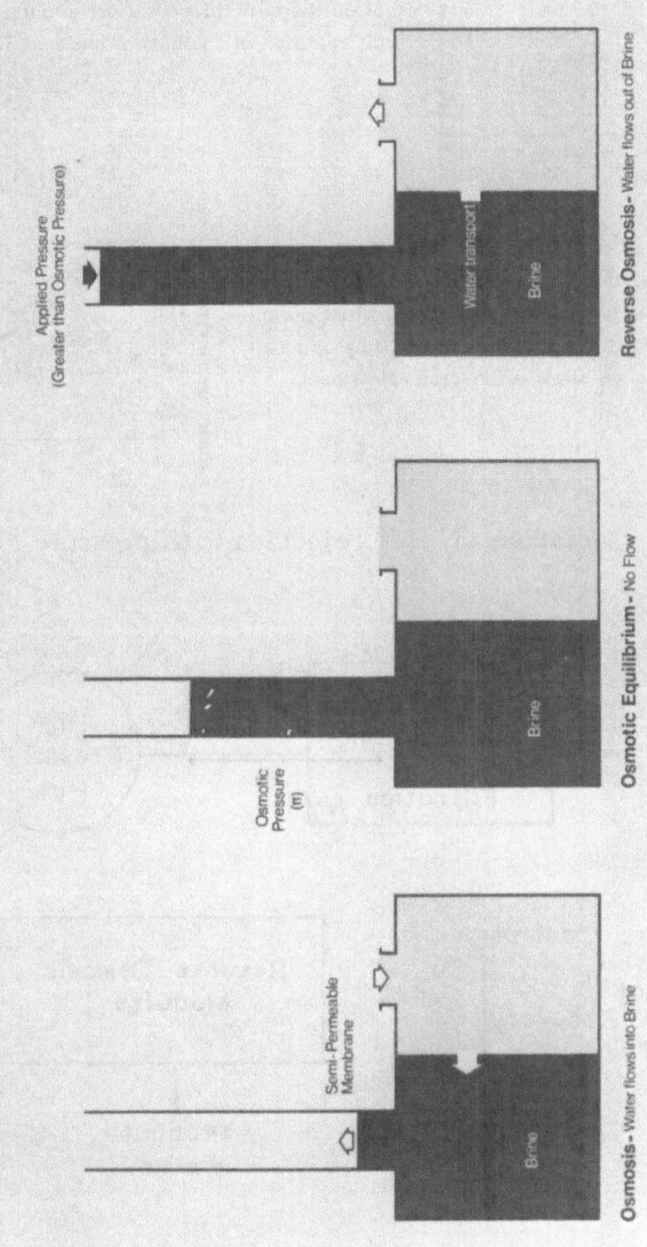

Principles of Osmotic Flow

Extract from Oxford Dictionary: from Greek "Osmosis" = push (tending to) percolation and intermixture of fluids separated by porous septa

Osmosis - Water flows into Brine

Osmotic Equilibrium - No Flow

Reverse Osmosis - Water flows out of Brine

Figure 1

46

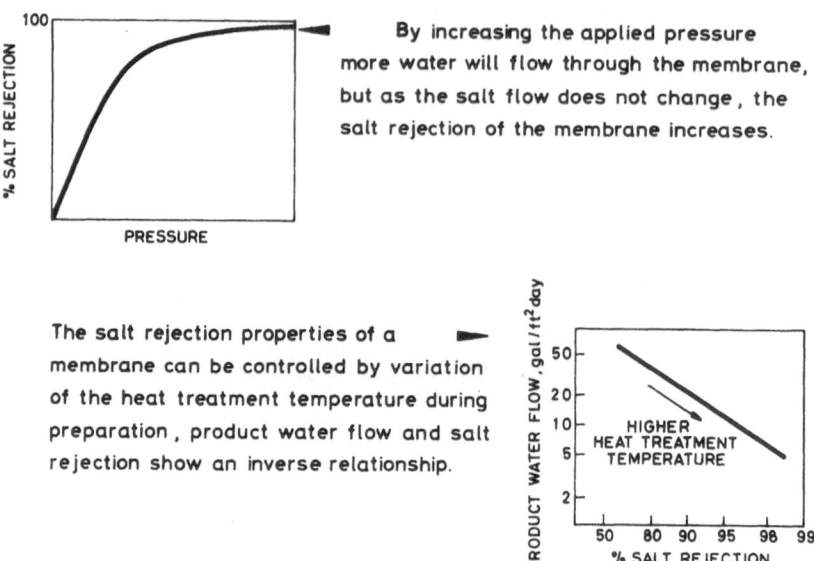

By increasing the applied pressure more water will flow through the membrane, but as the salt flow does not change, the salt rejection of the membrane increases.

The salt rejection properties of a membrane can be controlled by variation of the heat treatment temperature during preparation, product water flow and salt rejection show an inverse relationship.

Fig. 2.   Variation of salt rejection with pressure

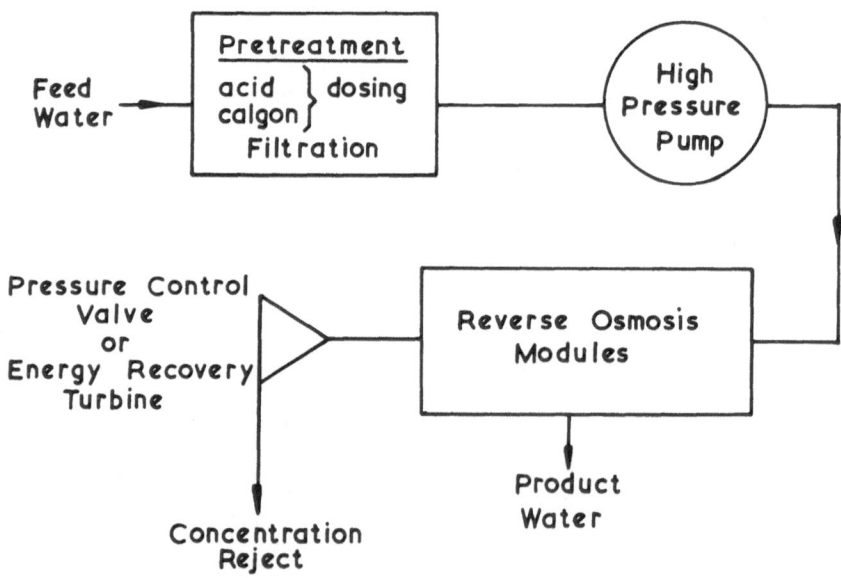

Fig. 3.   Diagram of a reverse osmosis plant

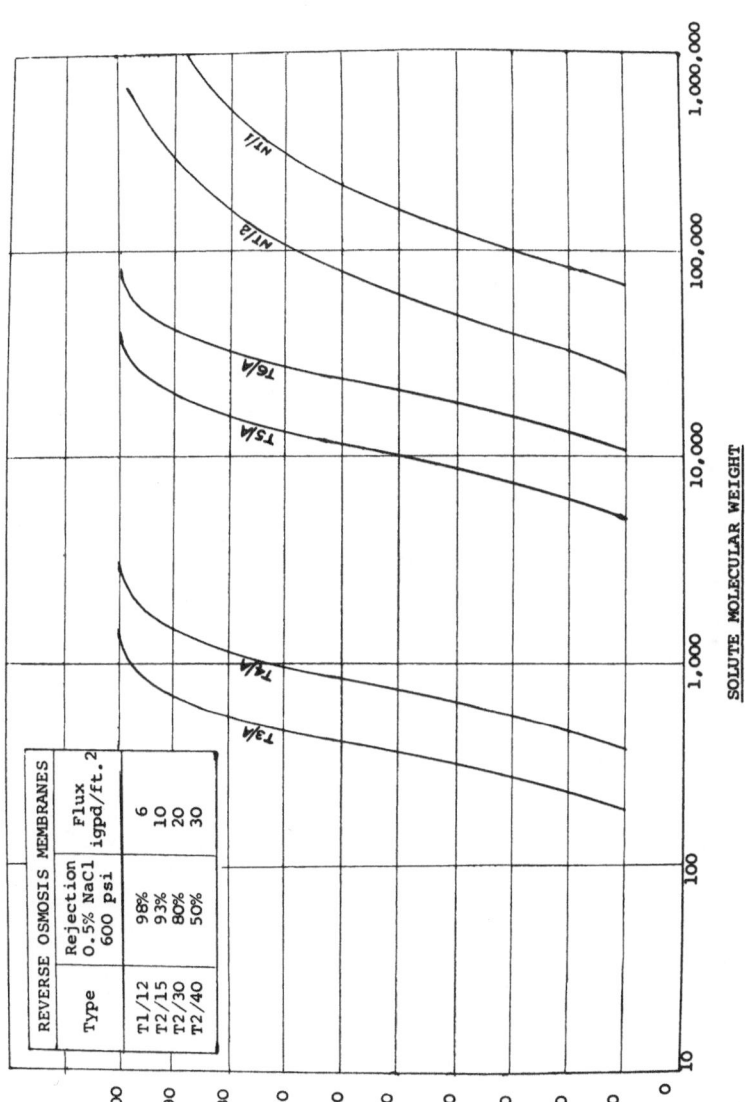

Figure 4 Characteristics of Ultrafiltration Membranes

MEMBRANES-FORMATION, STRUCTURE AND MECHANISM OF OPERATION

D. C. Sammon

Chemistry Division, AERE, Harwell,
Didcot, Oxon, U.K.

ABSTRACT. The development of asymmetric membranes of cellulose acetate is reviewed and the importance of each step in the formation process is discussed along with the possible mechanisms of these processes. Hollow fibres and composite membranes are briefly considered and various postulated mechanisms of salt rejection are discussed. These lead to suggested criteria to be applied in the development of new rejecting polymers.

Asymmetric ultrafiltration and reverse osmosis membranes have much in common and the range of available types of the former is outlined.

1. INTRODUCTION

A membrane can be defined as a barrier between two phases. Sometimes the barrier is totally impermeable as in a protective coating but more often it is semipermeable, that is permeable to some species and impermeable to others. The walls of living cells are semipermeable membranes and the operation of our kidneys is governed by various types of membrane. The immense field of biological membranes is a fascinating one which is outside the scope of this lecture though there is a great deal that we, who make artificial membranes, will have to learn before we can achieve the separations that nature manages so easily.

In addition to a membrane that is permeable to certain species only, a force is needed to cause the species to move. Various separation processes rely on membranes; four are listed in table I along with the driving forces used. Other processes

Table I

Membrane processes

Process	Driving force	Membrane permeable to
Dialysis	concentration difference	solutes below a fixed size
Electrodialysis	potential difference	anions or cations
Reverse osmosis	hydrostatic pressure	water
Ultrafiltration	hydrostatic pressure	water and solutes below a fixed size

are used but we will be concerned only with the last two, the pressure-driven processes. Practical membranes are not truly semi-permeable and this limits the degree of separation obtainable. Practical membranes must also allow enough material to go through for a separation process to be economically viable and this requirement has been met by the development of asymmetric membranes. The formation, structure and mechanism of operation of such membranes is the subject matter of this lecture.

2. OSMOTIC MEMBRANES

Before considering asymmetric membranes it is useful to review the early work on membranes which are permeable to water and impermeable to solutes. Such membranes can be used to demonstrate the phenomenon of osmosis and to measure osmotic pressures.

Most historical accounts of osmotic membranes start with experiments in 1748 by the Abbe Nollet who used a pig's bladder to separate water and ethanol. Only the former passed through the membrane but there is no indication that separation of an alcohol-water mixture was attempted or achieved. The flow of water only through animal membranes was found to occur for aqueous solutions in general and the term osmosis was used to describe the spontaneous flow of water through a membrane into a solution. The term semi-permeable was suggested by van't Hoff for such membranes. Inorganic membranes were made by Graham and by Traube in the middle of the nineteenth century and those made from copper ferrocyanide approximated closely to the ideal semi-permeable membrane. These precipitate membranes were flimsy until Pfeffer produced a robust form by causing the precipitation

to occur within the pores of a porous pot. van't Hoff used
Pfeffer's extensive measurements of osmotic pressure in his
well-known theory of solutions.

At that time two ideas were prevalent on the nature of these
membranes; some considered them as very fine sieves while others
postulated that the semipermeable character was due to water
dissolving in and diffusing through the membrane. Solutes did
not dissolve in the membrane and there was some evidence that
copper ferrocyanide membranes absorbed much greater amounts of
water than of sucrose. As an analogy one can consider layers
of three liquids in a tube. The liquids are ether at the top,
water in the middle and chloroform at the bottom and, since
adjacent pairs are incompletely miscible, well defined layers
are formed. However ether does dissolve to some extent in the
water and thus gradually diffuses through the water layer into
the chloroform. This last-named liquid does not dissolve in the
water and cannot reach the ether layer. Thus the water layer
is semipermeable and there is ample evidence that solution and
diffusion in membranes is of considerable relevance for reverse
osmosis.

Tinker[1] obtained some evidence in support of the sieve
theory by measuring the crystallite size of various precipitates
and calculating the size of the opening between close-packed
crystallites. He obtained good correlation between the size of
the openings and the degree of semipermeability of the membrane
made from the precipitate. A similar close-packed assembly of
aggregates of polymer molecules has been suggested to account for
the properties of reverse osmosis membranes.

Until about 1960 osmotic effects were mainly of interest to
physical chemists though, of course, they were readily obvious
in the plasmolysis of biological cells. Osmotic membranes were
not used in separation processes though ultrafiltration membranes
had been developed by Elford and Ferry in the 1930's.

3.   REVERSE OSMOSIS MEMBRANES

The stimulus came from the US Department of the Interior
which was concerned about inadequate water supplies and set up
the Office of Saline Water to study and develop possible solutions.
As well as conventional processes such as distillation the novel
idea of reverse osmosis was considered. In the spontaneous process,
water flows into the solution and a hydrostatic pressure can be
built up, but if a pressure is applied to the solution in
excess of the osmotic pressure the direction of water flow is
reversed - hence the name reverse osmosis. The osmotic pressure
of sea water is about 23 atmospheres, for brackish water it is

much less and the pressure used is governed by the need to obtain an economically useful flow through the membrane.

The precipitate membranes of the physical chemist are appropriate for measuring osmotic pressure but the flow rate is too low and the membranes are unsuitable in other ways. Attention was turned to polymeric films and of the many available types tested only one, cellulose acetate, was potentially useful. We have tested films from many sources by exposing them to salt solution at 67 atmospheres and the usual result is no flow - an impermeable film. Occasionally, as with gelatin, a fair amount of water passes through with unaltered salt content - not semi-permeable. Cellulose acetate lies between; a commercial film, 25µ thick, under our test conditions, gave a flux of 1/100 m/d with a 50-fold reduction in salt content. This permeation rate is too low but could be increased by making the film thinner, although appreciably thinner films are difficult to make and to handle.

French work in the 1930's had shown that magnesium perchlorate could be used to induce pore-formation in films of cellulose acetate. Loeb and Sourirajan developed a process for making an asymmetric membrane in which the bulk was highly porous but the top skin was dense and had properties similar to those of thick films of the polymer. The top skin is called the active layer and is usually less than 1µ in thickness. The rest is the porous layer which can be greater than 100µ in thickness and it offers little resistance to the passage of water and salt. The development of this asymmetric membrane has made reverse osmosis a commercially viable process. The permeation rate is increased about a hundredfold and adequate salt rejection is obtainable for many applications.

4. FORMATION OF ASYMMETRIC MEMBRANES

Asymmetric membranes of cellulose acetate are made from a solution of that polymer in a good solvent such as acetone with the addition of a pore-forming agent such as aqueous magnesium perchlorate[2] or formamide[3]. From this viscous solution the membrane is made in the following steps:-

i) casting - a layer of uniform thickness (150-250 µ) is spread out on a glass plate or porous substrate;

ii) evaporation - this layer is exposed to air for a period ranging from a few seconds to a few minutes;

iii)   gelation – the layer on its support is immersed in
water, usually for more than one hour;

iv)   heat treatment – the membrane so formed is immersed
in water at temperatures ranging from
60–90°C for a few minutes.

   The membranes are robust films somewhat thinner than the
cast layer. Depending on the composition of the casting solution,
they range from almost transparent to opaque. The membranes are
not readily damaged by handling. Details of the optimum conditions for membranes made using inorganic or organic pore-forming
agents or modifiers are given in table II.

Table II

Membrane production details

		Inorganic modifier	Organic modifier
Solution composition (weight %)	cellulose acetate acetone water magnesium perchlorate formamide	22.2 66.7 10.0 1.1 –	25 45 – – 30
Temperature (^{o}C)	casting evaporation gelation heat treatment	–5 to –10 –5 to –10 0 to 2 65 to 85	room room 0 to 2 65 to 85

   A great deal of work has gone into discovering the best
inorganic and organic modifiers and optimising the conditions of
formation. Factors governing the choice of components for the
casting solution will be dealt with later on. Qualitatively the
role of the four steps in membrane formation can be described
as follows:-

i)   casting – this simple process can be used to give flat sheet
membrane using the conventional doctor-blade technique. These
membranes normally come off polished supports, such as glass,
during the gelation step. It is possible to cast membranes
directly on to a porous support structure whether this be a flat
sheet or the inside or outside of a tube.

ii) evaporation - here the volatile acetone is lost from the surface layers and this contributes to the formation of the active layer. The permeate flux falls if the evaporation period is increased mainly due to the active layer increasing in thickness. The amount of salt passing through the membrane decreases faster than the permeate flux and thus the rejection factor increases. Too thin an active layer probably is somewhat incomplete.

iii) gelation - immersion in water solidifies the membrane. It does play a role as far as the active layer is concerned but the most obvious effect is the inhomogeneous precipitation of the bulk of the polymer giving rise to the porous structure. Gelation in water at room temperature results in a much weaker, brittle membrane. Additives can be used in the gelation bath to alter the properties of the membrane.

iv) heat treatment - the 'as-cast' membrane has little ability to reject salt and heating in water improves the salt rejection at the expense of permeate flux. As higher temperatures are used the performance of the membrane approaches that of a dense film of cellulose acetate of thickness equal to that of the active layer. The process can thus be used to adjust the performance of the membrane to meet the requirements of any particular application. Temperatures below $60^{\circ}C$ are ineffective in altering the properties whereas temperatures much in excess of $90^{\circ}C$ result in considerable loss of flux due to changes in the porous backing. The range of combinations of flux and rejection available in heat treatment depends on the composition of the casting solution.

Early work had shown that a cellulose acetate with $2\frac{1}{2}$ of the 3 hydroxyl groups replaced by acetyl gave the best performance. The recent trend has been to use more highly acetylated material and this has necessitated the formulation of different casting solutions.

5. MECHANISM OF FORMATION

The role of the modifier is to produce porosity in the bulk of the membrane and this has been clearly demonstrated in a study where the formamide content was systematically varied[4]. As the formamide content was increased from zero, the thickness, water content and water flux also increased. Electron micrographs indicated that the membranes were non-porous in the absence of modifier; with increasing amounts of formamide the closed-cell porosity which resulted in very low water flux gave way to the open-cell structure and high flux characteristic of the Loeb-Sourirajan membrane. The results were also discussed in terms of the effect of formamide on the solvent-power of acetone. The formation of the pores results from the separation of the polymer solution into polymer-rich and polymer-poor phases due to loss of acetone and addition of water.

The three component polymer solution is complex enough before the addition of the fourth component, water. An alternative approach was to study membranes formed from cellulose acetate dissolved in a range of single solvents. Each solution contained 20% of polymer and was cast, evaporated for less than 1 minute and gelled for 1 hour at $0^{\circ}C$. The results are given in table III and demonstrate the variations in structure due simply to varying the single solvent[5]. Membranes made from solutions of cellulose acetate in acetic acid are not greatly inferior to those produced from the solutions detailed in table II. These various structures of all-skin, no skin and skin on top or bottom were then correlated with the amount of water needed to precipitate polymer from the solvent and with the direction and magnitude of flow of solvent out and water into the cast solution during gelation. The directions of flow are summarised as follows:-

acetone and dioxane - net flow is from polymer solution to
water;

acetic acid and triethyl phosphate - no flow detectable;

dimethyl formamide and dimethyl sulphoxide - net flow is from
water to solution.

The porosity is likely to be fixed by the polymer content at the moment of precipitation. Since solutions from the first two solvents become more concentrated due to solvent loss, a lower porosity will result than for the other four solvents. The skin on the membrane made from acetic acid is probably due to the volatility of this solvent (at least with respect to solvents such as triethyl phosphate). The non-volatile nature of the last two solvents accounts for the absence of a top skin but the reason for the bottom skin is not clear. Neither, too, are the factors governing the direction of flow.

Thus the mechanism of membrane formation is beginning to be understood at least for single solvent systems. Solutions used in membrane manufacture are not binary, thus the mechanism will be complex though the basic features are likely to be. similar.

6.   CHOICE OF SOLVENTS AND MODIFIERS

There is now sufficient evidence to show that cellulose acetate is not unique in its ability to be formed into asymmetric

Table III

Properties of membranes cast from 20 weight % cellulose acetate, evaporated for < 1 minute and gelled in ice-water

Casting solvent	Biebrich Scarlet dye penetration into:		Flow rate* 10^4 (ml/cm^2.sec)		Salt rejection* (%)		Water content % weight	Structure of membrane
	Air side surface	Glass side surface	Air side in contact with salt solution	Glass side in contact with salt solution	Air side in contact with salt solution	Glass side in contact with salt solution		
Acetone	−	−	0.07	**	82	**	40	Homogeneous, very dense
Dioxane	−	−	0.3	0.3	92	89	71	Homogeneous, dense
Acetic acid	−	+	2.4	4.0	60	23	73	Skin, air side
Triethyl phosphate (TEP)	+	+	77	69	None	None	75	Homogeneous, porous
Dimethyl formamide (DMF)	+	−	190	18	None	77	78	Skin, glass side
Dimethyl sulfoxide (DMSO)	+	−	>280	38	None	25	78	Skin, glass side

*Measured at 40 atm operating pressure.
**Practically identical to the performance of the air side surface.

membranes. Ultrafiltration membranes of this type have been available for some time and recently ethyl cellulose[6] and polyamides[7] have been made into asymmetric reverse osmosis membranes. As new salt-rejecting polymers become available, appropriate solvents and procedures will have to be identified for them and it would be useful to have some guiding principles to replace the time-consuming empirical approach.

The general requirement is that a solvent or solvents must be identified and that these should be miscible with water since that is the usual (but not essential) gelation medium. The use of solubility parameters has been of great value in formulating solvent systems for paints and varnishes and this approach was used in the development of asymmetric membranes from ethyl cellulose. The energy of evaporation of a solvent is required to overcome contributions from three types of force or bond namely dispersion and polar forces and hydrogen bonds. These three components are useful in the prediction of the solvent power of mixtures hence the term solubility parameter for each of the three contributions and for the total effect. The contributions can be calculated from various physico-chemical parameters of the solvents and thus each solvent or mixture of solvents can be represented as a point on a three-dimensional plot with the three components as axes. For any polymer, all the solvents that dissolve it occur clustered together in a solubility zone in the three-dimensional plot and thus the prediction of additional solvent systems is facilitated.

This method has been little used as yet and, in any case, gives information on equilibrium effects only whereas kinetic effects are of considerable importance in membrane formation. In addition it is not yet clear whether the ideal solvent mixture should be near or far from the boundary of the solubility zone or what the relationship there should be between the position of the solvent and of the gelation medium which, of course, is outside the zone. However, as more desalting polymers become available some of these questions should be answered.

7.   STRUCTURE OF MEMBRANES

A number of techniques have been used to elucidate the structure of asymmetric membranes. The most direct approach is to use electron microscopy; transmission techniques give most information about the active layer while the scanning microscope is better for studying the rest of the membrane[8]. Three layers can be identified and are designated A, B and C. The A-layer is the active layer and no structure is visible in micrographs of membrane cross-sections. With increasing evaporation period the A-layer increases in thickness and then decreases as the thickness of the B-layer increases. This latter has a laminar structure but its transport properties are not clear. The C-layer is obviously porous though the porosity may be uniformly distributed or may be quite inhomogeneous due to the presence of pear-shaped vesicles which may extend through an appreciable fraction of the thickness of the membrane.

Some evidence of a nodular structure in the active layer is seen in micrographs of shadowed surface replicas[9]. The active layer may thus be a close-packed assembly of aggregates of cellulose acetate chains. The diameter of these aggregates is about 18.5 nm and it is concluded that water and salt are transported through the much smaller intervening gaps. Supramolecular aggregates do occur in solutions of cellulose acetate and thus the structure and properties of the active layer would appear to be derived from the state of the polymer in the casting solutions. Altering solution conditions to decrease the size of the aggregate has been shown to produce improvements in salt rejection due, presumably, to smaller gaps between the aggregates[10]. This recalls the work of Tinker mentioned earlier. Further evidence on the structure of the salt rejecting layer will be presented in the section on mechanisms of desalting.

The porous C-layer has been considered as simply a support for the thin active layer and it was assumed to offer little resistance to the flow of water or salt through it. It contains 60 to 70% of water and yet has to be strong enough to resist compaction by the pressures employed in reverse osmosis. The hydraulic resistance of the active layer is considerably increased by the heat treatment process while that of the porous layer is little changed and we have found that the porous layer contributes less than 10% to the total hydraulic resistance of high rejection membranes. For lower heat treatment temperatures, that is lower rejection membranes, the porous layer contributes significantly to the overall resistance. Changes in the hydraulic resistance of the membrane on prolonged exposure to high enough pressures can be correlated with changes in resistance of the porous layer and changes in thickness. The ability to resist compaction is dependent on the composition of the casting solution.

The Loeb-Sourirajan has to be highly asymmetric in order to produce a high-enough flux but for hollow fibres a much lower degree of asymmetry is necessary. These fibres are typically 50μ outside diameter and 25μ inside and for them a much lower permeate flow per unit area is acceptable. Indeed too high a flow could not be utilised since the pressure drop along the interior of the fibre would be excessive - these fibres have a skin on the outside and are strong enough to withstand external pressure of the magnitude used in reverse osmosis. Thus thicker active layers and lower porosity in the support are acceptable and can be produced either by gelation in water or by thermal gelation[11].

The asymmetric membranes considered so far have the same polymer in both active and porous layers but it is possible to make a useful composite membrane where the thin active layer of one polymer is deposited on a porous support made separately and usually from another material. In principle one could separately optimise each layer and produce a membrane of performance superior to that of the asymmetric membrane. Such composite membranes have been made by depositing, on a porous cellulose acetate-cellulose nitrate support, a layer of cellulose acetate less than 0.1μ in thickness[12]. The support is made under carefully controlled conditions to ensure the optimum pore structure on the surface and is coated by dipping it into and withdrawing it from a dilute solution of polymer. A masking layer can be applied to the porous support prior to coating. These membranes have stable, long-term fluxes of 1 m/d and over 99.5% rejection of salt from sea water at 102 atmospheres. Composite membranes from other polymers have also been reported[13].

8. MECHANISM OF SALT REJECTION

All the work outlined earlier on the production and structure of reverse osmosis membranes has been extended by studies on the mechanism of salt rejection. A thorough understanding of this is of importance in improving membranes made from cellulose acetate but it is also of considerable relevance in the search for new desalting polymers.

The solution-diffusion mechanism of transport through the membrane has received considerable support. Both water and salt are assumed to dissolve in the polymer and diffuse through. The permeability of each species can be calculated from measurements of uptake and diffusion coefficient on dense films. The results so obtained can be used to calculate rejection factors under reverse osmosis conditions and these are compared with experimental values. Both sets of figures for different solutes are presented in table IV[14]. Though the trends are similar, calculated RF's are generally higher than observed ones. Some but not all, of the difference can be ascribed to tiny flaws in the membranes. Nonetheless, it is clear that the solution-diffusion mechanism does go a fair way towards accounting for the observed results. Thus salt rejection can be ascribed to the low uptake and diffusion of salt in the membrane and any factors that decrease these parameters, when compared with those for water, would be expected to result in increased salt rejection. Increasing the acetyl content of the polymer is one way of achieving this result.

Table IV

Comparison of calculated and observed rejection factors (RF)

Solute	Observed RF	Calculated RF
NaCl	50–90	150
NH_4Cl	35	61
NaH_2PO_4	1400	1100
$NaNO_3$	28	42
$NaHCO_3$	140	560
Urea	1.8	5.7
H_3BO_3	2.2	5.3

Water and salt are dissolved in the polymer but the solution-diffusion hypothesis does not predict how these are distributed. They could be grouped together in pores or spread out uniformly. Another mechanism requires the existence of pores; it is called the preferential sorption-capillary flow mechanism[15]. Due to negative adsorption (i.e. depletion) of salt at a water-air interface a very thin layer of pure water exists at that interface. If the same were true for the interface between water and a porous membrane, then pure water could be drawn off through the pores if these had a diameter less than twice the thickness of the pure water layer. Bigger pores would result in water of lower purity being transported. Thus the essential requirement in this hypothesis is to have pores of appropriate diameter. Rejection by porous membranes has also been ascribed to dielectric repulsion experienced by ions in the pores[16].

The size of pores in dense cellulose acetate can be measured by a number of techniques which all give fairly similar results. A clear correlation exists between decreasing pore diameter and increasing salt rejection[17]; for appreciable salt rejection pore diameters have to be in the range 0.6 to 1.0 nm. It should be noted that the assumptions inherent in the various techniques are invalid for such small pores though the results are compatible with the good rejection of sucrose (diameter 1.0 nm) and the poor rejection of urea (diameter 0.4 nm). Pores of these diameters can scarcely be considered as cylindrical holes. The thickness of cellulose acetate chains is 1.0 nm and, in the polymer, gaps of comparable dimensions must occur between the chains[18]. Thus the pore must be seen as made up of a number of such gaps.

Regardless of the mechanism proposed it is obvious that the amount and ordering of the water in the polymer is of crucial importance. With too much water there would be little difference in the solution inside and outside the membrane and hence little rejection. With too little water the permeate flux would be unacceptably low. In general, salt rejection does increase with decreasing water content and it has been the aim of those developing new membrane materials to produce a system where water content can be systematically varied. However water content is not the sole criterion; the ratio of amounts of bound and free water is important as is the extent to which water molecules are clustered together. It is in this area of the way that water is distributed in the polymer that advances in understanding will be needed in order to design new polymers with desalting properties.

9. ULTRAFILTRATION MEMBRANES

These membranes have much in common with those described so far. Asymmetry is required, not so much for increasing the flux, but for confining the separative properties to the surface skin in order to prevent blocking of pores. These membranes have fairly well defined pores and can be made by casting techniques similar to those used for cellulose acetate. Indeed a range of ultrafiltration membranes can be made from cellulose acetate solutions of somewhat different compositions to those appropriate for reverse osmosis. Ultrafiltration membranes can be produced from a range of polymers though little has been published on the factors responsible for control of pore size and degree of asymmetry. There are many patents relating to production of such membranes. The mechanism of operation is simply sieving and a range of pore sizes can be obtained. The most important characteristic of these membranes is the molecular weight cut-off i.e. the minimum size of solute that is rejected by the membrane. Flat sheets of various specifications are available from various manufacturers as are tubes which usually have the skin on the inside.

The chemistry of the polymer is of prime importance for reverse osmosis; indeed the properties of the active layer approximate to those of the dense material when in equilibrium with water. For ultrafiltration membranes the chemical constitution is of less importance and the properties of dense films offer no guide to ultrafiltration potential. Ability to form a porous structure with the right characteristics is the essential feature.

10. CONCLUSION

Asymmetric membranes are proving most useful in a range of applications yet our understanding of the mechanisms of formation and operation is still far from complete. There are ample opportunities for further work and as we develop a more detailed picture we should be able to produce improvements in performance.

REFERENCES

1. F. Tinker, Proc. Roy. Soc. A92, 1916, p2.
2. S. Loeb & J. W. McCutchan, Ind. Eng. Chem. Prod. Res. Dev. 4(2) 1965 p114.
3. S. Manjikian, Ind. Eng. Chem. Prod. Res. Dev. 6(1) 1967 p23.
4. R. E. Kesting & A. Menefee, Koll. Z. u Z. für Polym 230(2) 1969 p341.
5. R. Bloch & M. A. Frommer, Desalination 7(2) 1970 p259.
6. H.K. Lonsdale & H.E. Podall 'Reverse Osmosis Membrane Research' Plenum Press 1972 p61.
7. H.K. Lonsdale & H.E. Podall 'Reverse Osmosis Membrane Research' Plenum Press 1972 p253.
8. G. J. Gittens, P.A. Hitchcock & G. E. Wakely, Desalination 12(3) 1973 p315.
9. R. D. Schultz & S. K. Asunmaa Rec. Prog. Surf. Sci. 3, 1970 p291.
10. B. Kunst & S. Sourirajan, J. Appl. Polym. Sci. 14, 1970 p1983.
11. H. K. Lonsdale & H. E. Podall 'Reverse Osmosis Membrane Research' Plenum Press 1972 p331.
12. R. L. Riley, G. R. Hightower & C. R. Lyons, Appl. Polym. Sci. Symposia No. 22, 1973 p255.
13. L. T. Rozelle, J. E. Cadotte, B. R. Nelson & C. V. Kopp Appl. Polym. Sci. Symposia No. 22, 1973 p223.
14. H. K. Lonsdale, B. P. Cross, F. M. Graber & C. E. Milstead J. Macromol. Sci.-Phys. B5(1) 1971 p167.
15. S. Sourirajan, 'Reverse Osmosis' Logos Press Ltd., 1970 p2.
16. S. Sourirajan, 'Reverse Osmosis' Logos Press Ltd., 1970 p163.
17. M. N. Sarbolouki & I. G. Miller, Desalination 12(3) 1973 p341.
18. P. Meares, J. B. Craig & J. Webster 'Diffusion Processes' (ed. J. N. Sherwood et al) 2 (1971) p609.

THE PERFORMANCE OF MEMBRANES - TEST METHODS AND RESULTS

D.C. Sammon

Chemistry Division, AERE, Harwell,
Didcot, Oxon, U.K.

ABSTRACT. Techniques and equipment are reviewed for testing reverse osmosis membranes either as flat sheets or in the forms used in plant. The performance under varying conditions of pressure, temperature and solute concentration is discussed and the behaviour with a wide range of solutes is surveyed. Factors affecting the life of membranes are considered.

The different requirements for ultrafiltration membranes are noted, particularly the importance of hydrodynamic factors which control gel-layer formation.

1. INTRODUCTION

The 'performance' of a reverse osmosis membrane is of crucial importance in assessing suitability in an envisaged application yet a brief survey of the literature would indicate that no universal way of expressing performance has emerged. Indeed it is unlikely that any simple way will emerge since potential applications embrace a wide range of variables. In the desalination field one can go from one extreme of the production of potable water from sea water in a single pass to the other of simply softening brackish water supplies. The former requires 99.5% rejection of NaCl and usually a pressure of 60 atm. or more whereas small scale units for the latter could operate at 3 - 6 atm. and only reject 50% of NaCl, but, of course, a larger percentage of calcium and magnesium salts. It is unlikely that one set of test conditions would be suitable for assessing the potential of membranes for both extremes.

However one is not always faced with such widely separated extremes and some compromise is often possible. In addition it is possible to predict performance under conditions other than those used for the test and this is clearly of value in comparing results from different sources. The main portion of this lecture will be devoted to describing how membranes perform under a range of conditions of pressure, temperature, solute concentration, solute type and elapsed time. Before coming to that survey we must be clear what we are trying to measure and how we can do it. Most of the discussion will be concerned with cellulose acetate membranes since the major part of the published information relates to them.

2.   BASIC EQUATIONS

Under any given set of conditions one is usually interested in measuring two properties namely how much water comes through the membrane and what is the solute content of that water. These measurements are easily made if one has the appropriate test equipment and are simply related to the basic membrane parameters, the water and salt permeability constants by the equations given earlier viz:-

$$F_w = A(\Delta P - \Delta \pi) \tag{1}$$

$$F_s = B(c_m - c_p) \tag{2}$$

F_w and c_p are measured, c_m is often assumed equal to c_f (or can be calculated from c_f) and $\Delta \pi$ is either small compared to ΔP or can be obtained from c_m (via c_f) and c_p using tabulated data[1]. $\Delta \pi$ can also be calculated from F_w and F_p which is the permeate flux in the absence of solutes.

$$F_p/F_w = \Delta P/(\Delta P - \Delta \pi) \tag{3}$$

Though c_p is measured, the parameter quoted is usually percentage rejection or rejection factor. The former is normally used in the technical literature while the latter, though used less often is more readily converted to F_s

$$F_s = F_w c_p \tag{4}$$

$$= c_f \times F_w/RF \tag{5}$$

and, for constant c_f $\alpha \; F_w/RF \tag{6}$

If we can assume that $c_m = c_f$ then

$$B = F_w/(RF - 1) \qquad (7)$$

Equations (2) and (4) - (7) have no pressure term and in many cases B is independent of pressure as would be expected from a simple solution-diffusion model of salt transport. The experimental evidence is that B is not always independent of pressure but then there is ample evidence that the solution-diffusion mechanism is not the only mechanism for transport of salt across the membrane. If equations (1) and (2) held under all conditions then the constants A and B would define the performance of the membrane, at least for a given solute and at a given temperature. For a more complete definition the temperature dependence of A and B would be needed as well as the relationships between B values for different salts.

The above treatment is clearly not totally rigorous and no doubt would not satisfy the theoretical experts. One obvious deficiency is the use of concentrations instead of activities but the emphasis here is more on the overall picture rather than the fine detail. There are various assumptions inherent in these equations and one has to remember this particularly in trying to compare measurements made under widely differing conditions.

The simplest comparison one can make is between membranes tested under the same conditions. Here the variable can be annealing temperature or composition of casting solution and we have found that graphs of log F_w against log RF permit comparisons to be made (under conditions where $P \gg \Delta\pi$). For membranes annealed at different temperatures a performance line is obtained as shown in figure I and this line can be used as a reference standard for comparing membranes from different casting solutions. Those with F_w and RF values above the line are better, those below are worse.

Leaks cause obvious problems. It is fairly easy to obviate leaks past seals but flaws in the membrane remain a problem. Fortunately small leaks are often plugged by colloidal species in the feed and this process is one contribution to the settling down period observed prior to fairly steady performance being obtained.

The symbols presented here have been in use at Harwell over a period of years but others appear in the literature. In Table I the units used here are compared with those used by Sourirajan and co-workers[2] who have made an extensive contribution to the literature on reverse osmosis. Different units are also used but these are readily interconverted.

The above discussion has been concentrated on the performance of economically useful membranes. Similar considerations apply to the evaluation of thin polymer films which are usually homogeneous in contrast to the asymmetry of most membranes. For these films the tendency is to use the fundamental parameters A and B to a greater extent than is the case for membranes.

To sum up, F_w and c_p are measured but must be converted to A and B in order to understand how the membrane will function under different conditions.

3. TEST EQUIPMENT

The simple requirements are a membrane and a pressurised feed along with some method of measuring F_w and c_p. The membrane must be adequately supported and the support must prevent damage to the membrane and yet must not contribute significantly to the hydraulic resistance of the system. The final requirement stems from the need to know c_m rather than c_f or at least to ensure that $c_m \approx c_f$. This requires in most cases a reproducible movement of the feed with respect to the membrane surface and is achieved either by pumping the feed solution or, in the case of flat membranes, stirring the feed. It is common to have one pump to produce pressure and flow though applying a pressure from a gas cylinder and pumping or stirring has been used. Sheet membrane is normally supported on a porous plate commonly of woven or sintered metal with sometimes filter paper interposed to counter roughness or coarse structure of the support.

Many designs of test cell have been published[3]. It is possible to make useful measurements in an unstirred cell[4] but since c_m changes with time F_w and c_p will also vary and the experimental measurements are not so directly related to A and B as is the case in a system where solution at constant c_f is pumped at pressure past the surface of a membrane. Mechanical stirring is used but here c_f will vary with time. The most common system is a recirculating loop with total volume large compared to the amounts of permeate collected to make measurements of F_w and c_p. Thus c_f remains virtually constant. The various designs differ depending on whether the flow past the membrane surface is laminar or turbulent. Under ideal conditions c_m is calculable from the dimensions and the volume flowrate of the feed. Otherwise conditions are chosen so that $c_m \approx c_f$ or c_m is derived empirically. We have used a recirculation loop system as shown in figure II[5] and in the cells turbulence is induced by directing the flow in a zig-zag pattern across the entire circular membrane area. A cross section of the cell is shown in figure III. In these cells conditions can be chosen so that $c_m < 1.1 \times c_f$.

A radial flow cell has been described[3d] in which the flow is laminar and c_m can be calculated. A dead spot occurs at the central inlet zone and problems at inlet and outlet points are a common feature of many designs.

For membranes on the inside of tubes the flow conditions are much more amenable to mathematical treatment and figure IV[6] shows how c_m varies with linear velocity. Membranes on the outside of tubes are frequently operated in a bundle as is also the case for hollow fibres and with these latter the whole problem is much more complex[7].

A complete discussion of concentration polarisation and the derived relationships between c_m, c_f and other parameters is beyond the scope of this lecture but one must be aware of such problems in interpreting test results. There is an extensive literature on this subject but a number of aspects are drawn together in Merten's book which though several years old still is of considerable value.

Measurements of F_w and c_p present no problems. The permeate is collected for a known time and the weight or volume is measured. c_p (and c_f) is measured either by non-specific methods such as conductimetry or refractometry or by any appropriate analytical technique.

As far as the actual equipment is concerned the materials of construction must be compatible with the solutions used. Corrosion products can be deposited on the membrane and cause a decrease in F_w. Much depends on the level of accuracy sought. In short term tests the corrosion of stainless steel by chloride solutions is not of much importance but we have found it impossible to measure reliably the small changes in performance with time in our stainless steel equipment (mainly type 316) if chloride solutions are used.

Choice of pumps, valves etc. depends on availability though reliability of pumps often leaves a lot to be desired.

4.   EFFECTS OF PRESSURE, TEMPERATURE AND CONCENTRATION

4.1 Effect of pressure

Equation (1) does express accurately the pressure dependence

of membrane flux in the absence of changes in A with time.
Changes in A with time will be discussed later in detail but
it is sufficient here to note that such pressure-induced
changes are permanent and A fairly soon reaches a quasi-steady
state. Thus for membranes that have been in use for some time
or for membranes that have been exposed previously to a higher
pressure A can be assumed to be constant. At low enough pressures
this is true for unused membranes and this is shown in the
F_W - P graphs in the literature. At low pressures the plot is
linear but with increasing pressure F_W increases less than
linearly in many examples since A is decreasing with time.

4.2 Effect of temperature

The effect of increasing the temperature is to increase A
and for various salts B increases by almost the same amount thus
the rejection factor is scarcely changed. For flow through a
set of pores of large enough dimensions the temperature dependence
of hydraulic permeability would be the same as the dependence
found for viscosity. However the observed result is that A
increases slightly less rapidly than viscosity decreases[9].
There is also some evidence that changes in A and B depend on
tightness of the membrane[10]. The overall result is that the
water flux is almost doubled by a 25°C rise in temperature.

4.3 Effect of concentration

Increasing the solute concentration causes a decrease in
F_W and in RF. These effects are most marked when $\Delta\pi$ is an
appreciable fraction of ΔP. The overall result is that less
water goes through the membrane and c_p rises more rapidly than
c_f since there is now less water to dilute the salt passing
through. There is no evidence that A is affected by solute
concentration for most solutes unless, of course, the solute
interacts with the membrane as happens with some organic
species. Such changes in A are usually irreversible. There
is not much evidence that B is appreciably affected by salt
concentration and small changes can only be detected if c_m is
known with sufficient accuracy. With a solution-diffusion
transport model it is possible that either solubility or diffusion could be a function of salt concentration but there is
little if any experimental evidence here. In some cases there
is a degree of coupling between salt and water fluxes and thus
decreases in F_W as $\Delta\pi$ increases would cause an apparent
decrease in B. There is some evidence that coupled flow increases
slightly with salt concentration[11].

Almost all the information on solute concentration relates

to simple salts; there is very little on other solutes.

5. TIME EFFECTS

Physical and chemical instability of membranes leads to changes in A and B. Fouling - the deposition of material on the membrane surface - causes changes in F_w and c_p and thus apparent changes in A and B. One normally attempts to prevent the occurrence of fouling but this is not always possible in laboratory tests and even less so in real applications.

5.1 Effect of pressure

Polymers such as cellulose acetate creep under applied pressure and a great deal of work has been done on measuring changes in A as a result of exposure to pressure. These changes are small and are becoming smaller as improved membranes become available. Such small changes are easily masked by other effects, particularly fouling. Hydrolytic effects are also possible but can be virtually eliminated by appropriate choice of pH. The effects are small and, in order to measure them with any degree of certainty, measurements must be made under carefully controlled conditions. We have found it advisable to keep the temperature constant to within 0.2°C while making measurements of F_w and c_p and to within 0.5° at other times. Pressure must likewise be carefully controlled but fouling is the principal problem. Others[3d] have found it necessary to use a once-through system - no recirculation - with equipment of Monel to limit corrosion due to chloride and periodic cleaning was also necessary.

We adopted the different approach of using a salt which was less corrosive namely sodium nitrate though this resulted in the growth of organic slimes. These, however, were eliminated by intermittently dosing with formaldehyde. Thorough cleaning between experiments was also practised. The total elimination of fouling is difficult to prove but, using these methods, no deposits were visible. Furthermore, as techniques were improved results became much more reproducible and the pattern of changes in A for different membranes became less random and more systematic. We have some evidence for the presence of colloidal species in the feed solution but either these have a highly reproducible effect or they are of little consequence in our tests.

Most of the work reported has been concerned with changes in A only but we have measured changes in A and B. It has become customary[12] to present these changes as a log-log plot of F_w versus time and such plots are linear. What is true for F_w is also true for A and we have demonstrated a similar relationship for B (and F_S). Thus it is possible to characterise each membrane under a given set of conditions by the negative slopes of the log-log plots of A and B versus time and these are given the symbols m_A and m_B. Typical plots are shown in figures V – VII in which the range of linearity covers four orders of magnitude. There is additional evidence that the linearity extends to much shorter and to longer times. The values of m_A and m_B for two types of membrane are shown in figure VIII. The difference between the L-7 (cellulose acetate-acetone-formanide) and the blend W-16 membranes is obvious and demonstrates the superiority of the latter. m_A increases with pressure (and with temperature) and A-value of membrane. For m_B the picture is not so simple but the relevance is more obvious in figure IX. For $m_B/m_A > 1$, RF will increase with time and this is the normal result with the ratio approaching unity as A increases.

The relevance of the m_A values is shown in table II. A value of 0.01 would be quite acceptable whereas 0.07 would mean that F_w would be almost halved in a year. The linearity of the log-log plot means that the membranes 'settle-down' fairly quickly. With m_A = 0.01, F_w decreases by 7.4% in one month and by only another 2.3% to the end of one year. To put it another way, 5% loss of flux from one hour to one week is followed by a further 5% loss in 168 x 168 hours, – about 3 years.

Changes in F_w with time are permanent being little affected by periods at lower pressure. Transient effects are sometimes observed for a short period after application of pressure. For changes in B the position is not so clear though somewhat similar effects are seen. For these membranes, A at 200 psi is virtually constant and this value and the appropriate value of m_A can be used to calculate A for any time at constant pressure. The non-linearity of plots of F_w versus P can readily be understood when one takes into account these pressure-dependent time effects. The changes in A are largely due to partial collapse of the porous backing whereas the causes of the changes in B are still not clear.

5.2 Chemical changes

The most important chemical instability of cellulose acetate is towards solutions of high and low p_H where hydrolysis results in the loss of acetyl groups. This in turn results in increases in A and B. Results have been reported on the hydrolysis of cellulose acetate powder and these are summarised in figure X[13]. The rate constant is at a minimum at p_H 5.0 and increases by a factor of 10 for an increase in temperature of 23°C and a change in p_H of about 1 unit (for values outside the range p_H 4-6). The resultant effect on the membrane is less well quantified though some information available indicates that log A and log B increase linearly with time. B increases more rapidly than A and thus the rejection factor decreases.

The rate of hydrolysis will be governed by the p_H within the active layer and this 'local' p_H will be different from that in the external solution since electrolytes are excluded from the swollen polymer. This 'local' p_H will depend on the species present as well as on the external p_H. In a carbonate-bicarbonate buffer, decrease in p_H is the result of decreasing the carbonate to bicarbonate ratio and since the uptake of the former into the polymer is less than that for the latter the local p_H should drop. For a sodium hydroxide solution the local p_H will also be less than the external value but there is no reason why the local p_H should be the same for the buffer as for the other solution. The precise meaning of 'local' p_H is not clear and there is no experimental evidence to quantify the effects postulated.

The initial effect of hydrolysis is to increase A and B but the increase in A slows down and then A decreases as the weakened porous structure begins to collapse. The net result as far as practical applications are concerned is that the p_H has to be kept to less than p_H8 (and presumably more than 2) if the lifetime is not to be shortened. Many natural waters have a p_H of around 8 and thus life may be limited particularly at higher operating temperatures.

Organic solutes can also cause degradation of cellulose acetate membranes but there is little quantitative information. The aromatic nylon membranes used in hollow fibres can be operated up to p_H 10[14] but there is little published information on these membranes. They are, however, sensitive to small quantities of chlorine (< 1 ppm). Other more resistant membranes have been described[15].

5.3 Fouling

Fouling is the deposition of material on the membrane surface and results in a decrease in F_w, with, on occasion, changes in RF. The concentration of any rejected species is higher at the membrane surface than in the feed and this can result in the solubility limit of a sparingly soluble salt being exceeded. Precipitation does occur in this way but a much more common effect is due to colloidal species where aggregation can result in a layer being formed on the membrane.

Depending on the nature of the species such aggregation can be reversible or irreversible. The relationship between F_w and time is the same as for pressure dependent effects namely a linear relation between $\log F_w$ and $\log$ time[16].

Decreases in RF as well as F_w would be expected and further decreases in RF might occur due to increased concentration polarisation. However decreases are not always observed because the fouling layer enhances the rejection of the membrane either by blocking flaws or in some other way.

Fouling is a problem in practical applications and also in laboratory studies as outlined earlier but so far few systematic investigations have been made which would enable one to predict how much fouling would be produced by a given species. One problem is the reproducibility of properties of some colloidal species. At the moment a given feed solution has to be tested empirically to measure how much fouling is produced. Increase in the feed velocity past the membrane reduces the amount of fouling observed.

6. REJECTION OF DIFFERENT SOLUTES

Early studies were almost all confined to inorganic salts but now a wide range of organic solutes has been studied. For cellulose acetate a number of factors have been identified as important in the rejection of solutes. These are:-

charge on solute
ability of solute to form hydrogen bonds
size of ion or molecule

6.1 Rejection of electrolytes

Here the size of the charge is of prime importance; divalent ions are rejected to greater extent than monovalent which in turn, are better rejected than uncharged species. Thus the order of increasing rejection would be:-

$$NaCl < MgCl_2 \text{ and } Na_2SO_4 < MgSO_4$$

Within each charge type, however, considerable differences exist and a number of rejection series have been reported:-

$$NaI < NaBr < NaCl < NaF \text{[17]}$$

$$NH_4Cl < KCl < CsCl < RbCl < NaCl < LiCl \text{[18]}$$

To illustrate actual performance figures the results obtained with various solutes and membrane are given in table III[19]. The flux figures are included to show how constant they are and hence the comparison of % R values is valid. The position of bicarbonate is uncertain but since the p_H in such experiments is not quoted it is likely that appreciable quantities of other species may have been present if the p_H was not carefully controlled. A detailed study[20] of the system CO_2 - $NaHCO_3$ - Na_2CO_3 indicated a slight negative rejection of CO_2 and a rejection of $NaHCO_3$ close to that for NaCl. The rejection of Na_2CO_3 was, of course, higher.

Clearly one must be aware of the dissociation equilibria when considering ions of acids with more than one stage of dissociation or ions of salts that are not completely dissociated.

The series are of little use for quantitative predictions and two more quantitative relationships can be used. We use the simple relationship:-

$$\log(RF-1)_1 / \log(RF-1)_2 = \text{constant}$$

This holds for one set of conditions i.e. pressure and temperature and is only useful if osmotic pressure effects can be neglected. A more general relationship holds namely:-

$$\log B_1 = k \log B_2 + k' \tag{8}$$

This equation has been found to hold with salt 1 being NaCl and salt 2 either Na_2SO_4 or $MgSO_4$ for B values extrapolated to zero pressure[9]. The same relationship had been reported earlier for B values for other salts[5].

The limitation on the validity of these equations is the same one as has been mentioned namely that B values are not always independent of pressure. Rejection factors of several thousand can be obtained for Na_2SO_4 and $MgSO_4$ and obviously a small flaw in the membrane or ineffective sealing around the membrane could swamp the small amount of true transport occurring.

6.2 Rejection of mixtures of salts

In many applications the feed contains a mixture of salts and it is of importance to be able to predict the overall salt rejection as well as that for individual species. Studies have been reported for simple mixtures[21][5] as well as for naturally occurring waters. The general finding is that those ions which are better rejected when present in single salts are rejected to an even greater extent when mixed with ions of less well rejected salt. For the latter, rejection in the mixture is reduced. Some success has been achieved in predicting rejections of individual ions in mixtures.

6.3 Other membranes

For the Dupont hollow fibres much less has been reported but ionic charge has the same relevance as for cellulose acetate. For charged membranes very different rejection series would be expected since the sign of the charge is as important as the magnitude and calculation of rejection from ion-exchange data is possible[22]

6.4 Rejection of organic solutes

This topic is of importance for two reasons namely estimating the rejection of organic species in waste waters processed by reverse osmosis and also assessing the possible use of the process in treating solutions of organic compounds either to concentrate or to separate solutes.

For the first the low rejection of urea and much higher rejection of sugars (higher than for NaCl) has been noted by various groups and can be understood in terms of the difference in molecular size. More surprising, however, is the low and sometimes negative rejection of phenol which must be ascribed to a high degree of interaction between this solute and cellulose acetate.

In a series of three papers[23] the separation of (1) aldehydes, ketones, ethers, esters and amines; (2) organic acids; and (3) hydrocarbons in aqueous solutions has been studied and correlated with physico-chemical parameters of the solutes. A wealth of information is presented and any attempt to summarise would have to be highly selective, and hence inadequate.

With many organic solutes considerable differences between F_p and F_w are observed[24] and these cannot be explained in terms of concentration polarisation and osmotic pressure. The changes are rapidly reversed when the solution is replaced by water and this effect is ascribed to adsorption of the solute on or in the membrane.

Mixtures of organic solutes seem to behave in a different way from mixtures of salts[25] but then maintaining charge neutrality is not required with the former. For pairs of acids rejections of components in mixtures lay between the values for the single solute system whereas, for a mixture of three alcohols, all the solutes were rejected to roughly the same extent from the mixture.

Some work has been reported on the separation of mixtures of organic compounds[26]. Separation factors are low but, for alcohols, increase with the differences in length of the hydrocarbon chain. For dissimilar solutes higher separations are possible. For example, with a mixture of ethanol and n-heptane, a feed containing 40 mole % of the former would yield a permeate containing 70 mole %.

The whole area of non-aqueous reverse osmosis has been little studied, and, of course, is much beset by irreversible effects of the organic species on the cellulose acetate membranes.

7. ULTRAFILTRATION

Some features of ultrafiltration are identical to those of reverse osmosis but several important differences can be recognised. With high molecular weight solutes, osmotic pressure can usually be neglected. Much lower pressures are normally used < 6 atmospheres compared to > 30 atmospheres for reverse osmosis. Probably the most important feature is the controlling effect that concentration polarisation produces.

The basic measurements are water flux and molecular weight cut-off and concentration polarisation problems are much more severe. Indeed it is general to measure F_p rather than F_w when characterising membranes since for a solute that is rejected F_w can decrease considerably.

Concentration polarisation results in the formation of a gel layer on the membrane surface and, on attempting to force more water through by increasing the pressure, the layer becomes thicker and eventually no increase in F_w is observed. A typical F_w versus P graph is shown in figure XII [27] where the three regions are clearly shown. In the first F_w is proportional to P and the membrane resistance controls the flux. Then there is a transition region leading to the constant flux region with the gel layer controlling the flux. Increasing the flow of solution past the membrane delays the onset of this third region.

In this region where the gel layer is the controlling

feature the flux varies inversely with the log of the solute concentration:

$$F_w = K \log c_g/c_f \qquad (9)$$

The flux reaches zero when the solute concentration reaches the value at which gelation occurs.

The proportionality constant K can be evaluated[28], using the theories developed for heat transfer, for laminar or turbulent flow and good agreement between theory and practice is obtained for macromolecules. For colloids, fluxes can be several orders of magnitude higher than the theoretical predictions and this is attributed to the pinch effect where the flow of solution causes the colloids to be concentrated some distance from the membrane.

Increase of temperature normally increases F_w but on occasions compensating effects can occur as with surfactant solutions where increasing the temperature is accompanied by an increase in the critical micelle concentration[29].

8.   CONCLUSION

The aim was to describe how membranes behave under a wide variety of conditions and to include quantitative relationships where possible. It will have been quite clear that some areas are much better understood than others with the problem of pressure-dependent solute flux being one which is not at all well quantified. Some detail has been included but much more is available from the references in the printed text. The overall picture has been presented since it would have been impossible to present the detail relevant to all the applications of membrane processes.

REFERENCES

1. S. Sourirajan, 'Reverse Osmosis', Logos Press Ltd., 1970, P.552.
2. S. Sourirajan, 'Reverse Osmosis', Logos Press Ltd., 1970, P.255.
3. a) S. Sourirajan, 'Reverse Osmosis', Logos Press Ltd., 1970, P.26.
   b) S. Sourirajan, 'Reverse Osmosis', Logos Press Ltd., 1970, P.36.
   c) R. McKinney, Anal. Chem. 41 (11) 1969, P.1513.
   d) J.P. Agrawal, C.R. Antonson and N.W. Rosenblatt, Desalination 11 (1) 1972, P.71.
4. R.J. Raridon, L. Dresner and K.A. Kraus, Desalination 1 (3) 1966, P.210.
5. T.D. Hodgson, Desalination 8 (1) 1970, P.99.
6. J.S. Johnson and J.W. McCutchan, Desalination 10 (2) 1972, P.154.
7. C. Chen and C.A. Petty, Desalination, 12 (3), 1973, P.281.
8. U. Merten, 'Desalination by Reverse Osmosis' M.I.T. Press 1966, P.161.
9. E. Glueckauf and D.C. Sammon, 3rd Int. Symp. Fresh Water from the Sea 2, 1970, P.397.
10. J.S. Johnson, J.W. McCutchan and D.N. Bennion, UCLA Report Eng-7139, 1971, P.111.
11. J.C. Osborn, & D.N. Bennion, Ind. Eng. Chem. Fund. 10, (2), 1971, p.273.
12. R.L. Riley, H.K. Lonsdale and C.R. Lyons, J. Appl. Polym. Sci. 15, 1971, P.1267.
13. U. Merten, 'Desalination by Reverse Osmosis' M.I.T. Press, 1966, P.151.
14. Dupont sales literature.
15. L.T. Rozelle, J.E. Cadotte, B.R. Nelson and C.V. Kopp, Appl. Polym. Symposium No.22, 1973, P.223.
16. D.G. Thomas and W.R. Mixon, Ind. Eng. Chem. Proc. Des. Dev. 11 (3), 1972, P.341.
17. U. Merten, 'Desalination by Reverse Osmosis' M.I.T. Press 1966, P.136.
18. M.F. Re and D.N. Bennion, Ind. Eng. Chem. Fund., 12 (1) 1973, P.69.
19. S. Sourirajan, 'Reverse Osmosis', Logos Press Ltd., 1970, P.243.
20. C.E. Milstead, A.B. Riedinger and H.K. Lonsdale, Desalination, 9 (3), 1971, P.217.
21. S. Sourirajan, 'Reverse Osmosis', Logos Press Ltd., 1970, P.391.
22. E. Hoffer and O. Kedem, J. Phys. Chem., 76 (24), 1972, P.3638.
23. a) T. Matsuura and S. Sourirajan, J. Appl. Polym. Sci. 16, 1972, P.1663.

78

b)   T. Matsuura and S. Sourirajan, J. Appl. Polym. Sci. 17, 1973, P.3661.

c)   T. Matsuura and S. Sourirajan, J. Appl. Polym. Sci. 17, 1973, P.3708.

24.   W.A. Duvel, T. Helfgott and E.J. Genetelli, A.I.Ch.E. Symp. Ser. 68 (124) 1971, P.250.

25.   C. Peri, P. Battisti and D. Setti. Lebensm-Wiss. Technol, 6 (4), 1973, p.127.

26.   S. Sourirajan, 'Reverse Osmosis, Logos Press Ltd., 1970, P.409.

27.   M.C. Porter and A.S. Michaels. Chemtech Jan 1971 P.56.

28.   M.C. Porter, Ind. Eng. Chem. Prod, Res. Dev. 11 (3), 1972, P.234.

29.   R.B. Grieves, D. Bhattacharyya, W. Schomp and J.L. Bewley A.I.Ch.E.J. 19 (4) 1973, P.766.

Terms and Symbols

Symbol	Description	
F_p	water flux in absence of solutes)	L/T eg gall/ft^{2}d or m/d
F_w	water flux in presence of solute)	
c_f	solute concentration in feed)	moles/L^3 eg molar or ppm
c_p	solute concentration in permeate)	
c_m	solute concentration at membrane)	
RF	rejection factor c_f/c_p	
%R	percentage rejection $100\ (c_f-c_p)/c_f$	
F_s	salt flux (moles/L^{2}T)	
A	membrane constant for water	
B	membrane constant for salt (L/T)	
P	pressure eg atmosphere, psi	
π	osmotic pressure	
c_g	concentration at which UF solute gels	

Table I

Comparison of Symbols

Symbols used here	Symbols used by Sourirajan
F_p	PWP
F_w	PR
F_s	N_A
A	A
B	$D_{AM}/K\delta$
%R	fx100

Note: The dimensions of each pair are the same but various units can be used.

Table II

Relevance of m_A values

m_A	F_w at time 't' relative to value at 1 hour			
	168 hours 1 week	730 hours 1 month	2200 hours 3 months	8800 hours 1 year
0.01	0.95	0.936	0.926	0.913
0.03	0.858	0.821	0.794	0.761
0.07	0.699	0.630	0.583	0.530

Table III

Flux and Rejection for Various Salt Solutions[19]

Solute	Membrane M1		Membrane M4	
	F_w	%R	F_w	%R
$MgSO_4$	45.8	99.9	163.1	98.7
Na_2SO_4	45.0	99.9	156.5	96.4
$BaCl_2$	44.8	99.8	157.7	81.2
$CaCl_2$	44.8	99.6	158.0	76.0
$MgCl_2$	44.7	99.5	157.8	77.1
LiCl	45.3	99.3	163.8	62.2
NaCl	45.3	99.2	163.9	64.4
KCl	45.3	99.0	164.2	55.4
NH_4Cl	45.4	98.6	164.4	49.6
$LiNO_3$	45.4	98.6	162.9	62.5
$NaNO_3$	45.4	97.6	164.1	54.5
$K\ NO_3$	45.3	98.1	165.4	42.9

Test Conditions 100 atmospheres, solution molality 0.05. F_w is in arbitrary units.

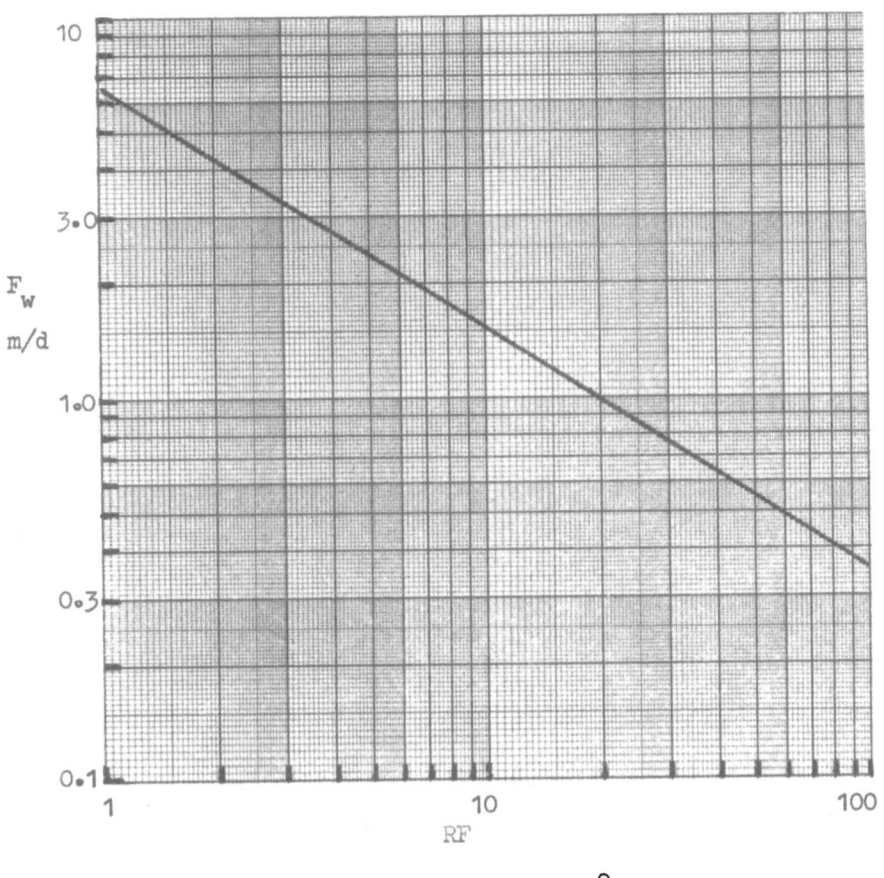

Test conditions   42 atm., 25°C
                         1170 ppm NaCl

Figure I   Plot of Flux versus Rejection Factor

82

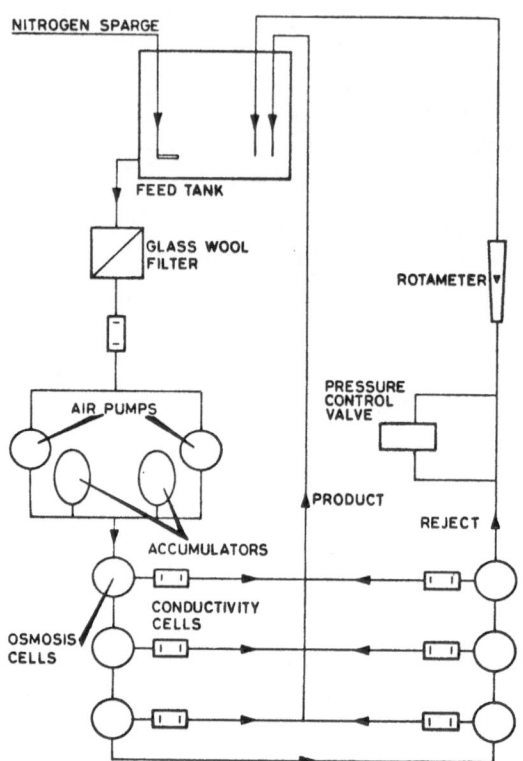

Figure II   Flow Diagram of Test Equipment

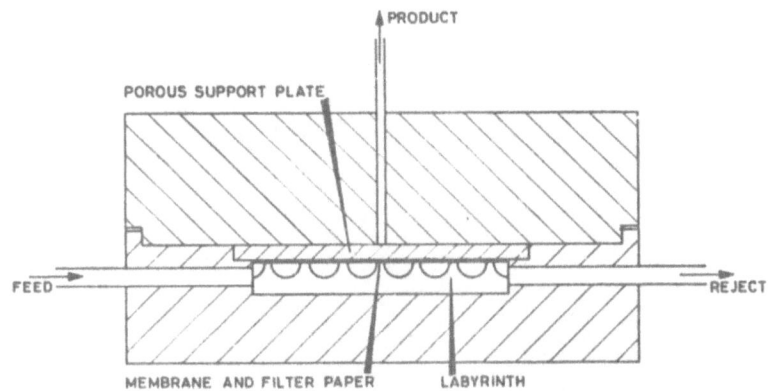

Figure III Cross-section of Test Cell

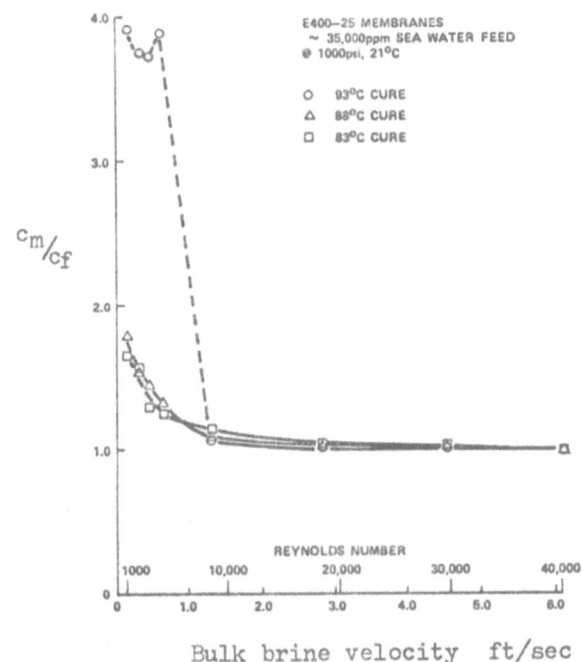

Figure IV Effect of Brine Velocity on c_m/c_f

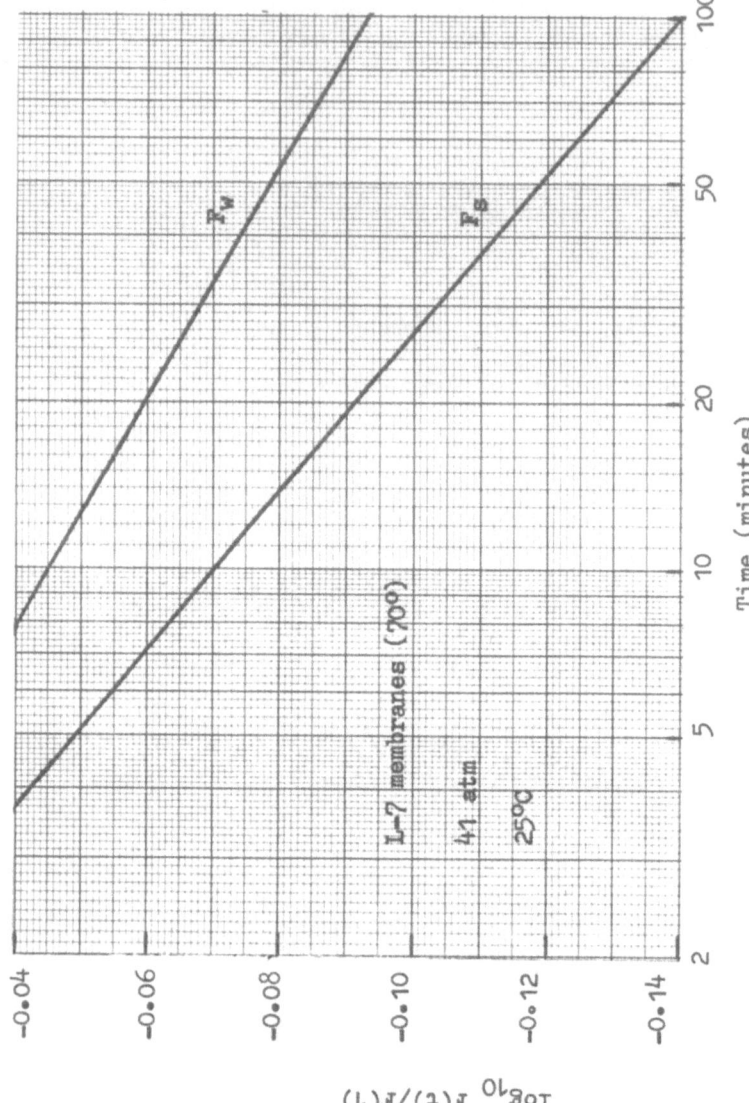

Figure V Change in F_w and F_s with time – up to 100 minutes

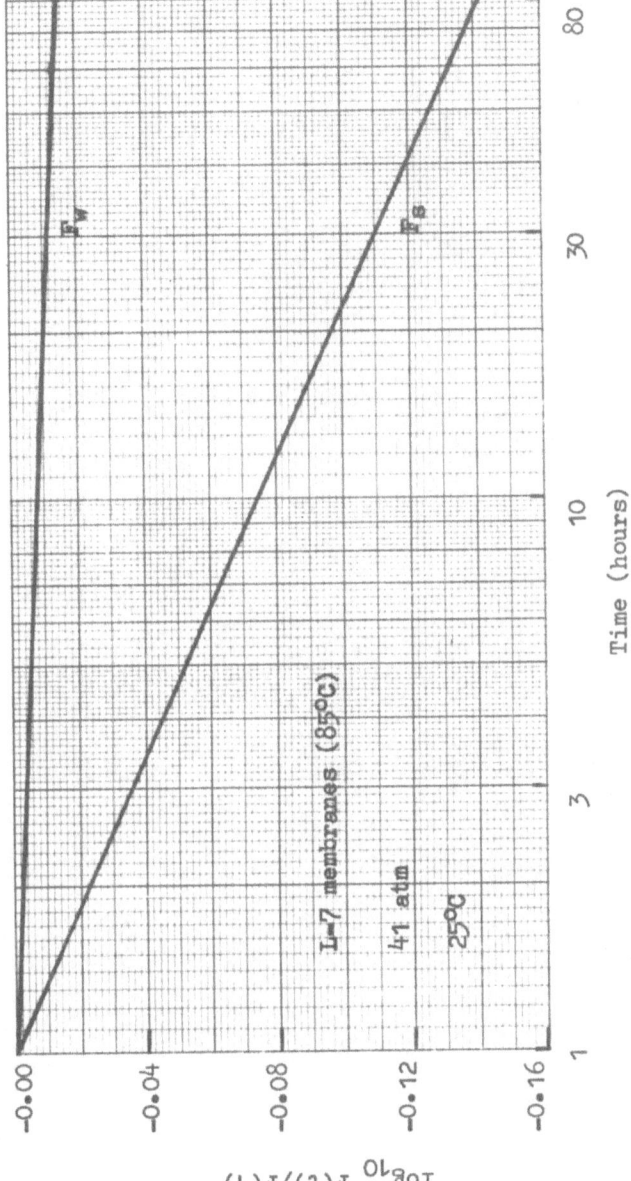

Figure VI   Change in F_w and F_s with time – up to 80 hours

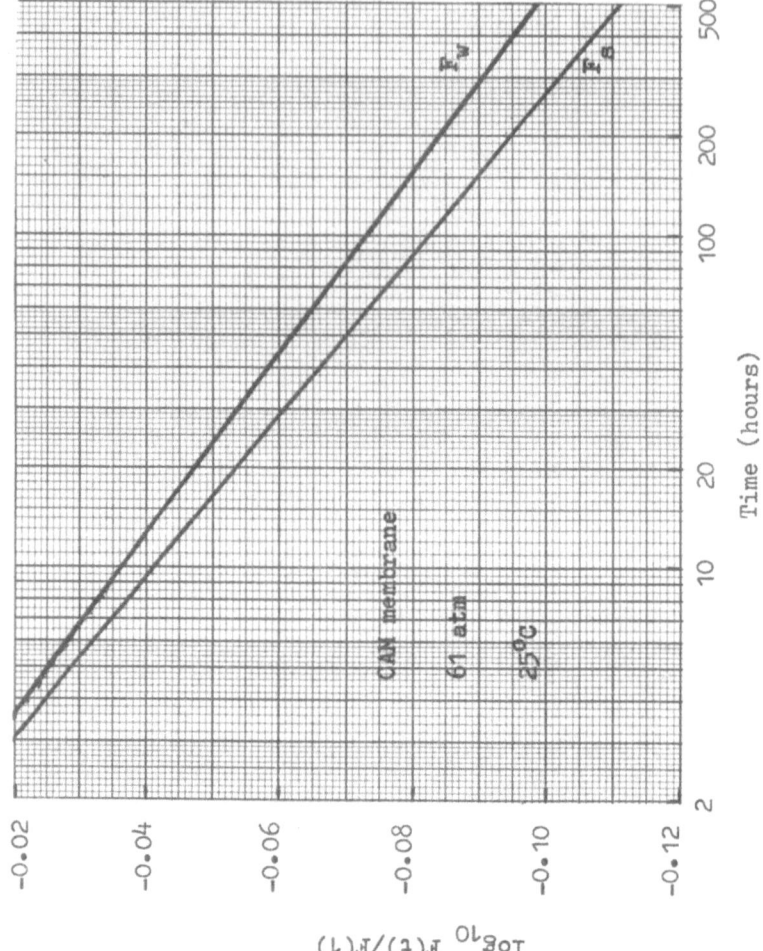

Figure VII   Change in F_w and F_s with time – up to 500 hours

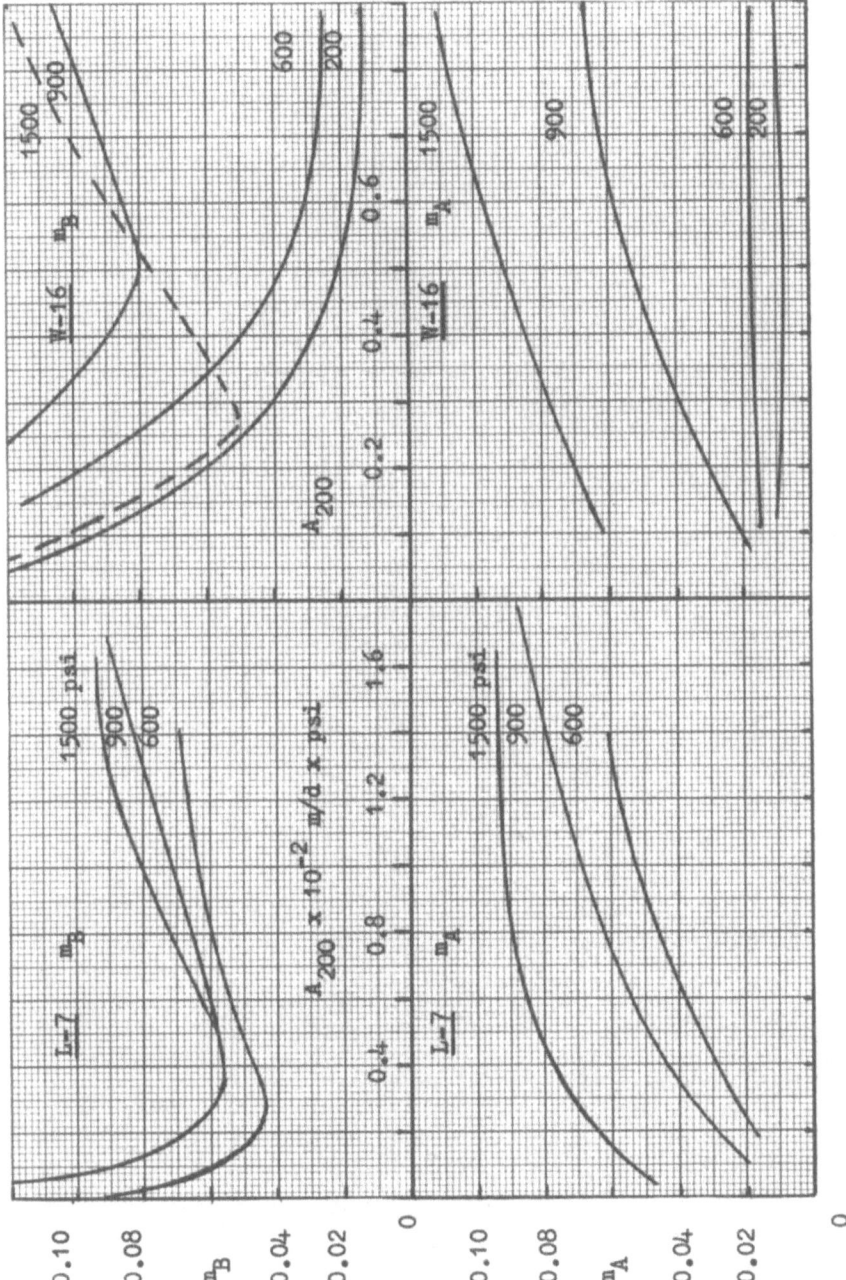

Figure VIII m_B and m_A versus A_{200}

88

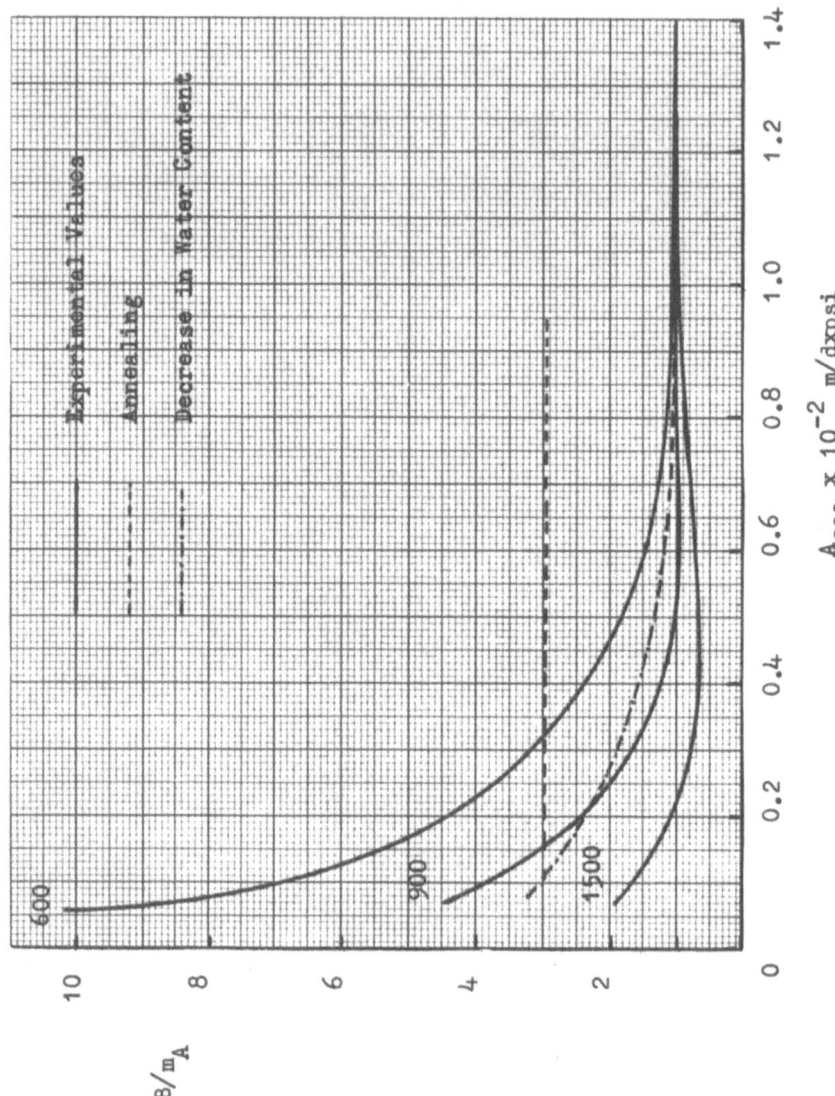

Figure IX m_B/m_A versus A_{200} for L-7 membranes

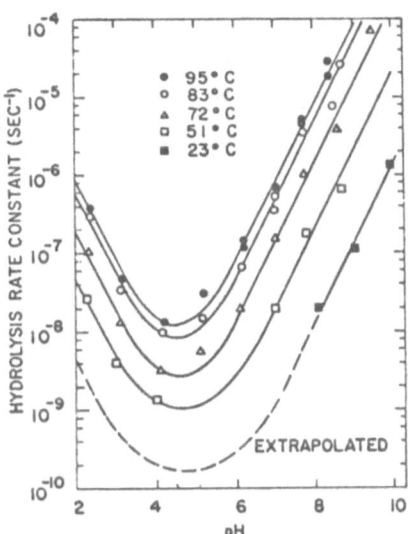

Figure X   Rate constants for hydrolysis of cellulose
acetate powder

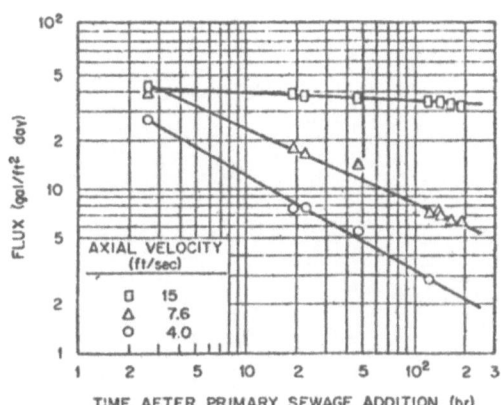

Figure XI F_W versus time in presence of fouling

90

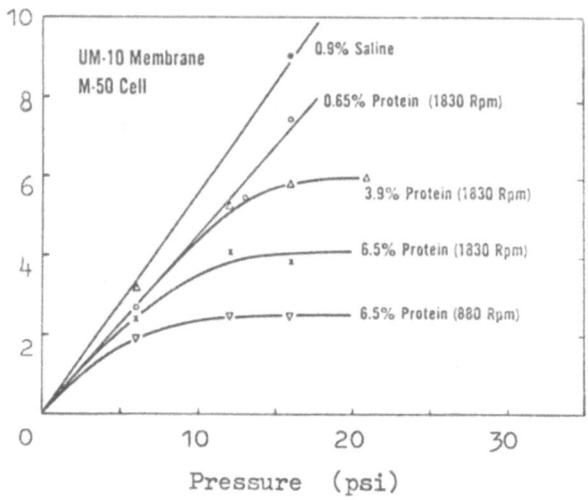

Figure XII F$_w$ versus Pressure in Ultrafiltration

SORPTION MECHANISMS

K.H. Lieser

Department of Inorganic and Nuclear Chemistry
Technische Hochschule Darmstadt

ABSTRACT. The lecture gives a survey of sorption
mechanisms and includes the following sections:
1. Introduction (The different groups of sorption:
Absorption, Adsorption on external surfaces, sorption
on internal surfaces and in pores, sorption by ion
exchange); 2. Absorption, 2.1. Gases in liquids,
2.2. Gases in solids; 3. Adsorption, 3.1. External
surfaces, internal surfaces and pores, 3.2. Adsorption
isotherms, 3.3. Adsorption energies, 3.4. Adsorption rates;
4. Adsorption on external surfaces, 4.1. Mathematical
treatment of the adsorption isotherms, 4.2. Examples
of physical and chemical adsorption from the gas phase,
4.3. Examples of physical and chemical adsorption from
the liquid phase, 4.4. Hydrolytic adsorption;
5. Sorption on internal surfaces and in pores,
5.1. Interpretation of sorption curves, 5.2. Examples
of porous systems, 5.3. Molecular sieves, 5.4. Layer
compounds, 5.5. Cage compounds; 6. Sorption by ion
exchange, 6.1. Distribution coefficients, separation
factors and selectivity, 6.2. Kinetics of ion exchange,
6.3. Organic ion exchange resins, 6.4. Inorganic
substances

1. INTRODUCTION

The term "sorption", introduced 1909 by Mc Bain embraces
different kinds of phenomena: It includes "adsorption",
that is the condensation of gases on free surfaces or
the fixation of solutes from a solution on the surface

of solids as well as "absorption", that is the uptake
of gas molecules or of molecules or ions from a solution
within the mass of an absorbing phase.

In all cases of sorption there is an interaction
between the molecules or ions which are beeing sorbed
and the molecules or ions of the sorbent. The forces of
interaction may be more of physical nature or of chemical
nature. They may be weaker and of the type of van der
Waals forces or stronger such as chemical bonds.

It seems reasonable to subdivide the phenomena of
sorptioninto four different groups:

1. Absorption, that means penetration of molecules
into the mass of the sorbent which may be a liquid or a
solid.

Absorption may be considered as a process of
dissolution. The sorbent represents the solvent and the
molecules absorbed are distributed at random within the
solvent.

Examples are: Absorption of molecular iodine or of
noble gases in organic solvents and absorption of hydrogen
in metals.

2. Adsorption on external surfaces. In this case
molecules from a gaseous phase or from a solution or
ions from a solution are found on external surfaces of
solid particles. They may be fixed at the surface by
physical forces or by chemical bonds. Accordingly we find
low or relatively high values for the heat of adsorption.
In the first step a monolayer is formed.

Examples of physical adsorption on external surfaces
are adsorption of relatively inert gases such as the noble
gases or nitrogen on non porous solids such as aluminium
oxide, graphite, ionic crystals or metal foils.

The determination of surface areas by the BET method
is based on this kind of phenomena.

Examples of chemisorption on external surfaces are
the uptake of carbon monoxide gas on the surface of
transition metals such as iron or nickel, the fixation
of molybdate from a solution on the surface of aluminium
oxide or the separation of pertechnetate ions on the
surface of cadmium sulfide.

In the chemisorption new chemical bonds are formed
and the resulting species may be best described in many
cases as surface compounds.

Furthermore the selectivity is much greater for
chemisorption than for physical adsorption. Therefore
selective separations are possible by chemisorption.

3. Sorption on internal surfaces and in pores. In
this group we find the same kind of phenomena as in group
2,but on the internal surfaces or within the pores of
porous solids,additional phenomena such as capillary

condensation within the pores and occlusion of molecules or ions just fitting into the porous system.

There is a great variety of examples in this group. Representative sorbents are silica gel, iron hydroxide or aluminium hydroxide gel and activated carbon. More complex examples are the clay minerals, such as montmorillonite, which form layer compounds and exhibit intercrystalline swelling, the molecular sieves containing pores of well defined structure and diameter which are able to take up and to bind rather firmly such molecules which are fitting well into these pores and finally the clathrates or cage compounds.

4. Sorption by ion exchange. Ion exchange is also a chemical process. In ion exchange reactions one kind of ions is taken up from a solution in exchange to another kind of ions contained in another phase which is generally a solid, in some cases a liquid (liquid ion exchange). The organic ion exchange resins are macromolecular compounds such as polystyrole which contain functional groups, mainly $-SO_3H$ or $-COOH$ groups for exchange of cations or $-NR_3OH$ groups for exchange of anions. Inorganic ion exchange materials may belong to different classes of compounds. Examples are hydrated oxides, phosphates, hexacyanoferrates, clay minerals, zeolites or sparingly soluble ionic compounds.

In the following chapters 2 and 3 absorption and adsorption processes will be treated first to point out the different principles. Absorption will be presented in a rather short and compact form because absorption phenomena are less complex and in general well known. Chapters 4, 5 and 6 will then be devoted to adsorption on external surfaces, sorption on internal surfaces and in pores and sorption by ion exchange.

2. ABSORPTION

2.1 Gases in liquids

All gases dissolve to some extent in liquids. If the interaction between the gas molecules and the molecules of the solvent is weak, small values for the solubility are to be expected. In Table 1 solubilities of some gases in water are listed. Solubilities are in most cases given in form of Bunsen's absorption coefficient. (Volume of gas (at $0°C$ and 1 at pressure) dissolved by unit volume of solvent under a partial pressure of 1 at at the temperature of the experiment). They are proportional to the partial pressure of the gas considered (Henry's law) provided that no chemical reaction is involved.

Gas	0°C	25°C
He	0,00901	0,0835
Ne	0,0110	0,098
Ar	0,054	0,0306
Xe	0,237	0,109
Rn	0,49	0,19
H_2	0,0208	0,0170
O_2	0,0471	0,0273
N_2	0,0227	0,0144
CH_4	0,0534	0,0291
C_2H_6	0,0941	0,0373
C_2H_2	1,70	0,91
CH_3Cl	-	1,90
CH_3Br	-	3,06
CO_2	1,676	0,746
H_2S	4,53	2,20
HCl	517	431
NH_3	1300	680

Table 1. Solubility of some gases in water (Bunsen absorption coefficient, Ncm^3/g)

In the case of oxygen and hydrogen for instance no reaction with the water molecules takes place and the solubility is rather low. The ammonia molecule, however, is a dipole and is therefore hydrated by water molecules. In addition it reacts with water according to the following equation

$$NH_3 + H_2O \rightleftharpoons NH_4^+ + OH^-$$

The solubility of ammonia in water is therefore high and Henry's law is not obeyed.

In Table 2 the absorption coefficients for the noble gases krypton and xenon in some organic solvents are given. Noble gases are monotomic and inert molecules. Solubilities are therefore generally low.

The data listed in Table 2 have been important for a study on the possibility of separation of radioactive krypton or xenon by washing the off-gas with organic solvents. As the partial pressures of these noble gases in the off-gas are very low the washing with organic solvents is not an efficient way in order to separate the noble gases.

It may be assumed that the transition of the gas

molecules from the gas phase into the liquid phase at
the interface gas/liquid is fast compared with the
migration within the liquid phase. Then the diffusion
of the gas molecules within the liquid phase will be
the rate determining step in attaining equilibrium.

Solvent	Air	Krypton	Xenon
n-heptane	1,6	8,0	22,5
benzene	0,49	2,62	-
toluene	0,44	3,87	14,5
p-xylene	0,75	4,02	16,2
fluorbenzene	1,35	3,39	-
trifluorbenzene	1,58	4,49	-
perfluorkerosene	5,03	11,0	-
kel-F-oil	2,66	9,1	-
benzylcyanide	0,34	1,31	-
benzonitrile	0,39	1,61	-
triethylamine	2,46	6,36	-
aniline	0,17	-	2,04

Table 2. Solubility of krypton and xenon in some organic
solvents [millimoles gas per mole solvent at
1 at].

2.2 Gases in solids

Very detailed investigations have been done on the absorption of hydrogen gas by metals. Palladium, uranium
and the lanthanide metals show high solubilities for
hydrogen forming metallic hydrides. As found by
Sieverts (1) the solubility is proportional to the
square root of the hydrogen pressure. That means that
the hydrogen molecules are split into hydrogen atoms
at the surface of the metal. The hydrogen atoms dissociate
into protons and electrons. The protons are found in
interstitial sites in the metal lattice and the electrons
join the electron gas as indicated by conductivity and
magnetic measurements.
    There are many other examples for the absorption of
gases from solids, which can not be referred here in
detail.
    In most cases the absorption of gas molecules by a
solid is a slow process because it is determined by the
diffusion of the molecules within the solid. Diffusion
coefficients in solids, however, are very small at room
temperature. If finely dispersed solids are used, the
rate of absorption is appreciably enhanced.

3. ADSORPTION

3.1 External surfaces, internal surfaces and pores

It is obvious that a solid will possess a large surface area if it consists of fine particles. There is an inverse relationship between the specific surface S (the surface area of 1 g of solid) and the particle size r

$$S \sim \frac{1}{r} \tag{1}$$

For an idealized case when the particles are cubes of equal size with an edge length a the specific surface is

$$S = \frac{6}{\rho a} \tag{2}$$

when ρ is the density of the solid. For actual powders made up of particles of different sizes and irregular shapes the relationship is more complicated but equation (2) gives still the order of magnitude.

The primary particles of a substance sometimes stick together to form aggregates - the secondary particles. The aggregation may be effected by condensation reactions as in the case of silica gel or hydroxides or the primary particles may become cemented together especially under the influence of temperature and pressure. In these aggregates the surface will be largely internal, they will possess a pore volume. The particles of a precipitate may have a wide variety of shapes and sizes. If they are heated they undergo sintering and their specific surface diminishes. In Fig. 1 as an example the specific surface of a barium sulfate precipitate is shown as a function of time as determined by BET measurements (2). First the specific surface is increasing due to disaggregation of the precipitate at lower supersaturation, then it is decreasing due to recrystallization under the influence of the solution. A number of other examples could be presented, mainly showing the decrease of specific surface area under the influence of temperature caused by recrystallization processes.

A large specific internal surface area may arise not only by aggregation of primary particles, but also by removal of parts of the solid in such a manner as to leave pores. The increase of specific surface of partially graphitized carbon by controlled burning provides an example. In this case the combustion occurs preferably along channels in the interior of the carbon which become enlarged and lengthened. The activation of wood charcoal by steam treatment probably occurs in a similar manner.

Another way in which solids of high specific internal surface area are produced is the thermal decomposition of solids by reactions of the type

Solid A $\longrightarrow$ Solid B + gas

The thermal decomposition of nitrates or carbonates provides examples.

The distinction between external and internal surfaces is somewhat arbitrary. It seems reasonable to include into the internal surface the walls of all cracks, pores and cavities which are deeper than they are wide. The surface of any solid is hardly ever truly plane on an atomic scale. Nearly always there are cracks or fissures present some of which may penetrate very deeply and therefore contribute to the internal surface. A wide range of porous solids have an internal surface greater by several orders of magnitude than the external surface, the total surface thus being predominantly internal. On the other hand very fine powders of numerous substances have a large external surface and a small or even negligible internal surface. When such particles aggregate to secondary particles, however, a great part of the external surface becomes converted into internal surface and a pore system is developed.

The pore systems of solids are of many different kinds, they may vary greatly both in size and in shape. For practical purposes the width of the pores is of greatest interest and it has become convenient to classify pores according to their average diameter. Pores with a diameter less than 20 Å are called micropores, those with a diameter greater than 200 Å macropores and those with diameters between 20 and 200 Å intermediate pores. Each range of pore size is associated with a characteristic adsorption behaviour.

3.2 Adsorption isotherms

The amount adsorbed per gram of solid adsorbent c_a depends on the equilibrium pressure p of the gaseous adsorbate (or the concentration c of the adsorbate in a solution respectively), the temperature T, the nature of the adsorbate A and the nature of the solid adsorbent S

$$c_a = f(p, T, A, S) \tag{3}$$

c_a may be measured in suitable concentration units, grams, moles or cm^3 (N.P.T.) gas per g sorbent. For a given adsorbate A, a given adsorbent S and a fixed temperature T equation (3) simplifies to

$$c_a = f (p)_T \tag{4}$$

If the adsorbate is a gas below its critical temperature the following form is more useful

$$c_a = f (p/p_o)_T \quad , \tag{5}$$

p_o being the saturation vapour pressure of the adsorbate. Equations (4) and (5) are expressions of the adsorption isotherm.

In the case of physical adsorption the majority of the thousands of adsorption isotherms measured may be grouped into five classes according to the classification originally proposed by Brunauer, Deming, Deming and Teller (3) but now commonly referred as the Brunauer Emmett and Teller (4) or BET classification. These types are shown in Fig. 2. Type I may be approximated by the Freundlich or the Langmuir equation. Types II and IV can be used to calculate the specific surface of the adsorbing solid. Type IV possesses a hysteresis loop and can be evaluated for making an estimate of the pore size distribution. From isotherms of type III and V on the other hand it is not possible so far to make any reliable estimate of either specific surface or pore size distribution.

If chemisorption is dominating an adsorption isotherm of the type shown in Fig. 3 is observed. It has a sharp knee. A plateau, which denotes completion of the monolayer, is reached at relatively low pressures or concentrations. If chemisorption is followed by physical adsorption the adsorption isotherm is bending upwards at higher pressures or concentrations as in Type II, Fig. 2.

3.3 Adsorption energies

Adsorption energies fall into two main groups, energies of physical adsorption and energies of chemical adsorption.
The forces bringing about physical adsorption are relatively weak and include the so-called van der Waals forces which may be of the type of dispersion forces (5) or forces due to permanent dipoles of the adsorbed molecules. The dispersion forces have been characterized by London (5); they originate through the changing electron

density in an atom, which induces a corresponding
electrical moment in the neighbour atom and so leads
to an attraction. The potential energy U(r) of two
isolated atoms separated by a distance r is given by

$$U(r) \sim - \frac{\alpha_1 \alpha_2}{r^6} \tag{6}$$

where α_1 and α_2 are the polarizabilities of the two
atoms. In addition there is a repulsion force

$$U(r) \sim \frac{1}{r^{12}} \tag{7}$$

The resulting potential energy U(r) is plotted in Fig. 4.
In order to apply these ideas to the problem of
adsorption it is necessary to summarize the individual
interactions of each atom of the adsorbing surface
with each atom of the molecules adsorbed. Because the
potential energy falls off very rapidly with distance
only a limited number of atoms have to be considered
and the result is not very different from the curve shown
in Fig. 4. The equilibrium position of the adsorbed
molecule with respect to the surface is given by the
minimum in the potential energy curve and is about
equal to the diameter of a single molecule. The value
of U at this minimum represents approximately the
energy of adsorption which is of the order of 1 kcal
per mole.
    The exact value of U will of course change to
some extent with the position on the surface. If the
system potassium chloride/argon is taken as an example
the following values for the adsorption energy are
found: 1,59 kcal/mole on a position just between 2 K^+
and 2 Cl^- ions, 1,31 kcal/mole on a position between
1 K^+ and 1 Cl^- ion, 1,45 kcal/mole on the centre of
a K^+ ion and 1,27 kcal/mole on the centre of a Cl^- ion.
The surface of the sorbent can in that way be covered
by contour lines of equal adsorption energy. The
molecules adsorbed may move freely on the surface along
these lines, but from one position on the surface to
another one there may exist a barrier which can be
surmounted without difficulty if the average thermal
energy per mole which is given by RT exceeds the height
of the barrier. For the adsorption of argon on potassium
chloride the theoretical analysis leads to the following
results: At room temperature the argon molecules are
freely mobile over the whole surface. A single layer

of molecules is adsorbed relatively strongly, whereas subsequent layers are held more weakly.

Dispersion forces are common to all matter and will always be present. If the sorbent is ionic in nature and the molecules of the adsorbate possess permanent dipoles, then electrostatic forces will contribute. The electrostatic contribution to the adsorption potential is given by

$$U(\mu) = - F \mu \cos \theta \qquad (8)$$

where μ is the dipole moment of the absorbed molecule, F is the strength of the electrostatic field and θ is the angle between the axis of the dipole and the field. For polar molecules such as water, ammonia or alcohols $U(\mu)$ assumes appreciable values of about 3-6 kcal/mole. Those molecules are therefore strongly adsorbed on inorganic salts and oxides.

In recent years the importance of the presence of a permanent quadrupole in adsorbate molecules such as CO_2 has also been recognized (6). The quadrupole contribution $U(Q)$ calculated for CO_2 on NaCl amounts to 4,4 kcal/mole.

Even if the adsorbed molecule or ion has no permanent dipole it will acquire an induced moment of the magnitude $F\alpha$, where α is again the polarizability of the molecule or ion, when placed in the electrostatic field of the strength F produced by the solid. The contribution to the interaction energy is given by

$$U(\alpha) = - \frac{F^2\alpha}{2} \qquad (9)$$

Generally the contribution amounts only to some percent of the total adsorption energy. The same formula applies if the absorbed molecules are polarized by the electrical double layer formed between the free electrons at a metal surface and the corresponding inner positive charge. With certain adsorbates such as saturated hydrocarbons on metals this electrostatic contribution may actually constitute the major part of the total adsorption energy (about 5 kcal/mole).

Dispersion forces will distinctly favor adsorption in crevices and pores of the solid because the adsorbed molecules are then close to a larger number of atoms of the solid. Adsorption should occur preferentially in the smallest pores. If the electrostatic contribution is considerable and positive and negative ions alternate on the surface, however, the picture is different. Within

narrow pores the alternating positive and negative
charges will tend to compensate each other so that the
electrostatic field will be weaker within the pores.

Chemical adsorption is characterized by the fact
that chemical bonds are formed between the sorbent and
the adsorbate. These results in the formation of a chemical
compound on the surface of the sorbent which can best
be described by the term surface compound.

An important practical example is the chemisorption
of oxygen by metals resulting in the formation of an
oxide on the surface. The properties of the surface are
determined by the physical and chemical properties of
the oxide. If the oxide forms a dense film, further
oxidation is hindered as in the case of Al and Zr. If
the oxide film is permeable further oxidation is
promoted, as in the case of iron. On the surface of
noble metals, however, oxygen is chemisorbed at lower
temperatures forming surface oxides which are decomposed
again at higher temperatures according to the decomposition temperature of the oxide.

An example of the formation of a surface compound in
a system solid/solution is the sorption of thorium by
silica gel at about pH3 (7). The hydroxo complexes of
thorium which are present in solution under these conditions react with the surface of silica gel forming
a surface compound corresponding to thorium silicate.
This surface compound shows some typical reactions. It
is able to react with other groups, for instance phosphate
ions can be fixed.

In any case the chemical and physical behaviour is
determined by the properties of the surface compound
formed.

For chemical adsorption in general the heat of
adsorption is much greater than for physical adsorption.
Whereas the heat of physical adsorption for gases rarely
exceeds twice the latent heat of condensation, the heat
of chemisorption is mostly several-fold greater than
the latent heat. Generally it is higher than 10 kcal/
mole and may even exceed 100 kcal/mole, for instance
for the chemisorption of oxygen on tungsten and other
metals.

3.4 Adsorption rates

Physical and chemical adsorption also differ in general
in the adsorption rate. Chemisorption frequently requires
an activation energy and so proceeds at a limited rate
which increases rapidly with rise in temperature. The
potential energy for chemical adsorption U_c and for

physical adsorption U_p are given in Fig. 5. A molecule
from the gas phase or from the solution may first be
physically adsorbed (heat of physical adsorption q_p).
Than the activation energy E or in total the energy
E + q_p is necessary for chemical adsorption (heat of
chemical adsorption q_c). The rate of chemical adsorption
only becomes measurable above some minimum temperature.
It is therefore often found that the amount adsorbed
increases with temperature. The situation is illustrated
by Fig. 6, which shows an adsorption isobar, that is
the amount adsorbed at a given pressure plotted against
temperature. ABB' is the equilibrium curve for physical
adsorption and C'CD that for chemisorption. At low
temperatures the rate of chemisorption is so low that
chemisorption is not observed. At B this rate is large
enough to produce an appreciable contribution to the
total measured adsorption. Because of the positive
temperature coefficient of the rate of chemisorption the
amount absorbed increases from B to C. At C this rate
is fast enough to enable equilibrium to be reached
during the period of the measurement. The chemical
adsorption is now predominant and the physical adsorption
is negligible.

Chemical adsorption is not always an activated
process. That depends on the point of intersection of
the potentials U_c and U_p in Fig. 5. Hydrogen for instance
is adsorbed rapidly by a tungsten filament both at room
temperature and at liquid air temperature to give a
saturated layer even at very low pressures. Fast chemisorption of gases such as hydrogen, nitrogen, carbon
monoxide, carbon dioxide and ethylene occurs on many
metals.

4. ADSORPTION ON EXTERNAL SURFACES

4.1 Mathematical treatment of the adsorption isotherms

The first equation for an adsorption isotherm which was
an empirical one, is named after Freundlich

$$c_a = ac^b \qquad (10)$$

when c is the concentration of the adsorbate in a solution
and a and b < 1 are constants. Instead of the concentration
c the partial pressure of a gas may be used as well.
A curve with b = 0,3 is shown in Fig. 7. At low concentrations the slope of the curve is infinite. This is not

in agreement with physical adsorption, where at low concentrations $c_a \sim c$. At high concentrations equation (10) gives no saturation value or plateau which is in contradiction to the monolayer concept of chemical and physical adsorption.

Langmuir (8) regarded the surface as an two dimensional arrangement of adsorption sites and postulated when a molecule from the gas phase strikes an empty adsorption site, it condenses there, remains attached to the site for a mean period of time and re-evaporates. He derived the following equation for the adsorption isotherm:

$$c_a = c_m \frac{B \cdot p}{1 + B \cdot p} \quad , \qquad (11)$$

where c_m is the monolayer capacity and B is a constant at constant temperature. The Langmuir adsorption curve is plotted in Fig. 8. At low pressure equation (11) gives $c_a \sim p$ as postulated by the model of physical adsorption. At high pressures saturation is obtained corresponding to the monolayer concept of chemical and physical adsorption. In equation (11) the concentration c in a solution may be used instead of the partial pressure of a gas.

Langmuir refered to the possibility that the evaporation-condensation mechanism might apply to second and higher molecular layers as well; but he focussed his attention on those cases where adsorption was restricted to a monolayer.

In 1938 Brunauer,Emmett and Teller (4) extended the Langmuir mechanism to second and higher molecular layers because experimental evidence had made it increasingly likely that multilayer physical adsorption occurred frequently. The Brunauer,Emmett, Teller (BET) equation has the form

$$c_a = c_m \frac{c(p/p_o)}{(1 - p/p_o)(1 + (c-1)p/p_o)} \qquad (12)$$

where c_m is again the monolayer capacity and

$$c = \frac{a_1 \nu_2}{a_2 \nu_1} e^{(E_1 - L)/RT} \qquad (13)$$

ν_1, ν_2 are the oscillation frequencies in the first and the second layer, a_1, a_2 are the respective condensation coefficients which allow for the possibility that only a proportion and not the whole of the incident molecules may condense on an empty site. E_1 is the heat of absorption in the first layer and L is the latent heat of condensation.

$v_1 e^{-E_1/RT}$ is the number of molecules evaporating from a given site per second. Equation (12) has been derived under the assumptions that the heat of adsorption in all layers above the first is equal to the latent heat of condensation and that the ratios v_i/a_i in all layers above the first are identical. A further assumption was that when p becomes equal to the saturated vapour pressure p_0 the adsorbate vapour condenses as an ordinary liquid on the adsorbed film so that the number of molecular layers becomes infinite on the surface.

For a limited number n of layers the following modified equation is obtained

$$c_a = c_m \frac{c(p/p_0)}{1-p/p_0} \cdot \frac{1-(n+1)(p/p_0)^n+n(p/p_0)^{n+1}}{1+(c-1)(p/p_0)-c(p/p_0)^{n+1}} \qquad (14)$$

which gives the Langmuir equation (11) by putting n=1 (the parameter B in equation (11) is replaced by (c/p_0)). Some curves of c_a/c_m against p/p_0 are plotted in Fig. 9 for different values of c. If $c > 2$ a curve of Type II as given in Fig. 2 results. For $c < 2$ a curve of Type III in Fig. 2 is obtained. The shape of the knee depends on the numerical value of c, becoming more sharp as the value of c increases, that is as the value of the net heat of adsorption E_1-L increases.

The BET equation is most often applied in the form

$$\frac{p}{c_a(p_0-p)} = \frac{1}{c_m \cdot c} + \frac{c-1}{c_m \cdot c} \frac{p}{p_0} \qquad (15)$$

which can be obtained from equation (12) by rewriting. According to this equation a straight line should result when $p/c_a(p_0-p)$ is plotted against p/p_0, with slope $s=(c-1)/c_m c$ and intercept $i=1/c_m c$. The solution of the two simultaneous equations gives c_m and c:

$$c_m = \frac{1}{s+i} \qquad (16)$$

$$c = \frac{s}{i} + 1 \qquad (17)$$

In general for low relative pressures p/p_o straight
lines are obtained as shown in Fig. 10 (4).
    From the monolayer capacity c_m obtained from
equation (16) the specific surface area is calculated
by the relationship

$$S = \frac{c_m}{M} N \cdot A_m \cdot 10^{-20} \left[m^2 g^{-1} \right] \qquad (18)$$

Here c_m is the monolayer capacity in grams of adsorbate
per gram of solid, N is the Avogadro number $(6,023 \cdot 10^{23})$
and A_m is the area occupied per molecule of adsorbate
in the completed monolayer in square Ångstrom units
$(10^{-20} m^2)$. By this method, the BET method, specific
surface areas are determined by physical adsorption
preferably of nitrogen molecules at low temperatures.
For nitrogen adsorption A_m has the value $16,2 \, Å^2$. The
BET method of surface determination is applicable for
adsorption isotherms of Type II. For curves of Type III
the evaluation does not always yield reasonable values.

4.2 Examples of physical and chemical adsorption from
    the gas phase

In the case of physical adsorption the shape of the
adsorption isotherm depends on the nature of the
adsorbent as may be seen from Fig. 11 (9) where the
adsorption isotherms for n-pentane on different adsorbents
are shown. The adsorption isotherms on graphitized carbon,
on barium sulfate and on silica are of Type II whereas
the curve for the adsorption on liquid water is of Type
III. By evaluating the curves it is found that the value
of A_m for pentane, that is the area occupied per molecule
of adsorbate, varies between 53 and 66 $Å^2$. A_m is found
to be a function of c and therefore of the net heat
of adsorption. For higher values of c lower values of
A_m are found. That means that the orientation of molecules
on the surface and heats of adsorption may be different
for different sorbents.
    Some examples of BET evaluation are shown in Fig.
12 (10). The plot of $p/x(p_o-p_o)$ against p/p_o gives
straight lines for adsorption of benzene at 20^oC, n-hexane
at 20^oC and nitrogen at -195^oC on graphitized carbon.
    The formation of the monolayer may be seen by

plotting the differential heat of adsorption as a
function of the quantity adsorbed (Fig. 13 (11)).
While the monolayer is formed the differential heat of
adsorption remains constant at a high value. When the
monolayer is completed and subsequent layers are
formed the differential heat of adsorption drops to a
lower value which corresponds nearly to the latent heat
of condensation. The measurements of Fig. 13 were taken
with a carbon sample which was graphitized by heating
at 2700°C. By this treatment a homogeneous surface was
obtained. If the surface is not homogeneous, but contains
centres of lower and higher activity as in the case of
non-graphitized carbon (Fig. 14 (12)) the highly active
centres are occupied first resulting in high values of
the differential heat of adsorption at the beginning.
The differential heat of adsorption is diminishing as
centres of lower activity are occupied. For inhomogeneous
surfaces the sharp drop of the differential heat of
adsorption at $c_a = c_m$ is therefore not pronounced (Fig.14)
    Well known examples of chemisorption are the chemi-
sorption of hydrogen, oxygen, carbon monoxide and others
on the surface of metals. In the case of hydrogen it is
important to distinguish between adsorption (cf. 2.2).
Pioneer work in the field of chemisorption of hydrogen
and oxygen was done by Langmuir (13)(8). The chemisorption
of oxygen by a metal is the first step in the oxidation,
a layer of bulk oxide ultimately being formed. The
thickness of the layer depends on the nature of the metal
and the temperature. But even at room temperature the
oxygen layer is of the order of 20-30 Å. The heat of
chemisorption of oxygen varies over a wide range, from
70 kcal/mole for Pt to 194 kcal/mole for W. Oxides such
as ZnO, MgO, Fe_2O_3 or Cr_2O_3 also adsorb oxygen. Carbon
monoxide is adsorbed by transition metals like Fe, Ni
and others with the formation of CO-metal bonds similar
to those found in metal carbonyls. Chemisorption of
this kind is very important with respect to catalysts
and their activity. It will not be discussed here in
greater detail.

4.3 Examples of physical and chemical adsorption from the liquid phase

From a liquid phase we have a competitive adsorption of
the solvent and the solute. In many cases adsorption
isotherms are observed which show a sharp knee and a
clearly defined plateau, Fig. 15. The monolayer capacity
is then given by the height of the plateau. Curves
of this type are also often encountered for dyestuffs

as solutes, Fig. 16. The use of dyestuffs for the determination of specific surface is quite common, mainly
on account of the relative ease of determination of
the amount adsorbed. The molecules of dyestuffs are
large so that the dispersion forces per molecule will be
relatively strong, furthermore the molecules usually
contain polar groups which will enhance the adsorption
on a polar adsorbent. Thus the desired preferential
adsorption of the solute is often present. It is difficult,
however, to assess the proper value of A_m because of the
large size and complicated shape of dyestuff molecules.

In general different types of adsorption isotherms
are being observed due to the different solid-solute,
solute-solute and solid-solvent interactions. The solute
molecules may be oriented on the surface in different
ways. The monolayer not always exists of tightly packed
and vertically oriented molecules of the solute. It may
contain solvent molecules in appreciable quantity.
Moreover the solute molecules may be clustered into
micelles.

For surface determination nitrophenol is recommended.
It has some advantages as compared with methylene blue
which is often used.

Chemisorption is usually characterized by a steep
rise of the adsorption isotherm at low concentrations
and a long plateau over a wide range of solute concentration corresponding to the formation of a monolayer.
Chemisorption is found for some species on aluminium
oxide, for instance for tellurate and molybdate ions.
These are probably bound by oxygen bridges, as shown
in Fig. 17. Reaction of that kind are highly selective
and may be used for selective separations.

Polar molecules such as water are adsorbed rather
strongly on the surface of oxides, hydrated oxides,
hydroxides or ionic crystals and bound by dipole-dipole
or ion-dipole interactions. In water as solvent these
compounds are covered with a film of water molecules;
they are hydrated. This hydration of the surface is
rather a chemical reaction and the heat of adsorption
is relatively high. Water films formed in this way are
also immobile at temperatures near $0°C$. This was found
by measuring isotopic exchange reactions at the surface
of ionic crystals as a function of temperature. The full
exchange capacity at the surface was reached in aqueous
solutions at temperatures of about $40°C$.

In mixtures of two solvents the two kinds of molecules compete in physical or chemical adsorption. If
ionic crystals such as strontium sulfate are suspended
in water-methanol mixtures for instance, the water
molecules are preferably adsorbed on the surface of the
crystals. For those mixtures

usually the apparent adsorption is plotted against
the mole fraction. The curves obtained are called
composite isotherms. Examples are given in Fig. 18
(14) and Fig. 19 (14). In the upper curves the composite
isotherm is shown. The apparent adsorption c'_a is given
by the expression

$$c'_a = \frac{n_A \, \Delta \, x_A}{m} \quad \left[\frac{moles}{g} \right] \tag{19}$$

where n_A is the mole number of the component A originally
present, Δx_A is the change in mole fraction of component
A by adsorption and m the mass of the solid in grams.
From the composite isotherm the individual isotherms
for the components A and B may be calculated. The
individual isotherms are shown in the lower curves of
Fig. 18 and Fig. 19.

4.4 Hydrolytic adsorption

The term hydrolytic adsorption was first used by
Fricke (15). External as well as internal surfaces are
involved in this kind of adsorption, which occurs
mainly in aqueous solutions.
        Sorbents showing hydrolytic adsorption all have
-OH groups on the surface. This is the case for all
hydrated oxides; for instance silica gel, SiO_2 powder,
glass powder, aluminium oxide, ferric oxide, MnO_2, TiO_2
and many other compounds.
        Adsorbates, on the other hand, must show hydrolysis
so that hydrolytic adsorption may occur. The easily
hydrolysable cations are those of oxidation states
3, 4 and 5. Examples of the oxidation state 3 are Fe^{3+},
Al^{3+} and the Lanthanide ions. The hydrolysis of these
ions starts at pH 3. Ions of the oxidation state 4,
such as Ti^{4+}, Zr^{4+}, Th^{4+} begin to hydrolyse at lower
pH, around pH 2 while ions of the oxidation state 5,
such as V^{5+}, Nb^{5+} and Pa^{5+} show hydrolysis even in
medium concentrated acids.
        Hydrolysis is a two step reaction. In the first step,
which is fast, hydroxo ions and hydroxides are formed,
for instance

$$Fe^{3+} \longrightarrow FeOH^{2+} \longrightarrow Fe(OH)_2^+ \longrightarrow Fe(OH)_3 \tag{20}$$

$$\longrightarrow pH$$

The progression of this reaction depends on the pH. At
higher pH the equilibria are shifted to the right. The

second step of the reaction is a slow one. It consists
of the condensation of the species formed by the first
reaction to species of higher molecular weight. Examples
are:

$$[HO-Fe-OH]^+ + [HO-Fe-OH]^+ \rightarrow [HO-Fe-O-Fe-OH]^{2+} \quad (21)$$

or

$$[HO-Fe-O-Fe-OH]^{2+} + HO-Fe-OH \rightarrow$$
$$\underset{\underset{OH}{|}}{}$$

$$[HO-Fe-O-Fe-O-Fe-OH]^{2+} \qquad\qquad (22)$$
$$\underset{\underset{OH}{|}}{}$$

These condensation reactions may also occur on the
surface of a sorbent carrying -OH groups, for instance

$$\underset{\diagup}{\overset{\diagdown}{-}}Si-OH \quad + \quad HO-Fe-OH \rightarrow \underset{\diagup}{\overset{\diagdown}{-}}Si-O-Fe-OH \qquad (23)$$
$$\qquad\qquad\qquad\qquad\underset{OH}{|} \qquad\qquad\qquad\underset{OH}{|}$$

silica gel   on the surface of
                                silica gel

The hydrolytic adsorption reactions on the surface are
slow chemical reactions and are to be classified as chemisorption. They are restricted, however, to hydroxo
complexes or hydroxides of low degree of condensation.
Species of higher molecular weight are adsorbed in a
much lower amount. If the sorbent has a pore structure,
molecules of higher molecular weight are not able to
enter the pores. There is also a possibility to separate
small particles from big ones, for instance those of
colloidal dimensions. The small ones are able to enter
the pore system and are sorbed, while the big ones which
are not able to enter the pores, are not sorbed.
    In weak alkaline solutions cations of the oxidation
state 2 also show hydrolysis, for instance Cu^{2+}, Pb^{2+},
Ni^{2+} and others. Under these conditions cations of the
oxidation state 2 may also be adsorbed by condensation
reactions similar to those shown in equation (23).
    In any case of hydrolytical adsorption the species
formed on the surface may be described as surface compound
These surface compounds are in many cases unsoluble such
as the silicates of heavy metals in neutral or alkaline
solutions.
    If other ions are present in solution which are able
to act as ligands in forming complexes, such as cyanide,

carbonate, oxalate or fluoride ions, the complexing reaction competes with the hydrolysis reaction. An example is the following reaction

$$Fe^{3+} + 3 \; Ox^{2-} \; \rightleftharpoons \; \left[Fe(Ox)_3\right]^{3-} \tag{24}$$

Therefore, if oxalate ions are present in sufficient concentrations, hydrolytic adsorption of Fe^{3+} does not occur.

By making use of complex formation and of the dependence of hydrolysis on pH, selective separations by hydrolytic adsorption may be obtained if the proper values of pH and (or) the proper complexing agents in suitable concentrations are applied.

5. SORPTION ON INTERNAL SURFACES AND IN PORES

5.1 Interpretation of sorption curves

If the sorbent is porous and has an internal surface the thickness of the adsorbed layer on the walls of the pores is limited by the width of the pores. The form of the isotherm is modified correspondingly. Instead of a Type II isotherm (Fig. 1) a Type IV isotherm is found and instead of a Type III isotherm a Type V isotherm. A Type IV isotherm is shown in Fig. 20. It resembles the Type II isotherm in having a point of inflection at the low pressure end, but instead of approaching the line $p/p_o = 1$ asymptotically it turns towards the pressure axis again giving a branch which is nearly horizontal. In addition Type IV isotherms have a hysteresis loop, the amount adsorbed being greater at any given relative pressure along the desorption branch HLF than along the adsorption branch FGH. This loop is reproducible provided that during the adsorption run point H was reached. If desorption commences from some intermediate point on the loop such as X the loop will follow the path XY.

Type IV isotherms are quite common, and are found expecially amongst the porous hydroxides (xerogels) such as silica gel or iron hydroxide. From an analysis of a Type IV isotherm it is possible, provided microporosity is absent, to estimate the specific surface of the solid and to arrive to some conclusions as to pore size distribution. In the pores having diameters ranging from tens to hundreds of Ångstrom units capillary condensation occurs giving rise to branch HLF of the isotherm. Branch

JK', when present, is ascribed to capillary condensation
in macropores or in the interstices between the grains
of the solid.

Along the low pressure branch DEF monolayer adsorption takes place on the walls of the pores in the
same manner as on the external surface of non-porous
solids. This process is reversible and there is no
hysteresis. At a point between E and F or actually at
point F the monolayer is completed. The BET method of
surface determination which was explained in 4.1 can
therefore be applied also for Typ IV isotherms using
the branch DEF or DEFG.

Concerning the events along branch FGH it is assumed
that a multilayer is gradually built up, again just
as on non-porous solids. The multilayer grows in thickness
as pressure increases till at H the pores are completely
filled with adsorbate. Thereafter adsorption, being
confined to the outside of the grains, increases very
slowly and the flat branch HJ results. When adsorbate
is withdrawn from the system the desorption branch HLF
is traversed. The liquid within any given pore begins
to evaporate from a meniscus which stretches across the
pore, as soon as the equilibrium pressure in the system
has fallen to a critical value p, given by the Kelvin
equation

$$\ln \frac{p}{p_o} = - \frac{2 \gamma V}{rRT} \cos \phi \, , \qquad (25)$$

where p_o is the saturation vapour pressure at the temperature T, γ is the surface tension, V is the molar
volume of the adsorbate in liquid form, r is the radius
of the pore, R is the gas constant per mole and ϕ is
the angle of contact between the liquid and the wall
of the pore. This equation plays a central role in
calculations of pore size distribution. It is usual to
make the simplifying assumption that $\phi = 0$ which gives

$$\ln \frac{p}{p_o} = - \frac{2 \gamma V}{rRT} \qquad (26)$$

This equation permits for a given liquid and known values
for γ and V the calculation of a numerical value of r for
the cylindrical radius corresponding to any value of
p/p_o. The capillaries which bring about condensation at
the left hand or low pressure end of the hysteresis loop
are very fine. Thus the radius corresponding to point F
of Fig. 20, where $p/p_o = 0,25$, is $r \approx 6$ Å.

The situation of non-cylindrical pores of different shape, V-shaped pores and of pores having only one opening has been discussed in detail and procedures for the evaluation of the pore size distribution from the hysteresis loop of adsorption isotherms of Type IV have been worked out. The procedures can not be described here in detail. An example is given in Fig. 21 (16) where the pore size distribution of silica gel has been determined from the adsorption isotherm of nitrogen at -195°C.

It may be mentioned that pore size distribution may also be determined by other methods, for instance by the mercury porosimeter.

5.2 Examples of porous systems

The pores present in sorbents may be divided into different categories: Open-ended cylindrical pores, pores formed by the capillary space between parallel plates or open slit-shaped capillaries, tubular or ink-bottle pores, pores between spherical particles and others. Some of these categories have characteristic adsorption isotherms as shown in Fig. 22.

Examples of sorption between parallel plates is furnished by some clay minerals such as montmorillonite and vermiculite. The crystals of these minerals have a layer structure, the layers being formed by a two dimensional aluminosilicate lattice. Between these layers there are cations such as Na^+, K^+, Ca^{2+} which may be exchanged against other cations. Water molecules are also found between the layers in variable amounts. An increase in the water content causes intercrystalline swelling which is a property typical for these compounds. Other plate-like compounds are found amongst the basic salts.

Xerogels such as silica gel, aluminium oxide hydrate, iron oxide hydrate or other compounds gained by precipitation resemble to some extent a packing of more or less spherical particles held together by chemical bonds. Active carbon and charcoal may be described as solids containing pores of approximately cylindrical shape and different diameters. By heating recrystallization occurs and larger crystals of graphite are formed, the mean diameter of the pores increases and the specific surface area decreases.

Uniform pores, however, and those having a geometrically ideal shape are very rarely found in nature.

5.3 Molecular sieves

Molecular sieves are compounds with a rather uniform
pore system which is formed by the structure of these
compounds. The molecular sieves belong to the class
of the zeolites. These are found in nature and prepared
artificially as well. The zeolites represent a very
large group of crystals containing pores in the form of
a network of intersecting channels. They are able to
form complexes with a great variety of molecules, inorganic
or organic, as well as with various salts, the molecules
or ions being taken up into the network of channels.
Because of their differing pore and channel geometries
they find a range of applications as selective sorbents
and are also used as selective ion exchangers.

All naturally occuring zeolites are aluminosilicates
in which the exchangeable cations are predominantly
Na, K, Ca and Ba. Synthetic zeolites are of the same
general structure and may contain other ions such as
Li, Rb, Cs, Tl, NH_4, Sr and alkylammonium ions $NH_{4-n}R_n^+$
as well. The replacement by other ions leads to modifi-
cations of the molecular sieve properties. Up to about
50 % of the volume of the crystals may consist of pores.

The zeolites found in nature include analcite,
natrolite, chabazite, gmelinite, levynite, mordenite,
faujasite and a great number of other compounds. Examples
of synthetic zeolites are the Linde Molecular Sieves.
The basic structural unit in all zeolites is the tetra-
hedron SiO_4 or AlO_4. These units are linked together
to a network which is negatively charged due to the
isomorphous replacement of Si by Al. The network charge
is neutralized by an equivalent number of interstitial
exchangeable cations. The frameworks characteristic for
the zeolites may be described in terms of polyhedra
constructed from the groups of tetrahedra mentioned above.
Thus eight tetrahedra form a cubic unit, twelve
tetrahedra a hexagonal prism and so forth. The resulting
framework of polyhedra brings about the porous structure
which has been studied in detail by Barrer (17). Fig. 23
(18) and Fig. 24 (19) give some examples of pores found
in molecular sieves. The free dimensions of the apertures
in the channel system compared with the dimensions of
molecules determine the ability of molecules to pass into
the pores. As molecules are not hard spheres and the
rings of tetrahedra are not entirely rigid there is some
overlap in dimensions which is responsible for the energy
barrier observed. Thus Linde Sieve 5A having a minimum
aperture of 4.3 Å is occluding argon molecules at $-183^{\circ}C$
very rapidly whereas at the same temperature the rate
of occlusion in Levynite with a minimum aperture of

3.2 Å is very slow. The cation distribution within the crystals is also important for the rate of sorption.

A classification of some molecular sieves according to the critical size of the pores given by the minimum aperture is shown in Table 3.

Ca-mordenite		Group 1: He, Ne, Ar, CO, H_2, O_2, N_2, NH_3, H_2O
Levynite	3.8	
		Group 2: Kr, Xe, CH_4, C_2H_6, CH_3OH, CH_3CN, CH_3NH_2, CH_3Cl,
Na-mordenite		CH_3Br, CO_2, C_2H_2, CS_2
Linde Sieve 4A	4.0	
		Group 3: C_3H_8, $n-C_4H_{10}$, $n-C_7H_{16}$, C_2H_5Cl, C_2H_5Br, C_2H_5OH, $C_2H_5NH_2$, CH_2Cl_2, CH_2Br_2, CHF_2Cl, CHF_3, CH_3J, B_2H_6,
Ca-chabazite		CF_4, C_2F_6, CF_2Cl_2, CF_3Cl,
Linde Sieve 5A	4.9	$CHFCl_2$
Ba-zeolite		
Gmelinite		Group 4: SF_6, iso-C_4H_{10}, $CHCl_3$, $CHBr_3$, CHJ_3, $(CH_3)_2CHOH$, $n-C_3F_8$, $n-C_4F_{10}$, B_5H_9, $(CH_3)_3N$, $(C_2H_5)_3N$, $C(CH_3)_4$, $C(CH_3)_3Cl$, CCl_4, CBr_4, $C_2F_2Cl_4$ C_6H_6, $C_6H_5CH_3$, Cyclohexane, Methylcyclohexane, Thiophene, Furan, Pyridine, Dioxane, $B_{10}H_{14}$, Naphthalene, Quinoline
Linde Sieve 10X		
		Group 5: 1,3,5 triethylbenzene
Linde Sieve 13X	Å	

Table 3. Classification of some molecular sieves

The guest molecules are listed also according to their size. Further subdivision within the major groups is possible.

Another important property of molecular sieves is their saturation capacity. This is given by the number of guest molecules which are accomodated per cavity. These numbers vary with temperature. Some examples are given in Table 4. Sorption isotherms in some cases qualitatively resemble the Langmuir isotherm, in many other cases they show several steps and (or) hysteresis. As an example the isotherms for the sorption of CF_4 in Na-faujasite are presented in Fig. 25 (20).

Compound	Largest cavities	Guest molecules per cavity
Chabazite	20-hedra (6 x 8-rings 2 x 6-rings 12 x 4-rings)	12-14H_2O ~7.7NH_3 ~6Ar,N_2,O_2 ~4.9CH_3NH_2 ~4.3CH_3Cl ~3.1CH_2Cl_2 ~2.0I_2
Linde Sieve A	26-hedra (6 x 8-rings 8 x 6-rings 12 x 4-rings)	~29H_2O(25+4) 19-20NH_3 14-16Ar,N_2,O_2 ~15H_2S ~12CH_3OH ~10SO_2 ~9CO_2 ~5.5I_2 ~5.4n-C_3H_7OH ~4n-C_4H_{10}
Faujasite	26-hedra (4 x 12-rings 4 x 6-rings 18 x 4-rings)	~32H_2O(28+4) 17-19Ar,N_2,O_2 ~7.5I_2 ~7.8CF_4 ~6.5SF_6 ~5.8C_2F_6 ~5.6 cyclopentane ~5.4 benzene ~4.6 toluene ~4.5n-C_5H_{12} ~4.1 cyclohexane ~4.1 perfluorocyclobutene ~3.5n-C_7H_{16} ~3.4C_3F_8 ~2.9n-C_4F_{10} ~2.3 perfluoro-Me-cyclohexane

Table 4. Capacity of molecular sieves for guest molecules

5.4 Layer compounds

Zeolites (molecular sieves) and layer compounds are able to form inorganic inclusion complexes. Many layer lattices

possess the property of being able to expand in the
presence of appropriate fluids and of intercalating
the molecules of these fluids. These layers appear in
a variety of natural layer minerals such as micas,
chlorites, vermiculites and a number of clay minerals
such as montmorillonite. Each layer is a threefold one,
in the upper and lower layers the SiO_4-tetrahedra are
linked together to rings of six. These layers are also
linked via Si-O-Al-bonds to some of the oxygens
surrounding the aluminium in octahedral coordination
in the central layer as shown in Fig. 26. Replacement
of Si by Al in the tetrahedral layers or replacement of
Al by Mg, Fe^{II} or others in the octahedral layers result
in anionic charges which are compensated by equivalent
numbers of cations such as Na, K, Mg, Ca, Ba between
the triple layers. The ions may be more or less hydrated.

The structure of the layer compounds has been
investigated by Barrer (17) and others while the intercrystalline swelling and ion exchange has been studied
in detail mainly by Weiss and Hofmann (Literature cited
in (17)). The extent of swelling depends in the first
instance on the number of charges per unit area on the
silicate layers. Water does not swell very highly charged
or uncharged layer silicates. The most highly swollen
structures are those with a small layer charge and
monovalent interlayer ions promote swelling more than
divalent ions. In some cases unlimited swelling is
observed leading to complete separation of the layers.

Intercalation in natural montmorillonites and some
other clay minerals is observed for water and other polar
molecules such as NH_3, CH_3OH, C_2H_5OH, pyridine, glycol
and others. The intercalation isotherms are usually
continuous and demonstrate hysteresis between sorption
and desorption.

By means of ion exchange a great range of modified
layer lattice silicates may be prepared which show
interesting powers of intercalation. When Na or Ca ions
are replaced by organic ions the compounds may become
hydrophobic and other types of neutral molecules,mainly
organic compounds, may be intercalated giving rise to
a great variety of sorption properties.

Taking advantage of these properties layer compounds
may be used either as inorganic ion exchangers or after
appropriate treatment as sorbents for organic molecules.

Layer compounds are not only found in the class of
aluminosilicates which have been described above such
as the clay minerals,montmorillonites or the vermiculites.
Similar properties are exhibited by other classes of
compounds,for instance the uranium micas with the general
formula $(M^+,1/2M^{2+})xH_2O \left[UO_2XO_4 \right]$ (X=P,As,V), the

trititanates consisting of layers of TiO_6 octahedra, complex cyanides such as $Ni(CN)_2 \cdot NH_3 \cdot C_6H_6$, basic salts and the graphite compounds. All these compounds have a layer structure and are able to form inclusion complexes by sorption.

5.5 Cage compounds (clathrates)

The molecular sieves contain a network of intersecting channels accomodating the guest molecules sorbed. In the layer compounds the guest molecules are arranged between the layers. In the cage compounds the guest molecules are found in isolated cavities of the lattice. Cage compounds can therefore not be formed by sorption into a given solid because the cavities are not accessible from outside. They are obtained only if the host lattice is formed in the presence of the guest molecules. The term cage or clathrate compounds is used to describe the molecular imprisonment of the guest molecules.

Well known examples are the gas or liquid hydrates. In the structure of these compounds the pentagonal dodecahedron composed of twenty water molecules is an essential unit. The water molecules are held together by hydrogen bonds. The dodecahedra are arranged as shown in Fig. 27. The diameter of such a dodecahedron is about 6.9 Å. Furthermore there are small voids between the dodecahedra which are able to accomodate small molecules additionally. In Table 5 the properties of some gas hydrates are given. The dimensions of four kinds of cavities are compared with the sizes of different molecules in Fig. 28.

Guest molecule	Dissociation pressure at $0^{\circ}C$ (atm)	Decomposition temperature at 1 atm $(^{\circ}C)$
Ar	105	-42,8
CH_4	26	-29,0
Kr	14,5	-27,8
CO_2	12,3	-24
C_2H_2	5,7	-15,4
C_2H_6	5,2	-15,8
Xe	1,5	-3,4
H_2S	0,96	+0,35
CH_3Cl	0,41	+7,5
SO_2	0,39	+7,0
Cl_2	0,33	+9,6
CH_3Br	0,25	+11,1

Table 5. Properties of some gas hydrates

Other types of cage compounds are the hydroquinone clathrates. They are found with SO_2, CO_2, H_2S, HCl, HBr, CH_3OH, HCOOH, Ar, Kr, Xe and others as guest molecules. The percentage of cavities occupied by guest molecules varies between about 30 and 99 %.

The crystal structures of all clathrates are alike in that they have enclosed spaces which may be occupied or not by the trapped molecules. In gas hydrates the bonds between the guest molecules and the host are mainly of the van der Waals type. This holds also for the hydroquinone clathrates. Stronger bonds such as atomic or ionic bonds are rarely found between host and guest molecules. Hydrogen bonds and van der Waals bonds are the operating forces, too, in holding together the lattice of the host molecules in the gas hydrates and the hydroquinone clathrates.

6. SORPTION BY ION EXCHANGE

6.1 Distribution coefficients, separation factors and selectivity

In the processes of absorption and adsorption considered so far a certain amount of substance is taken up by a sorbent. In many cases these processes could be described by solubilities or adsorption isotherms, respectively. In ion exchange reactions, however, equivalent amounts of ions are taken up and given off. Adsorption isotherms are not applicable any more.

As already mentioned, ion exchange is a chemical process. The exchange of cations is usually described by an equation of the following kind:

$$nA^{m+} + m\overline{B^{n+}} \rightleftharpoons m\overline{B^{n+}} + nA^{m+} \qquad (27)$$

where n and m are the respective charges of the ions A and B and the bar means that the ions are bound in the solid. A similar equation may be written for anion exchange.

Ion exchange is usually characterized by the distribution coefficient. For reaction (27) the equilibrium constant is given by the expression

$$\frac{\left[\overline{B^{n+}}\right]^m \left[A^{m+}\right]^n}{\left[\overline{A^{m+}}\right]^n \left[B^{n+}\right]^m} = K \qquad (28)$$

The distribution coefficient K_d is the ratio of B in the solid form, that is in the ion exchange substance, and in the solution

$$K_d = \frac{[\overline{B^{n+}}]}{[B^{n+}]} \qquad \left[\frac{liter}{g}\right] \qquad (29)$$

As $[\overline{B^{n+}}]$ is usually expressed in moles/g and $[B^{n+}]$ in moles/liter, the dimension of K_d comes out to be liter/g.

The ratio of the distribution coefficients for two different ions is called separation factor

$$SF(A/B) = \frac{K_d(A)}{K_d(B)} \qquad (30)$$

Separation factors are a measure of the selectivity. If the distribution coeff. for one special ion is much higher than that for all other ions we have a high selectivity for that special ion. Thus the selectivity for an ion A may be defined by the equation

$$S = \log K_d(A) - \log K_d(others) \qquad (31)$$

Selectivities $S > 2$ are of great importance for practical separations and may be reached in some cases.

It should be pointed out that the concept of distribution coefficients, separation factors and selectivity which has here been applied to ion exchange reactions may also be used for other sorption processes. It has many advantages for practical purposes.

Taking as an example a simple ion exchange reaction between protons and metal ions M^{n+}

$$n\overline{H} + M^{n+} \rightleftharpoons \overline{M} + nH^+ \qquad (32)$$

it may be seen that the slope of the curve for the logarithm of the distribution coefficients plotted against pH should be n. The equilibrium constant for reaction (32) is

$$\frac{[\overline{M}][H^+]^n}{[\overline{H}]^n[M^{n+}]} = K \qquad (33)$$

and the distribution coefficient

$$K_d = \frac{[\overline{M}]}{[M^{n+}]} = K \frac{[\overline{H}]^n}{[H^+]^n} \tag{34}$$

$$\log K_d = \log K [\overline{H}]^n - n\log [H^+] \tag{35}$$

$$\log K_d = \log K' + npH \tag{36}$$

The concentration of the hydrogen ions bound in the solid has been considered constant which is the case as long as only a small amount is exchanged.

6.2 Kinetics of ion exchange

If the ion exchange takes place on external surfaces it is a fast reaction - except ion exchange reactions on the surface of ionic crystals which in general need an activation energy. Ion exchange on internal surfaces and in pores, however, is always a slow reaction, the rate depending on the size of the grains and the size of the pores. In the case of organic resins two kinetic processes are of importance which are illustrated in Fig. 29, film kinetics and gel kinetics. Film kinetics involve the diffusion of the ions within the Nernst diffusion film adhering to the resin particles and gel kinetics the diffusion of the ions within the pore system of the resin particles. Film kinetics depend on the thickness of the Nernst diffusion film and the temperature Characteristic half times of film kinetics are of the order of 10 seconds. Gel kinetics, on the other hand, depend on the size of the resin particles and the size of the pores, as already mentioned, and also on the temperature. For Dowex 50 x 8 200-400 mesh (particle size 0.04-0.08 mm) at room temperature half times of about 3 minutes are found, for the same type of resin but with a particle size of 0.15-0.3 mm (50-100 mesh) the half time is about 6 minutes. With the so called macroporous resins of the same particle size half times have been measured which are somewhat shorter.
    If the functional groups of the ion exchange resin are restricted to the surface of the resin particles gel diffusion is no more the rate determining step but film diffusion (21). The exchange rates for a normal resin and a resin carrying functional groups on the surface only can be seen from Fig. 30. A similar result is obtained if very small resin particles with a particle

size of the order of 0.01 mm are taken. In this case
film kinetics again become rate determining. The
situation is illustrated in Fig. 29.

In the case of inorganic ion exchange particles
the kinetics depend also on the grain size, the pore
system and the temperature. For most inorganic ion
exchange substances the diffusion within the grains
or particles is a very slow process so that the exchange
rate is characterized by long half times of the order
of ten minutes or more. Only if the exchange reaction
is restricted to the surface of the solid particles
fast exchange rates are observed.

6.3 Organic ion exchange resins

Many ion exchange substances consist of a matrix, a
functional group and exchangeable ions. Examples of this
kind are the ion exchange resins in which the matrix
is an organic macromolecule like polymerized styrole
linked by vinyle benzene groups. The functional groups
may be in the case of cation exchange resins $-SO_3H$ or
$-COOH$ - groups. The first are strongly acid, the second
weakly acid. In the case of anion exchange resins
$-NR_3OH$, $-NHR_2OH$, $-NH_2ROH$ or $-NH_3OH$ groups are used as
functional groups. The basicity is decreasing from
$-NR_3OH$ to $-NH_3OH$. Macromolecular organic compounds with
other functional groups have been synthesized with the
tendency to introduce groups of high selectivity, such
as complexing or chelating ligands.

Distribution coefficients for some cations on a
Dowex 50 resin are shown in Fig. 31. They are plotted
against the pH. The slope of the curves corresponds to
the charge of the cations according to equation (36).
The differences in the distribution coefficients for
the different cations are not very pronounced or, in
other words, the selectivity of organic resins is not
very high.

The kinetic behaviour of organic resins has already
been discussed in greater detail in 6.2.

For selective separations functional groups of
higher specifity are needed. Complexing, or even better,
chelating agents are very suitable as functional groups.
But synthesis of compounds containing such groups in
easily accessible positions and in high concentrations
is not an easy task for the chemist.

6.4 Inorganic substances

In most inorganic substances exhibiting ion exchange
matrix and functional group are identical. Compounds
found in nature are the zeolites and the clay minerals.
They consist of an aluminosilicate lattice which may be
arranged in form of layers, as in the clay minerals,
or in form of a three-dimensional network with pores,
as in the zeolites. The aluminosilicate lattice carries
negative charges equivalent to the amount of aluminium
atoms in the lattice. The negative charges of the matrix
are compensated by an equivalent number of cations which
can be exchanged. The synthetic molecular sieves have
similar properties as the zeolites. Zeolites and molecular sieves also work on the same principle. Other
important synthetic inorganic compounds are hydrated
oxides, phosphates and hexocyanoferrates. Hydrated oxides
such as $TiO_2 \cdot xH_2O$ or $ZrO_2 \cdot xH_2O$ are able to exchange to
a certain amount H^+-ions, in certain cases also OH^--ions,
both from their water molecules, to other cations or
anions. Mainly, however, these compounds show adsorption
without giving off an equivalent amount of ions. Phosphates and even more hexacyanoferrates such as titanium
phosphate, zirconiumphosphate, titaniumhexacyanoferrate
and many other hexacyanoferrates to a much greater extent
exchange H^+-ions to other cations. The H^+ ions are
mainly contained in the acidic part of the compound, for
instance in the form of $Ti(OH)[\underline{H}Fe(CN)_6] \cdot xH_2O$ or
$Ti[\underline{H}PO_4]_2 \cdot xH_2O$. The exchangeable H^+-ions are underlined.
    Finally ionic compounds such as $BaSO_4$ or $AgCl$ also
exhibit ion exchange properties. $BaSO_4$ for instance is
able to exchange Ba^{2+} ions to Sr^{2+} ions and $AgCl$
exchanges Cl^- ions to I^--ions. Here we have no matrix
and no functional group but only ions which may be exchanged. In general, formation of the less soluble compound
is preferred in these ion exchange reactions, for instance
the formation of AgI from $AgCl$ and I^--ions so that the
reaction is a quantitative one in this case.
    Two examples are given for illustration. In Fig. 32
log K_d values plotted against acid concentration are
shown for titanium hexacyanoferrate (22). The high selectivity for cesium ions may be seen from this figure. It
is assumed that the cesium ions fit excellent into the
interstices between the octahedral groups of the hexacyanoferrate ions.
    Another example is the exchange of Cl^--ions against
I^--ions in silver chloride, a reaction which proceeds
qualitatively until the total amount of $AgCl$ present is
transformed into AgI (23)

$$AgCl + I^- \longrightarrow AgI + Cl^- \tag{37}$$

Similar reactions occur if Ag^+-ions are present in form of $AgNO_3$ or if they are sorbed on zeolites or molecular sieves by ion exchange. The silver ions also react quantitatively with molecular iodine if water is present:

$$6\ AgCl + 3\ I_2 + 3\ H_2O \rightarrow 5\ AgI + AgIO_3 + 6\ HCl \tag{38}$$

In inorganic ion exchange rather high selectivity is obtained in many cases. The reactions involved are generally more complex than for resins. The rate of exchange, however, is slow for most inorganic substances. Higher rates of exchange are reached by use of finely powdered material and application of pressure, for instance in a high pressure apparatus. This could be shown recently (24).

LITERATURE

A. K. Hauffe u. S.R. Morrison, Adsorption, Walter de Gruyter, Berlin u. New York, 1974.
B. S.J. Gregg and K.S.W. Sing, Adsorption, Surface Area and Porosity, Academic Press, London and New York, 1967.
C. J.J. Kipling, Adsorption from Solutions of Non-Electrolytes, Academic Press, London and New York, 1965.
D. D.W. Breck, Zeolite Molecular Sieves, John Wiley & Sons, New York, 1973.
E. R.M. Barrer, Inorganic Inclusion Complexes in Non-stoichiometric compounds, Edited by L. Mandelcorn, Academic Press, New York and London, 1964.
F. K. Dorfner, Ionenaustauscher (Eigenschaften und Anwendungen), 2. Aufl., Walter de Gruyter, Berlin 1964.
G. C.B. Amphlett, Inorganic Ion Exchangers, Elsevier Publ. Comp., Amsterdam-London-New York, 1964.

REFERENCES

1. A. Sieverts, Z. Metallkde 44, 152 (1938).
2. Ph. Gütlich u. K.H. Lieser, Z.physik.Chem. N.F. 46, 257 (1965).
3. S. Brunauer, L.S. Deming, W.S. Deming and E. Teller, J.Amer.Chem.Soc. 62, 1723 (1940).
4. S. Brunauer, P.H. Emmett and E. Teller, J.Amer.Chem. Soc. 60, 309 (1938).
5. F. London, Z.Physik 63, 245 (1930); Z.physik.Chem. 11, 222 (1930).
6. L.E. Drain and J.A. Morrison, Trans.Far.Soc. 49, 654 (1953).
7. H. Kautsky u. H. Weßlau, Z.Naturforschg. 9b, 569 (1954).
8. I. Langmuir, J.Amer.Chem.Soc. 38, 2221 (1916).
9. A.V. Kiselev and Y.A. Eltekov, Int. Cong. Surface Activity II, 228, Butterworth, London (1957).
10. A.A. Isirikyan and A.V. Kiselev, J.phys.Chem. 65, 601 (1961); 66, 210 (1962).
11. L.G. Joyner and P.H. Emmett, J.Amer.Chem.Soc. 70, 2353 (1948).
12. R.A. Beebe, J. Biscoe, W.R. Smith and C.B. Wendell, J.Amer.Chem.Soc. 69, 95 (1947).
13. I. Langmuir, J.Amer.Chem.Soc. 34, 1310 (1912).
14. C.G. Gasser and J.J. Kipling, J.phys.Chem. 64, 710 (1960).
15. R. Fricke u. H. Schmäh, Z.anorg.Chem. 255, 253 (1948).

16. K.S.W. Sing and J.D. Madeley, J.appl.Chem. 3 (1953).
17. R.M. Barrer in Non-stoichiometric compounds, Edited by L. Mandelcorn, Academic Press, New York and London, 1964.
18. R.M. Barrer and I.S. Kerr, Trans.Far. Soc. 55, 1915 (1959).
19. R.M. Barrer and S. Wasilewski, Trans.Far.Soc. 57, 1140 (1961).
20. R.M. Barrer and P.J. Reucroft, Proc.Roy.Soc. A 258, 431 (1960).
21. M. Skafi u. K.H. Lieser, Z.Anal.Chem. 249, 182 (1970).
22. J. Bastian u. K.H. Lieser, J.inorg.nucl.Chem. 29, 827 (1967).
23. K.H. Lieser u. W. Hild, Z.anorg.allg.Chem. 350, 237 (1967); Radiochim.Acta 7, 74 (1967).
24. K.H. Lieser u. T. Baumgartner, unpublished results.

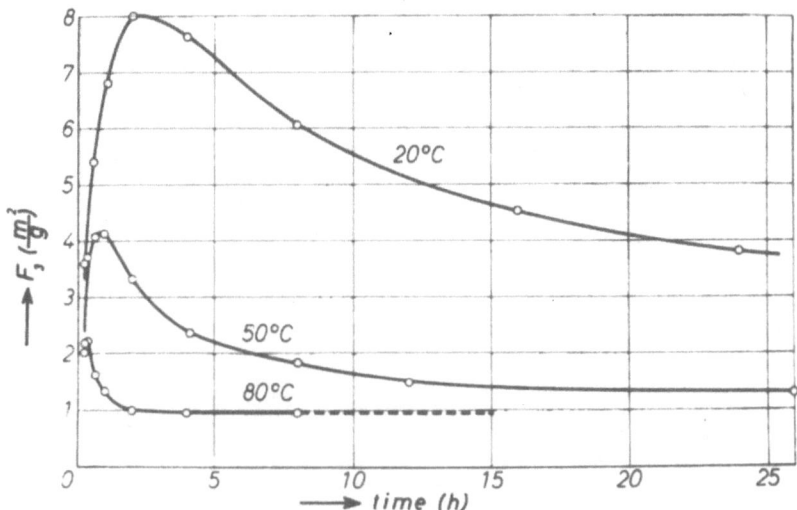

Fig. 1. Specific surface of BaSO$_4$ after precipitation
at different temperatures as a function of time   e

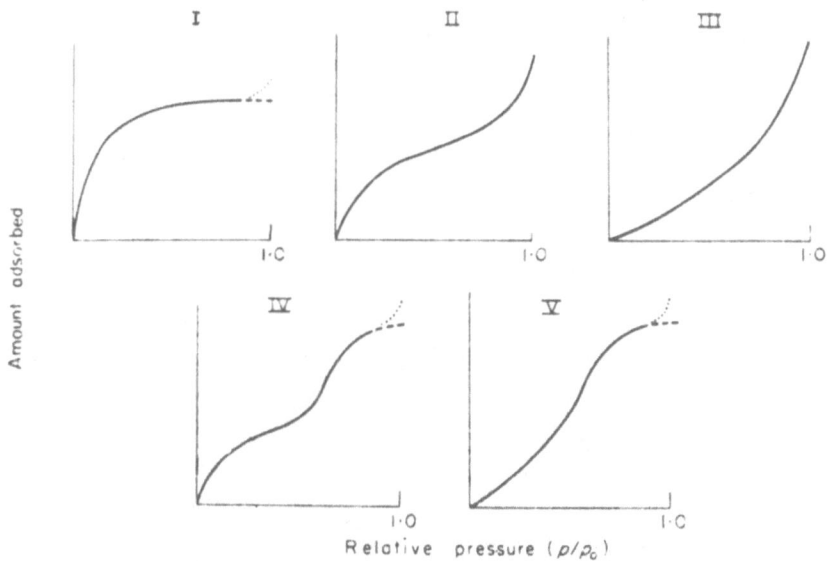

Fig. 2. Different types of adsorption isotherms (BET
classification)

126

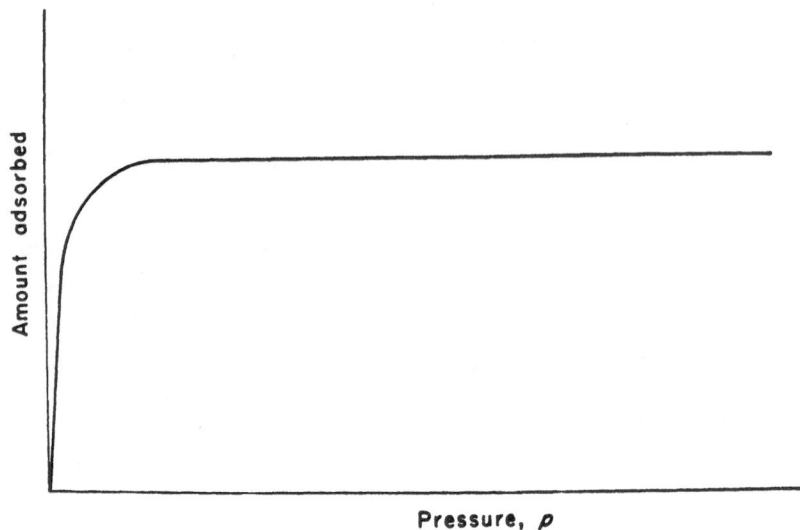

Fig. 3.   Isotherm for chemisorption

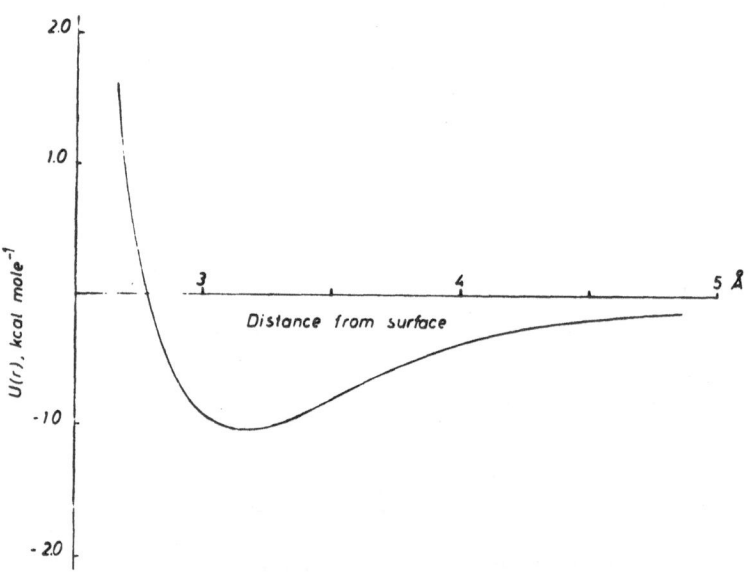

Fig. 4.   Potentional energy for physical adsorption

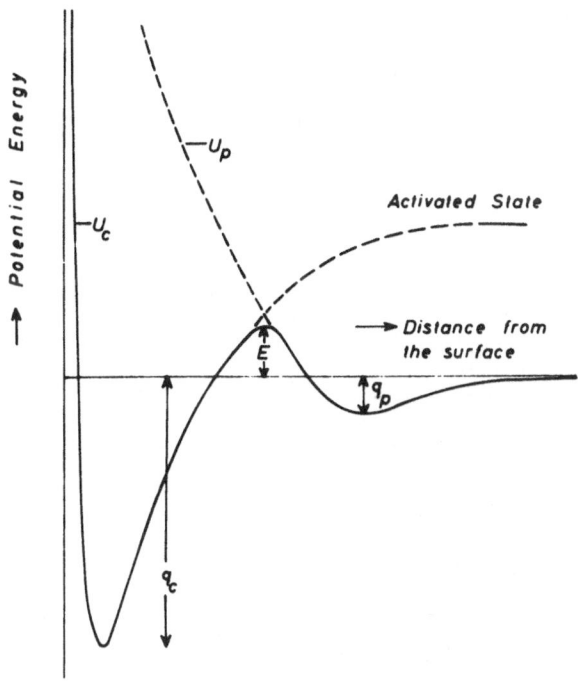

Fig. 5. Potential energy for chemical adsorption (U_c) and for physical adsorption (U_p)

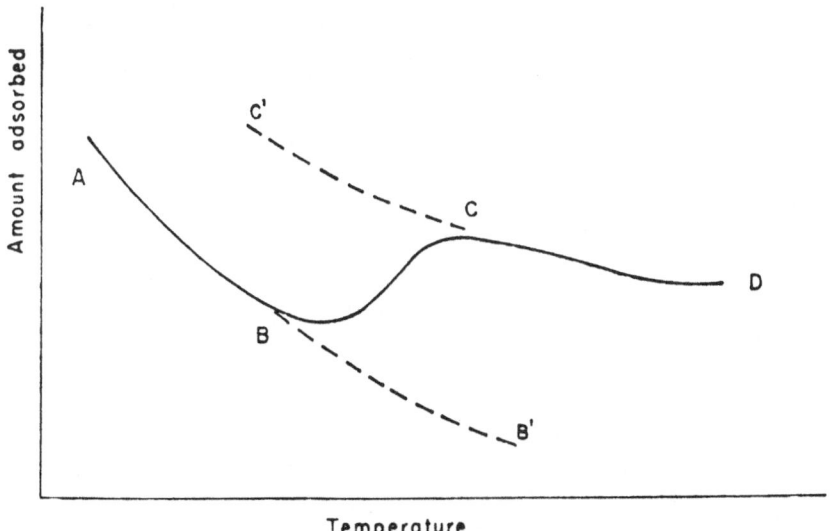

Fig. 6. Adsorption isobar showing physical and chemical adsorption

128

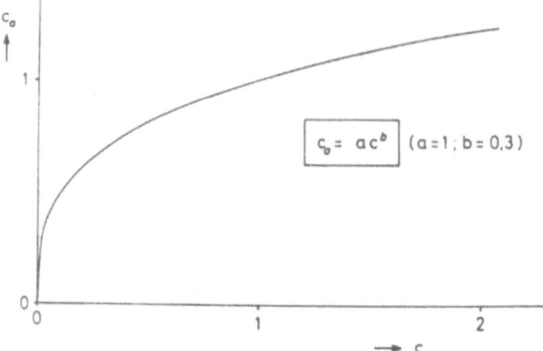

Fig. 7. Freundlich isotherm

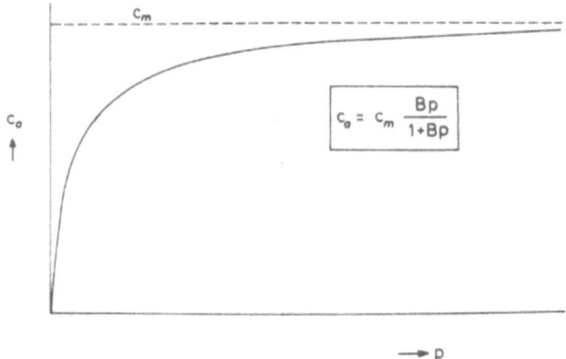

Fig. 8. Langmuir isotherm

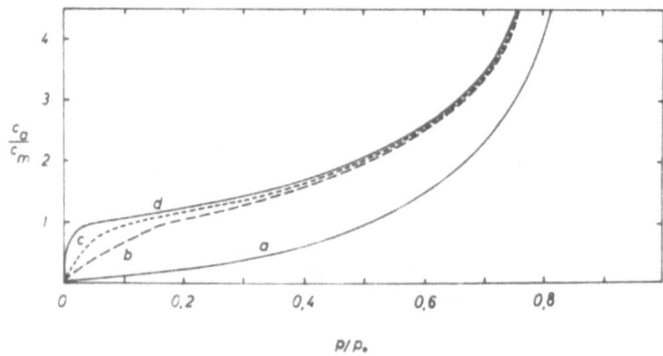

Fig. 9. Curves of c_a/c_m against p/p_0 calculated from the BET eq.

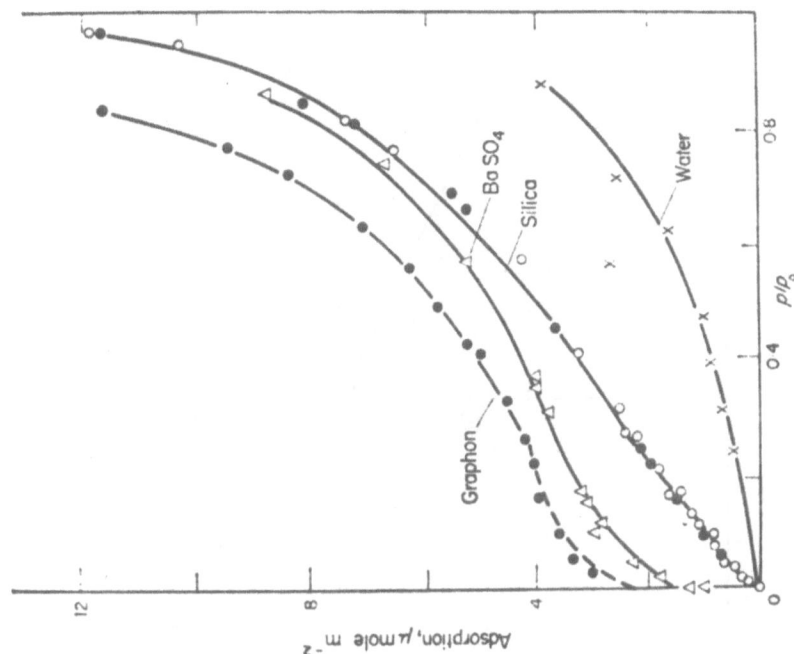

Fig. 11. Adsorption isotherm of n-pentane on different adsorbents

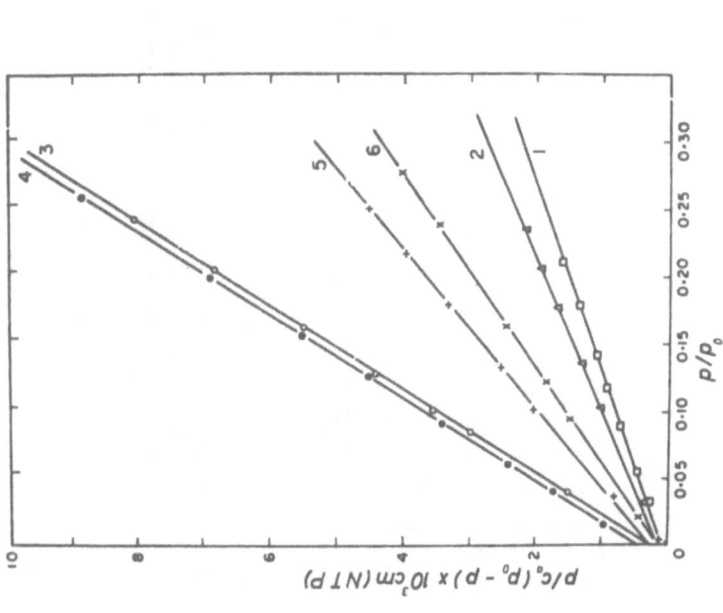

Fig. 10. Adsorption of nitrogen at -193°C on a number of catalysts: 1. Unpromoted Fe, 2. Al_2O_3-promoted Fe, 3. $Al_2O_3/$ K_2O-promoted Fe, 4. Cu, 5. Chromium oxide gel, 6. silica gel

130

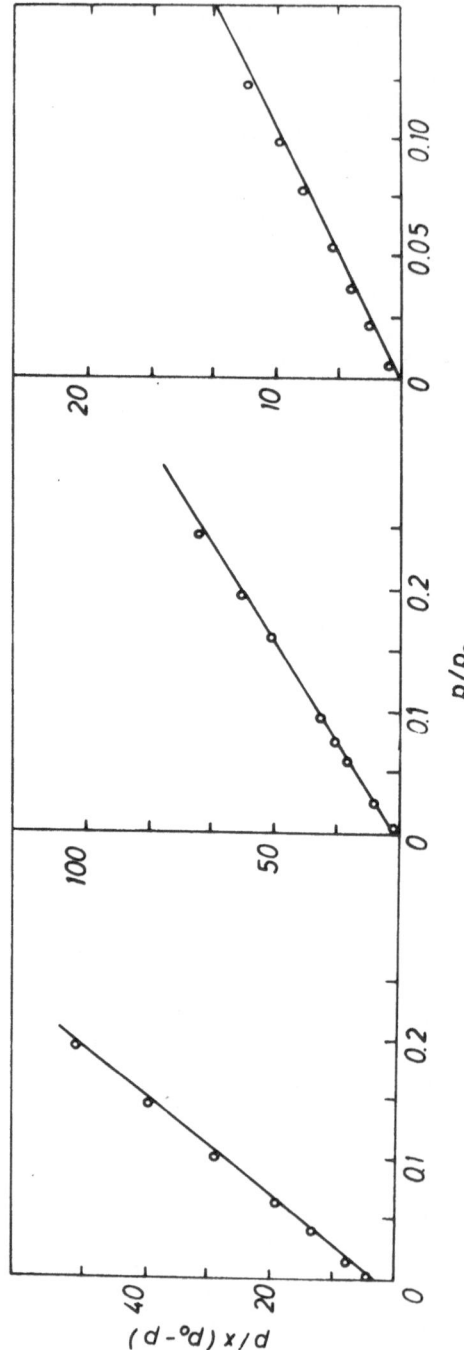

Fig. 12. $p/x(p_0-p)$ against p/p_0 for adsorption of benzene at $20^\circ C$, n-hexane at $20^\circ C$ and nitrogen at $-195^\circ C$ on carbon black, graphitized at $3100^\circ C$. $x = c_a$ in μ moles per square meter.

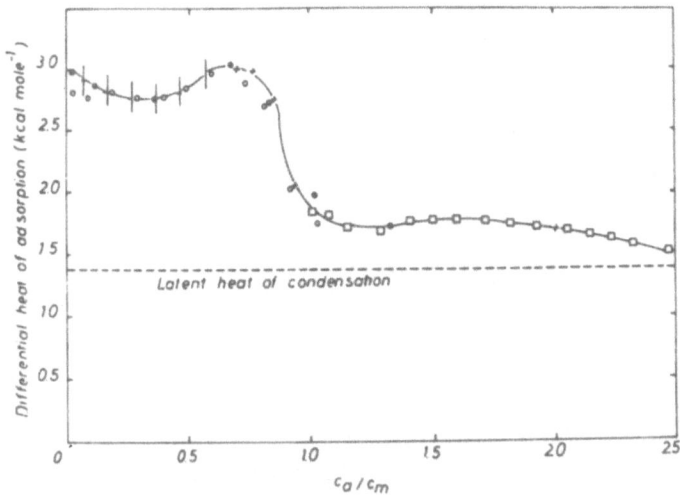

Fig. 13. Differential heat of adsorption of nitrogen on carbon
black, graphitized at 2700°C, showing the formation
of a monolayer

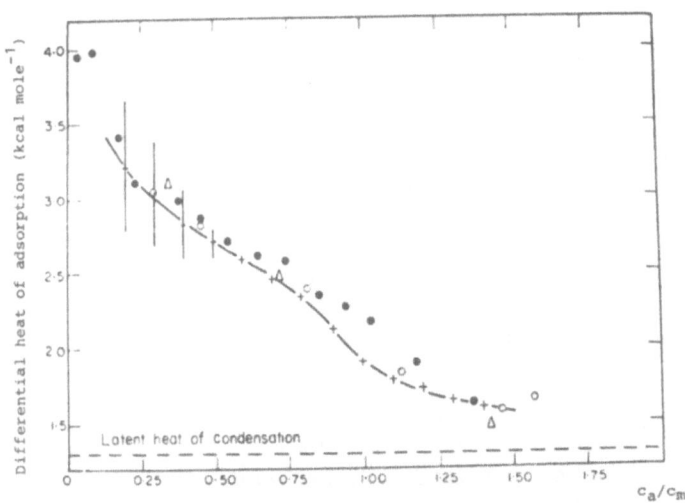

Fig. 14. Differential heat of adsorption of nitrogen on
carbon black before graphitization

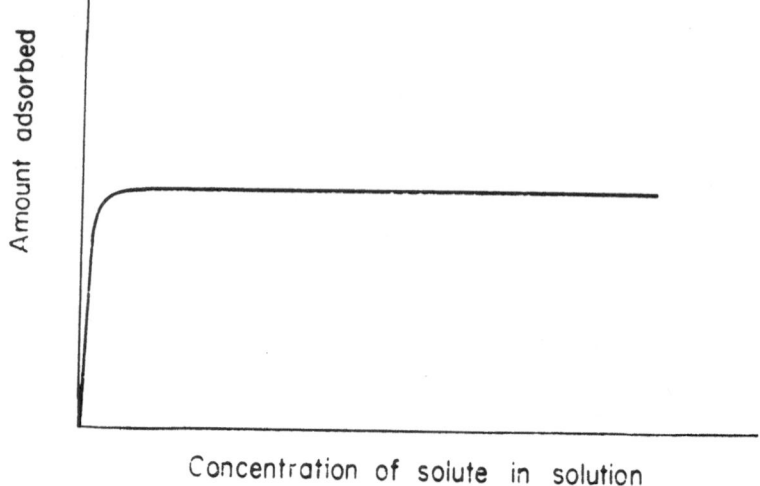

Fig. 15.   Adsorption isotherm for a solute from solution

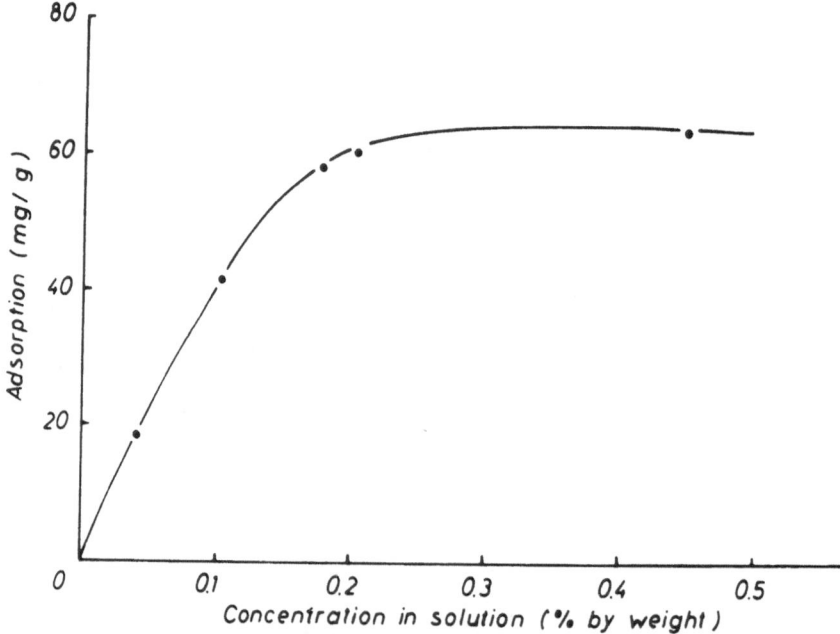

Fig. 16.   Adsorption of methylene blue from aqueous solution
by spheron 6

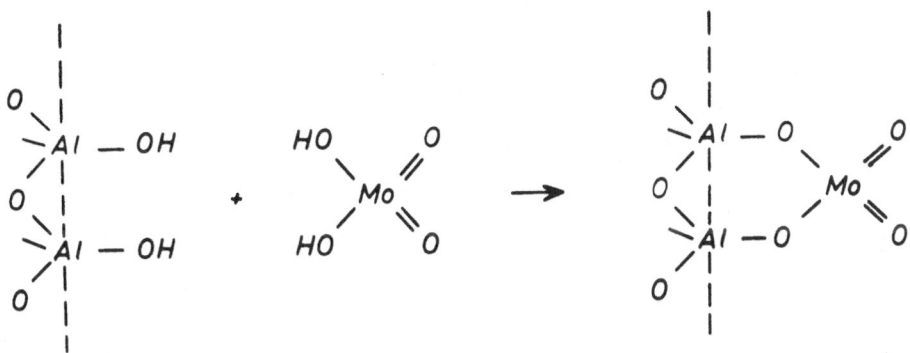

Fig. 17. Chemisorption of molybdate on aluminium oxide

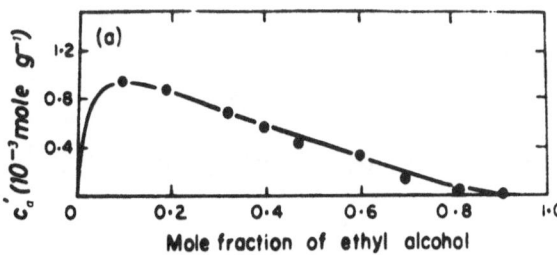

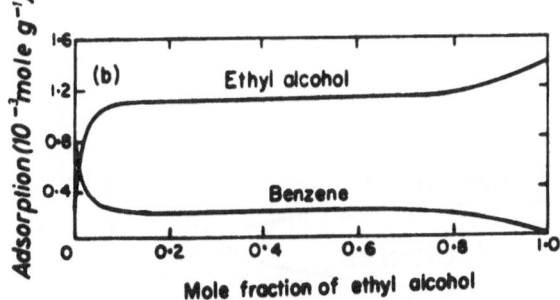

Fig. 18. (a) Measured isotherm for the adsorption of ethyl alcohol from its solution in benzene on activated Al_2O_3 (apparent adsorption c_a')

(b) Individual isotherms of ethyl alcohol and benzene

134

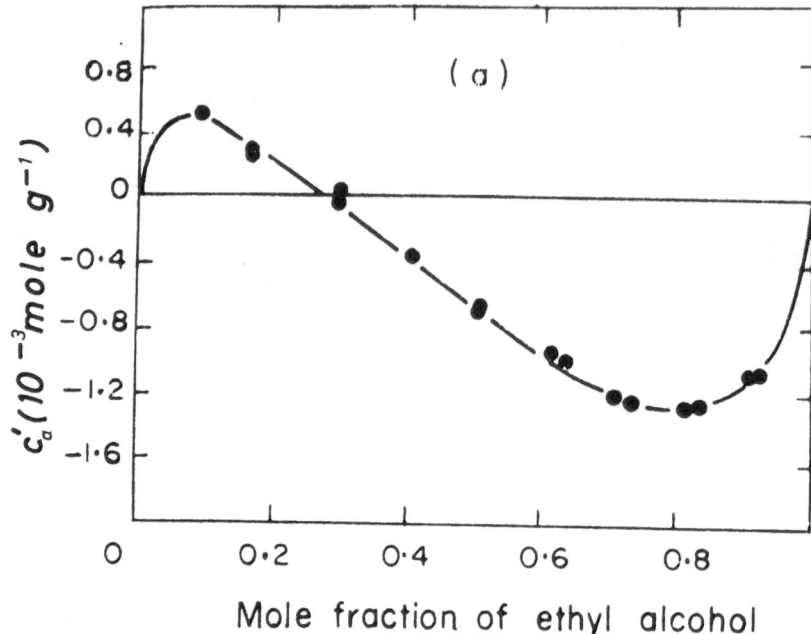

Mole fraction of ethyl alcohol

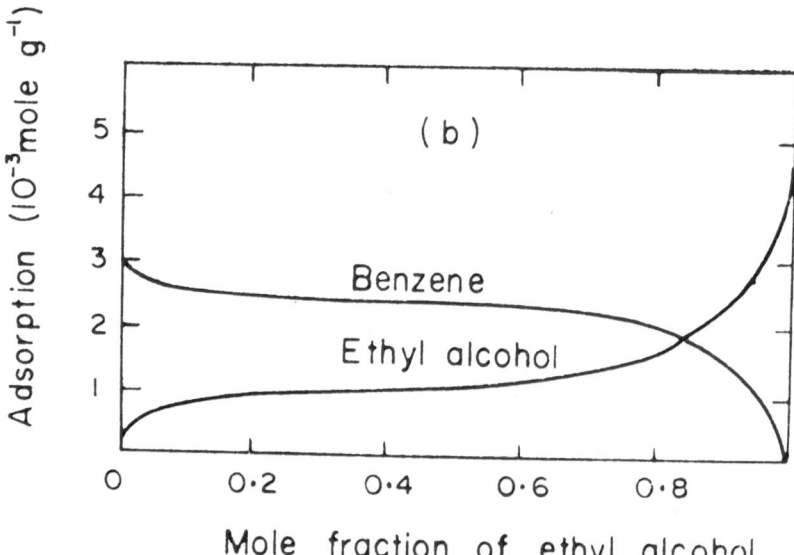

Mole fraction of ethyl alcohol

Fig. 19. (a) Measured isotherm for the adsorption
of ethyl alcohol from its solution in
benzene on charcoal

(b) Individual isotherms of ethyl alcohol and
benzene.

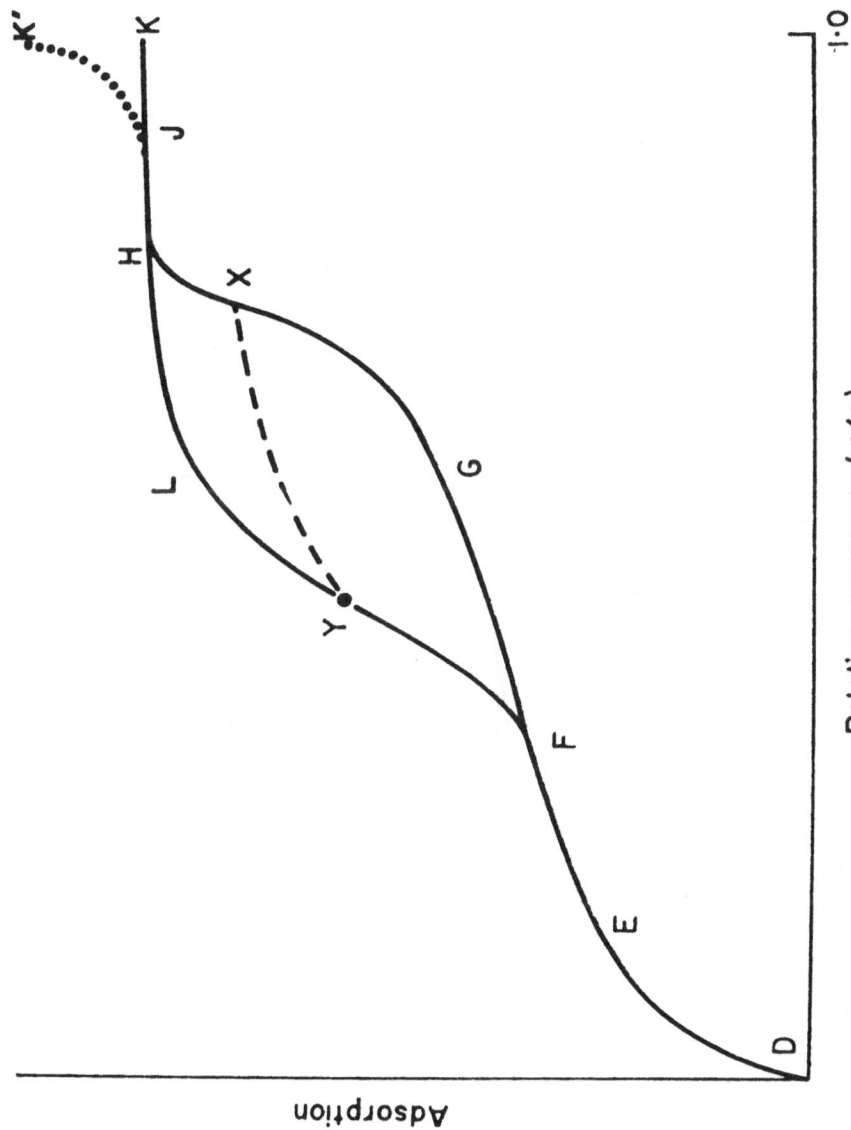

Fig. 20. Interpretation of a Type IV isotherm.

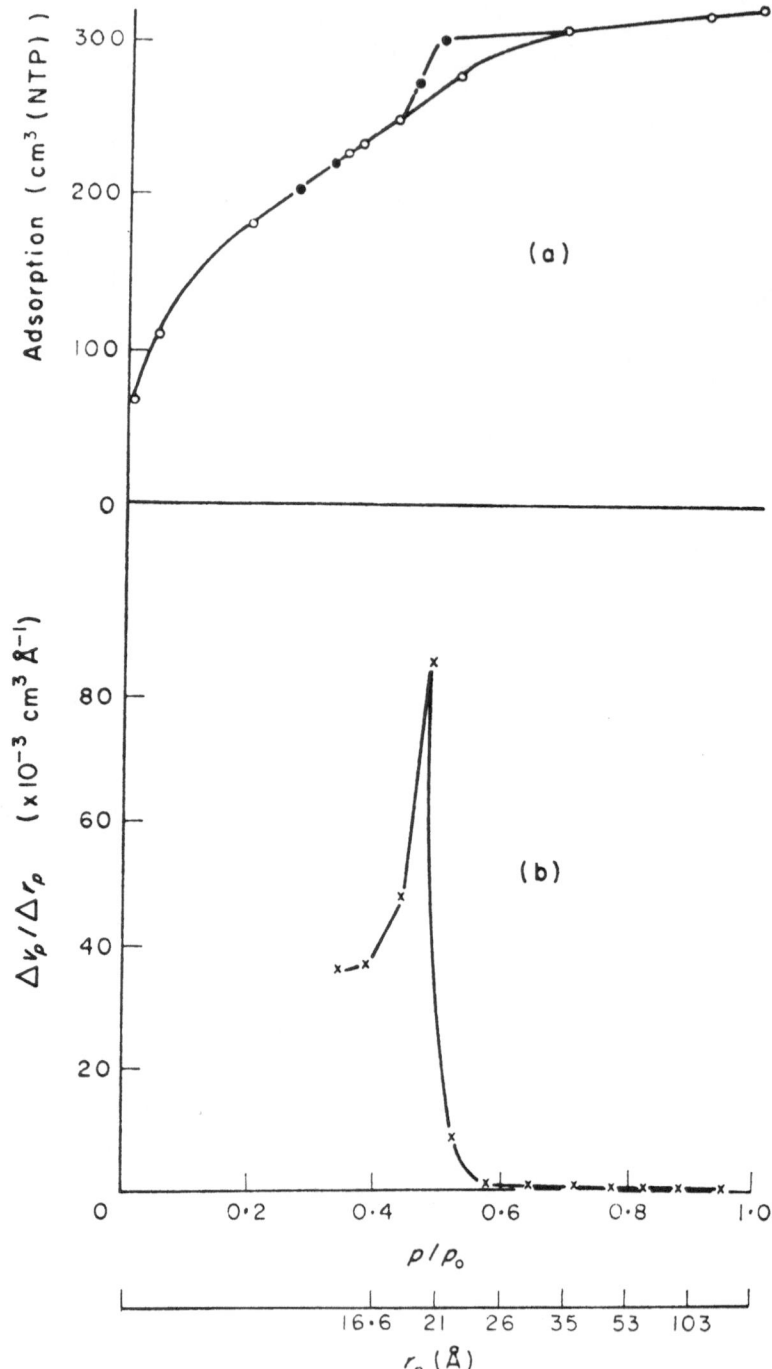

Fig. 21. Adsorption isotherm of nitrogen at
         -195°C on silica gel (curve a) and pore
         size distribution of silica gel (curve b).

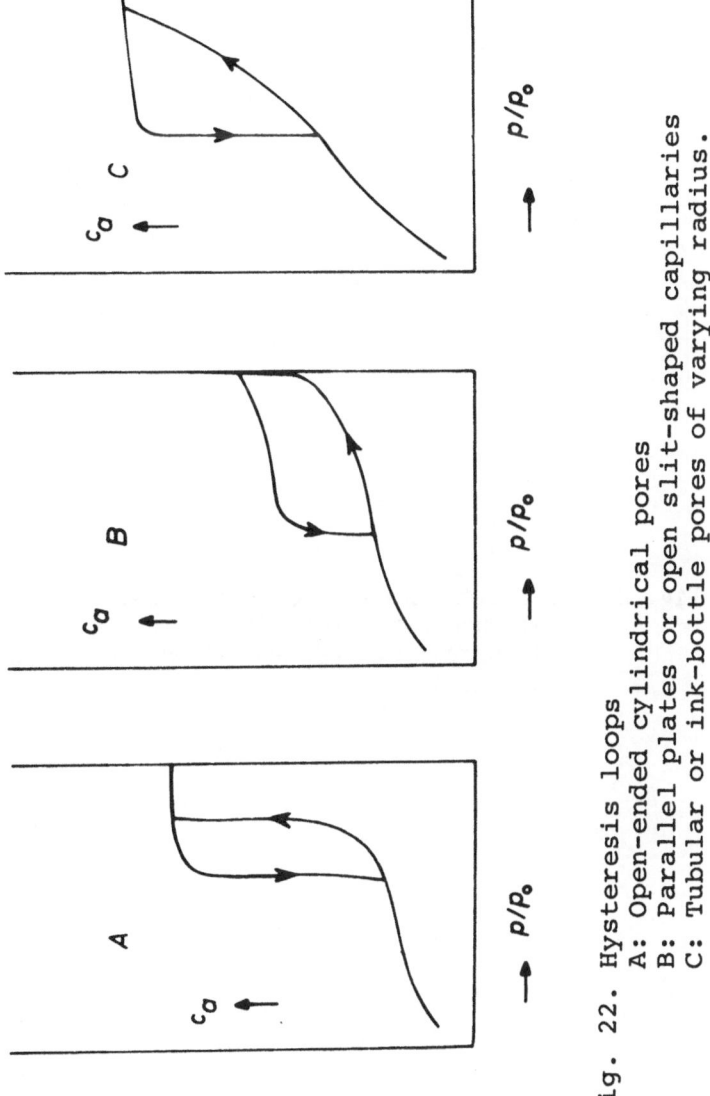

Fig. 22. Hysteresis loops
A: Open-ended cylindrical pores
B: Parallel plates or open slit-shaped capillaries
C: Tubular or ink-bottle pores of varying radius.

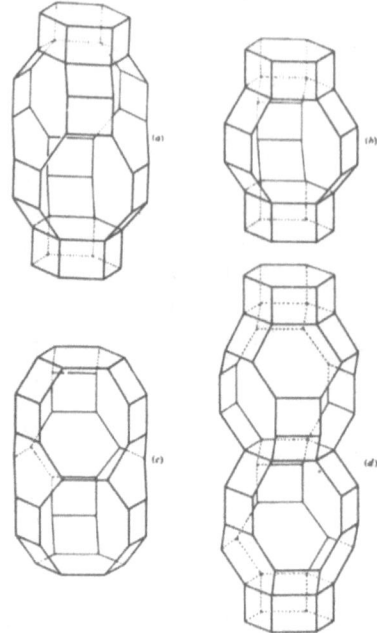

Fig. 23.   Cavities in different zeolites.
(a) Chabazite, (b) Gmelinite, (c) Erionite,
(d) Levynite.   (Si and Al on each corner, 0 on
the edges).

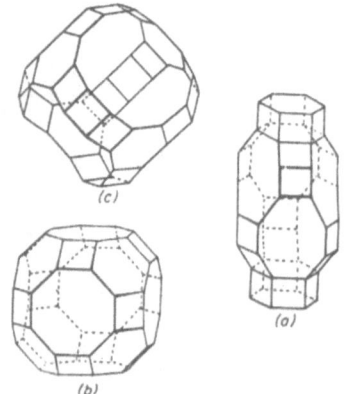

Fig. 24.   Cavities in (a) Chabazite, (b) Linde Sieve A,
(c) Faujasite or Linde Sieves X and Y.

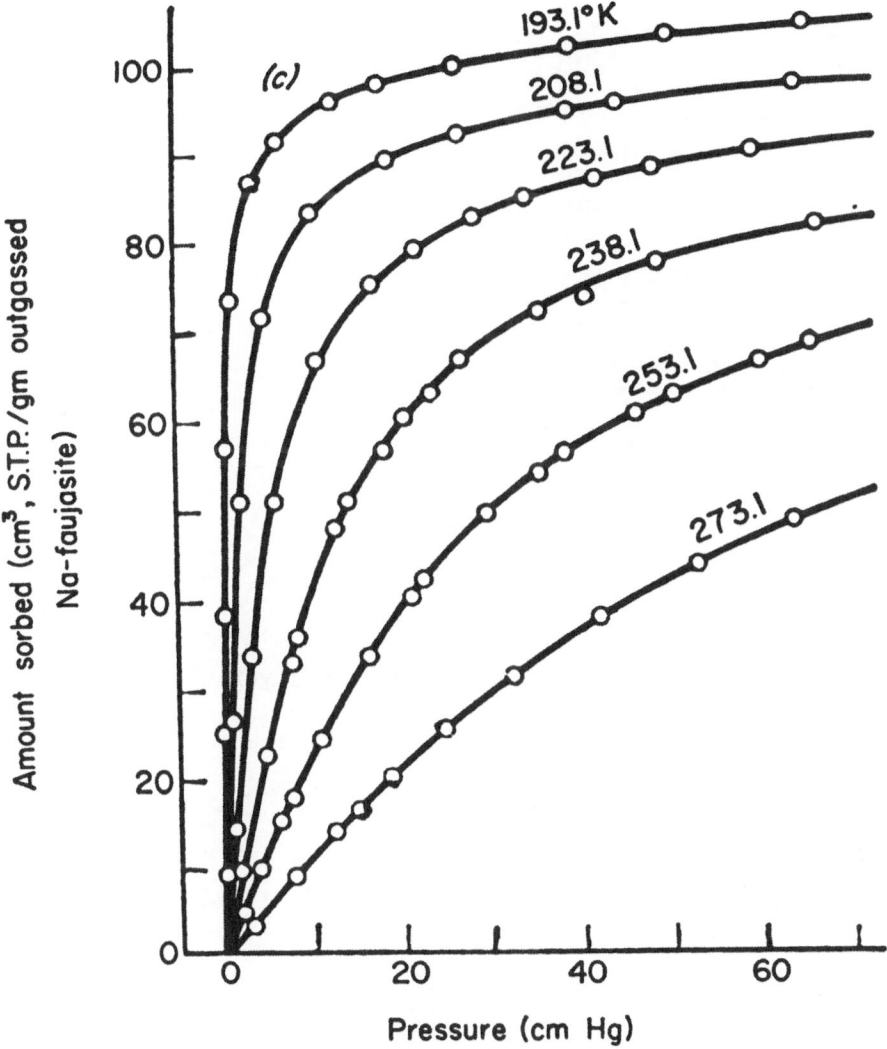

Fig. 25. Sorption of CF_4 in Na-faujasite at different temperatures.

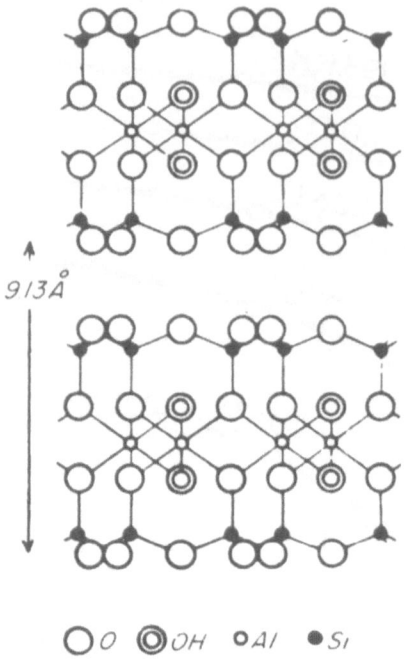

$9/3\overset{\circ}{A}$

○ O ◎ OH ○ Al ● Si

Fig. 26. Cross section of aluminosilicate layers
(pyrophyllite).

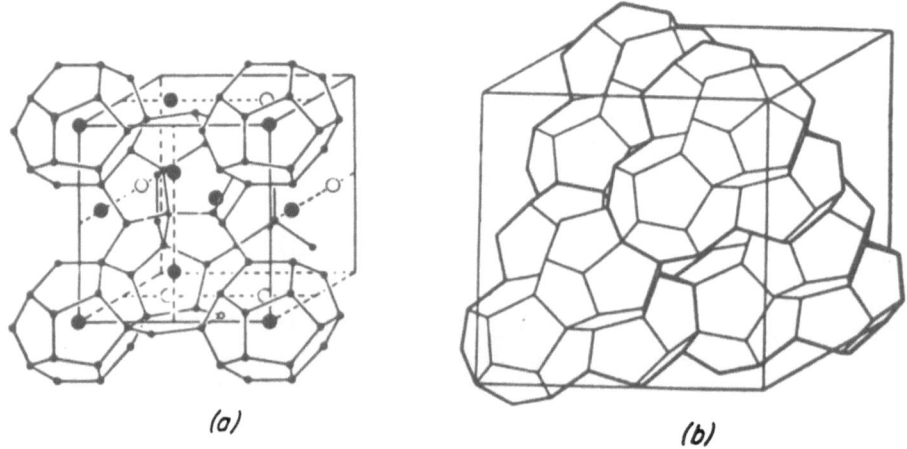

(a)   (b)

Fig. 27. Stacking of dodecahedra in hydrates
(a) Type I, (b) Type II

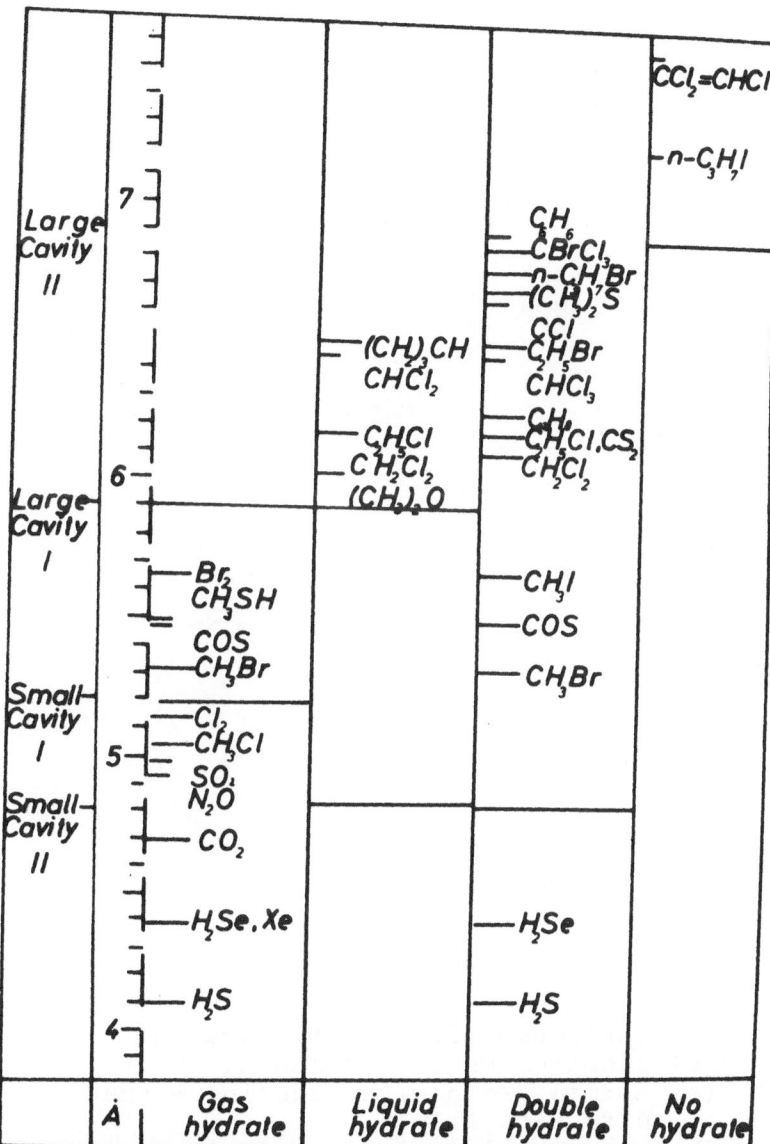

Fig. 28. Comparison of molecular diameters
of molecules which are forming
hydrates with the diameters of the
cavities.

142

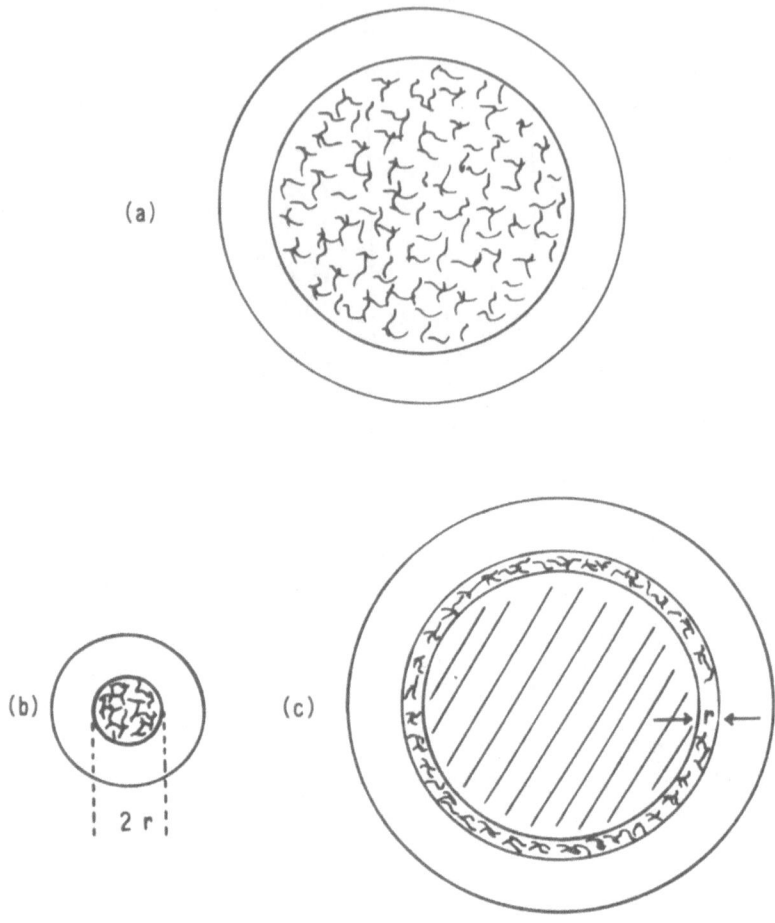

Fig. 29. The influence of film kinetics and
        gel kinetics on ion exchange.
        (a) normal grain with adhering Nernst
            diffusion film
        (b) very small grain ($r \approx 10$ μm)
        (c) grain of normal size but carrying
            the functional groups on the
            surface only ($r \approx 0.01 - 10$ μm)

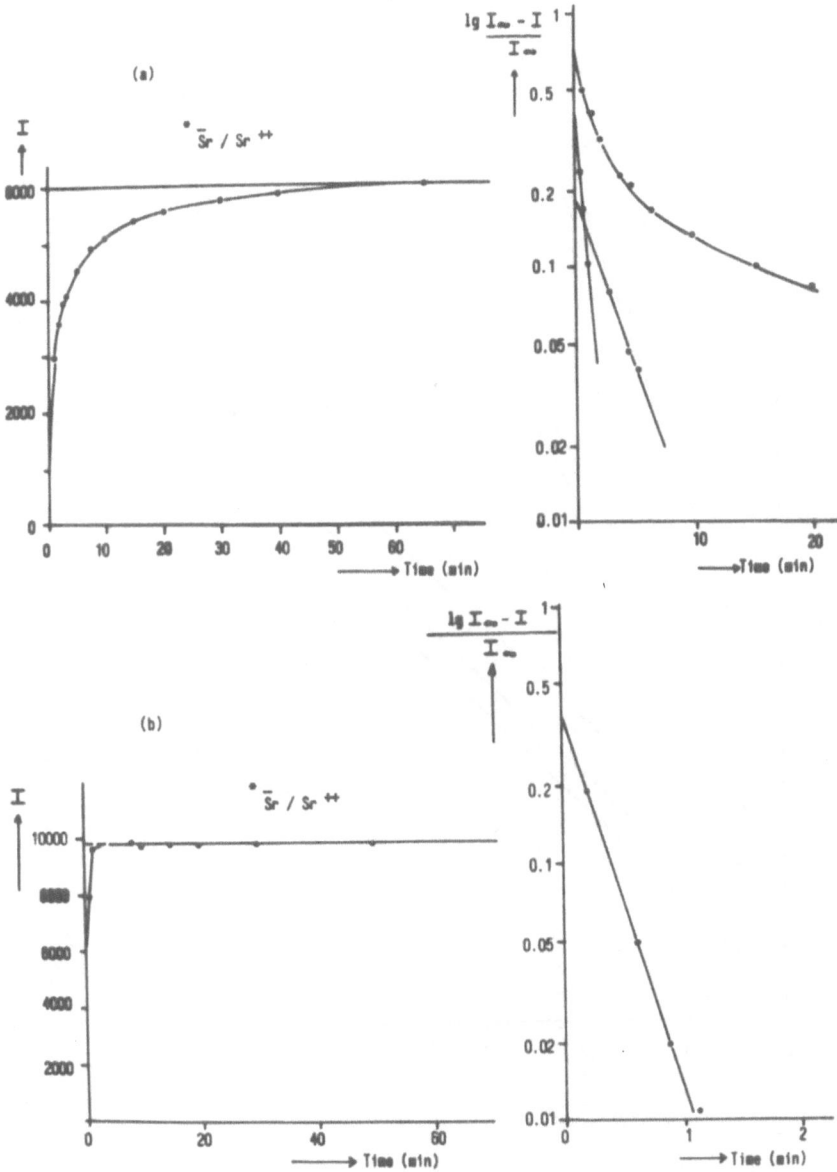

Fig. 30. Kinetics of ion exchange on resins of
normal grain size
(a) thoroughly sulfonated (cf.(a) in Fig.29)
(b) superficially sulfonated (cf.(c) in
Fig.29)

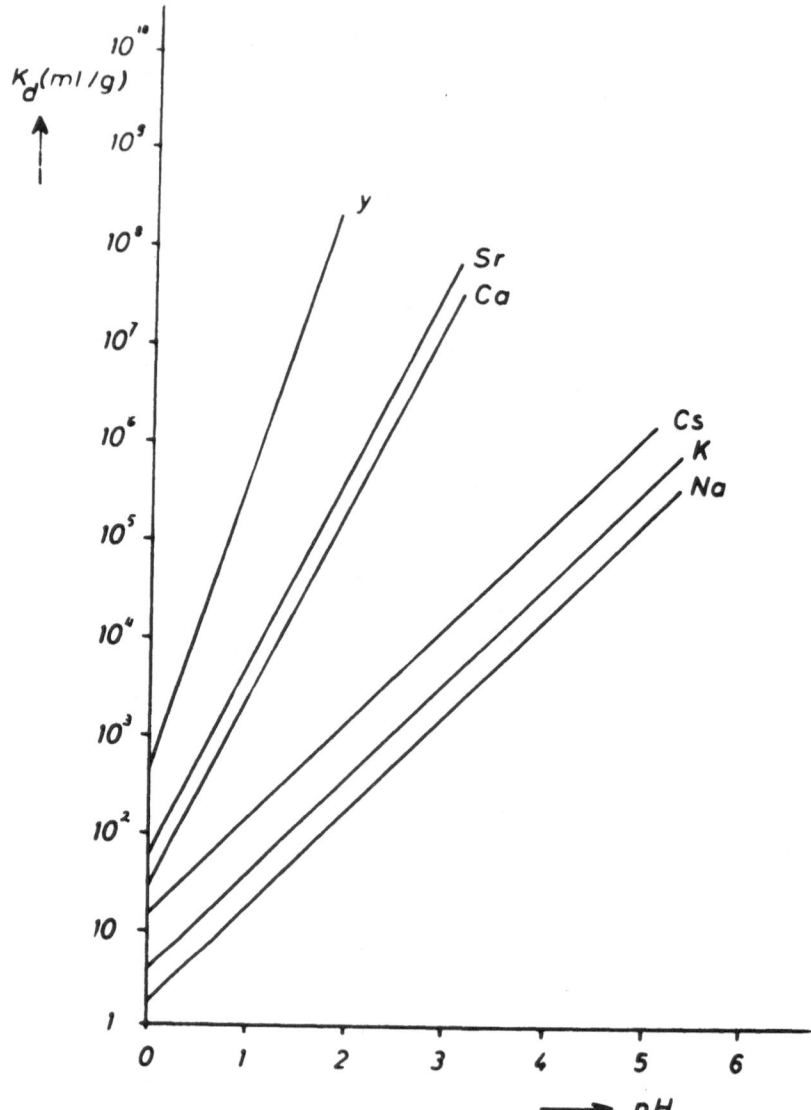

Fig. 31. Distribution coefficients for different
cations on organic ion exchange resins
(Dowex 50)

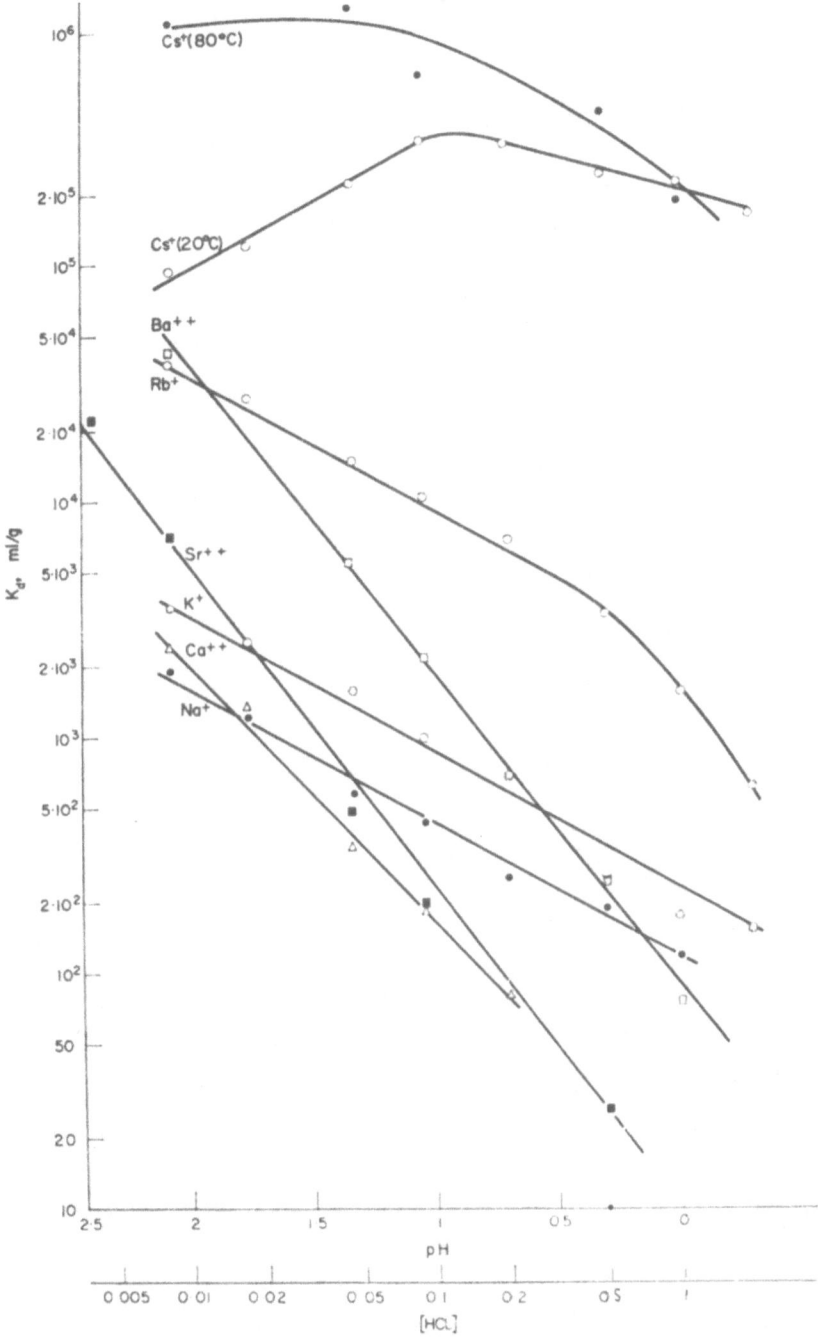

Fig. 32. Distribution coefficients for different
cations on titanium hexacyanoferrate

ADSORPTION ON ACTIVATED CARBON*

Walter J. Weber, Jr.

Professor of Environmental and Water Resources
Engineering and Chairman, Water Resources Program,
College of Engineering, The University of Michigan,
Ann Arbor, Michigan, U.S.A.

1. INTRODUCTION

Activated carbon has been indicated as the sorbent of choice
for most applications in water and effluent treatment. The
reasons are several fold. Activated carbon has been demonstrated
to have good sorption characteristics for most organic compounds
of interest in these fields. This is attributable to its dual
properties of large surface area per unit weight (and bulk volume)
and high degree of surface activity. It can be produced with
relative ease and at reasonable cost from a number of different
raw materials and with a variety of surface properties to meet
the requirements of specific applications. Finally, activated
carbon -- particularly granular activated carbon -- can be
efficiently regenerated for multiple reuse as a sorbent. Despite
exhaustive searches for alternatives, no other material has been
found which combines the desirable properties of a sorbent for
water and effluent applications as effectively as does activated
carbon.

2. PRINCIPLES OF ADSORPTION

Adsorption occurs in large measure as a resultant of forces
active within phase boundaries, or surface boundaries. These
forces result in characteristic boundary energies. Classical

---

*Lecture 2. Session II. NATO Advanced Study Institute: Scientific
Aspects of Sorption and Filtration Methods for Gas and Water
Purification, Fauske, Norway, 23-29 June 1974.

chemistry defines the properties of a system by the properties of its mass; for surface phenomena the significant properties are those of the surface or boundary.

Pure liquids tend to reduce their free surface energy through the action of surface tension. A large number of soluble materials (e.g., detergents) can effectively alter the surface tension of a liquid. A material which is active at surfaces will decrease the tension at the surface of a liquid by virtue of its movement to the surface. Migration to the surface or boundary results in a net reduction of the work required to enlarge the surface area, the reduction being proportional to the concentration of sorbate at the surface. Hence the energy balance of the system favors the adsorptive concentration of such surface-active substances at the phase interface. The tendency of an impurity to lower the surface tension of water is referred to as hydrophobicity; that is, the impurity "dislikes" water.

Adsorption of a dissolved impurity from water onto activated carbon may result from the hydrophobicity of the impurity, or it may be caused by a high affinity of the solute for the carbon. For the majority of systems encountered in water and effluent treatment, adsorption results from a combination of these factors.

The solubility of a substance in water is significant; solubility can be thought of as the chemical compatibility between the water and the solute. The more hydrophilic the substance the less likely it is to be adsorbed. Conversely, a hydrophobic sustance will more likely be adsorbed.

In the context of solute affinity for the solid, it is common to distinguish between three types of adsorption. The affinity may be predominantly due to: 1) electrical attraction of the solute to the sorbent (exchange adsorption); 2) van der Waals attraction (physical or ideal adsorption); or, 3) chemical reaction (chemisorption or chemical adsorption).

Many adsorptions of organic substances by activated carbon result from specific interactions between functional groups on the sorbate and on the surface of the sorbent. These interactions may be designated as "specific adsorptions." It is possible for specific adsorptions to exhibit a large range of binding energies, from values commonly associated with "physical" adsorption to higher energies involved in "chemisorption." The adsorptive interactions of aromatic hydroxyl and nitro-substituted compounds with active carbon, for example, are specific adsorption processes resulting from the formation of donor-acceptor complexes of the organic molecule with surface carbonyl oxygen groups, with adsorption continuing after these sites are exhausted via complexation with the rings of the basal planes of the carbon microcrystallite.[1]

Adsorption results in the removal of solutes from solution and their concentration at a surface, to such time as the amount of solute remaining in solution is in equilibrium with that at the surface. This equilibrium is described by expressing the amount of solute adsorbed per unit weight of sorbent q_e, as a function of C, the concentration of solute remaining in solution. An expression of this type is termed an adsorption isotherm.

Several types of isotherms may be observed. The most common relationship between q_e and C obtains for systems in which adsorption from solution leads to the deposition of an apparent single layer of solute molecules on the surface of the solid. Occasionally, multimolecular layers of solute may be adsorbed. Then, for adequate description of the phenomenon, resort must be had to adsorption models which are somewhat more complex.

The Langmuir adsorption model has been set forth for description of single-layer adsorption, whereas the Brunauer, Emmett, Teller (BET) model represents isotherms reflecting apparent multilayer adsorption. Both equations are limited by the assumption of uniform energies of adsorption on the surface. Graphically the Langmuir isotherm has the form shown in Figure 1, and the most common BET isotherm has the form shown in Figure 2; in both drawings the saturation concentration of the solute in the solution at a given temperature is represented by C_s. The BET isotherm, the more generally applicable of the two, reduces to the Langmuir model when the limit of adsorption is a monolayer.

Both the Langmuir and the BET isotherms may be deduced either from kinetic considerations or from the thermodynamics of adsorption.[2-4] The latter derivations are somewhat more

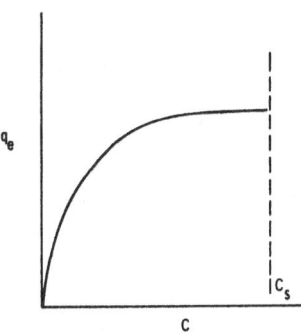

Fig. 1. Typical isotherm for Langmuir adsorption pattern.

150

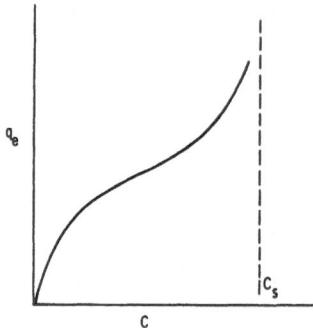

Fig. 2. Typical isotherm for BET adsorption pattern.

sophisticated, though less intuitive, than the kinetic treatments since fewer assumptions are involved (e.g., the balancing of forward and reverse rate processes according to some assumed mechanism).

The Langmuir treatment is based on the assumption that maximum adsorption corresponds to a saturated monolayer of solute molecules on the sorbent surface, that the energy of adsorption is constant, and that there is no transmigration of sorbate in the plane of the surface. The BET model assumes that a number of layers of sorbate molecules form at the surface and that the Langmuir equation applies to each layer. A further assumption of the BET model is that a given layer need not complete formation prior to initiation of subsequent layers; the equilibrium condition will therefore involve several types of surfaces in the sense of number of layers of molecules on each surface site.

For adsorption from solution with the additional assumption that layers beyond the first have equal energies of adsorption the BET equation takes the simplified form

$$q_e = \frac{BCQ^{\circ}}{(C_s - C)[1 + (B - 1)(C/C_s)]} \qquad (1)$$

in which C_s is the saturation concentration of the solute, C is the measured concentration in solution at equilibrium, Q° is the number of moles of solute adsorbed per unit weight of sorbent in forming a complete monolayer on the surface, q_e is the number of moles of solute adsorbed per unit weight at concentration C, and

B is a constant expressive of the energy of interaction with the surface. The Langmuir isotherm is

$$q_e = \frac{Q^\circ bC}{(1 + bC)} \tag{2}$$

in which b is a constant related to the energy or net enthalpy of adsorption, and all other symbols have the same significance as in Equation 1. Equation 1 reduces to Equation 2 if b is set equal to B/C_s, C is taken as negligibly small compared with C_s, and B is taken as much greater than 1.

For very small amounts of adsorption, that is, when $bC \ll 1$, adsorption is proportional to the final concentration of adsorbate in solution, yielding a linear adsorption relationship

$$q_e = Q^\circ bC \tag{3}$$

For large amounts of adsorption

$$bC \gg 1 \quad \text{and} \quad q_e \approx Q^\circ \tag{4}$$

When adsorption is in accord with the Langmuir equation the total capacity of the sorbent for a sorbate is given by the limiting value of q_e as C approaches C_s and is equal to the value of Q°. Assumption of a value for the surface area covered per molecule then allows computation of the active specific surface area of the sorbent. The specific area, Σ_s, can be determined by the relationship

$$\Sigma_s = Q^\circ N_{Av} \sigma^\circ \tag{5}$$

in which σ° is the area per molecule, N_{Av} is Avogadro's number, and Q° is expressed in moles of sorbate per unit weight of sorbent.

There is no similar "total capacity" for adsorption which follows the familiar BET type II pattern illustrated in Figure 2, for q_e tends asymptotically toward infinity as C approaches C_s. However, Q° has the same significance as in the Langmuir equation and gives a limiting capacity that can be used for computation of specific active surface areas. Often the value of Q° is found to correspond closely to the inflection point of the q_e versus C plot.

At concentrations less than that at which adsorption equal to Q^o is reached, the equilibrium capacity of the sorbent is less than the limiting value. Efficient utilization of a sorbent requires that conditions of operation be such that full use be made of the equilibrium capacity, at least, if not of the full limiting capacity.

Although the basic assumptions explicit in development of the Langmuir equilibrium model for adsorption are not met in most adsorption systems of concern for water and effluent treatment, the Langmuir equation is often found useful for description of equilibrium data for such systems. Rarely, for example, does a value of Q^o developed for adsorption of an organic water or effluent component on activated carbon represent a true monolayer capacity. Nonetheless, Q^o does represent a practical limiting capacity for adsorption. It is also unlikely in such a system that the sorption energies are uniform over all sites on the carbon surface, or that sorbed molecules do not undergo movement in the surface. Such deviations from the basic assumptions of the Langmuir model do limit interpretation of values for Q^o and b in terms of absolute surface areas and sorption free energies. They do not, however, negate the value of the Langmuir equation for quantification and mathematic representation of observed equilibrium relationships. As long as its restrictions and limitations are clearly recognized, the Langmuir equation can be used for describing equilibrium conditions for adsorption and for providing parameters (Q^o and b) with which to quantitatively compare adsorption behavior in different sorbate-sorbent systems, or for varied conditions within any given system.

One other adsorption isotherm equation, the Freundlich or van Bemmelen equation, has been widely used for many years. This equation is a special case for heterogeneous surface energies in which the energy term, b, in the Langmuir equation varies as a function of surface coverage, q_e, strictly due to variations in heat of adsorption. The Freundlich equation has the general form

$$q_e = K_F C^{1/n} \tag{6}$$

where K_F and n are constants, and n > 1.

The value of K_F is roughly an indicator of sorption capacity and n of adsorption intensity. The Freundlich equation generally agrees quite well with the Langmuir equation and experimental data over moderate ranges of concentration, C. Unlike the

Langmuir equation, however, it does not reduce to a linear adsorption expression at very low concentrations, nor does it agree well with the Langmuir equation at very high concentrations, since n must reach some limit when the surface is fully covered.

The adsorption isotherm is useful for representing the capacity of an activated carbon for adsorbing organics from a waste, and in providing description of the functional dependence of capacity on the concentration of pollutant. The steeper the isotherm, the more effective is the activated carbon; that is, the sharper the rise of the isotherm to a given ultimate capacity as concentration increases, the higher will be the effective capacity at the concentration level desired for the treated water. Experimental determination of the isotherm is routine practice in evaluating the feasibility of adsorption for treatment, in selecting a carbon, and in estimating carbon dosage requirements. The Langmuir and Freundich equations provide means for mathematic description of the experimentally observed dependence of capacity on concentration. The adsorption isotherm relates to an equilibrium condition, however, and practical detention times used in most treatment applications do not provide sufficient time for true equilibrium to obtain.

Rates of adsorption are thus significant, for the more rapid the approach to equilibrium, the greater is the fraction of equilibrium capacity utilized in a given contact time. There are essentially three consecutive steps in the adsorption of materials from solution by porous sorbents such as activated carbon. The first of these is the transport of the adsorbate through a surface film to the exterior of the adsorbent ("film diffusion").

Film, penetration, boundary layer, and other theories have been postulated to explain mass transfer in the region separating a turbulent bulk solution and a solid surface. However, the fluid mechanics of this region are not well defined. Boundary layer theory accounts for a velocity distribution and is more realistic than film theory, which assumes a laminar film surrounding the particle. The term "film diffusion" will be used in this lecture to generally describe the resistance to mass transfer at the surface of the particle. However, use of this term is not intended to imply the existence of a definable film nor is it meant to restrict treatment of the data to the film theory. The second of the three consecutive steps in sorption by porous sorbents, with the exception of a small amount of adsorption that occurs on the exterior surface of the sorbent after transport across the exterior film, is the diffusion of the sorbate within the pores of the sorbent ("pore diffusion"). The third and final step is adsorption of the solute on the interior surfaces bounding the pore and capillary spaces of the sorbent.

Consideration of rates at which interfacial tensions are lowered by chemical compounds representative of organic pollution materials gives indication that the adsorption process itself is probably not rate-determining, and that a much slower process must control the overall rate of uptake by porous carbon. Under certain operating conditions, transport of the sorbate through the "surface film" or boundary layer to the sorbent may be rate-limiting; if sufficient turbulence is provided, transport of the sorbate within the porous carbon may control the rate of uptake. One of the most significant factors to consider is, therefore, the nature of the step which controls the speed at which the reaction proceeds, in order that the process may be described in terms of appropriate rate expressions and rate parameters.

Certain properties of the sorbate are useful in determining the nature of the rate-controlling step. For example, if intra-particle transport determines the rate of reaction, the size and structure of an individual solute ion or molecule will affect this rate to the extent that it affects molecular mobility.

The rate-controlling step can also be characterized in part by the observed activation energy for the process. A study of the effect of temperature on rate, in addition to yielding information relative to optimum conditions of operation, permits evaluation of the activation energy and is, consequently, a further means for determining the nature of rate-limiting reactions.

For a process in which the overall rate is controlled by a strictly adsorptive reaction the variation of rate should be directly proportional to the concentration of solute, and for very simple diffusion the rate is expected also to be proportional to the first power of concentration. However, complex mathematic expressions for intraparticle transport indicate that the relationship between concentration and the rate of the reaction will not be one of direct proportionality. Since concentration affects a number of the parameters of these equations, it is not possible to predict an exact concentration-rate relation for this reaction. Qualitatively, if diffusion of solute within the pores and capillaries of the carbon limits the rate, the variation of rate with concentration is not expected to be linear, whereas a direct proportionality is anticipated for strictly adsorptive reactions. Thus the concentration-dependence of the rate of reaction may be used as a partial test of hypotheses regarding the nature of the rate-controlling step. The effect of the concentration of solute on the speed at which its sorptive uptake proceeds is also significant for any prediction of the most efficient manner in which adsorption can be utilized for removal of the solute from solution.

For processes in which the rate-limiting reaction is

adsorption on the exterior surfaces of the sorbent or transport through an external surface film, the rate is expected to vary as the reciprocal of the diameter of the sorbent particles for a given total weight of sorbent; this because the rate is in this case a first-order function of exterior surface area, which in turn is inversely proportional to particle diameter. Conversely, according to appropriate mathematical expressions for transport relationships, the rate of diffusion of solute into the pores of a particle will vary as the reciprocal of some higher power of the diameter of the particle. Variation of rate with particle size is then another method which is useful for the characterization of the rate-limiting step for a particular system. Particle size is an important consideration also for achieving optimum utilization of a sorbent in treatment operations.

The method by which the carbon is contacted with the water determines in large part which of the transport or reaction steps is rate-limiting. For a completely and vigorously mixed batch (CMB) reactor, pore diffusion may be rate limiting. For continuous flow systems (e.g., beds of carbon) film diffusion is usually rate limiting for normal flow rates of 2-8 gpm/ft^2 (81-326 l/min/m^2).[1]

3. COMPONENTS AND CONDITIONS

3.1 Adsorbent properties

Activated carbon is a generic term for a broad range of amorphous carbon-based materials so prepared as to exhibit a high degree of porosity and an extensive associated surface area. Hundreds of different commercial activated carbons with unique properties, and therefore different application suitability, exist.

The purification properties of carbons, or at least charcoal, have been known for over 30 centuries, but the first commercial application appears to have been the use of bone char for decolorization in the cane sugar industry in the 1780's, and bone char is yet the most commonly used sorbent in the cane sugar industry today.[5]

During the 19th century different relatively crude activated carbons were prepared from a variety of raw materials, but manufacturing problems, and the absence of a real need to find anything significantly better than bone char stifled product development. The first activated carbon preparation and use for purification of potable water occurred in 1862.[6]

In the late 1890's and early 1900's Ostrejko developed two improved processes for the manufacture of activated carbon.[7] The first involved the carbonization of vegetable substances impregnated with metallic chlorides; the second the activation of charcoal with carbon dioxide and steam at high temperatures. Virtually all activated carbons are yet made by one or the other of these two processes, either low temperature chemical activation or high temperature gaseous oxidation. The low temperature chemical activation involves chemical dehydration and charring of a carbonaceous raw material, and usually is carried out at temperatures between 200 and 650 °C.

Porosity is developed by the action of dehydrating chemicals -- normally phosphoric acid, zinc chloride and/or sulfuric acid -- on the cellulose structure of the starting material, or by the action of oxidizing gases generated in the process. In the high temperature oxidation of a previously charred carbonaceous substance, a porous structure is developed in the low surface area carbonaceous starting material by controlled oxidation at temperatures between 800 and 950 °C with steam, flue gas, or some other oxidizing gas mixture. Bone charcoal is manufactured by a different process, in which collagen and other carbon containing components of bone are carbonized at high temperatures in the absence of oxidizing gases to form carbon deposits within the hydroxyapatite structure of bone.

The properties of an activated carbon depend on the nature of the raw material used, the conditions under which carbonization is accomplished, the activation process and conditions, and post treatment of the product.

Surface areas usually range between 450 and 1500 m^2/g. Bone charcoal is an exception, having, in the freshly prepared form, an area of about 100 m^2/g char. Bone char, however, contains only 5 to 12% by weight of carbon, which accounts for 50% of its surface. The carbon present is truly in an activated state, having an area of between 400 and 1000 m^2/g carbon, depending on the composition of the char.

The surface areas of some typical carbons are shown in Table 1.

Surface area is normally determined by measuring the volume of nitrogen gas adsorbed at liquid nitrogen temperature (-195 °C) at various pressures. From this data one can calculate, using the BET equation, the monolayer coverage and the surface area. The actual area available for adsorption of a specific compound from water can be considerably less than the total surface area determined by nitrogen adsorption.

For gas adsorbing carbons, for example, most of the surface

Table 1. Surface areas of typical activated carbons

Name	Origin	Area m^2/g
Actibon S	Wood	850 - 900
Columbia G	Coconut shell	1100 - 1150
Columbia AC	Coconut shell	120u - 1400
Darco S51	Lignite	500 - 550
Darco KB	Wood	950 - 1000
Nuchar Aqua	Pulp mill residue	550 - 650
Nuchar C	Pulp mill residue	1050 - 1100
Nuchar WV-G	Coal	1000 - 1100
Pittsburgh CAL	Coal	1000 - 1100
Pittsburgh RB	Coal	1250 - 1400

area is in micropores, and these carbons have little capacity
for molecules too large to enter a pore less than about 20Å in
diameter. As another example, bone char has a surface area of
about 65 m^2/g. Typical commercial granular activated carbons
may have areas in excess of 1000 m^2/g; these, however, have only
between 5 and 10 times the capacity of bone char in a use such
as sugar decolorization. Based on total surface area alone they
should be over 15 times as active, but bone char has a much
lower proportion of its surface in micropores and thus more
readily available for sorption of the large color bodies. The
conclusion that must be reached then is that total surface area
is not by itself a satisfactory measure of available surface
for liquid phase applications. Rather, it is the distribution
of surface area as a function of pore size within the sorbent
which is important.

Pore volume distribution can be determined by a combination
of mercury porosimetry and gas adsorption-desorption measurements.
Gas carbons are characterized by a large percentage of pore volume
in micropores, the almost total absence of pore volume in transitional pores, and a secondary maximum in macropores. Conversely,
liquid phase carbons have a pronounced percentage of pore volume
in transitional pores together with capacity in both micro and
macropores. Thus, just as with total surface area, total pore
volume is not acceptable by itself as a measure of capacity.
Total pore volumes of gas and liquid phase carbons are similar,
but it is the pore volume in large pores which determines in
large measure the capacity of liquid phase carbons. Together,
total surface area and total pore volume thus give some measure
of the potential capacity of a carbon, which depends on the

distribution of area or volume with pore size, and the distribution of molecular sizes to be adsorbed.

The chemical properties of the surface of the activated carbon are also important in determining activity; that is capacity for a specific adsorbate. The chemical properties of the surface depend on the starting material, the activation process, and the conditions employed in activation.

Activated carbon can be considered to consist of essentially two types of surfaces, excluding contributions from inorganic impurities. The first are planar, non-polar surfaces, which comprise the bulk of the surface for most carbons. Adsorption on this surface would be largely of the van der Waals type. The second type of surface is comprised of the heterogeneous edges of the carbon planes, which make up the crystallites, whereon carbon-oxygen functional groups formed by oxidation in the manufacturing process are located. These groups, which include phenolic hydroxyl, alcoholic hydroxyl, carboxyl, n-lactone, f-lactone and chromene groups, enable activated carbon surfaces to undergo halogenation, hydrogenation, oxidation, and to act as a catalyst in many reactions.

The surface properties of different carbons can have profound effects on both rate and capacity for adsorption. The surface chemistry of active carbon has been a subject of much interest for some time, yet surprisingly little is known about the nature of the surface functional groups of this material. Recent work has provided an examination of the character of functional groups formed on active carbon under different conditions of activation, using the technique of multiple internal reflectance spectroscopy (MIRS) as a means for characterizing surface functional groups.[1] This technique will be discussed in detail in Lecture 1, Session III.

While the activity of a sorbent is related to its distribution of surface area and the chemistry of that surface, it must be recognized that activity or capacity is only one parameter which must be taken into account when selecting a carbon for a particular process. Other properties to be considered include hardness and head-loss characteristics for granular carbons, filterability and bulk density for powdered carbons, water solubility of impurities, and pH.

As already noted it is not possible to determine activity or capacity from basic carbon properties such as surface area, nor to relate activity for a water or effluent application to capacity for a reference sorbate, such as iodine or methylene blue. Activity or capacity must be determined directly on the system of interest. For granular carbons, hardness is probably

second in importance to capacity among properties to be considered
in carbon selection. Hardness determines, in large measure, the
loss on each adsorption-regeneration cycle. Losses result from
attrition on handling and burn-off during reactivation. For coal-
based carbons, losses of about 5% per cycle can be expected;
losses for softer carbons can be as high as 15%.

Head-loss or pressure drop in downflow columns and bed
expansion in upflow columns of granular carbon are determined in
part by particle size and size distribution; these properties
are therefore factors influencing design, installation and capital
costs. In general, the smallest size of particle that conditions
of efficient operation permit should be used, for this increases
the adsorption rate and thus reduces the size of the plant
required. For powdered carbon, which usually must be removed
from the treated water by filtration, filterability is affected
by particle shape, size and size distribution.

Bulk density is also important in the selection of a
powdered carbon, for it determines, to a large extent, the length
of the filtration cycle. Filterability is important, for poor
filtration results in the use of more carbon and filter aid, and
the need to provide a larger plant.

All carbons contain some soluble impurities depending on
the nature of the starting material, the activation process, and
final treatment. The amounts and the nature of the compounds
which can be tolerated will be governed by the purity requirements
of the treated water or effluent.

The "pH" of a carbon is actually the pH measured on a water
extract from the carbon. It is a function primarily of the nature
of the activation process; steam activation usually yields alkaline carbons, while activation with phosphoric acid, for example,
gives carbon pH values below 5. The pH of carbon can be modified
by washing. Steam activated carbons can be acid washed to give
products with pH values between 5 and 7, while phosphoric acid
carbons can be "neutralized" by caustic wash. The adsorption of
many solutes can be related to solubility, which in turn is
affected by pH. The optimum pH of the carbon, as well as of the
solution, can be determined only by experiment with the specific
water or effluent to be treated.

Regeneration is an important consideration in the use of
active carbon for water and effluent treatment. Detailed
discussion of this aspect of activated carbon lies beyond the
scope of this lecture. It should at least be noted, however,
that it is currently feasible to regenerate granular carbon by
conventional thermal techniques for at least 15 cycles of successive saturation and regeneration. The results of attempts to

develop efficient and economical means for chemical regeneration of carbon have thus far been disappointing.

3.2 Solute properties

In general, an inverse relationship between the extent of adsorption of a solute and its water solubility can be anticipated. The water solubility of organic compounds within a particular chemical class decreases with increasing chain length, because the compound becomes more hydrocarbon-like as the number of carbon atoms becomes greater. Thus, adsorption from aqueous solution increases as an homologous series is ascended, largely because the expulsion of increasingly large hydrophobic molecules from water permits an increasing number of water-water bonds to reform.

Molecular size is of significance if the adsorption rate is controlled by intraparticle transport, in which case the reaction generally proceeds more rapidly the smaller the absorbate molecule. It must be emphasized, however, that the rate dependence on molecular size can be generalized only within a particular chemical class or series of molecules. Large molecules of one chemical class may sorb more rapidly than smaller ones of another if higher energies (driving forces) are involved.[1]

Many organic compounds exist, or have the potential of existing, as ionic species. Fatty acids, phenolic species, amines, and many pesticides are a few materials having the property of ionizing under appropriate conditions of pH. Activated carbon commonly carries a net negative surface charge; further, many of the physical and chemical properties of certain compounds undergo changes upon ionization. Most observations point to the generalization that as long as compounds are structurally simple, sorption is at a minimum for charged species and at a maximum for neutral species. As compounds become more complex, the effect of ionization decreases. Studies of amphoteric compounds indicate an adsorption maximum at the isoelectric point, consistent with other observations that adsorption is at a maximum for neutral species. A polar solute will be strongly sorbed from a non-polar solvent by a polar sorbent, but will prefer a polar solvent to a non-polar sorbent. Polarity of organic compounds is a function of charge separation within the molecule. Almost any asymmetric compound will be more or less polar, but several types of functional groups tend to produce fairly high polarities in compounds Examples of these are hydroxyl, carboxyl, nitro, nitrile, carbonyl, sulfonate, and amine. Thus ethanol, C_2H_5OH, is polar, having an incremental negative charge on the hydroxyl and a corresponding positive charge on the ethyl group. Because solvation by water involves formation of a hydrogen bond from one of the positively charged hydrogens of the water to a group bearing more or less of a negative charge, along with some bonding in the reverse direction to the water oxygen, water solubility is expected to increase with increasing polarity. It therefore follows that adsorption decreases as polarity increases, even though active carbon is a polar sorbent.

Because hydrogen and hydroxide ions are sorbed quite strongly, the adsorption of other ions is influenced by the pH of the solution. Further, to the extent to which ionization of an acidic or basic compound affects its adsorption, pH affects adsorption in that it governs the degree of ionization. In general, adsorption of typical organic pollutants from water is increased with decreasing pH.

Adsorption reactions are normally exothermic; thus the extent of adsorption generally increases with decreasing temperature. The changes in enthalpy for adsorption are usually of the order of those for condensation or crystallization reactions, thus small variations in temperature tend not to alter the adsorption process in water and effluent treatment to a significant extent.

The organic components of a waste mixture may mutually enhance adsorption, may act relatively independently, or may interfere with one another. Mutual inhibition can be expected if the adsorption affinities of the solutes do not differ by several orders of magnitude and there is no specific interaction between solutes enhancing adsorption. Similarly, because the adsorption of one substance will tend to reduce the number of open sites and, hence, the "concentration" of adsorbent available, mutually depressing effects on rates of adsorption may be predicted.

It should be apparent from the foregoing discussion of the effects of solute character on adsorption that an analytical characterization of the impurities present is helpful to a thoughtful prediction of the effectiveness of activated carbon in water and effluent purification.

REFERENCES

1. Weber, W. J., Jr., Physicochemical Processes for Water Quality Control, Wiley-Interscience, New York, N.Y., 1972.
2. Langmuir, I., The adsorption of gases on plane surfaces of glass, mica, and platinum, Jour. Amer. Chem. Soc., 40, 1361, 1918.
3. Brunauer, S., Emmett, P. H. and Teller, E., Adsorption of gases in multimolecular layers, Jour. Amer. Chem. Soc., 60, 309, 1938.
4. Adamson, A. W., Physical Chemistry of Surfaces, Wiley-Interscience, New York, N.Y., 1967.

162

5. Deitz, V. R., _Bibliography of Solid Adsorbents_, National Bureau of Standards, Washington, D. C., 1944.
6. Libscombe, F., _British Patent 2887_, 1862.
7. Von Ostrejko, R., _British Patents 14,224 (1900); 18,040 (1900)_. _German Patent 136,792 (1901)_.
8. Abram, J. C., The characteristics of activated carbon, _Proceedings, Activated Carbon in Water Treatment_, Water Research Assoc., Medmenham, Marlow, Bucks, England, 1974.

GENERAL SCOPE OF ZEOLITE PROPERTIES

D. Barthomeuf

Institut de Recherches sur la Catalyse et
Université Claude Bernard, Lyon I- France

ABSTRACT. The general properties of zeolites are presented. The lecture includes: 1. Introduction; 2. Different kinds of zeolites; 3. Preparation; 4. Structure; 5. Porosity; 6. Activation before use; 7. Adsorption with geometric, interaction and kinetics effects
and adsorption of binary mixtures; 8. Catalysis with
nature of active sites and sieving properties and catalysis.

1. INTRODUCTION

The first definitive experiments on the separation of
mixtures using the dehydrated zeolite mineral chabazite as a molecular sieve was performed by Barrer in
1945.[1] In 1948 the first industrial research efforts
by Milton and his associates at Union Carbide Corporation resulted in the synthesis and the manufacture of
synthetic zeolite molecular sieves which did not exist
in nature.[2] This controlled synthesis was an important
research achievement.
The wide variety of applications of zeolites includes
separation and recovery of n-paraffins, catalysis of
hydrocarbon reactions, drying of refrigerants, separation of the components of air, carrying catalysts in
the curing of plastics and rubber, recovery of radioactive waste solutions, removing carbon dioxide and
sulfur compounds from natural gas, cryopumping, sampling air at high altitudes, solubilizing enzymes, separating hydrogen isotopes and the removal of atmos-

pheric polluants such as sulfur dioxide.
Some of these properties are related to the general
properties of zeolites while others, such as selective adsorption, stereospecific catalytic and ion-sieve
behavior, involve their molecular sieving ability.
General properties are due to the aluminosilicate
structure. Zeolites are crystalline, hydrated aluminosilicates of mono or polyvalent elements (mineral or
organic cations such as $(CH_3)_4N^+$). The zeolite structure consists of an infinitely extending three dimensional network of AlO_4 and SiO_4 tetrahedra linked to
each other by sharing all of the oxygens. Zeolites
may be represented by the formula :

$$M_{\frac{x}{n}} \; (AlO_2)_x \; (SiO_2)_y \; , \; w \; H_2O$$

where M is the cation of valence n, w is the number of
water molecules and the ratio y/x usually has a value between,1 and 5 depending upon the structure. Cations and water molecules are located in channels and
inter-connected voids of the framework. Adsorptive properties depend on the nature of the cations (their size, valence ...). The exchange of cations for protons
may be achieved directly with acid solutions or for
zeolites unstable in acidic media by using as an intermediate the ammonium cationic form. Heating results
in evolution of NH_3 and leaving the zeolite in the protonic form. The acidity is responsible for the catalytic properties of many zeolites. To illustrate the
dual character of zeolites, figure (1) shows the changes in acidity and adsorptive properties on increasing
the cation content. The acidity decreases while adsorptive monolayer capacity increases.

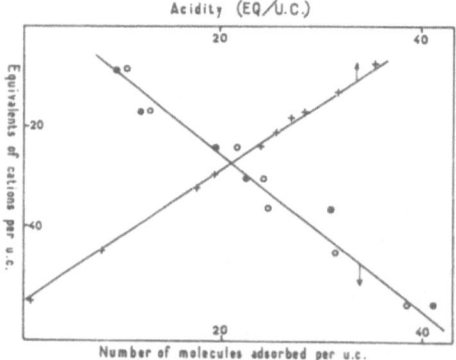

Fig. 1. Changes in acidity (+) or monolayer capacity
of adsorption for benzene (●) and cyclohexane (o) as
a function of sodium content in Y-type zeolites.

2. DIFFERENT KINDS OF ZEOLITES

There are 34 species of zeolite minerals and about
100 types of synthetic zeolites but only a few of these have practical use. In fact, the structure of the
zeolite must remain intact after activation- generally
dehydration. With many of the zeolites, the structure
partially collapses during dehydration.

3. PREPARATION

The study of zeolite synthesis is interesting both
from a fundamental and a commercial point of view. The
discovery of new zeolites with selective adsorptive or
catalytic properties would be very important. Therefore a large amount of work is being carried out on
this subject. Zeolites are synthetized in aqueous media at a relatively high pH. Starting materials may
be various : alumina and silica gels, alumino-silicate
gels of alkali or alkaline earths metals, clays, systems with organic cations such as the tetraethylammo-
nium ion. The cation plays a prominent structure -- directing role in zeolite crystallization. It has been
postulated that the cation stabilizes the formation of
structural subunits (or secondary building units SBU)
which are the precursors or nucleating species in crystallization. Further exchange of cations of the synthetized zeolites for other cations give a wide variety
of zeolites. In this way zeolites containing trivalent
ions (La^{3+}, Ce^{3+}...) are obtained.

4. STRUCTURE

The complexity of zeolite structure occurs because of
the various ways in which the tetrahedral groups AlO_4
and SiO_4 (or primary units) may link by the common
sharing of oxygen ions to form polynuclear complexes.
Several classifications of zeolites have been proposed.
The last one suggested by Breck[3] is based on seven
groups according to seven secondary building units.[4]
The zeolite structure is characterized by channels and
cavities. There are three types of channels - Type 1
(L zeolite for example) is a one-dimensional system
which does not permit intersections of the channels.
Type 2 is a two-dimensional intersecting channel system. Mordenite belongs to this class. Type 3 is a three
dimensional intersecting channel system (faujasite zeolite for example).

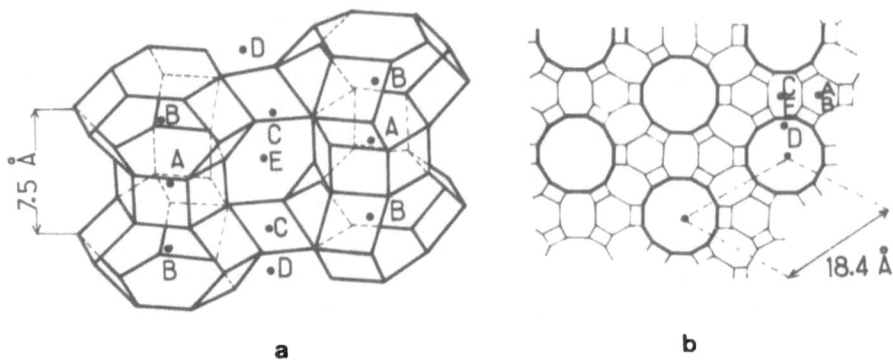

Fig. 2. L type structure (a) showing cancrinite cages and hexagonal prisms or (b) seen along C axis

As an illustration, the structure of zeolite L is presented in figure 2. The largest cavities are linked
through hexagonal prisms. The columns formed thereby
are joined by single oxygen bridges. Six columns cir-
cumscribe the main channels running parallel to the c
axis. A corresponding projection down the c axis shows
the channel apertures, of 7.1 Å diameter. The structure of zeolite L has been visualized by electron microscopy.[5] The structural configuration and distances are
observed on the electron micrographs of a L particle
seen along the c axis.

5. POROSITY

The characteristics of some usual zeolites are given
in table 1. The framework density d_f expresses the density of the aluminosilicate framework without the water
or exchangeable cations and is equal to one tenth the
number of tetrahedra per 1000 A. 3. The values are
slightly lower than density values. The water void volume is the amount of water contained in the fully hydrated zeolite. This volume is used to characterize the
zeolites. It may slightly change if a molecule other
than water is considered. The values vary for usual
zeolites from 0.28 to 0.50. The free apertures of channels or cavities is calculated from a hard sphere model

of the structure. The sizes may change as the temperature is raised or as the process of adsorption occurs. The next column gives indications of the largest molecule which can be adsorbed. There is of course a correlation between the free apertures and the size of the molecules. This size is evaluated by a kinetic diameter[6] (or collision diameter) which is the intermolecular distance of closest approach for two molecules colliding with zero initial kinetic energy. It has been shown that this kinetic diameter corresponds with the apparent pore diameter of various zeolites better than the Pauling diameter. The last column gives the type of channel present in the zeolite.

6. ACTIVATION BEFORE USE

Commercial zeolites possess of course the fundamental stability requirement during activation. At this stage it is interesting to consider a stabilization effect occurring during activation which has been described in the case of Y zeolites. The crystallinity of protonic Y type zeolites heated in a dry air flow starts to decrease near 700°C. When heated in contact with water vapor or / and ammonia at a temperature higher than 500-600°C, the thermal and hydrothermal stability is increased and the crystal structure is retained at 1000°C.[7,8]. It is supposed that tetrahedral aluminum is removed from the framework by hydrolysis with water vapor. The increase in the framework ratio Si/Al accounts for the high stability. It has to be noted that similar superstable zeolites are obtained by direct extraction of aluminum atoms from zeolites with chemical reagents.[9]

7. ADSORPTION

Among commercial adsorbents which exhibit ultraporosity, zeolites are particularly interesting because of their uniform pore size which is uniquely determined by the unit structure of the crystal. Hence they are very selective geometric adsorbents. Further, they adsorb molecules with a selectivity not found in other adsorbents because of specific interactions. Many investigators tried to explain the very particular character of adsorption in zeolites from a scientific approach. From the basic results of the work of Barrer and results of other workers, three main effects may be presented to account for adsorptive properties.

Table 1

Type of zeolite Unit cell formula	d Density g/cm^3	d_f Framework density g/cm^3	Water void volume cm^3/cm^3
A Na_{12} $(AlO_2)_{12}(SiO_2)_{12}$ $27H_2O$	1.99	1.27	0.47
chabazite Ca_2 $(AlO_2)_4(SiO_2)_8, 13H_2O$	2.05	1.45	0.47
erionite $(Ca,Mg,Na_2,K_2)_{4.5}$ $(AlO_2)_9(SiO_2)_{27}$, $27H_2O$	2.02	1.51	0.35
L K_9 $(AlO_2)_9(SiO_2)_{27}$, $22 H_2O$	2.11	1.61	0.32
mordenite Na_8 $(AlO_2)_8(SiO_2)_{40}, 24H_2O$	2.13	1.70	0.28
mordenite large port (zeolon) $Na_{8.7}$ $(AlO_2)_{8.7}(SiO_2)_{39.3}$, $24 H_2O$		1.7	0.28
T $Na_{1.2}K_{2.8}$ $(AlO_2)_4(SiO_2)_{14}$, $14.4 H_2O$	2.12	1.5	0.40
X $Na_{86}(AlO_2)_{86}(SiO_2)_{106}$, $264 H_2O$	1.93	1.31	0.50
Y Na_{56} $(AlO_2)_{56}(SiO_2)_{136}$, $250 H_2O$	1.92	1.25-1.29	0.48

Free apertures of main channels Å	Largest molecule adsorbed	Kinetic diameter	Type of channel
4.2	C_2H_4 at R.T O_2 at $-183°C$	3.9 and 3.6	3
3.7 x 4.2	n paraffins hydrocarbons	4.3	3
3.6 x 5.2	n paraffins hydrocarbons	4.3	3
7.1	$(C_4H_9)_3N$ and $(C_4F_9)_3N$ slowly at 50°	8.1	1
6.7 x 7	C_2H_4	3.9	2
6.7 x 7	C_6H_6	5.85	2
probably like erionite	n paraffin hydrocarbons	4.3	3
7.4	$(C_4H_9)_3N$	8.1	3
7.4	$(C_4H_9)_3N$	8.1	3

7.1 Geometric effect

In static conditions the amount of gas adsorbed and the specificity of adsorption depend largely on the respective size of the adsorbate molecule and of the zeolite channels.

One of the well known examples of molecular sieving is the adsorptive behavior of normal and branched chain paraffin hydrocarbon. For example Na A does not adsorb paraffins (normal or branched). The replacement of four sodium ions by two calcium ions permits the rapid diffusion of normal paraffin hydrocarbons into the zeolite channels and therefore their separation from branched-chain paraffins.

As to the capacity of adsorption, the water void volume expressed as the void fraction and given previously (table 1) for usual zeolites is reported in figure 3 as a function of the framework densities for a series of zeolites. The relation is linear. The known zeolite structures suggest that the maximum observed void fraction is about 0.5. Values of 0.6 have been postulated but stability factors may rule out the formation of such a zeolite. The void volume depends also on the nature and size of the adsorbate molecule. For instance for NaX it decreases from 0.51 (H_2O) to 0.50 (N_2), 0.45 (O_2), 0.42 ($n-C_5H_{12}$) and 0.38 (neopentane).
. The changes come from the sizes of the molecules and the different types of zeolite cages which adsorbate may enter. Due to their kinetic diameter, water molecules (2.65 Å) may enter sodalite cages (aperture 2.6 Å) while O_2 (3.46 Å) and larger molecules are

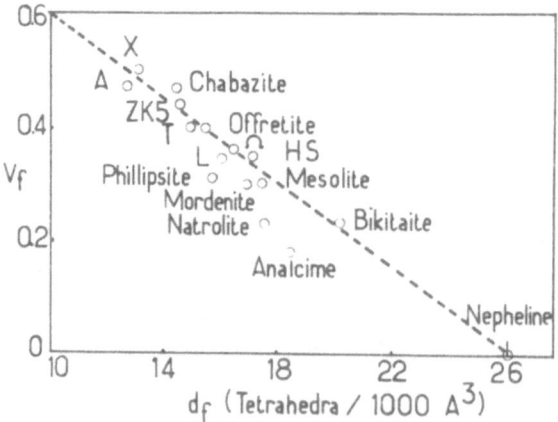

Fig. 3. Void fraction as a function of the framework density (10) Reprinted by permission of John & sons, Inc

excluded. Breck suggests that N_2 molecules (3.64 Å) do not enter the small cages and the large amount of nitrogen adsorbed might be due to interactions of molecules with cations in the large cavities. In that case the density of nitrogen would be higher than that of liquid nitrogen at its boiling point.[10] Figure 4 classifies the results of various workers and gives the effective pore diameter of zeolites in equilibrium adsorption. The size of this aperture can be determined from the kinetic diameter of molecules which are or are not adsorbed under a given set of conditions.

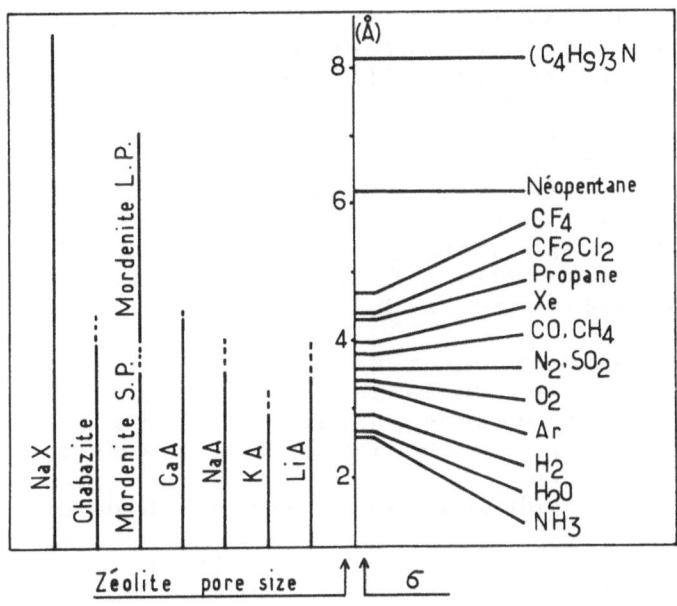

Fig. 4. Chart showing a correlation between effective pore size of various zeolites in equilibrium adsorption over temperatures of 77K to 420 K (indicated by ....) with the kinetic diameters of various molecules (10) Reprinted by permission of John Wiley & Sons, Inc.

Two conclusions may be drawn from figure 4 i). The dashed lines correspond to the increase in pore diameter as the temperature changes from 77 to 420 K. A drastic effect due to temperature is observed in the adsorption of Ar, O_2, N_2 on NaA. At very low temperature ($\sim$ -150°C) only O_2 is easily adsorbed. In the case of A, X and Y zeolites, at room temperature, polar molecules such as NH_3 do not enter the sodalite cage and adsorption equilibrium is attained rapidly. As the temperature is raised above 473 K, 5 hours are necessary to reach equilibrium, since NH_3 diffuses into the small cages. These results may be explained by several factors, such as a process of activated diffusion that is a function of temperature, or the effect of increase in temperature on the vibration amplitudes of the oxygen atoms in the structure, which implies an increase in the size of the apertures. ii) Figure 4 also shows that the various cationic forms of a zeolite have different pore diameters. In the case of type A zeolite the following sequence is obtained KA< LiA < NaA< CaA. X-ray diffraction studies confirm this sequence since the unit cell dimension of LiA is 12 Å compared to 12.3 Å for NaA. Several factors may account for the change in pore aperture with different cation, such as the size of cations, their location near the apertures, the number of ions, their valency as each bivalent ion is replaced by two monovalent ions. For instance results have been reported for offretite type zeolite.[11] This zeolite contains potassium and tetramethylammonium ions (TMA) which may be exchanged for protons.

1- K_2 $(TMA)_2$ adsorbs only H_2O and CO_2. The large TMA cations in both small cages and large channels reduce the void volume

2- K_3 TMA adsorbs some n-hexane indicating replacement of the TMA ions in the large channels by K^+ ions

3- K_2H_2 adsorbs cyclohexane (kinetic diameter 6 Å) indicating that K^+ ions in the large channels reduce the effective pore size

4- KH_3 adsorbs cyclohexane and m-xylene (kinetic diameter 6.8 Å).

The size of the channels may also be varied by preadsorption of molecules such as water, ammonia or N_2.

Another method of varying the pore size is steaming. For instance steaming NaA at 550°C for 25 minutes eliminates the slow adsorption of CHF_2Cl_2 and NaA may then be used as a drying for this refrigerant.

Finally as to the geometric effect, it must be mentioned that the zeolitic structure is not a rigid matrix. It is known that during the adsorption process of polar molecules the shape of the apertures is changed because of adsorbate-adsorbent interactions. Further it has been shown that the localization of the cations is also modified under conditions of adsorption. Cations Cu^{2+} for example may migrate from internal positions to sites in the large cavities when pyridine is adsorbed on Y zeolites and then modify the pore volume.[12] These dynamic properties of zeolites indicate that the adsorption capacity and selectivity effect may depend greatly on the adsorbate-adsorbent system.

7.2 Interaction effect

Interactions between adsorbate molecules and zeolite atoms are important parameters which explain many adsorption selectivity effects in zeolites. Experimental results allow enthalpies and entropies of adsorption to be easily calculated. The differential enthalpy of adsorption $\Delta \overline{H}_{ads}$ or isoteric heat of adsorption q^{st} ($q^{st} = - \overline{\Delta H}_{ads}$) depend upon the pressure and the adsorbate coverage. Integral enthalpy of adsorption $\Delta \widehat{H}_{ads}$ is given by

$$(1) \quad \Delta \widehat{H}_{ads} = \frac{1}{n} \int_{o}^{n} \overline{\Delta H}_{ads} \, dn$$

at constant temperature. Figure 5 gives an example of changes of heat of adsorption as a function of coverage

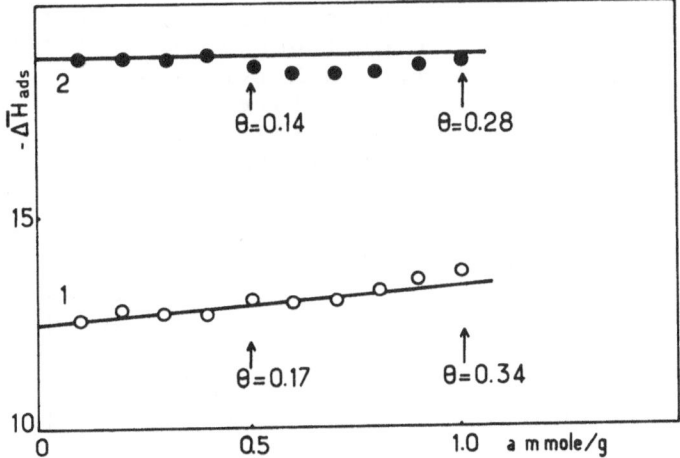

Fig. 5. Differential heat of adsorption of C_6H_6 (●) or cyclohexane (o) on NaY zeolite

The increase in $\overline{\Delta H}$ is attributed to interactions between the adsorbed molecules. The interactions depend on the zeolite (the type, nature and content of cation) and on the adsorbate. Figure 6 shows the changes in the initial enthalpy of adsorption of water vapour with cation radius.[13] The enthalpy of adsorption cation depends on cation size and on the amount adsorbed.

174

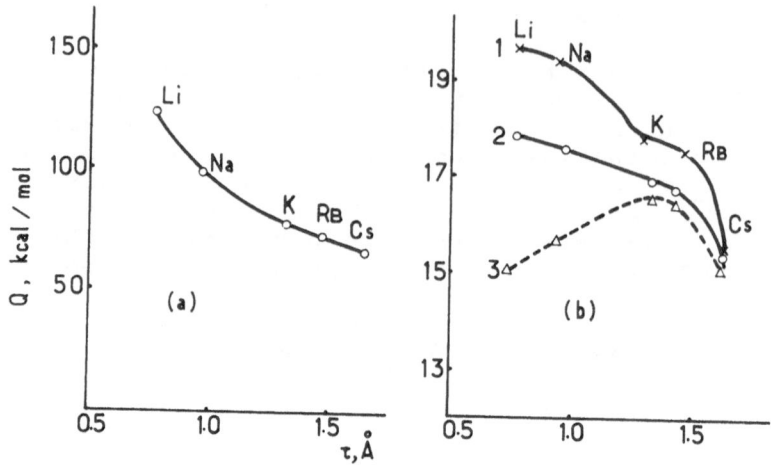

Fig. 6. Dependence on the radius of ions of alkali metals (a) of heat of hydration of these ions, and (b)
of heats of adsorption of water on X type zeolites for
different numbers of molecules per large cavity:
(1) 0.5 molecule, (2) 2 molecules, (3) 6 molecules.
Reprinted by permission of Chemical Society.(13).

Initial enthalpy of adsorption

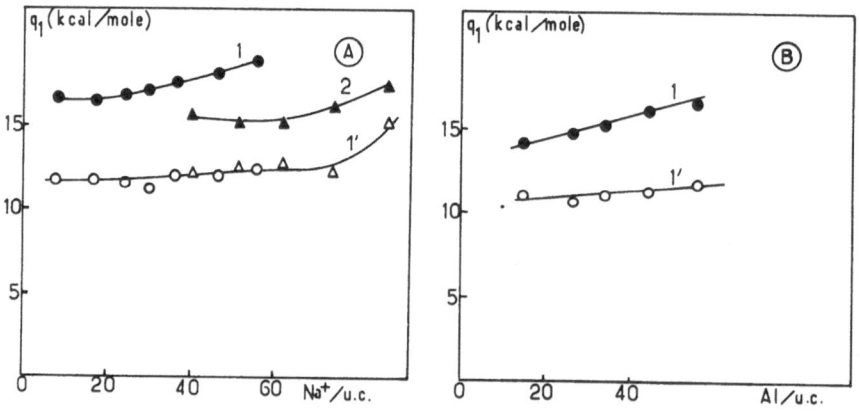

Fig. 7. Dependence of the initial enthalpy of adsorption of benzene (black points) and cyclohexane (open
points) on the number of sodium (A) or aluminum contents (B) in Y type zeolites (14).

Figure 7 indicates the dependence of the initial enthalpy of adsorption of benzene and cyclohexane upon
Na^+ content in X and Y type zeolites and upon Al con-

tent in aluminum-deficient Y materials.[14] Due to specific interactions of its π electron, benzene interaction is greater than that of cyclohexane. Furthermore X and Y type zeolites give similar results for
the saturated hydrocarbon, while the interaction of
the π electrons of benzene allow differences between
the two types to be detected. At high level of adsorption (coverage equals 1), see figure 8 which indicates
that the integral enthalpy of adsorption calculated
for a monolayer adsorbed in a unit cell is a linear
function of the cation (K⁺ or Na⁺) content. The slopes
of such lines give the integral enthalpy of adsorption
per cation. Such a value only depends on the type of
zeolite X or Y and on the hydrocarbon but not on the
cation content and nature.[14]

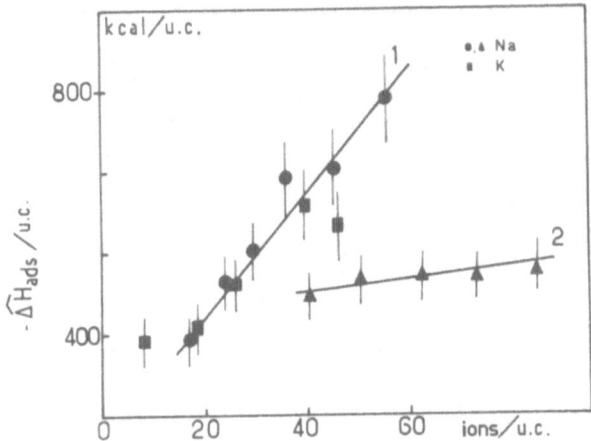

Fig. 8. Dependence of the integral enthalpy of adsorption per unit cell on the number of cation content
in Y zeolites (curve 1) and X zeolites (curve 2) (14).

Differences generally observed between low and high
coverage results give information on the degree of ho-
mogeneity of the inside of the zeolite relative to the
adsorbate concerned. Therefore results are in general
given for two different cation contents at zero and
high degree of coverage.
At low adsorption level a theoretical approach relate
the initial enthalpy of adsorption to several component interaction energies. At OK, $\overline{\Delta H}$ is given by

$$(2) \quad \overline{\Delta H} = \emptyset_D + \emptyset_R + \emptyset_P + \emptyset_{F-\mu} + \emptyset_{F-Q}$$

where $\emptyset_D$ and $\emptyset_R$ are dispersion and short range repulsion energies between the molecule and the zeolite, $\emptyset_P$ is the polarizing energy. The two other terms are attributed to electrostatic interaction, $\emptyset_{F-\mu}$ is the field dipole energy and $\emptyset_{F-Q}$ field gradient-quadrupole energy.

The specificity shown by zeolite adsorbents toward an adsorbate molecule is determined by one or more of these different type of interactions energies. Dispersion and repulsion energies exist in every case of adsorbents. The term $\emptyset_D$ may be very important where dipole and quadrupole interactions are absent. Polarization energy however is present only if the adsorbent is heteropolar. This case occurs in zeolites because of negative and positive ions of the structure which involve a local electrostatic field polarizing molecules. A relationship has been found between the polarisability of the adsorbed molecule and the initial enthalpy of adsorption on different types of zeolites.[15] For instance on Ca-A chabazite $\overline{\Delta H}$ increases nearly 4 fold as the polarisability increases 9 times from N_2 to C_2H_6. Molecules which have permanent dipole moments such as water and ammonia interact very strongly with zeolite electrostatic field and $\emptyset_{F-\mu}$ is important. This gives high values for $\overline{\Delta H}$. The term $\emptyset_{F-Q}$ is important for molecules which possess a quadrupole moment such as nitrogen and carbon dioxide. A specific interaction of π electrons of olefins with cations has also been demonstrated. It may be 25 % of the enthalpy of adsorption.

Therefore the initial enthalpy of adsorption depends greatly upon the type and value of the interactions implied. For instance, for argon the main interaction energies are due to dispersion and polarization forces whereas for nitrogen the quadrupole interaction is important, up to 50% of the total interaction.

All these results indicate that the selectivity of adsorption may be modified by methods which alter energy of interaction terms. Three main methods are :

- the preadsorption of small amounts of polar molecules such as water which is adsorbed strongly and selectively on the most energetic sites and will not be deplaced by an other molecule.

- the exchange of cation which affect not only the pore size as already mentionned but which also modify the local electrostatic field and the polarization effect. The charge, size, location of the cations is of importance

- the decationization or dealumination which alters the zeolite field and decreases the interactions of adsorbate molecules (fig 7).

7.3 Kinetic effect

The rate of adsorption is of very practical importance.
It depends on the diffusion of the adsorbate into zeolite. The model generally considered is that of a diffusing molecule which encounters a periodical potential field within the zeolite. The diffusion coefficient D (in cm^3/s) for an assembly of spherical or cubic particles and a small period of time is deduced
at a constant pressure from the equation :

$$(3) \quad \frac{Q_t - Q_o}{Q_\infty - Q_o} = \frac{2A}{V} \left(\frac{Dt}{\pi}\right)^{1/2}$$

where Q_o, Q_t, Q_∞ are amounts adsorbed at time $0, t, \infty$.
A is the external surface area of the particles in
cm^2/g and V is the volume of the crystals in cm^3/g.
This applies for small amounts adsorbed where Henry's
law is valid for the adsorption isotherms. The value
of D so obtained is an average over the range of adsorption and varies with the amount of adsorption, the
nature of the adsorbate, the type of zeolite (structure, cation ...) and the temperature. The effect of temperature allows the activation energy of diffusion E_a
to be calculated from the equation

$$(4) \quad D = D_o \, e^{-E_a/RT}$$

This energy arises from the repulsive interaction in
the case of molecules whose size approaches the pore
size since a potential barrier has to be passed at
each aperture.
The effect of zeolite structure has been studied in
the case of erionite and zeolite CaA. The rate of adsorption of n-paraffin hydrocarbons is much lower in
erionite than in CaA. In zeolite A n-paraffins can diffuse through the channel system from cavity to cavity
in a unidirectionnal manner whereas in erionite the
hydrocarbons must diffuse through a zigzag channel in
an irregular manner. Moreover a very specific effect
of diffusion has been described in erionite. Figure 9
shows the changes in diffusion coefficient of hydrocarbon in erionite as a function of the number of carbon atoms.[16] It is very surprising to find an increase
in D for hydrocarbons $> C_8$. This usual effect is related to the size of the unit cell.
A part from this very specific structure, it is gene-

rally observed that D decreases with the size of the adsorbate.

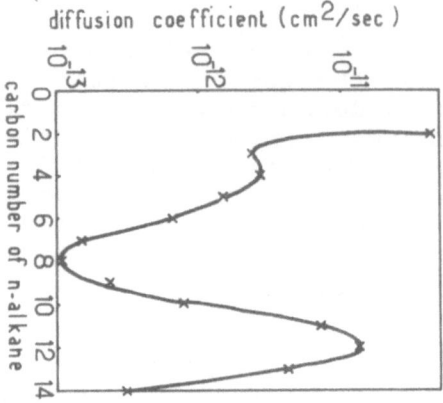

Fig. 9. Diffusion coefficients of n-alkanes in potassium T zeolite at 300°C. Reprinted by permission of Academic Press, Inc. (16).

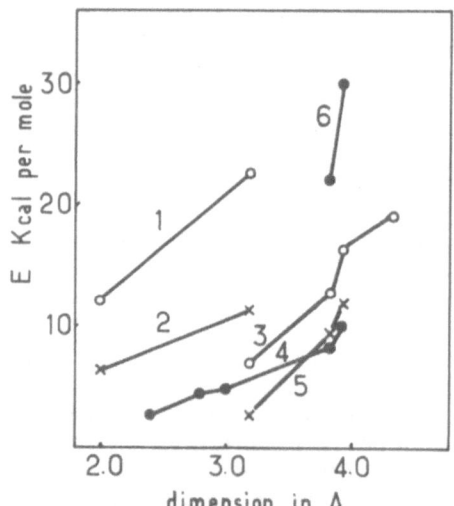

Fig. 10. Dependence of energy of activation for diffusion in several media upon the critical van der Waals dimension of the diffusing molecules
1: α-Tridymite, 2: Silica glass, 3: K-A (Linde Sieve 3A)
4: K-mordenite, 5: Levynite, 6: Basic sodalite
Reading from left to right, the diffusing molecules
are : He, H_2, O_2, N_2, Ne, Ar, Kr, and Xe (17)
Published in "MOLECULAR SIEVES ZEOLITES II"

Besides the influence of the geometry of zeolite structure and of molecules on adsorption rate, adsorbate-adsorbent interactions are of importance and are reflected in energy activation values. For instance figure 10 indicates a change in activation energy of diffusion with molecule size and zeolite nature.[17]

7.4 Adsorption of binary mixtures on zeolites

Relatively few fundamental results are known on the adsorption of gas mixtures on zeolites. If, owing to a sieving effect one component of the mixture of two gases is excluded or slowly adsorbed, a simple separation then occurs. But very often both gases are adsorbed and the mixture adsorption depend in interactions of adsorbed molecules with zeolites.
An approach considers the separation factor α as

$$(5) \quad \alpha = \frac{Y_a . X_g}{Y_g . X_a}$$

where X_a, Y_a are the mole fractions of the two adsorbates in the adsorbed phase and X_g, Y_g in the gas phase.
If mutual interactions between adsorbate molecules are neglected, as well as the perturbing influence of adjacent cations, one may assume that the composition of the adsorbed gas at equilibrium depends on the relative heats of adsorption of the two different species.
An equation of the following type has been derived

$$(6) \quad \alpha = K \exp \frac{\Delta q_{iso}}{RT}$$

where Δq_{iso} is the difference in isosteric heats of the two adsorbates. Data support the conclusion that in Δq_{iso} term the dipole and quadrupole moments of the molecules involved are very important. As an example, in the case of air adsorption in zeolite A, it is known that, due to its quadrupole moment nitrogen adsorbs on active sites and inhibits the adsorption of oxygen while pure oxygen is readily adsorbed and pure nitrogen excluded.
Many other examples of separation of binary mixtures may be related to differences in interactions energies of the two molecules with the zeolite. In a mixture of methane and CO the dispersion and polarization terms are higher for methane. However the dipole and quadrupole interactions of CO is larger and results in a o-

verall selectivity for CO over methane in the mixture. Because of interactions between surface sites and π electrons of adsorbate, the selectivity depends also upon the unsaturated bond. For example in a mixture of n-hexane - benzene, there is a strong selectivity at 97°C for benzene over n-hexane on KY zeolites.

8. CATALYSIS

The main difference between zeolite adsorbents and catalysts is that the adsorbent has to be an inert material while the catalyst must activate the adsorbed molecules. Therefore the latter must have catalytically active sites. The type of reaction catalyzed by zeolite depend upon the nature of these sites and also on the sieving properties of zeolites.

8.1 Nature of catalytic sites

The most important use of zeolites is the catalytic cracking. The active sites are acid centers which are created in the zeolite during the exchange of cations for protons. It is generally considered that protons come from zeolitic OH groups of which the hydrogen atom has a positive charge (figure 11). In fact not only protons are active sites but also Lewis acid sites. These sites are tricoordinated aluminum atoms which may arise from reaction given in figure 11.

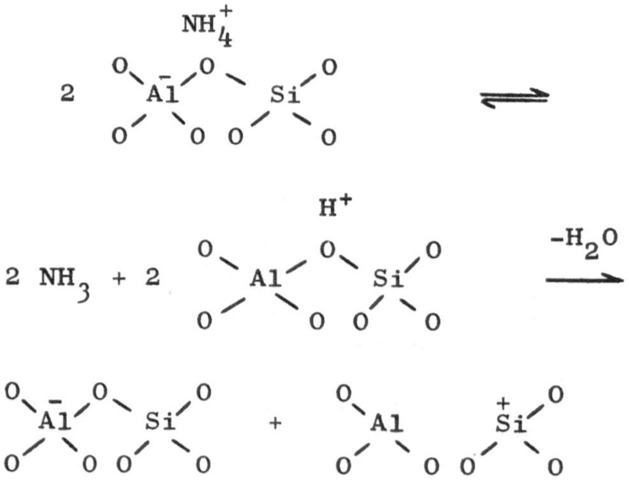

Fig. 11.

The hydrocarbon molecules form carbonium ions with the
two types of sites. Cracking requires strong acid sites and an interesting property of zeolites is that
they have such strong acidity. At the present time,
90% of cracking catalysts in the United States are based on zeolites.
Zeolites are also used as bifunctionnal catalysts. A
metal such as Pd or Pt is incorporated in zeolite. The
dispersion of metal may be very high, small aggreates
are formed ($\emptyset \sim 5$ to 6 Å) which contain only a few
number of metal atoms. The high metallic area promotes
a high activity of this metal. Hence zeolites may easily
hydrogenate unsaturated hydrocarbons. They are used
as reforming catalysts where both acidic properties of
zeolites and hydrogenating properties of the metal are
implied.

8.2 Sieving properties and catalysis

Since the external surface areas of zeolites is small
compared to the internal surface, such catalysts are
active only for molecules which may enter the channels
and cavities. The main effects which were shown to re-
gulate adsorption process are of course also operating
in catalysis since adsorption is a necessary step in he-
teregeneous catalysis. It is known for example that a-
cidic A type zeolite may dehydrate n-alcohols but not
large isoalcohols. An important process is selectofor-
ming, where the process of reforming is combined with
a shape selectivity property of erionite. Effluents
arising from a reforming unit pass over erionite catalyst which do not adsorb isoparaffins while n-paraffins are adsorbed. The elimination of n-paraffins from
the effluent increasesthe octane number of gasoline.
Erionite is used as a hydrocracking catalyst which
transforms selectively the n-paraffins into lighter hydrocarbons, particularly propane which is practically
important.

REFERENCES

1. Barrer, R.M., Separation of mixtures using zeoli-
    tes as molecular sieves. I. Three classes of mo-
    lecular sieves zeolites, J. Soc. Chem. Ind., 64,
    130, 1945.
2. Milton, R.M., Commercial development of molecular
    sieve technology, in Molecular Sieves, Society
    of Chemical Industry, London 1968.

3. Breck, D.W., _Zeolite Molecular Sieves_, John Wiley and Sons, New York, 1974, 47.

4. Meier, W.M., Zeolite structures, in _Molecular Sieves_, Society of Chemical Industry, London 1968.

5. Frety, R., Ballivet, D., Barthomeuf, D., Trambouze, Y., Examen au microscope électronique de la structure poreuse d'une zéolithe de type L, _C.R. Acad. Sci._, 275, ser. C, 1215, 1972.

6. Breck, D.W., _Zeolite Molecular Sieves_, John Wiley and Sons, New York, 1974, 636.

7. Mc Daniel, C.V., Maher P.K., New ultrastable form of faujasite, in _Molecular Sieves_, Society of Chemical Industry, London 1968.

8. Kerr, G.T., The intracrystalline rearrangement of constitutive water in hydrogen zeolite Y, _J. Phys. Chem._, 71, 4155, 1967.

9. Barthomeuf, D., Beaumont, R., X, Y, aluminum-deficient and ultrastable faujasite-type zeolites. III Catalytic activity, _J. Catal._, 30, 288, 1973.

10. Breck, D.W., Grose, R.W., A correlation of the calculated intracrystalline void volumes and limiting adsorption volumes in zeolites, in _Advances in Chemistry Series_, 121, Meier, W.M. and Uytterhoeven, J.B., American Chemical Society, Washington, 1973.

11. Aiello, R., Barrer, R.M., Davies, J.A., Kerr, I.S., Molecule sieving in relation to cation type and position in unfaulted offretite, _Trans. Faraday Soc._, 66, 1610, 1970.

12. Gallezot, P., Ben Taarit, Y., Imelik, B., X-ray diffraction study of cupric ion migrations in two Y-type zeolites containing adsorbed reagents, _J. Catal._, 26, 295, 1972.

13. Bezus, A.G., Kiselev, A.V., Sedlacek, Z., Pham Quang Du, Adsorption of ethane and ethylene on X-zeolites containing Li^+, Na^+, K^+, Rb^+ and Cs^+ cations, Trans. Faraday Soc., 67, 468, 1971.

14. Barthomeuf, D., Ha, B.H., Adsorption of benzene and cyclohexane on faujasite type zeolites, part I, _J.C.S. Faraday I_, 69, 2147, 1973. Part II, _J.C.S. Faraday I_, 69, 2158, 1973.

15. Barrer, R.M., Specificity in physical sorption, _J. Colloid Interface Sci_, 21, 415, 1966.

16. Gorring, R.L., Diffusion of normal paraffins in zeolite T. Occurence of window effect, _J. Catal._, 31, 13, 1973.

17. Barrer, R.M., Intracrystalline diffusion, in _Advances in Chemistry Series_, 102, Gould, R.F., American Chemical Society, Washington, 1971.

STUDY OF ZEOLITE AT THE LABORATORY

D. Barthomeuf

Institut de Recherches sur la Catalyse et
Université Claude Bernard, Lyon I. France

ABSTRACT. The differents physical and chemical methods
applied to zeolite study are presented. The lecture
includes: 1. Structure; 2. Adsorptive properties;
3. Surface state chemistry

1. STRUCTURE

X-rays diffraction is used to determine the crystalline structure of zeolites, the pores apertures and the
theoretical pore volumes from unit cell parameters...
Besides this very important application other information is obtained from X-rays studies. Changes in crystallinity and unit cell constants may be studied as a
function of pretreatment of zeolites, extent of cation
exchange or aluminum removal. They are useful to know
both from fundamental and practical aspects. From the
intensity of diffracted lines related to cations located in the very definitive sites in zeolites it is possible to evaluate the occupancy factor of cations in
the different sites. Changes in intensity allow the
variations in cation population of sites to be studied.
Very recent results have shown the cation migration
upon adsorption and desorption of hydrocarbons.
Infrared spectroscopy has been recently applied to
structural studies of zeolites. Vibrations of
O - (Al, Si) - O bonds and vibrations of secondary
building units, D-6, D-8, D-12 rings, give absorption
bands in the $1300-350$ cm^{-1} region. The intensity and

184

frequency of these bands give information on changes
in zeolite framework upon different treatments and
adsorption.
Electron microscopy is a useful tool giving important
information on the shape of particles, their size,
the heterogeneity of a sample. Moreover recently, lattice images of the framework have been obtained where
lattice parameters may be directly measured. Such images may be observed either directly on crystallites
correctly oriented under the electronic beam or on
thin sections of particles.

2. ADSORPTIVE PROPERTIES

The adsorptive properties of zeolites are studied with
the same methods as those of other micropores adsorbents. The properties studied may be classified in two
main types, amount adsorbed and energy of adsorption.
The first kind is related to the adsorption capacity.
It includes adsorption isotherms, pore filling, diffusion coefficient, surface area... The methods used
in that case are thermogravimetric analysis (TGA), vo-
lumetry and chromatography. All of these techniques allow isotherms to be determined. In the case of chromatography it is necessary to be sure that adsorption
is rapid enough to reach equilibrium during the time
the adsorbate goes through the zeolite.

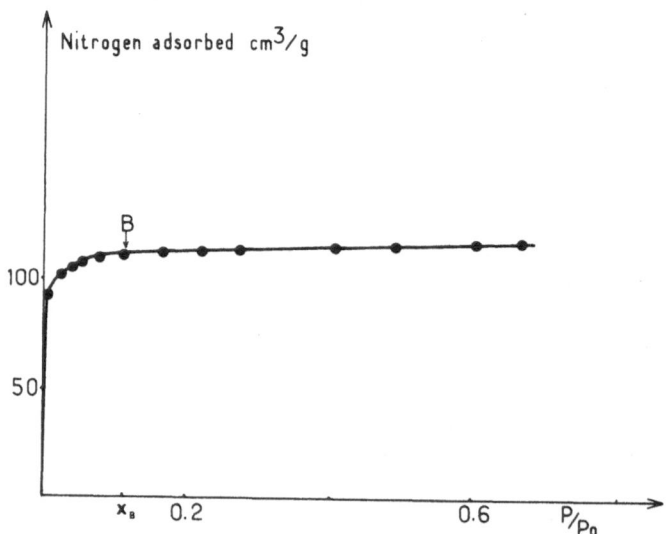

Fig. 1. Isotherm of nitrogen adsorption at 77 °K on
a L zeolite.

Zeolites exhibit the type I isotherm (figure 1). These adsorption isotherms do not exhibit hysterisis, as do isotherms on many other noncrystalline, microporous solids. Several attempts have been made to derive a standard adsorption isotherm which would apply to the adsorbed phase in zeolites under all conditions but until now no universal adsorption equation exists. At very low pressures, the amount adsorbed is proportional to pressure P(Henry's law). Besides the classical Langmuir or Volmer equations, other models have been applied to adsorption in zeolites. Barrer proposes an equation based upon a virial equation by the application of solution thermodynamics :[1]

$$(1) \quad \frac{\pi}{CRT} = 1 + A_1 C + A_2 C^2 + \ldots + A_n C^n$$

where π is a pressure and C the concentration of adsorbate in zeolite A_i are contants.
An other virial equation [2] has also been used by Kiselev et al at low pressure P and amount adsorbed n :

$$(2) \quad P = K'_1 n + K'_2 n^2 + K'_3 n^3 + \ldots$$

K'_i are constants, K'_1 is Henry constant. The equilibrium constant of adsorption is $1/K'_1$ which may be easily calculated.
The Polanyi-Dubinin potential theory[3] has been found to apply in many instances to zeolites. The volume occupied by the adsorbed phase W, at a given temperature, T, and pressure P, is given by the equation

$$(3) \quad W = W_o \exp \left[- \frac{B}{\beta^2} (T \log_e \frac{P_s}{P})^2 \right]$$

where W_o is the pore volume, B a constant characterising adsorbent and β is an affinity coefficient. W_o and B can be calculated.
More recently Loughlin and Ruthven have derived a new isotherm equation from the application of statistical thermodynamics.[4] This was applied to the adsorption of low molecular weight hydrocarbons on zeolite A.
As to surface areas determination, BET equation is not applicable to zeolites because of the small size of pores. In order to obtain the monolayer capacity the "point B" method may be used.[5] On the isotherm of figure 1 this adsorbed volume is x_B. Pore filling is obtained from the volume adsorbed at saturation.
The energy of adsorption which is the second aspect of the adsorptive properties may be studied by a direct

186

or indirect method. Hence calorimetry and microcalori-
metry have been used to evaluate initial, isosteric
and integral heats of adsorption of different gases,
hydrocarbons, bases, ... on zeolites. Heats of adsorption are also obtained from a family of isotherms by
the Clausius Clapeyron equation at a constant volume
n adsorbed :

$$(4) \quad \overline{\Delta H}_{ads} = -(\frac{\delta \ln P}{\delta T})RT^2$$

where $\overline{\Delta H}$ is differential enthalpy of adsorption.
Changes in entropy during adsorption $\overline{\Delta S}_{ads}$ are obtained from the equation

$$(5) \quad \overline{\Delta S}_{ads} = RT \ln \frac{P_o}{P} + \frac{\overline{\Delta H}_{ads}}{T}$$

Integral values $\widehat{\Delta H}$ and $\widehat{\Delta S}$ are obtained by integration of differential values.

3. SURFACE STATE CHEMISTRY

Differential thermal analysis is an important tool in
studying adsorbents. Figure 2 gives a schematic DTA
curve showing low temperature endotherms caused by water or ammonia desorption and high temperature exo-
therms due changes in structure or loss of structure.

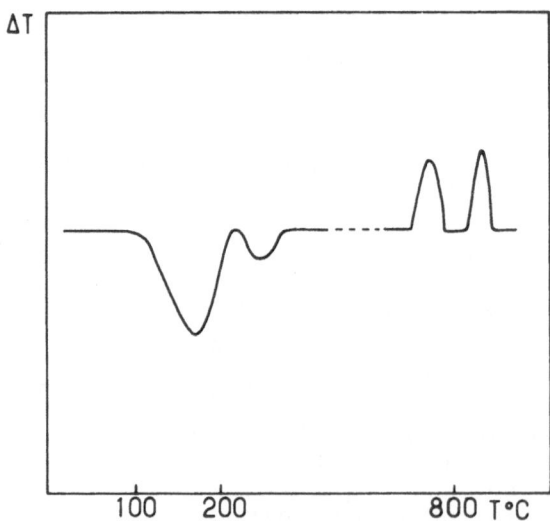

Fig. 2. Typical DTA curve of zeolite

In the case of zeolites containing organic ions such
as TMA (tetramethylammonium), DTA curves obtained detect decomposition of this ion.

Infrared spectroscopy is used to study both the surface state and the adsorbate phase. Compressed disks
of zeolites are inserted in a quartz sample-holder
which is introduced into an i-r cell and spectra are
recorded. Zeolites are characterized by structural hydroxyl groups which give rise to i-r absorption bands
in the range 3500-3800 cm^{-1}. Changes in intensity and
frequency of the bands upon several treatments detect
variations in the number and the environment of OH
groups. Some of these groups are acidic. For instance
it has been shown that adsorption of pyridine decreases the intensity of the bands at 3540 and 3640 cm^{-1}.
I-r spectroscopy has often been applied to evaluate
the Bronsted or Lewis acid character of zeolites surfaces by adsorption of bases. In the case of adsorption of acetylene on Ni-zeolites, i-r spectroscopy gives information on the state of the adsorbed hydrocarbon (monomer or trimer).

ESR gives information on oxidant and reductor conters
of surfaces. Reactants such as perylene (electron donor) or tetracyanoethylene (electron acceptor) become
paramagnetic after adsorption on zeolites and are used
to evaluate the number of such centers. The adsorption
of NO on copper zeolite has been studied by ESR. The
spectra allows characterization of active copper ions.
Generally speaking, ESR give information on the valence and environment (hence location) of the ions.

NMR is used to study the proton in acidic zeolites and
has shown that protons jump from sites to sites very
rapidly.

Chemical titration of acidity with bases in gaseous or
liquid phase give the number and the strength of acid
centers. Many methods have been used for this purpose.
They give results which are correlated with catalytic
properties of zeolites.

The catalytic active sites are studied by catalytic
tests. In the laboratory, generally microreactors are
used which allow the concentration of sites, their selectivity and stability to be studied.

REFERENCES

1.   Barrer, R.M., Coughlan, B., Influence of crystal
      structures upon zeolitic carbon dioxide. I. Iso-
      therms and selectivity, in Molecular sieves, So-

188

ciety of Chemical Industry, London, 1968.

2. Bezus, A.G., Kiselev, A.V., Sedlacek, Z;, Pham Quang Du, Adsorption of ethane and ethylene on X-zeolites containing Li^+, Na^+, K^+, Rb^+ and Cs^+ cations, <u>Trans. Faraday Soc.</u>, 67, 468, 1971.

3. Dubinin, M.M., Astakhov, V.A., Description of adsorption equilibria of vapors on zeolites over wide ranges of temperature and pressure, in <u>Advances in Chemistry Series</u>, 102, Gould, R.F., American Chemical Society, Washington 1971.

4. Ruthven, D.M., Loughlin, K.L., Sorption of light paraffins in type-A zeolites, <u>J.C.S. Faraday I</u>, 4, 696, 1972.

5. Gregg, S.J., Sing, K.S.W., <u>Adsorption Surface Area and Porosity</u>, Academic Press, London and New York, 1967, 54.

IMPREGNATED ADSORBENTS

F.A.P. Maggs

Chemical Defence Establishment, Porton Down, Salisbury, Wiltshire, England

ABSTRACT. The mode of action of impregnated adsorbents is discussed and a number of illustrative examples are given.

INTRODUCTION

Impregnated adsorbents are generally used in situations where physical adsorption fails to provide the required protection, and this occurs when the solid/gas interaction is not strong. One crude assessment of this interaction is provided by mutual forces between adsorbate molecules, an interaction reflected in a number of properties, such as boiling point. We may generalise by saying that for gases of boiling point lower than ca. $50^{\circ}C$ physical adsorption is insufficient to allow the proper functioning of a filter bed. The malfunction of the filter may take three forms - (i) the amount adsorbed is low, so that the penetration time is too short to be of practical use; (ii) a dose may be weakly adsorbed, and then subsequently desorbed with continued passage of air; (iii) a dose may be safely adsorbed and the use of the canister discontinued for a period, during which the adsorbate is redistributed through the bed; in subsequent use the vapour is immediately desorbed from the effluent end of the respirator canister.

In such cases an enhancement of the binding force may be attempted either by using a filtering medium with which the vapour reacts chemically, or by the addition of impregnants to an adsorbent.

In the former case, some limitation on the extent of the reaction is imposed by the reaction occurring only at the external surface of the granule, unless this contains large pores permitting gas-phase diffusion (surface diffusion to the interior is clearly not possible). In the case of impregnated adsorbents, the impregnant forms a reactive species on the internal surface, and chemi-sorption occurs; the nature, extent and accessibility of the base adsorbent may therefore be as important as the impregnant properties. The term chemi-sorption is used here in a broad sense, implying that electron-transfer has occurred.

Several modes of action by which protection is given by such adsorbents may be distinguished:-

(i) The adsorbate vapour may interact chemically to give an involatile product which remains on the surface.

(ii) The adsorbate/solid interaction may yield a volatile but non-toxic product, such as carbon dioxide, nitrogen, or water.

(iii) The adsorbate may form a complex with the solid or the impregnant.

In many cases the solid/gas interaction has its counterpart in bulk phases; often, however, the reaction at the surface proceeds much more rapidly either because of the extended surface available, or because of polar or orientation effects at the surface. Catalytic action may also occur – a desirable process, since the filter should then have a long life, although in practice catalyst poisoning usually intervenes. Some examples of these divergent reactions follow.

EXAMPLES

(a) Catalytic action

A familiar example of catalytic air purification is provided by the removal of CO by aerial oxidation over hopcolite. This material is composed of a mixture of MnO_2 and CuO; silver is sometimes added, and may even replace the copper.

Water vapour is found to inhibit the catalytic oxidation of CO to CO_2 over hopcolite; the use of dried gases allows an indefinitely long period of use of the hopcolite. If the filter is operated at temperatures below ca. $0^{\circ}C$, the catalytic product, CO_2, remains adsorbed and the activity of the catalyst is severely reduced. The catalyst deactivated by water vapour may be regenerated by vigorous drying (> $200^{\circ}C$); (in high concentrations of CO, the heat of reaction may raise the catalyst temperature

sufficiently to reduce very considerably the effect of moisture).
If the hopcolite is operated at a high temperature (e.g. 500°C),
water vapour no longer inhibits the reaction, and the life of
the catalyst is only terminated by deposition of carbon from
atmospheric organic contaminants, by granule degradation, or by
very slow sintering.

The catalytic nature of the oxidation is amply demonstrated
by the behaviour described above: a) long life in the absence of
inhibitors; b) inhibition by chemisorbed water; c) inhibition by
products. The practical aspects of art are also exemplified in
the preparation and use of hopcolite; whilst the composition of
the catalyst must clearly be dependent on the activity towards
the desired reaction, the pore structure must be developed to
provide ready access of the reactants (carbon monoxide and
oxygen) to an extensive surface. Again, the granular material
must be robust with respect to attrition, size degradation, and
temperature changes; any inert binding material must not clog
the porous structure.

We may also note in passing that charcoals, both as such or
as a support medium, are of industrial importance as catalysts
in manufacture of vinyl chloride and acetate, fluorocarbons,
sulphuryl chloride, and phosgene.

(b) Chemical Interaction

The impregnant may interact chemically with the contaminant
vapour, and many examples of this type of filter filling are
known.

(i) Hydrogen Sulphide, Copper Impregnants. The uptake of
hydrogen sulphide by a charcoal treated with a copper salt
is far greater than that shown by the unimpregnated charcoal.
The copper may be added by merely spraying the charcoal with
a copper salt, or it may be incorporated in the raw material
before charcoal manufacture. A solution-treated charcoal
is improved by heating. The marked temperature rise
accompanying the uptake of hydrogen sulphide shows that
strong chemisorption occurs.

Copper impregnation also enhances the protection given by
charcoal or silica gel against ammonia; cobalt salts are
also effective.

(ii) Oxidation of Nitric Oxide. The oxidation of nitric oxide
to nitrogen dioxide (which is more strongly adsorbed on
charcoal than nitric oxide) may be brought about by passing
the contaminated air through granules treated with an
oxidising agent, such as potassium permanganate.

(iii) Conversion of Hydrogen Cyanide to Carbon Dioxide.
This system, which will be discussed in a later paper,
provides an example of the conversion of the toxic contaminant to a relatively non-toxic gas, carbon dioxide.

The interaction of hydrogen cyanide with copper is complex,
and depends to a large extent on the presence of oxide and
hydroxide groups on the metal surface. Under suitable
conditions a charcoal containing copper effects the
conversion of carbon dioxide and nitrogen. Note that
cyanogen may be formed if cupric salts are present on the
charcoal - a condition clearly to be avoided.

(iv) Alkyl Halide Removal. A number of low molecular weight
alkyl halides which are of industrial importance are known
to be undesirable atmospheric contaminants (e.g. methyl
bromide, vinyl chloride, methyl chloride, methyl iodide).
Physical adsorption is unacceptably low, and impregnation
with tertiary nitrogen compounds has been found to be
effective, presumably through the formation of quaternary
amine salts.

e.g. $R_3N + MeI \longrightarrow R_3N^+ MeI'$

Cyclic amines are found to be particularly effective: the
order of increasing effect with respect to methyl iodide
has been reported (Collins, Taylor and Taylor[1]) as
pyridine < morpholine < piperazine < tetraethylene diamine.
Initial experiments have shown tetraethyl diamine to be
effective as a charcoal impregnant for the removal of
methyl bromide.

The same workers also found an inorganic impregnant,
potassium iodide, was effective; some evidence of ion
exchange was noted (by using radioactive methyl iodide),
and this opens the possibility of other interactions when
the amines form the impregnant.

Whilst considering the alkyl halides, one may also mention
the value of tertiary amines in enhancing the protection
against cyanogen chloride. Presumably the von Braun
reaction occurs (better known when cyanogen bromide is
used)

$R_3N + ClCN \longrightarrow R_3NCN^+ Br'$

Whether further breakdown occurs (with ring opening in
the case of cyclics such as piperidine) is not clear.

Several interesting points arise from the impregnations with amines. Collins, Taylor and Taylor note that different charcoals responded to very different extents to impregnation with a given amine, and that other adsorbents, such as alumina, silica gel, offered no protection against MeI when impregnated. It would appear that not only is a large internal surface a necessity, but that this must be associated with the correct sub-microscopic pore structure. It may be suggested that not only the pores should be large enough to allow the impregnant to cover the surface, but that blockage of the pore by the impregnant should be avoided. It also seems possible that the type of surface can be influential; for instance, the value of the tertiary nitrogen may be impaired if this is attached to the surface by other than dispersion forces. The presence of metal impurities may also be relevant if, for example, an initial hydrolysis of the alkyl halide is required before reaction with the amine.

EFFECT OF IMPREGNATION ON PHYSICAL ADSORPTION

Impregnation of the charcoal by addition after charcoal manufacture, whilst enhancing considerably a specific protection, may lead to a reduction of the non-specific uptake of other vapours. This is illustrated in the following Table, which shows the effect on the uptake of carbon tetrachloride, of the addition, by solution spraying, of various amounts of an inorganic salt; (the samples were dried before test).

EFFECT OF IMPREGNATION ON ADSORPTION OF CARBON TETRACHLORIDE

% Salt, w/w	Weight adsorbed at penetration	Weight adsorbed at equilibrium
0	19.4	24.1
2	19.2	24.2
4	18.5	23.8
6	17.7	22.8
8	17.0	21.6
10	15.9	19.8

A similar effect will be observed in treating charcoals with amines. If the adsorbent is also required to function as a general adsorbent, a compromise between this and the effective amount of impregnant must be reached. In cases where the impregnant may be incorporated during the charcoal manufacture (as with copper) the impregnant becomes effective without causing a loss of physical adsorption capacity.

REFERENCES

1   Collins, D.A., Taylor, L.R., Taylor, R., 9th Air Cleaning
    Conference, Boston, 1966.

MEASUREMENT OF FUNDAMENTAL SURFACE PROPERTIES: MULTIPLE INTERNAL
REFLECTANCE SPECTROSCOPY*

Walter J. Weber, Jr.

Professor of Environmental and Water Resources
Engineering and Chairman, Water Resources Program,
College of Engineering, The University of Michigan,
Ann Arbor, Michigan, U.S.A.

1. INTRODUCTION

For analysis of the fundamental surface properties of
sorbents it would often be advantageous to obtain spectrophoto-
metric data relative to the surface functionality, or surface
chemistry, of these materials. However, most sorbents do not
readily lend themselves to routine transmission measurements
because they are usually characterized by one or more of the
following properties: insolubility; a high extinction coefficient;
and high light-scattering properties.

A number of previous efforts to obtain infrared spectra for
carbon black, activated carbon, and coal have involved transmission
measurements on prepared KBr pellets or Nujol mulls.[1-3] Although
some information regarding structure of the bulk materials has
been gained in this manner, such techniques have proved rather
unsatisfactory for identification of the surfaces of carbon
materials because of invariably poor resolution. Ergun has shown
that the extinction coefficient of graphite in the infrared spectral
range is very high, approaching that of a metal.[4] The average
extinction coefficient, k, is about 0.66 in the visible region
and little variation is observed through the short wavelength
region of the infrared. Common organic compounds or functional
groups exhibit extinction coefficients approximately two orders
of magnitude smaller than that of graphite, and therefore transmit

* Lecture 1. Session III NATO Advanced Study Institute:
  Scientific Aspects of Sorption and Filtration Methods for Gas
  and Water Purification, Fauske, Norway, 23-29 June 1974

sufficient radiation to give infrared spectra of reasonable contrast. From our own spectral studies of activated carbon, it appears that the bulk extinction coefficient of this material is approximately that of graphite. On this basis it is reasonable to conclude that light cannot be transmitted through particles of carbon unless they are extremely thin, in fact, light of 5-μ wavelength will decay to 1% of its initial value after passing through 3.7 μ of graphite. Thus, infrared light incident upon most microcrystals of activated carbon will be totally absorbed, unless it hits at a sufficiently grazing angle to allow the light to be reflected from the particle. The same is true for many other sorbent materials.

In view of the magnitude of the extinction coefficient for sorbents like activated carbon it would appear that the attempts reported in the literature to obtain transmission spectra of these materials in KBr or Nujol have actually involved measurement of a complicated type of mixed transmission and reflection spectrum. This type of spectrum results from a combination of forward scattered radiation and radiation which misses the particles entirely, as illustrated in Figure 1, as well as radiation which passes through extremely small particles. The resulting spectrum might be best referred to as a "diffuse reflectance" spectrum. This approach to examining the surface of a sorbent such as activated carbon is complicated by the fact that light losses due to scattering are inversely proportional to the fourth power of the wavelength, producing huge scattering losses at shorter wavelengths.

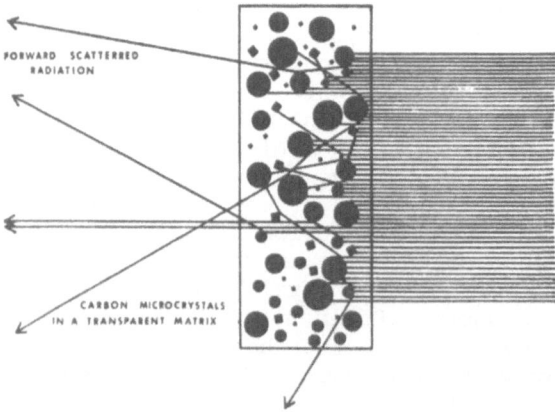

FORWARD SCATTERED RADIATION

CARBON MICROCRYSTALS IN A TRANSPARENT MATRIX

Fig. 1. Schematic diagram of the forward scattering process obtained when strongly absorbing carbon particles are included in KBr or Nujol, and infrared light is transmitted through the matrix.

Multiple internal reflection spectrometric (MIRS) techniques allow all of the incident light to interact with the sample, without losses due to scattering.[5-6] Thus it is possible to obtain spectra of high contrast and resolution with which to examine the nature of surface structures of active carbon. The general theory behind the use of MIRS has been discussed by Harrick.[6-7] Work in our laboratories has involved the use of MIRS to measure spectra of activated carbon, and in particular to obtain spectrometric evidence of the nature of functional groups on the activated carbon surfaces. The experimental details necessary to obtain such information are presented here along with the results of some typical studies on the effects of adsorption of p-nitrophenol on the spectra of activated carbon.

2. EXPERIMENTAL

A Perkin-Elmer 621 infrared spectrophotometer has been employed in our work along with a Wilks Model 12 IRS attachment (Wilks Scientific Corp., South Norwalk, Conn.). It has been found necessary to purge the 621 for two to four hours with dry, CO_2-free air because it is difficult to ensure equal path lengths in both the reference and sample beams with the Wilks Model 12 in place. Atmospheric absorptions will otherwise interfere with the sample absorptions in the 20X scale expansion mode employed. The Wilks Model 12 attachment is used in the double beam mode with 45° germanium crystals 2mm in thickness, 52.5mm in length, which provide 25 reflections in each beam. Scrupulous attention must also be paid to the cleanliness of the surfaces of these crystals. They must be freshly polished and employed in no other application than that for which they are intended. Germanium adsorbs many organic compounds, forms a set of surface oxides, and is generally very susceptible to fouling. While solid activated carbon without any labile compounds associated does not pose a problem, adsorbed organic molecules, such as the p-nitrophenol used in the typical studies discussed herein, are readily transferred to the germanium surface in sufficient quantity to pose an interference at high levels of signal amplification. Polishing with jeweler's rouge between each run is the minimum cleaning required. Because germanium is relatively inert to organic solvents, as well as to mild acids and bases, chemical cleaning is preferable to that provided by mechanical abrasion alone. Spectrometrically pure graphite (from Ultra Carbon Corp., Bay City, Michigan) is used as the reference material. It is assumed that this material effectively cancels most absorptions due to the graphite-like bulk structure of activated carbon. An overall advantage in using a graphite reference is that it nearly balances out the very high bulk extinction coefficient of the sample. The magnitude of the signal expected to result from absorption of light by a functional group on the carbon surface can be estimated as follows. The complex refractive index of an absorbing medium is given by

$$\hat{n} = n + ik$$

where k is the extinction coefficient and n is the observed refractive index.[8] The energy lost to the absorbing medium in attenuated total internal reflection is given by

$$\int_0^\infty P_x dx = \int_0^\infty \frac{nk(E_x^2)}{\lambda} dx = \frac{nk(E_0^2)dp}{\lambda} \tag{1}$$

where P_x is the power absorbed per unit volume at a distance x into the medium, (E_x^2) the time average of the square of the amplitude of the electric vector at x, and dp is the depth of penetration[7],[9] into the absorbing medium. Calculation of dp according to Hansen,[9] assuming that the carbon is covered with a 50 Å thick layer of organic molecules gives:

$$\frac{\text{Energy absorbed with organic layer}}{\text{Energy absorbed by pure graphite}}$$

$$\frac{n_{org}k_{org}(50 \text{ Å}) + n_{GR}k_{GR}dp}{n_{GR}k_{GR}dp}$$

$$\cong 1.0004 \tag{2}$$

After 25 reflections, the ratio is $(1.0004)^{25} \cong 1.01$. The change in reflectance is about 1%. For this reason, it is necessary to use the maximum scale expansion of the instrument (20X) in order to detect this small signal on top of the large bulk absorption. It is not possible to use a different MIRS crystal material due to the high refractive index (2.2) of the carbon, as well as the desirability for obtaining spectra out to 10 μ.[4],[7],[10]

The graphite reference has a 1-μ particle size, and the carbon samples are ground and passed through a 325-mesh screen. The carbon is washed with triply distilled water, and that portion which decants off after a reasonable settling time is discarded. The range of particle sizes thus selected is about 10-40 μ. The particle size is dictated by experimental considerations; the surface coverage on the MIRS crystal is determined by packing considerations, and it was necessary to analyze the solutions in the p-nitrophenol adsorption studies by colorimetric methods.

Because of some difference in the extinction coefficients for graphite and activated carbon, and because the evanescent wave depth of penetration increases with wavelength, the spectra exhibit a well-defined sloping base line. The spectra reported here have been corrected for the base line.

3. RESULTS AND DISCUSSION

    As an example, the results of a study on a sample of Nuchar
C-115, a lignin based carbon obtained from West Virginia Pulp and
Paper Co., are reported here. Spectra were obtained from: (i) the
sized, washed, and dried (to constant weight at 200 °C) original
material, (ii) a portion of (i) which was used for obtaining an
adsorption isotherm of *p*-nitrophenol and subsequently dried at
200 °C, and (iii) a portion of (i) which was subjected only to
the pH 2 solution of HCl used in (ii) to stabilize the neutral
species of *p*-nitrophenol. The original sample was also subjected
to elemental analysis (Spang Microanalytical Laboratory, Ann
Arbor, Mich.) and found to contain 92.1% carbon, 1.3% hydrogen,
0% nitrogen, and 1.1% sulfur, with an ash content of 0%. The
amount of oxygen present was determined in this laboratory by
triton activation, giving 12.2% oxygen.[11] (The higher value for
the oxygen content is probably due to tightly bound H_2O_2.)

    Figures 2 and 3 show 5X scale expansion spectra of both
samples. Some of the absorption peaks observed in these spectra
correspond to those observed in the diffuse reflectance spectra
of Garten and Weiss,[1] Brown,[2] and Friedel and Quieser.[3] Because
internal reflection spectra, in the case of particulate matter

Fig. 2. IRS spectrum of Nuchar c-115 (5X scale expansion).

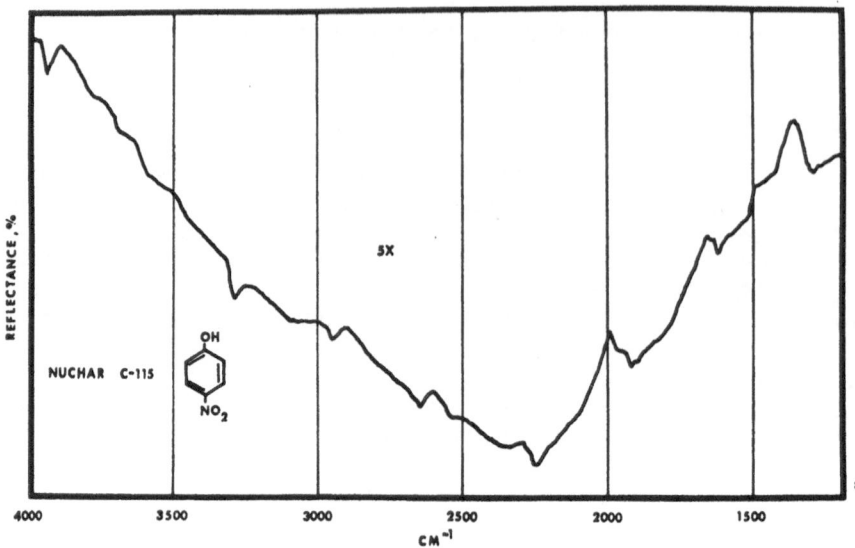

Fig. 3. The IRS spectrum of Nuchar C-115 after adsorption of 1.6 mequiv *p*-nitrophenol from pH 2 solution (5X scale expansion)

significantly larger than the depth of penetration of the evanescent wave, are representative of surface structures, this supports the contention that the spectra obtained by previous workers are actually from diffuse reflectance, containing a large amount of forward scattered light, rather than transmission alone.

The absorption of 3300 cm^{-1} can be seen in both samples (i) and (ii), and the size of the band in sample (ii) shows some increase over that in (i). These results support assignment of this band to phenolic —OH, but upon thorough drying of sample (ii) (not shown), this band is observed to disappear. Although the latter fact suggests that the absorption is caused by tightly bound water, this would require the presence of some chemical mechanism which disposes of the —OH hydrogen on the *p*-nitrophenol upon adsorption.

Broad bands at 2050-1800 cm^{-1} and 1250-1150 cm^{-1} in Figure 2 are not present after adsorption of *p*-nitrophenol, nor do they show up in the spectrum of the sample (iii) which was exposed only to the pH 2 HCl solution. The loss of the first band is shown in Figure 4, and of the latter in Figure 5. The appearance of the two bands together before exposure to acid, and their disappearance after adsorption suggest either an interaction with the acid or with the *p*-nitrophenol. Investigation of sample (iii) showed that the higher energy band was eliminated by acid

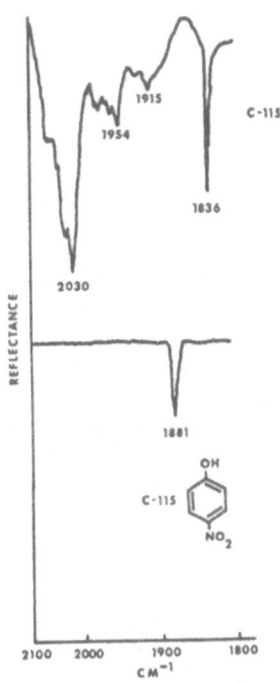

Fig. 4. 2100-1800 cm^{-1} region, top spectrum of Nuchar C-115 before treatment, bottom spectrum obtained after adsorption of p-nitrophenol (20X scale expansion).

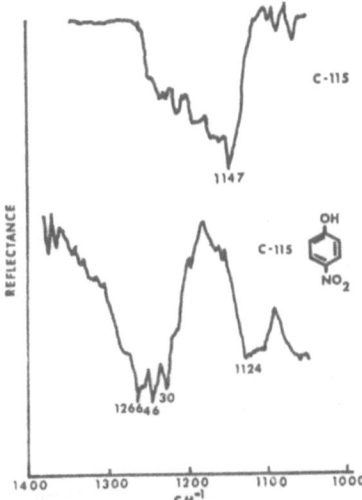

Fig. 5. 1400-1050 cm^{-1} region, top spectrum of Nuchar C-115 before treatment, bottom spectrum obtained after adsorption of p-nitrophenol (20X scale expansion).

202

treatment. Acid hydrolysis of a lactone group would produce the
spectral change observed. The fate of the lower band escaped
detection because of experimental difficulties. It is conceivable
that the lower band represents the C—O—C of an ester, and the
higher band the C=O of the same ester. The position of this
carbonyl absorption is high (2030 cm^{-1}), and represents a 600-cal/
mole shift over the energy of an ordinary lactone carbonyl group
(1820 cm^{-1}).

Figure 6 shows the detail around 2700 cm^{-1}, attributable to
a C—H bond shifted to lower energy than normal, quite probably
as a result of its proximity to the carbonyl groups observed.
This peak is unaffected by the adsorption of either acid or
p-nitrophenol. Figure 7 shows the strong peak occurring at
1600 cm^{-1}. In previous investigations,[1] this has been attributed
to ring vibrations of the graphite-like bulk structure, but in
light of what has been presented here, it is more likely that it
represents a vinyl-like C—C functional group. The strong peak at
1836 cm^{-1} in sample (i) is observed to shift to 1865 cm^{-1} in
(iii) (not shown), and to shift to 1881 cm^{-1} in the p-nitrophenol
isotherm sample (Figure 4). This band is attributed to another
type of carbonyl C O group, but the overall functional group
cannot be readily identified because all of the C=O groups on the
surface studied exhibit unusual vibrational energies. The three
peaks observed in Figure 5 after adsorption are caused by ring
vibrations in the p-nitrophenol, and the C—O of the phenolic O—H

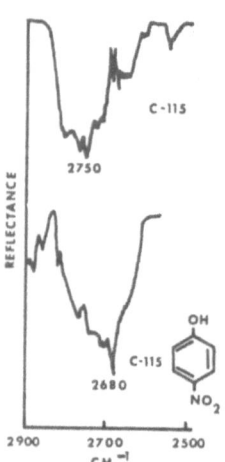

Fig. 6.   2700 cm^{-1} absorption band due to C—H in close proximity
to carbonyl group; note band is not affected by adsorption of
p-nitrophenol (20X scale expansion).

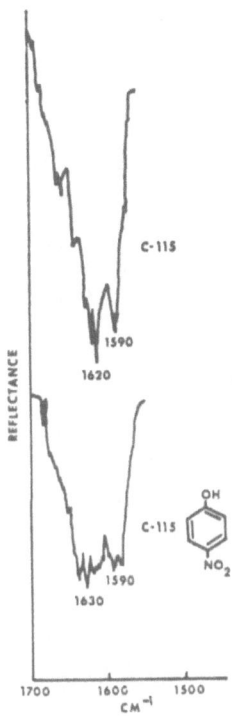

Fig. 7. 1600 cm^{-1} band due to vinyl-like C—C absorption; note that absorption band is unaffected by adsorption of p-nitrophenol (20X scale expansion).

on the p-nitrophenol causes broadening of the 1250 cm^{-1} peak in that group.

A detailed study of the spectra of a variety of different activated carbons and carbon blacks and the effects of various adsorbates on their surface structure is in progress. It is hoped that these studies will indicate which functional groups are involved in chemisorption processes and will shed light on the nature of the bonding involved. Correlations of the surface composition of functional groups and effectiveness in adsorption processes with the preparation conditions are being studied. As noted in Lecture 2, Session II, these studies have already indicated that the adsorptive interactions of aromatic hydroxyl and nitro-substituted compounds with activated carbon are specific adsorption processes resulting from the formation of donor-acceptor complexes of the organic molecule with surface carbonyl oxygen groups, with adsorption continuing after these sites are exhausted via complexation with the rings of the basal planes of the carbon microcrystallite.

REFERENCES

1. Garten, V. A., Weiss, D. E. and Willis, J. B., Aust. J. Chem., 10, 295, 1957.
2. Brown, J. K., J. Chem. Soc. (London), 744, 1955.
3. Friedel, R. A. and Queiser, J. A., Anal. Chem., 28, 22, 1956.
4. Ergun, S., in Chemistry and Physics of Carbon, Vol. 3, Philip L. Walker, Ed., Marcel Dekker, Inc., New York, 1968, p. 45.
5. Fahrenfort, J., Spectrochim. Acta, 17, 698, 1961.
6. Harrick, N. J. and Riederman, N. H., Spectrochim. Acta, 21, 2135, 1965.
7. Harrick, N. J., Internal Reflection Spectroscopy, Interscience, New York, 1967.
8. Stratton, J. A., Electromagnetic Theory, McGraw-Hill Book Co., New York, 1941, p. 490.
9. Hansen, W. N., J. Opt. Soc Amer., 58, 380, 1968.
10. Wilks Scientific Corporation, South Norwalk, Conn., Internal Reflection Spectroscopy, Vol. 1, 1965.
11. Mattson, J. S. and Mark, H. B., Jr., Department of Chemistry, University of Michigan, Ann Arbor, Mich., unpublished results, 1968.

DETERMINATION OF PARAMETERS IMPORTANT IN DYNAMIC ADSORPTION

J. Medema

Chemical Laboratory TNO, Rijswijk, The Netherlands

ABSTRACT. From a model describing the adsorption process in a dynamic system it is possible to derive which parameters are important
in an adsorption process. Attention is paid to the determination
of the size of the adsorbent particles, the porosity of the bed
and the porosity of the particles. With respect to the porosity
several methods for the determination of the pore size distribution are discussed. An attempt has been made to calculate various
diffusion parameters influencing the adsorption process, such as
axial dispersion, diffusion in the macro pores and diffusion in
the micro pores. The rate of adsorption and the equilibrium adsorption constant is dealt with. Finally the discussed methods are
applied to the dynamic adsorption of benzene onto dry charcoal:
the results are in agreement with the experimentally found adsorption.

1. INTRODUCTION

In general it is rather easy to establish the adsorption of a gas
onto an adsorbent in a static system (1) . The quantity adsorbed
(σ) is related to the residence time of a molecule on the adsorbent surface (t), the number of collisions of the gas molecules
with the surface per unit time and unit surface area (n) and the
total surface area (o)

$$\sigma = t.n.o.$$

According to De Boer (1) the time a molecule spent on the surface
is directly related to the heat of adsorption according to the
equation

$$t = t_o . e^{Q/RT}$$

In which Q is the heat of adsorption and t_o a vibration constant comprising the entropy change in the act the adsorption. On the basis of this fundamental equation it is possible to derive more practical equations such as the Langmuir, Freundlich and even the BET isotherm equation.

The adsorption of a flowing gas in fixed bed adsorbers is subject to much more complicated equations. In this case one is forced to set-up the differential equations describing the subsequent steps in the adsorption process. Generally these differential equations are too complicated to permit an analytical solution. Two different lines of appraoch remain: first, the introduction of simplifications to make the desired analytical solution possible, and second, a numerical solution of the differential equations. The latter line of approach may become inevitable when the former approach yields equations which fall short to describe the adsorption process due to oversimplifications.

In the past several approximative solutions have been given (2) , but they all suffer from oversimplification and, even more important, the consequences of the approximations made can hardly be derived from a comparison of the approximative solution with experiment. Therefore, it is worthwhile to incorporate as many parameters as possible. However, to obtain a solution from the complicated differential equations it is necessary to subject the equations to Laplace transformation. The solution is then expressed in so-called Laplace co-ordinates. However, the inverse transformation to normal co-ordinates is extremely difficult and in the case of complicated differential equations impossible. Fortunately, it may serve to arrive at expressions for the so-called statistical moments which are directly related to important characteristics of the adsorption process, such as the inflection point of the adsorption front and the width of the front.

2. ADSORPTION MODEL

In our laboratory we are working with active charcoal consisting of cylindrical pellets. The pellets are composed of small particles which are microporous. The micro pores give rise to a large surface area. The channels between these particles are much larger and we refer to these as macro pores. Openings between the pellets in the bed will named voids. The adsorption on this adsorbent might proceed in the following steps:
1. In the air stream passing through the bed axial dispersion (Eddy diffusion) occurs as a result of the non uniformity of the air stream in the voids between the pellets. By axial diffusion the adsorbate is transported from the gas phase to the

diffusion the adsorbate is transported from the gas phase to the pellets.

2. The adsorbent pellets are supposed to be surrounded by a stagnant film of air through which the gas molecules have to diffuse in order to enter the macro pores.

3. The adsorbate diffuses into the pellets under the influence of a concentration gradient in the macro porous system.

4. Near the walls of the macro porous system adsorption on the surface, or rather in the mouths of the micro pores, may take place.

5. Adsorbate molecules move deeper into the micro pores by means of surface diffusion. Adsorption as well as desorption, taking place during the various steps, are assumed to follow simple first order rate-laws.

Step 5 is of particular importance because it describes the way in which the equilibrium is reached between gaseous molecules and adsorbed molecules. In our model the driving force is not the difference in equilibrium amount adsorbed and a lower lying point of the adsorption isotherm but the difference in covered and non-covered surface. This implies that the equilibrium adsorption is not reached through equilibrium points on the isotherm but straight from zero to the equilibrium amount adsorbed.

In step 3-5 a division is made in the macro porous and micro porous system. This is necessary when the gas phase diffusion in the macro porous system is much faster than other types of transport, for instance diffusion along the external and internal surface. If these latter processes are rate-determining steps 3-5 should be represented by one step comprising the diffusion along the surface. This step is similar to step 5 except that it holds now for the pellets instead of the smaller particles.

Step 1 and 2 are in fact two parallel processes. Both transfer the adsorbate from the gas stream to the outer surface of the pellet. Therefore, these steps must be add in parallel instead of in series as is done with the other steps.

Each of the five steps can be represented by a differential equation and a set of boundary conditions. By applying the appropriate mathematics a solution in Laplace co-ordinates is obtained. From this solution the equation for the inflection point of the adsorption front and the equation for the width of the front or the Height equivalent to a Transfer Unit may be derived (3). In fact these equations are surprisingly simple in view of the considerable complexity of the model. The mean of the adsorption front assumes the form:

$$\mu_1 = \frac{L}{V} + \frac{L}{V} \; Eg \; (\; 1 + E_p H) \tag{1}$$

Taking into consideration the remarks about the various steps the HTU becomes

$$HTU = \frac{1}{\frac{v}{2D_g} + \frac{3E_g k_g}{vd_p}} + \frac{vd_p^2}{30E_g D_p} + \frac{vd_m}{3E_g E_p k_p} + \frac{vd_m^2}{30E_g E_p HD_m} \qquad (2)$$

or provided that surface diffusion is rate controlling

$$HTU = \frac{1}{\frac{v}{2D_g} + \frac{3E_g k_g}{vd_p}} + \frac{vd_p^2}{30E_g E_p HD_m} \qquad (3)$$

The meaning of the various symbols is given in Table I.

From these equations a list of parameters which are supposed to be important in dynamic adsorption can be made up. In the next sections it will be tried to calculate these parameters or to find measuring methods for them.

symbols	units	description
Table I		
Parameters influencing the adsorption of a fixed bed		
d_m	m	diameter of the micro porous particles
d_p	m	diameter of the pellets
E_g	–	porosity parameter of the bed
E_p	–	porosity parameter of the pellets
D_g	$m^2 s^{-1}$	coefficient of axial dispersion
D_m	$m^2 s^{-1}$	diffusivity in the micro pores
D_p	$m^2 s^{-1}$	diffusivity in the macro porous system
k_g	$m s^{-1}$	mass transfer coefficient from the stream to the pellets
k_p	$m s^{-1}$	rate constant of adsorption from macro to micro pores
v	$m s^{-1}$	real velocity of the carrier gas in the bed
L	m	bedlength
H	–	equilibrium constant, adsorption capacity parameter

3. BED AND PARTICLE DIMENSIONS

The bedlength, L, is a parameter which can easily be measured.
For the determination of the pellet size it must be remembered
that the model assumes spherical particles. When pellets of other
shape are used an effective diameter must be used. This effective
diameter can be derived in two ways: at first it can be calculated according to the assumption of Wilke and Hougen (4) that the
outer surface area presented by the pellet must be equal to the
surface area of a hypothetical sphere, secondly it can be derived
from the hydraulic pore radius and the external porosity, which
can be found from pressure drop measurements at various flows.
We are using charcoal pellets with a diameter of 1 mm and a length
of 3 mm. According to the first method the effective diameter is
calculated.

$$d_p = d_{cylinder} \sqrt{2}$$

The pressure drop method will be treated later, it gives approximately the same result.

There are no well-defined methods to determine or to calculate d_m, the size of the particles in the pellets. Probably the
best way to obtain a reasonable value is to inspect the pellets
under a microscope and make a good estimate about the size of the
particles. Another method sometimes used is to crush the pellets
slightly and to determine the particle size of the powder. In the
case of non-porous particles which are compressed together the
particle size can be easily derived from the total surface area
minus the outer surface of the pellets. However, in this case
problems arise with the roughness factor of the particle surface.
Sometimes the particle size is equivalent to the crystallite size
which can be derived from diffraction experiments. However, in
general there is no exact method for the determination of d_m especially in the case of charcoal.

4. POROSITY PARAMETERS OF THE BED AND PARTICLES

The porosity of the bed is easily derived from the difference in
density the pellets assume in the bed and in mercury (5). The
porosity parameter E_g is defined as the ratio in bed volume and
void volume minus 1. It might be difficult sometimes to determine
the mercury density, an example of a handy apparatus is shown in
Fig. 1. The use of other more handsome liquids should be avoided
because even mercury penetrates into the cracks and larger pores
of the pellets giving a wrong density value and as a result a
wrong value of the porosity parameter. Therefore, an even better
way to determine the porosity is directly from a flow experiment.
According to Ergun (6) the pressure drop ΔP per unit bedlength L

is related to the space velocity v_o according to:

$$\frac{\Delta P}{Lv_o} = A + Bv_o \qquad (4)$$

In which A and B are constants both containing the pellet diameter and the porosity. From a plot of the left hand side of eq. (4), versus the space velocity v_o the slope and intercept yield two equations with two unknowns and the porosity as well as the particle diameter can be found. For the charcoal it was found that the effective diameter is about 1.5 times the diameter of the cylindrical pellets. This value is in good agreement with the one calculated. The agreement between the mercury porosity and the pressure drop porosity is even better. The difference amounted to 10%.

The real flow in the bed v, is defined as the space velocity divided by the bed porosity, or the volume flow divided by the product of bed porosity and cross section of the bed. Once knowing E_g it is easy to find v.

As with the diameter the porosity of the pellets is much more difficult to obtain from direct methods such as density or pressure drop methods. One can try to do measurements on single pellets (7) but a much more elegant method is to use eq. (1) for a gas with H = 1.

$$\mu_1 = \frac{L}{v} + \frac{L}{v} E_g (1 + E_p H)$$

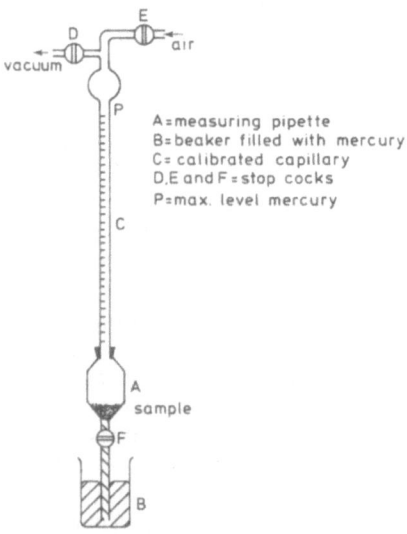

A = measuring pipette
B = beaker filled with mercury
C = calibrated capillary
D, E and F = stop cocks
P = max. level mercury

Fig. 1 Mercury density pycnometer.

H is the ratio in concentration inside and outside the micro pores. The condition H = 1 is fulfilled when a non-sorbing gas is fed to the adsorber. In practice the experiments are carried out as follows. A charcoal bed is supplied with hydrogen carrier gas. Then a pulse of deuterium or helium is injected into the carrier gas stream. The time in which the centroid of the pulse pass through the column can be substituted in eq. (1), and E_p can be calculated from the other known quantities. In order to establish whether the gas is non-sorbing an experiment at a higher temperature needs to be performed.

The list of parameters set-up in the introduction is based on a model describing the adsorption in a system containing micro pores. In fact the whole adsorption is assumed to take place on the very large surface of the micro pores. A definite other model should be used when a considerable part of the adsorption takes place on the surface of larger pores (8) . In the context of this lecture it should go to far to give all the arguments for the discrimination between micro-, transition- and macro pores but it should be stressed that the transport mechanism in micro pores is mainly by surface diffusion and the macro pores are filled up mainly by flow through the gas phase and adsorption directly from the gas phase (9) . In the case of transition pores both mechanisms will work. Which mechanism actually works does not depends on the pore size only but also on the ratio in concentration adsorbed and concentration in the gas phase. The higher the adsorption the more important the surface diffusion. However, to be able to apply a given model the pore size distribution should be known.

Three methods are available to determine the pore size distribution. The method generally applied for larger pore systems is based on the penetration of mercury into the pores under pressure (10) . Fig. 2 shows an example of a measurement upto 30.000 atmosphere pressure. It is seen that only pores above a certain size are penetrated and the smaller pores cannot be measured in this way. Moreover, the concept underlying the method is not well-suited for the smaller pores even when the mercury pressure is increased above 30.000 atmospheres (11) .

A second method is based on the Kelvin equation (12) . In this equation the vapour pressure is corrected for the curvature of the surface. In small pores capillary condensation takes place at pressures below the saturated vapour pressure. Due to the surface tension the surface of the liquid in the pores will be curved. This gives rise to an additional energy factor when the molecules evaporate from the liquids in the pores. As a result the pores are emptied on desorption at a lower pressure then the pressure at which the filling with liquid is completed. This effect is known as the hysteresis of adsorption isotherms. The way in which the pore size distribution can be calculated from the Kelvin equation is described in detail in the literature (13) . The method

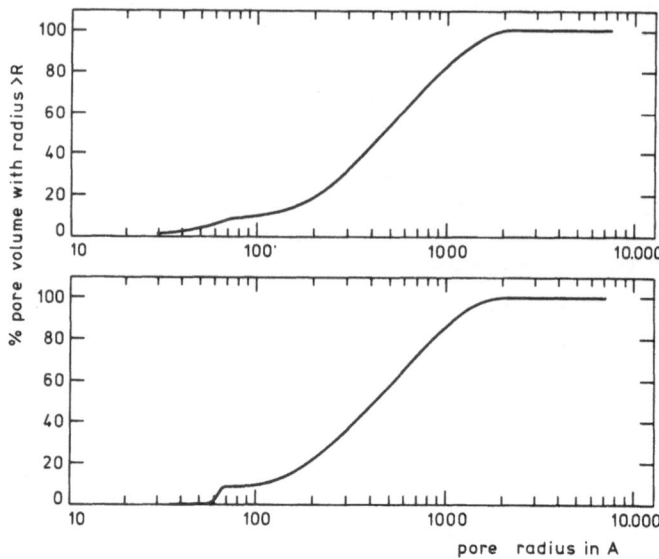

Fig. 2 Cumulative pore size distribution from mercury penetration.

is rather time consuming and as far as any check on the reliability of the method was possible it appeared to give reasonable result in the region of transition pores (30-100 Å) (14) . The application of the method to micro pores systems is highly questionable because the dimensions of the pores (10 Å) is comparable to the size of the molecules (4 Å) used and one can hardly speak of surface curvature when 2 or 3 molecules are involved.

The micro pores which are already filled in the low pressure region of the adsorption isotherm (the BET monolayer point corresponds to a pore width of approximately 8 Å) can be derived from a so-called V-t plot (15) . The quantity adsorbed is plotted versus the thickness of the adsorbed layer instead of the relative pressure as is normally done for isotherms. In order to convert the pressure axis into the thickness an adsorption isotherm of a non-porous solid is used. The thickness of the adsorbed layer on the non-porous solid is derived from the volume adsorbed and the monolayer capacity.

The method is demonstrated in Fig. 3. First the isotherm on the non-porous solid is measured and the surface area is calculated using the BET method. Next the volume adsorbed is converted into a thickness and a graph of the thickness versus the relative pressure is obtained. With the help of this graph the pressure axis of the isotherm of a porous solid is converted. A plot of results obtained with charcoal is shown in Fig. 4. The nearly horizontal part at t-values exceeding 6 Å indicates that mainly micro pores are involved here (5) .

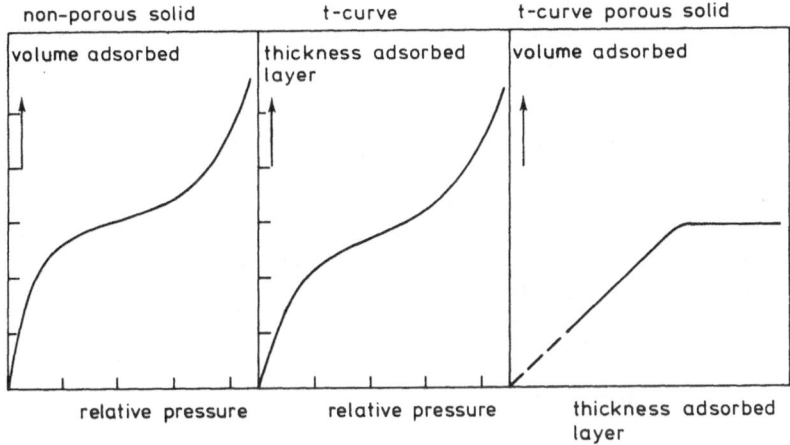

Fig. 3 Conversion of an isotherm in a so-called V-t plot.

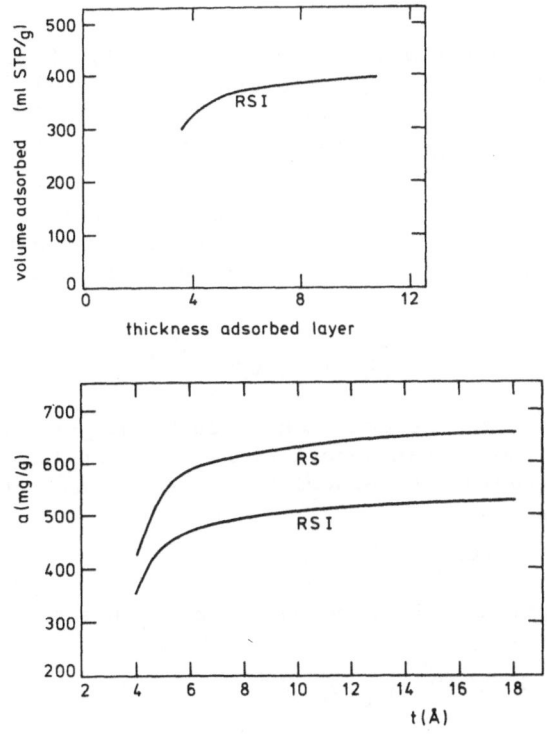

Fig. 4 V-t plot for nitrogen (top) and benzene (bottom) on charcoal.

A similar result can be obtained when the average pore radius is derived from the ratio of pore volume and surface area. When cylindrical shaped pores are assumed the relation between surface area and pore volume is (5) :

pore radius = 2x pore volume / surface area.

The pore volume directly follows from the densities in mercury and toluene as imbibition liquids, toluene will penetrate the pores whereas mercury will not. It has no use to apply much smaller molecules as imbibition liquids because pores accesible to relative large molecules such as benzene and toluene are of interest only. When this method is applied to charcoal it appears that this material has an average pore diameter of 10 Å.

5. DIFFUSIVITY PARAMETERS

To relate the coefficient of axial dispersion D_g to flow and particle size the following expression has been accepted (16) :

$$D_q = D_o/q + \lambda d_p v$$

The tortuosity factor q is in the order of 1.5-2 (17) and the contribution of D_o may be neglected when the particles and the flow are not too small. The coefficient λ amounts to 0.8 when Re (Reynolds) >10 and d_p >0.5 mm. If these conditions are fulfilled D_g can be calculated.

Thoenes (18) has investigated the mass transfer in packed beds and has formulated the dependence of k_g upon the velocity v and the hydraulic radius of the macro pores. He arrives at the relation:

$$k_g \, d_h/D_o = 0.7 \left(\frac{d_h v}{\nu} \right)^{1/2} (\nu/D_o)^{1/3}$$

which is similar to the relation given by Hougen and Watson (19). The factors ν and D_o can generally be found in the literature. The hydraulic radius is related to d_p and E_g according to

$$d_h = d_p/6E_g$$

For the adsorption of benzene on charcoal this expression leads to:

$$k_g = 1.15 \cdot 10^2 \, (v_o/d_p)^{1/2}$$

According to the kinetic theory of gases a diffusion coefficient is a function of the average velocity of the molecules $\bar{v}$

and the mean free path λ:

$$D = 1/3 \ \bar{v}.\lambda.$$

In pores where the dimensions are of the same order as the mean free path the collisions with the pore walls will happen more frequently than the collisions with each other. Therefore, the diffusion coefficient will change. Finally in very small pores the diffusion coefficient is dependent on the pore radius r and the average velocity of the molecules $\bar{v}$ according to:

$$D_m = \frac{2}{3} \ r.\bar{v}$$

In between these pore regions an over-all equation covering both effects is necessary (20). In the macro porous system the pores are very large (>10.000 Å) and the diffusion coefficient may be taken as the gas phase diffusion coefficient. In cases where the macro pores are smaller corrected equations should be used.

The actual values of the diffusion coefficient can be derived from measurements with single adsorbent pellets. From these measurements it can also be derived which type of diffusion mechanism is operative. In the literature there are only a few of these experiments described (21) and it is therefore not possible to say how accurate the calculated diffusion coefficients are.

6. RATE OF ADSORPTION AND ADSORPTION CONSTANT

Uptill now we have been able to determine or to calculate all constants in the equations independently except for the adsorption constant H and the rate constant for the transport from the macro pores into the micro pores k_p. From eq. (2) it follows that k_p can be found from an experiment with a non-sorbing gas (H = 1). However, in that case all the uncertainties introduced in calculating or measuring the other constants become visible in k_p. Another way to deal with the problem is to investigate how the broadening of the breakthrough curve depends on d_m. Fortunately nature helps a little in solving this problem. In small pore systems with a large pore volume the total cross section of the pore mouths is large. As the rate of transport to the micro pores is directly related to this cross section the rate of transport is usually high (22). The term containing k_p may be neglected therefore. The situation around the evaluation of k_p is rather unsatisfactory, in practice it is often used for adjustment of the model. At the moment we are investigating other possibilities to say something reasonable about k_p. This is done by studying the initial breakthrough. It should go to far to give all the details, but one may imagine that in the initial breakthrough, which represents the first steps in the adsorption process the influence of the last

term in eq. (2), the diffusion into the micro pores is greatly
diminished. As was pointed out many times (22, 23, 24) the initial breakthrough is an exponential function of time. From our
differential equations this relation can be derived directly and
it was found that the diffusion in the micro pores does not influence the initial breakthrough. However, this is still in the
experimental state and at the moment we cannot confirm this idea.
Hereafter we will assume that the rate of transport to the micro
pores is fast, consequently the term containing k_p may be omitted.

In most models describing the adsorption from a gas flow onto
a fixed bed adsorber a linear, quadratic or Langmuir isotherm driving force for the adsorption is assumed. The linear driving force
is justified by assuming that the adsorption takes place according
to a linear isotherm (25) . In fact equilibrium between the concentration in the gas phase and quantity adsorbed is assumed for
every concentration. So during the rise of the concentration in
the adsorption front the amount adsorbed follows the adsorption
isotherm. If the adsorption takes place in this way it is easily
seen that H is a constant only in the case of a linear isotherm.
Unfortunately most isotherms happen to be curved and H is not a
constant any longer. This introduces a considerable complexity in
the differential rate equations and uptill now only approximated
solutions for these cases are given.

Looking at the actual adsorption in a micro porous system it
is very unlikely that the establishment of equilibrium will follow
the adsorption isotherm. Once adsorption has taken place in the
pore mouth, the equilibrium between the quantity adsorbed in the
pore mouth and in the gas phase is rapidly established. The pore
mouths are of the dimensions of a few molecules and a monolayer
adsorption narrows the opening to such an extent that contact between gas phase and surface inside the pores is lost. The total
quantity adsorbed is now reached by diffusion from the pore mouth
into the interior of the pores.

In the first case equilibrium is attained through a gradual
rise in concentration during which every part of the surface is
in equilibrium with the gas phase. In the second case there is a
steep rise in the gas phase concentration from zero to final. A
small part of the surface is in equilibrium with this concentration. The final equilibrium is attained by gradually filling up
the non-covered surface area. The important implication of this
adsorption mechanism is that the adsorption capacity is always
consumed in a linear way and the capacity constant H is a real
constant for one concentration. Moreover, the constant can be derived directly from the adsorption isotherm as it is simply the
total quantity adsorbed at the concentration used.

7. APPLICATION OF THE MODEL

At this point we have methods available to obtain all the necessary constants for the calculation of the mean of the breakthrough curve and the broadening. In order to see whether our model and assumption are correct the model is applied to the adsorption of benzene from air onto dry active charcoal. For the calculation of the mean, bedlength L, real velocity v and E_g are determined in the way described before. The term E_pH is derived from the adsorption isotherm by means of the relation:

$$W = \frac{C_o}{\rho_p} E_p H$$

In which W is the quantity adsorbed in kg adsorbate per kg adsorbent at the concentration C_o and ρ_p is the density of the adsorbent. If these values are inserted in eq. (1) the following inflection points in the breakthrough curves are calculated under various conditions. In Table II these inflection points are compared to the actual observed values. The agreement is satisfactorily.

By comparing experiments with different bedlength, different velocities, different porosities and different particle sizes, it was found that all parameters appear in the right way in eq. (1).

It is far more difficult to obtain equally well fitting data for the broadening of the curve. First a decision has to be made which of the two equations for the HTU is applicable in the case of benzene adsorption onto charcoal. If the values for the velocity, the pellet size, the bed porosity and the gas phase diffusion coefficient are inserted in the third term of eq. (2) it appears that the calculated broadening is much larger than the experimentally observed one. As a result eq. (2) does not describe the broadening of the breakthrough curve and eq. (3) must be tried. These equation consists of two terms, the first term describes the contribution of external mass transport to the broadening and the second term gives the broadening due to transport limitations in the interior of the pellets. The external HTU contribution can be calculated rather exactly. The axial dispersion term amounts to $1.6\ d_p$. The transfer term becomes approximately $8.v^2.d_p^{3/2}$. Which of the two terms predominates depends on the velocity. In our case the velocities were such that the axial dispersion term has a minor influence only.

The second part of equation 3a can be easily evaluated from the adsorption isotherm (E_pH), the bed porosity E_g, the real velocity v, the pellet size d_p and the diffusion constant in the pores D_m from the equation given by Wheeler (20) :

$$D_m = 2/3.r.\bar{v}$$

Table II

Comparison of calculated and experimental values of t_m

No. exp.	L (m)	ϕ_v 1/h	C_o mg/l	calcul. t_m (sec)	expt. t_m (sec)	% deviation
40	0.030	300	42.7	1070	1070	0.0
59	0.015	2000	3.85	640	640	0.0
61	0.015	2000	2.85	830	800	− 3.8
76	0.060	600	5.15	6800	6700	− 1.7
89	0.060	1000	4.05	4800	4700	− 2.1
108	0.030	1000	4.22	2350	2430	− 3.7
111	0.060	600	4.05	6800	6740	− 0.9
112	0.015	600	4.05	1700	1720	+ 1.2
115	0.030	600	4.05	3400	3420	+ 0.6

In order to see whether equation 3a describes the broadening of the breakthrough curve satisfactorily the velocity dependence and the pellet size dependence was evaluated from experimental
curves. The relation according to the model is:

$$\text{HTU} = a\, v^{1/2} d_p^{3/2} + b\, v d_p^{2} \tag{3a}$$

In Fig. 5 the HTU/v is plotted versus $1/\sqrt{v}$. Straight lines
are obtained indicating that the velocity is incorporated in the
right way in eq. 3a. Moreover, the constants appearing in eq. 3a
can be evaluated from this graph and it appears that there is a
satisfactory agreement between the experimentally observed and
the theoretically calculated values. This agreement is shown in
Table III.

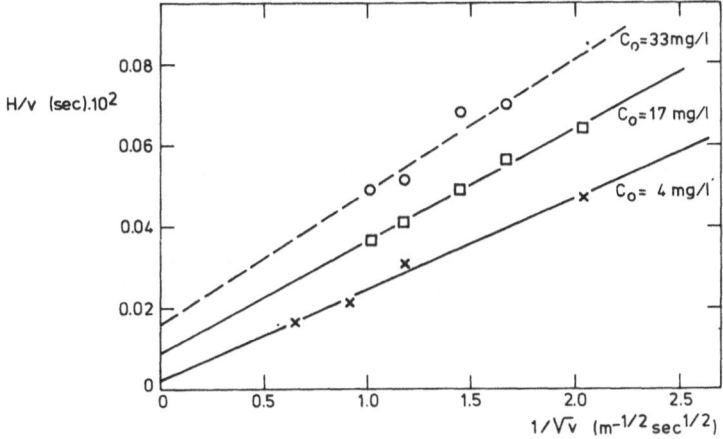

Fig. 5 Correlation between HTU and velocity at three concentrations.

Table III
Comparison of experimentally observed and calculated values for the constants in eq. (3a).

c_o mg/l	a $(sec^{1/2}\,cm^{-1})$ cal.	obs.	b $(sec\,cm^{-2})$ cal.	obs.	
33	$5.7 \cdot 10^{-2}$	$6.5 \cdot 10^{-2}$	$1.6 \cdot 10^{-2}$	$1.6 \cdot 10^{-2}$	Fig. 5
17	5.7	6.1	$0.9 \cdot 10^{-2}$	$0.9 \cdot 10^{-2}$	Fig. 5
4	5.7	5.5	$0.26 \cdot 10^{-2}$	$0.27 \cdot 10^{-2}$	Fig. 5
4	5.7	5.2	$0.26 \cdot 10^{-2}$	$3.0 \cdot 10^{-2}$	Fig. 6

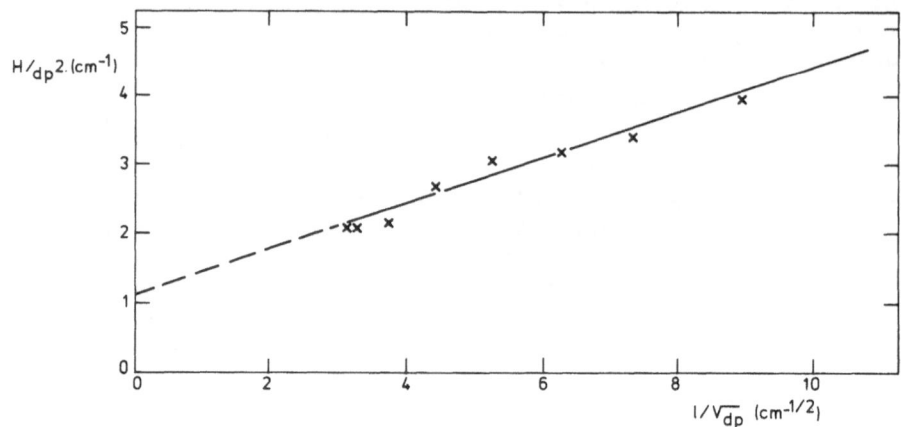

Fig. 6 Correlation between HTU and pellet size.

In Table III the values evaluated from the pellet size dependence shown in Fig. 6 are given also. The charcoal had adsorbed some water from the ambient in this case and therefore there is an additional transport limitation in the interior of the pores. The value of D_m had to be corrected. But it appears that also in this case the pellet size is incorporated in the right way and that the contribution of the external term to the broadening is approximately the same.

8. CONCLUSIONS

If a model for the adsorption in a fixed bed is assumed, which in particular accounts for the influence of the micro porosity of the adsorbent, it is possible to arrive at a breakthrough curve which can be calculated from independent found parameters. In Fig. 7 a calculated and observed breakthrough curve are compared. Besides the external mass transfer limitations the transport inside the pellets is governed mainly by surface diffusion as appeared from the validity of eq. 3a. This is probably due to the high adsorptive power of charcoal for benzene. Transport along the surface is faster than transport through the gas phase.

REFERENCES

1. de Boer, J.H., The Dynamical Character of Adsorption, Clarendon Press, Oxford, 1953.
2. Bohart, G.S., Adams, E.G., J. Am. Chem. Soc., 42, 523 (1920)
3. van Dongen, R.H., Stamperius, P.C., Report Chem. Lab. TNO, 1974 no. 6.
4. Wilke, C.R., Hougen, O.A., Trans. Am. Inst. Chem. Engrs., 41, 445 (1945).
5. van Aken, J.G.T., Thesis, Delft, 1968.
6. Kramers, H., Fysische transportverschijnselen, lecture book, Delft, 1961.
7. Carman, P.C., Raal, F.A., Flow of Gases through Porous Media, Butterworths, London, 1956.
8. Carman, P.C., Trans. Faraday Soc., 222A, 109 (1954).
9. Haul, R.A.W., Naturwissenschaften, 41, 255 (1954).
10. Ritter, H.L., Drake, L.C., Ind. Eng. Chem. Anal. Ed., 17, 782 (1945).
11. Gregg, S.J., Sing, K.S.W., Adsorption Surface Area and Porosity, Academic Press, London, 1967 p 184.
12. Thomson, W.T., Phil. Mag. 42, 448 (1971).
13. Gregg, S.J., Sing, K.S.W., Adsorption Surface Area and Porosity, Academic Press, London, 1967 Chapter 3.
14. Dubinin, M.M., et. al., Russ. J. Phys. Chem., 34, 959 (1960).
15. Lippens, B.C., Thesis, Delft, 1961.
16. Kramers, H., Niet-stationaire kolomprocessen, Lecture book, Delft, 1960.
17. Carman, P.C., Flow of Gases through Porous Media, Butterworths, London, 1956, p 48.
18. Thoenes, D., Thesis, Delft, 1956.
19. Hougen, O.A., Watson, K.M., Chemical Process Principles, III Kinetics and Catalysis, John Wiley, New York, 1943 p 985.
20. Wheeler, A., Adv. in Catalysis and related subjects III, 1951 p 266.
21. Barrer, R.M., Diffusion In and Through Solids, Mac Millan, New York, 1941.
22. Canjar, L.N., Kostechi, J.A., Physical adsorption processes and principles, Chem. Eng. Progress, Symposium Series 74, vol 63, 1967.
23. Danby, C.J., et. al., J. Chem. Soc., 1946, 918.
24. Wheeler, A., Robell, A.J., J. Catalysis, 13, 299 (1969).
25. Perry, J.H., Chemical Engineering Handbook, 4th ed., Mac Grawhill, New York (1963).

EMPIRICAL TESTS AND CHEMICAL ANALYSES FOR THE CLASSIFICATION AND
QUALITY CONTROL OF COMMERCIAL SORBENTS

John E. Urbanic

Research & Development Department
Calgon Corporation, Subsidiary of Merck & Co., Inc.
Pittsburgh, Pennsylvania   U.S.A.

INTRODUCTION

Commercial sorbents, particularly molecular sieves and activated carbon, have found a wide variety of uses in gas and water
purification. This has created a need for describing the quality
of the sorbents in a meaningful way so that the manufacturer can
control quality during production, and the user can choose the
type and quality of sorbent he desires for his particular application. Tests to determine the fundamental characteristics of
the sorbents, such as pore volume, surface area, and pore size
distribution, are usually too time-consuming to be carried out on
a routine basis. They also require sophisticated equipment and
skilled personnel which are not always available at the production plant or at the customer's facilities.

Direct measurement of a product's ability to adsorb a particular impurity of interest would of course be the ideal situation. In practice, however, a substitute adsorbate is chosen
and the test is carried out under controlled conditions familiar
to both buyer and seller so that both understand the meaning of
the empirical result. During the course of this lecture, a number of empirical tests will be described.

MOLECULAR SIEVES

Commercial molecular sieves are complex crystalline structures consisting of aluminum and silicon oxide tetrahedra. An
exchangeable cation (Na, K, and Ca) provides electrical balance
for the system. When water is removed from the crystal, the

structure becomes a series of cavities or "cages" connected by round pores of smaller diameters. Unlike activated carbon, the pore diameters are uniform and there is essentially no pore size distribution.

The general chemical formula of the molecular sieves are determined by conventional analytical techniques. X-Ray diffraction is used to determine the cage size. Type A molecular sieves have an Al to Si ratio of 1:1 and a cage diameter of 11.4 Angstrom units. Type X sieves have approximately twice the cage diameter and the Al to Si ratio is 1:1.25 or 1:1.5. The connecting pore diameters can be altered by changes in the Al to Si ratio or the nature and valance state of the exchangeable cation. Commercial sieves are produced with pores of 3, 4, 5, 8, and 9 Angstroms.

Molecular Probe Techniques

Adsorption tests are used to verify the pore size of the sieves. In these tests the adsorption capacities of the molecular sieves are determined under conditions of low partial pressures to eliminate the possibility of condensation on the external surfaces. The adsorbates used are pure compounds of known molecular diameter. This approach is commonly called the molecular probe technique.

Conventional volumetric or gravimetric adsorption methods are used to determine the adsorption capacities of the sieves.[1] The sample to be tested is heated to 350° C. for several hours to insure that all water is removed. This operation is frequently performed under vacuum in the same device that the adsorption measurement will be made. The sample is cooled in the absence of water vapor and evacuated. The adsorbate of interest is then introduced into the system and allowed to contact the sample. After the pressure in the system has stabilized and, in the case of a gravimetric system the sample has ceased gaining weight, the amount of adsorbate removed by the sieve is calculated and reported as a percentage of the sample weight. For pore characterization it is necessary to know the molecular dimensions of the adsorbate. A list of compounds with molecular dimensions and sieve types which will adsorb these compounds is given in Table 1. Naturally, sieve types with pores larger than the critical diameter of a molecule will adsorb that molecule; e.g., Type 13X will adsorb all the compounds listed.

The cleanness of the separation (adsorption of a molecule of one size and rejection of a molecule of a larger size) will determine the quality of the molecular sieve being tested.

TABLE 1

Molecular probe classification of molecular sieves

Molecular probe molecule	Critical diameter $\overset{\circ}{A}$	Sieve minimum pore diameter $\overset{\circ}{A}$	
ACETYLENE	2.4	3	(TYPE 3A)
CARBON MONOXIDE	2.8	3	
WATER	3.2	4	
AMONIA	3.6	4	(TYPE 4A)
HYDROGEN SULFIDE	3.6	4	
ETHYLENE	4.2	5	
ETHANE	4.4	5	(TYPE 5A)
METHANOL	4.4	5	
BUTANE TO $C_{22}H_{46}$	4.9	5	
ISOBUTANE TO ISO $C_{22}H_{46}$	5.2	8	
M-XYLENE	7.1	8	(TYPE 10X)
O-XYLENE	7.4	8	
TRIETHYLAMINE	8.4	9	(TYPE 13X)

Conventional adsorption tests are time-consuming and the devices for performing the tests usually can accommodate only one sample at a time. Landolt[2] describes a device to allow the measurement of as many as eleven samples at one time. In this system, the dry molecular sieve samples, in weighing bottles, are placed in a large adsorption chamber. The chamber and a series of adsorbate reservoirs are evacuated to 10^{-3} Torr. Adsorbate is admitted to the chamber through one of the reservoirs. Pressure is controlled at a preset value. Equilibrium is reached in one to three hours, depending on the adsorbate, after which the samples are removed and weighed to determine the amount adsorbed. Another multiple testing system is described by Roelofsen.[3]

Thermal Analysis

Borthakur[4] found that differential thermal analysis might be used as a tool for judgint the quality of molecular sieves. A large endotherm in the temperature range of 120° C. to 400° C. corresponded with the loss in weight due to elimination of zeolitic water. They also noted that the peak temperature of zeolitic water loss varied from 180° C. for a zeolite with 3.5 $\overset{\circ}{A}$ pores to 250° C. for a molecular sieve with 8 $\overset{\circ}{A}$ pores. By further analysis, based on the assumption that the water in the pores adheres uniformly to the zeolite frame, these authors deduced that the endotherm curve for zeolitic water loss should be smooth. Therefore, inflections in the curve could be due to presence of non-zeolitic materials or mixtures of zeolites. The presence of non-zeolitic structure was confirmed by X-ray studies.

ACTIVATED CARBON

Adsorption tests for activated carbon classification can be divided broadly into two types: those that measure adsorption from aqueous solution and those that measure adsorption from gas streams.

Liquid Phase Tests

Iodine. The adsorption of iodine from solution has been used as a measure of the relative adsorptive power of activated carbon. There are several versions of the tests recorded in the literature.[5,6] The method currently used in the Calgon Quality Assurance Laboratory will be discussed here.[7]

A representative sample of the activated carbon to be tested is pulverized so that 95% passes through a No.325 U.S. standard sieve. The pulverized sample is dried at 150° C. for at least three hours and cooled in a desiccator prior to testing. This sample preparation step is common to all liquid phase adsorption tests used at Calgon. A weighed portion (approximately one gram) of the dried, pulverized carbon is placed in a standard taper 250-ml iodine flask. Ten milliliters of 5% HCl are added to the carbon and the flask is swirled to wet the carbon--then heated to bring the mixture to boiling for exactly thirty seconds. This step is necessary to eliminate sulfides which would interfere with the test. When the flask is cooled, 100 ml of 0.1 normal iodine solution is added and the flask is stoppered and agitated for thirty seconds. The mixture is then filtered by gravity through a folded filter paper into a beaker. The first 20-30 ml of filtrate is discarded, a 50-ml aliquot of the remaining filtrate is pipetted into a 250-ml iodine flask and titrated with 0.1 N sodium thiosulfate solution using a starch indicator for the final end point. The milligrams of iodine consumed are calculated from the normality of the filtrate. The "iodine number" is based on the milligrams of iodine adsorbed per gram of carbon at equilibrium with 0.02 N iodine solution. If the normality of the filtrate is less than 0.035 N and greater than 0.008 N, a correction is applied to obtain the iodine number. If the normality of the filtrate is outside that range, the test is repeated with a greater or lesser amount of sample. Iodine number has been shown to correlate empirically with total surface area in pores greater than 10 Å.[8]

Phenol. The early uses of activated carbon in water purification concerned removal of phenols. Consequently, a test was devised to measure the relative efficiencies of activated carbons based on phenol adsorption. An early version, known as Baylis'

phenol value,[6] consisted of treating a 0.1 ppm phenol solution
with various dosages of the pulverized activated carbon under
test. After a two-hour contact time the carbon was removed from
the phenol solution by filtration and the residual phenol concentration of the filtrate was determined. The "phenol value" is
obtained from the isotherm by selecting the carbon dosage in ppm
required to remove 90% of the phenol from solution. This test
was very cumbersome and included distillation and concentration
steps prior to final phenol analysis. A modified version, known
as the "Colebaugh" modified phenol test, is reported by Hassler.[5]
In this test, a 200 ppm phenol solution and 30-minute contact
time are used. Residual phenol is determined by titration with
bromate-bromide solution to convert the phenol to tribromophenol.

Molasses. The tests described above measure the carbon's
ability to adsorb relatively small molecules such as those involved
in taste and odor. Experience has shown that they do not correlate with the carbon's ability to remove molecules such as the
color bodies in raw sugar solutions.

When activated carbon came to be used for the decolorization
of sugar syrup, its relative merits were judged in relation to a
molasses decolorization test already in use for classifying bone
char. The test is quite empirical but does provide a measure of
relative decolorizing power.

In the Calgon version of the method,[7] equal weights of a
standard carbon with an assigned molasses value of 400 and test
carbon are contacted with portions of molasses solution (diluted
black strap molasses solution which has a residual absorbance of
0.35-0.41 measured at 425 nanometers when treated with standard
carbon).

The carbon and molasses solution mixture is heated to boiling for 30 seconds and filtered by vacuum through a Whatman No. 3
filter paper coated with filter paper pulp. The first 20 ml of
filtrate is discarded and the absorbance of the remaining filtrate
is measured against distilled water at 425 nanometers and through
a 2.5 mm effective light path. Molasses number is calculated by
the following formula:

$$\text{Molasses No.} = \frac{400 \times \text{absorbance of std molasses solution}}{\text{absorbance of test solution}}$$

A molasses test described by Hassler[5] ratios the weight of
test carbon required to remove 90% of the color from the molasses
solution with the weight of standard carbon required to accomplish
90% color removal. The ratio is expressed as percent relative
efficiency. This method requires testing a range of dosages so
that 90% removal point can be obtained from the plot of the isotherm.

Methylene Blue and Other Dyes. The instability and non-uniformity of black strap molasses has prompted research to supplant the molasses test as a measure of decolorizing ability. A test using methylene blue dye is employed in judging activated carbons used in water purification applications. The method[7] is similar to other liquid phase tests. A general procedure for dyes is given by Hassler[5].

Methylene blue numbers correlate with surface area in pores greater than 15 Å in diameter, and molasses number roughly correlates with surface area in pores greater than 28 Å in diameter.[8]

Many activated carbon manufacturers quote molasses number, iodine number, and methylene blue number for the liquid phase carbons as a means of empirical characterization of their products.

Vapor Phase Tests

Carbon Tetrachloride Activity and Retentivity. Adsorption tests for vapor phase activated carbon generally measure the activated carbon's total capacity for adsorption of pure organic vapors from air. The carbon tetrachloride activity test is typical of this type of test. It is a good quality control test which can be performed rapidly with reproducible results. The apparatus for the carbon tetrachloride activity test consists of an air drying and pressure regulating section, a vapor saturation section, and a means of metering the vapor saturated air (see Figure 1) into a tube containing a known weight and volume of the activated carbon on test. In Calgon's carbon tetrachloride activity test,[7] a weighed sample tube such as that shown in Figure 2, is filled to the 10-cm mark at a rate of 9 ml/min with granular activated carbon and re-weighed to obtain the weight of the carbon. Carbon tetrachloride is placed in the vapor saturator in an ice bath. Dried air is passed through the saturator at a regulated pressure. The system is allowed to equilibrate for at least thirty minutes prior to the first analysis. At this time the filled sample tube is placed in a 25° C. constant temperature bath and attached to the vapor laden air metering system which is adjusted to a flow of 1,560 ml/min (one l/min/cm² velocity). After thirty minutes the sample is re-weighed to obtain the weight of carbon tetrachloride adsorbed. The sample is replaced on the apparatus for additional 10-minute periods until the weight change for the period is less than 5 milligrams. Carbon tetrachloride activity is expressed as percent by weight carbon tetrachloride adsorbed.

Vapor phase carbons generally have a minimum carbon tetrachloride activity specification.

The method described above will determine saturation capacity of the activated carbon under test for the adsorption of

FIGURE 1
ADSORPTION TESTING APPARATUS
For CCL$_4$ activity and retentivity testing

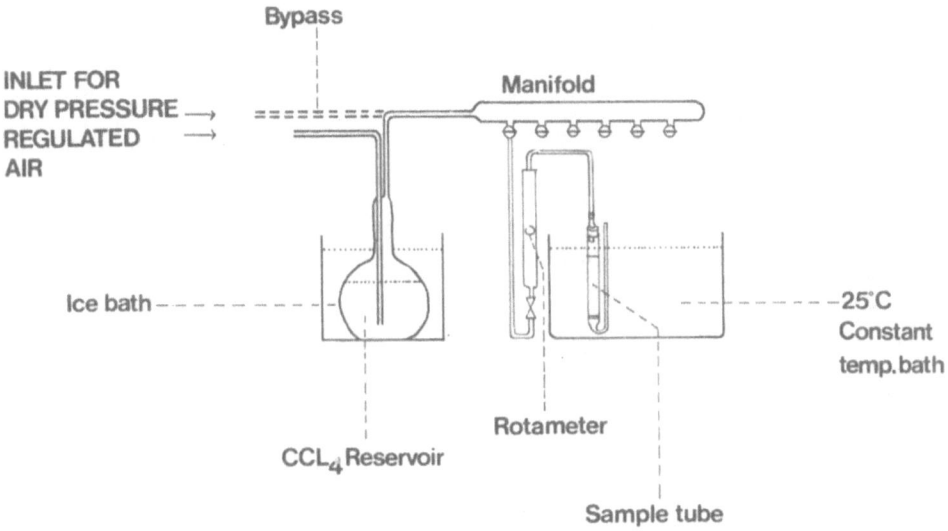

Bypass

INLET FOR
DRY PRESSURE →
REGULATED →
AIR

Manifold

Ice bath

CCL$_4$ Reservoir

Rotameter

Sample tube

25°C
Constant
temp. bath

volatile organic compounds at partial pressures. In some air
purification applications, the activated carbon is required to
adsorb organic compounds at extremely low partial pressures. The
carbon tetrachloride retentivity test[7] is a measure of the carbon's ability to retain the adsorbed carbon tetrachloride when
exposed to a stream of air devoid of the vapor. The test is run
on a sample of activated carbon already saturated with carbon
tetrachloride as above. The sample tube is returned to the apparatus and air is passed through the system, bypassing the carbon
tetrachloride vaporizer. Dried air at 780 ml/min is passed
through the sample for six hours. After this time the sample
tube is weighed and the amount still retained on the activated
carbon is calculated by difference. The retentivity property is
generally related to the ability of the activated carbon to
adsorb organic compounds at low concentrations in air.

Benzene. Benzene capacity can be obtained in a manner
similar to the carbon tetrachloride capacity. In fact, if carbon
tetrachloride activity is known, benzene capacity can be predicted by computing the ratio of the liquid density of the two
solvents at the same temperature of adsorption. At saturation

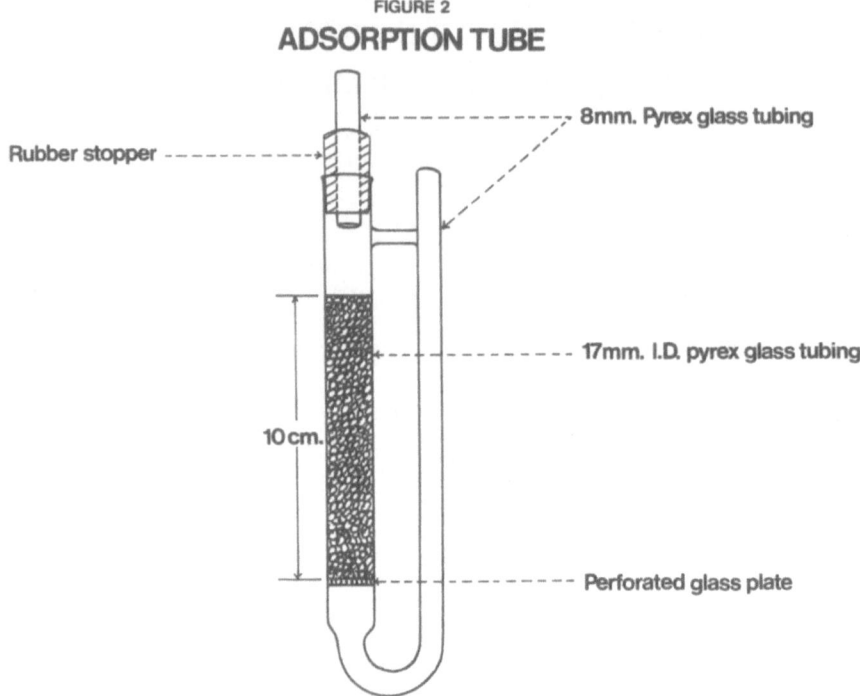

FIGURE 2
ADSORPTION TUBE

Rubber stopper

8mm. Pyrex glass tubing

17mm. I.D. pyrex glass tubing

10 cm.

Perforated glass plate

both compounds would occupy approximately the same pore volume.

In certain cases, knowing the saturation value or retentivity is not sufficient to describe the activated carbon property appropriate for a particular application. In poison gas tests, as will be discussed later, it is necessary to know the point at which the adsorbate will no longer completely adsorb all of the vapors. In other cases, when "in situ" desorption is used and complete desorption is not possible, it is important to know the amount desorbed since this represents the usable capacity or "working capacity" available for subsequent adsorption cycles.

Butane Working Capacity. Use of activated carbon in automobile evaporative loss devices is a case in point. The automobile companies needed a test method to insure that the activated carbon they purchased for the evaporative loss control canisters would have sufficient working capacity to be effective. The test devised to measure this parameter is called "butane working capacity". The apparatus used for this test, shown in Figure 3, is similar to the carbon tetrachloride test, except that a butane cylinder is used in place of a carbon tetrachloride vaporizer. An adsorption tube (Figure 2) is filled with dry carbon as before. The tube, with known weight and volume of carbon, is placed on the apparatus in a 25° C. water bath and butane flow

FIGURE 3

BUTANE WORKING CAPACITY TESTING APPARATUS

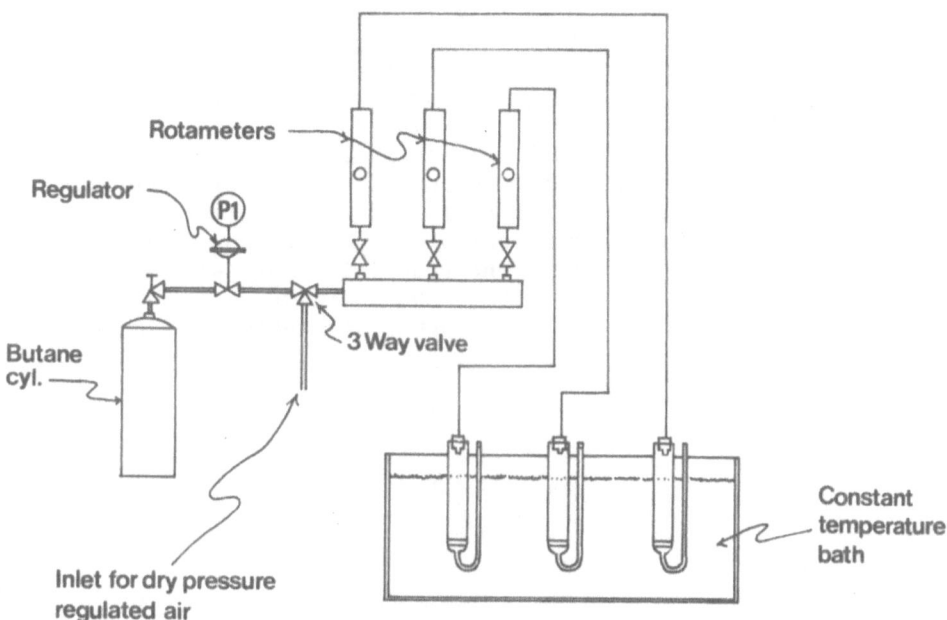

(approximately 250 milligrams per minute) is started through the carbon in upflow direction. After 15 minutes the sample is removed and re-weighed. Butane flow is stopped and air flow is adjusted to 10 bed volumes per minute (170 ml/min). The tube is attached so that the air flows in the opposite direction from the butane flow. After exactly 20 minutes the sample is removed and re-weighed. Butane working capacity is calculated in the follow-ing manner:

Difference in weight between the saturated weight and the weight after 20-minute air purge measures the amount desorbed in grams. This weight is divided by the sample volume and the result expressed as butane working capacity in grams of butane per 100 milliliters of activated carbon.

The test provides a control which assures adequate perform-ance in use. Development of this test is an illustration of the "use oriented" nature of activated carbon control tests. Butane was chosen as the adsorbate because it can be obtained in a pure form and it is a major constituent of gasoline vapor.

Impregnated Activated Carbons

Activated carbons manufactured for the control of poisonous
gases are usually impregnated with metals which react with some
of the gases the carbon is meant to protect against. These carbons are manufactured to minimum "service life" times; that is,
the time in minutes from the start of the test until a specified
concentration of the test gas is detected in the effluent

Chloropicrin Service Life. The accelerated chloropicrin
service life test[7] is used as a control test for the production
of Whetlerite carbon (carbons impregnated with copper, chromium,
and small amounts of silver). In this test, a stream of dry air
containing 47 milligrams of chloropicrin per liter is passed
through a 10-cm bed of carbon at 10 meters per minute linear
velocity until there is a detectable concentration of chloropicrin in the effluent stream. The result is expressed as service
time in minutes. The test tube shown in Figure 4 holds approximately 14 cubic centimeters of granular activated carbon when

FIGURE 4

SAMPLE TUBE AND FUNNEL — CHLOROPICRIN TEST

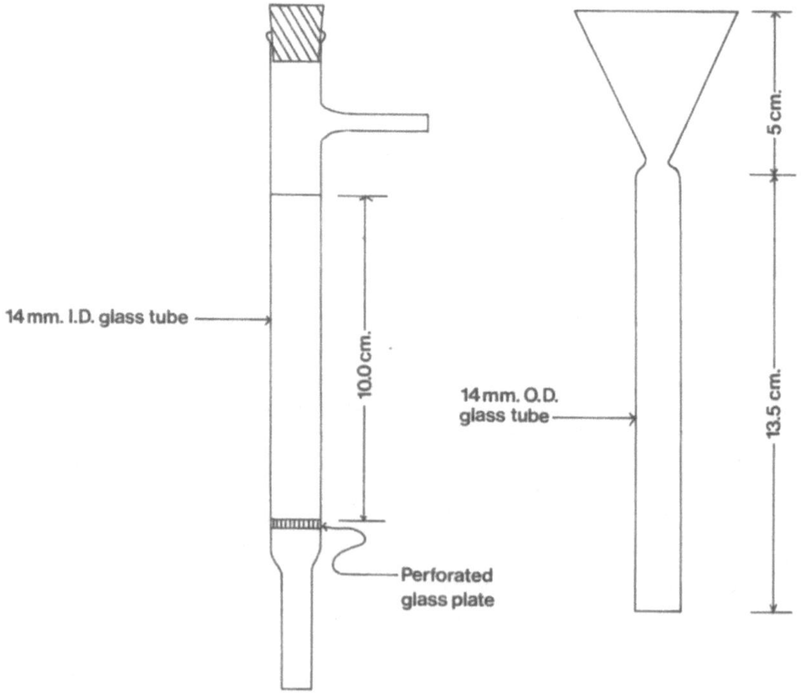

14 mm. I.D. glass tube

10.0 cm.

14 mm. O.D. glass tube

5 cm.

13.5 cm.

Perforated glass plate

filled to the 10-cm mark. The special funnel shown is employed when filling the tube to insure accurate filling. The rate of fill should be no faster than 9 cubic centimeters per minute. Once again the apparatus is very similar to the other gas adsorption apparatus as described, except for the tube design and the addition of a detection device for chloropicrin at the exit. This latter device consists of a heated quartz tube, and a liquid gas trap containing starch iodide indicator solution through which the gas exiting the sample tube is passed.

The sample tube is attached to the apparatus and appropriate flow set and time noted. At the first appearance of chloropicrin as indicated by the starch iodide solution, flow is stopped and the time noted. The actual concentration of chloropicrin in the air is determined by adsorbing the chloropicrin on activated carbon from a known volume of chloropicrin laden air. The observed service time is multiplied by a ratio of actual concentration, and the specific concentration of 47 milligrams per liter. The corrected value is the chloropicrin service life.

Chemical Analysis for Impregnants. When impregnated activated carbons are produced, it is necessary to control, within reasonable limits, the amount of impregnants deposited on the carbon surface. In the production of copper and chromium impregnated carbons, a polarographic method[7] is employed to determine the amount of these metals on the impregnated carbon. In this procedure a one-gram sample of carbon is ashed according to the procedure outlined in ASTM D-271. The ash obtained is fused with a mixture of sodium carbonate and sodium borate at 900° C. for thirty minutes. The melt is dissolved in a mixture of nitric acid and sulfuric acid. Sodium hydroxide solution is added until the solution is slightly alkaline. Sodium salicylilate and gelatin solution are then added and the solution is diluted to volume with distilled water. A portion of the ash solution is poured into the dropping mercury electrode of the polarograph and a polarogram is determined. Copper and chromium content of solution are calculated from the ratio of the wave heights of the sample and the wave heights of a standard solution of copper and chrome. The results are reported as a percentage of the original impregnated carbon weight.

The methods discussed in this lecture represent a brief survey of the adsorption techniques available for quality control and classification of molecular sieves and activated carbon. Limited as these techniques are, they play an important role in assuring uniform production for the manufacturer and continued performance for the customer.

As new uses are discovered for both molecular sieves and activated carbons, new tests may need to be devised. As exceptions become the rule with existing tests, these also will need to be revised or new tests developed.

REFERENCES

1. Young, D. M., and Crowell, A. D., Physical Adsorption of Gases, Butterworths, London, 1962.

2. Landolt, G. R., Method for the rapid determination of adsorption properties of molecular sieves, Analytical Chem., 43, 613, 1971.

3. Roelofsen, D. P., Analytical Chem., 43, 631, 1972

4. Borthakur, P. C., et al., Quality Assessment of molecular sieve zeolites by thermal analysis, J. Appl. Chem. Biotechnol., 23, 415, 1973.

5. Hassler, J. W., Activated Carbon, Chemical Publishing Co., New York, 1963, 329.

6. Baylis, J. R., Elimination of Taste and Odor in Water, McGraw-Hill, New York, 1935, 185.

7. Pittsburgh Activated Carbon Division, Calgon Corporation: Test Method No. 4, Determination of iodine number; Test Method No. 3, Determination of molasses number; Test Method No. 11, Determination of methylene blue number; Test Method No. 6, Determination of carbon tetrachloride activity and retentivity; Test Method No. 27, PS life; and Test Method No. 24, Polarographic determination of total copper and chromium impregnants.

8. Pittsburgh Activated Carbon Division, Calgon Corporation, Basic concepts of adsorption on carbon, Adsorption Handbook.

BIBLIOGRAPHY

Hersh, C. K., Molecular Sieves, Reinhold, New York, chap. 5.

MODELING, PILOT TESTS, AND CONTROL TECHNIQUES: NUMERIC METHOD
FOR PREDICTION AND DESIGN*

Walter J. Weber, Jr.

Professor of Environmental and Water Resources
Engineering and Chairman, Water Resources Program,
College of Engineering, The University of Michigan,
Ann Arbor, Michigan, U.S.A.

1. INTRODUCTION

Much of the time and expense in planning and designing
adsorption facilities is involved in predicting or forecasting
the operational dynamics of the adsorption process for a given
sorbent and water or effluent stream; this generally requires
extensive experimental pilot study. Both the time and expense
for such prediction can be minimized by a general modeling scheme
which is capable of describing the dynamics of adsorption processes
given certain basic information about the system of interest. At
a very minimum, a general modeling scheme will aid design of
programs to be carried out at the pilot level, and evaluation of
the effects of process and operational variables, thereby easing
the transition from pilot to full scale. In short, a general
modeling scheme should conserve both time and money, and ensure
more optimum full scale design and operation.

This lecture is a brief discussion of a numeric method for
adsorber design. The numeric model is termed MADAM I; Michigan
Adsorption Design and Applications Model - I.[1]

Because it is based on numeric solution techniques, MADAM I,
unlike many other models for adsorption systems, is not restricted
to simplified rate and equilibrium expressions to facilitate
analytic solution. Again because of its numeric base, MADAM I can

* Lecture 3. Session III NATO Advanced Study Institute:
Scientific Aspects of Sorption and Filtration Methods for Gas
and Water Purification, Fauske, Norway, 23-29 June 1974

accomodate the dynamic aspects of fluid dispersion, solids mixing, multisolute interactions, and biological growth on activated carbon surfaces, aspects which must be excluded because of mathematic complexity from models which are based on analytic solution techniques.

2. MODELING TECHNIQUES

In efforts to model adsorption and ion exchange processes, analytic solutions have been developed for a variety of conceptual models.[2-7] In each of these it was necessary to use simplified rate and/or equilibrium expressions (e.g., first or second order rate equations and linear or irreversible isotherm models) to describe the adsorption process in order to permit mathematic solution; most have dealt with description of only single-component systems. In the first attempt to model the adsorption behavior of complex mixtures of organic substances typical of industrial and municipal effluents, Usinowicz and Weber[8] applied the mathematic techniques outlined by Hiester and Vermuelen[2] and Thomas[6] for single-component systems to mixtures exhibiting a broad spectrum of sorption behavior. Because this modeling procedure employed an analytic solution technique and was therefore restricted to relatively simple rate and equilibrium expressions describing one-component adsorption, Usinowicz and Weber utilized a lumped-parameter approach. This type of model predicts a sigmoidal breakthrough curve. The experimental data with which Usinowicz and Weber tested the model could be adequately predicted for only about the first half of the effluent breakthrough profile. Thus, although the model previously had been shown by Keinath and Weber[4] to adequately predict and describe breakthrough curves for some single-component systems of interest in effluent treatment, its use in adsorber designs for heterogeneous systems (which normally exhibit complex non-symmetric breakthrough curves) was shown by Usinowicz and Weber to be limited.

Another significant shortcoming of the analytic approach is the lack of flexibility that would allow inclusion of process equations describing: 1) multicomponent interactions, such as hysteresis and competitive adsorption;[9,10] 2) biological activity, which increases adsorptive capacity;[11-13] 3) fluidization and solids mixing;[14] and, 4) liquid dispersion and channeling.[15] To increase flexibility, we have developed and tested in our laboratories several numeric techniques which can accomodate the mathematic complexity introduced by inclusion of the aforementioned factors. The numerical techniques developed here provide the first step toward a general model encompassing a more complete picture of dynamic adsorption processes in column operations.

3. MATERIAL BALANCE

The principle of mass conservation is central to the application of material balance relationships for analysis of process dynamics.[15] A material or mass balance on an adsorption bed, illustrated in Figure 1, results in a pair of partial differential equations (PDE's) for each solute entering the bed in the liquid phase and adsorbing on the solid phase. If the mass balance is complete and accurate, substitution of appropriate rate and equilibrium relationships and solution of the resulting system of PDE's should yield an exact description of the dynamic behavior of the adsorber -- in theory. If analytic solution techniques are used, severe mathematic constraints are imposed on the complexity of the equations used to describe the several process dynamics factors involved. Indeed, many important factors may have to be neglected.

If, on the other hand, the system is solved numerically, then virtually any dynamic complexity that is known to exist and that can be described mathematically may be incorporated into the conceptual model for the adsorber system.

Thomas[6] solved simplified forms of the equations shown in Figure 1 -- analytically -- for the following limited conditions: 1) second-order kinetics; 2) single-solute adsorption; 3) no dispersion (Peclet Number, $N_{Pe}=0$); and, 4) the Langmuir adsorption isotherm.

We have developed an implicit backward finite difference method (IBFDM) solution to these equations for several different and more realistic conditions. The conditions which will be illustrated here include: 1) three different rate expressions; 2) time-variant influent concentrations; 3) completely mixed solid phase with liquid phase gradients (a condition which does occur in expanded beds at high flow velocities); and, 4) a two-component solute system. Unlike an analytic solution, the numeric solution solves for the values of dependent variables at discrete points in the domain of the independent variables. The independent variables are calculated at the intersection points of an appropriate grid. To do this, one replaces the derivatives in the equations shown in Figure 1 with finite difference approximations, and an algebriac equation is obtained at each grid point.

4. COMPARISON OF THE NUMERIC AND ANALYTIC SOLUTIONS

The Langmuir equilibrium expression is used in all of the models considered here to provide a comparison between the numeric and analytic solution techniques.

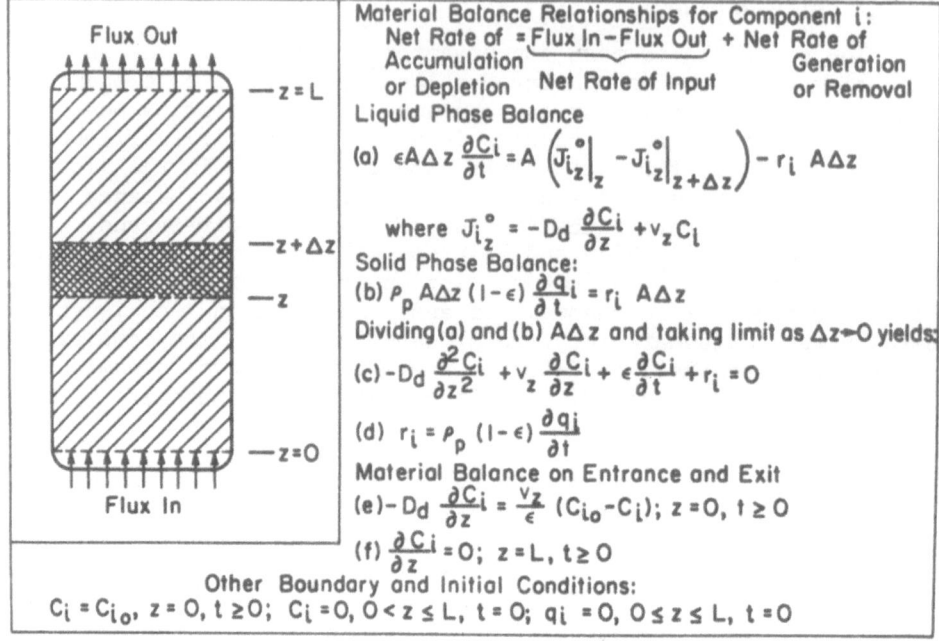

Fig. 1. Material balance relationships for an adsorber

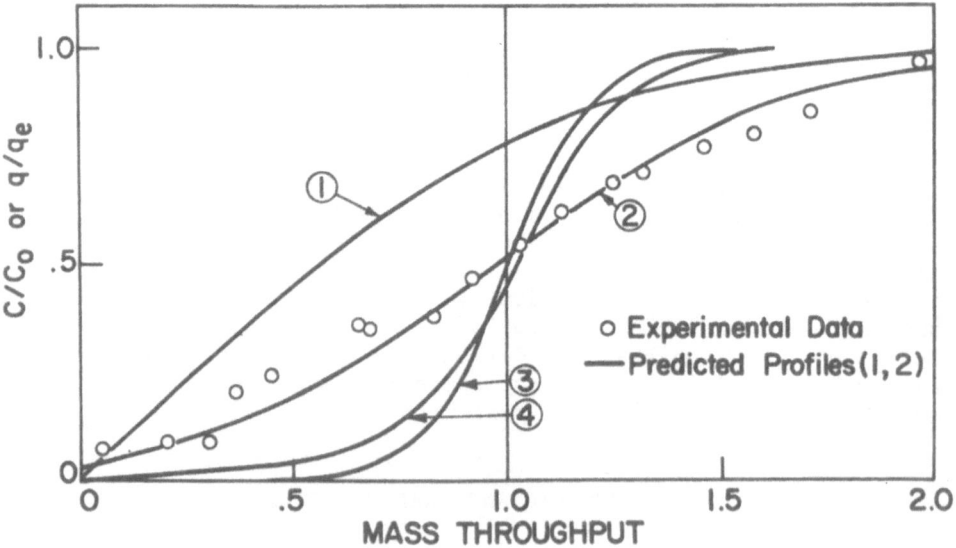

Fig. 2. Experimental and predicted breakthrough and adsorption curves for DNOSBP and o-nitrochlorobenzene.

Keinath and Weber[4] found the Hiester-Vermuelen[2] approach incorporating the Thomas[6] analytic solution adequate in describing single-component behavior for a wide range of adsorber parameters; different bed depths, flow rates, and influent concentrations. The adsorbate used for this evaluation, di-nitro, *ortho*, *sec*-butyl, phenol (DNOSBP, an herbicide) is unique in some of its properties, however. It belongs to a class of adsorbates -- usually small molecules -- for which adsorption rate is controlled by a single-step mechanism.

The particular rate expression utilized by Thomas in his analytic solution is based on a Langmuir adsorption model. A comparison between the analytic solution and the MADAM I numeric solution to the appropriate PDE's -- using, for consistency, the rate expression developed by Thomas -- is presented in Figure 2, curves 1 and 2. Curves 1 and 2 are, respectively, the ratio of effluent to influent liquid phase concentration and the ratio of the average solid phase concentration to the equilibrium solid phase concentration. Traces for the analytic and numeric solutions for identical adsorption kinetics can not be separately identified here because they fall virtually on top of one another along curves 1 and 2. For fifty evenly spaced points on the graph the standard deviation between the numeric and analytic solutions is 1/6% for C/C_0 and 1/4% for q/q_e. This validates the accuracy of the numeric solution. The computing time (machine time) for the numeric solution is approximately the same as for the analytic solution in this relatively simplified case. It is interesting that for this particular system there is no significant difference in the numeric prediction whether the solid phase is treated as fixed or completely mixed (no analytic solution exists for the completely mixed solid phase condition). This may be another reason why the analytic solution, with its fixed solid phase assumption, was found adequate by Keinath and Weber[4] for this expanded bed system with DNOBP, despite the fact that some solids mixing is known to occur in expanded beds. For other solutes the differences between fixed and completely mixed solid phases may be substantial. Curves 3 and 4 in Figure 2 illustrate such a difference for a single solute system, *o*-nitrochlorobenzene, reported by Usinowicz and Weber.[8]

Another solute that follows a single-step rate mechanism, film transport, throughout the breakthrough curve is p-toluene sulfonate. Curve 2 in Figure 3 is an experimental breakthrough curve measured by Weber[16] for this solute, and curve 3 in the same figure is the MADAM I numeric prediction of the breakthrough for the following conditions: a) no dispersion; b) fixed solid phase; c) rate of adsorption controlled by film transport.

Curve 1 in Figure 3 is experimental data[16] for a sulfonated alkylbenzene. This breakthrough profile cannot be described

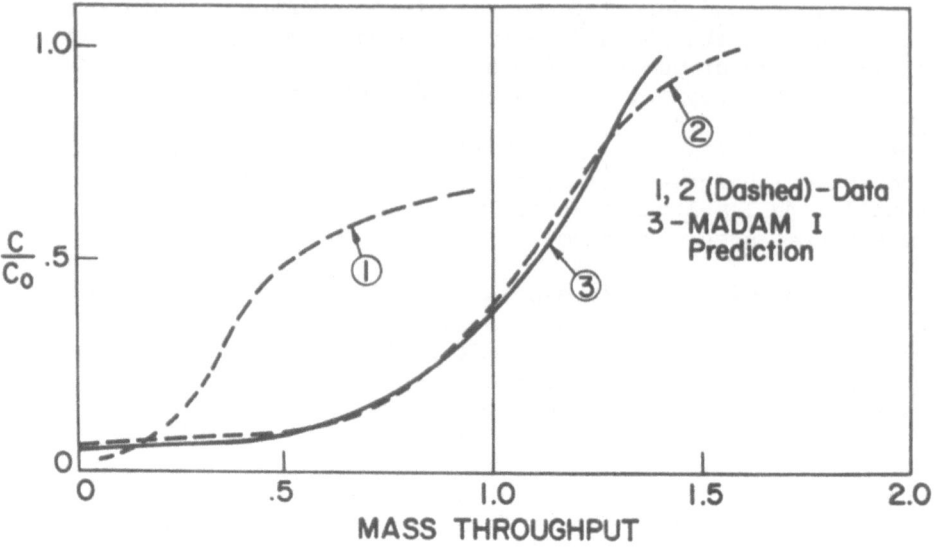

Fig. 3. Experimental and predicted breakthrough and adsorption curves for a sulfonated alkylbenzene and *p*-toluene sulfonate.

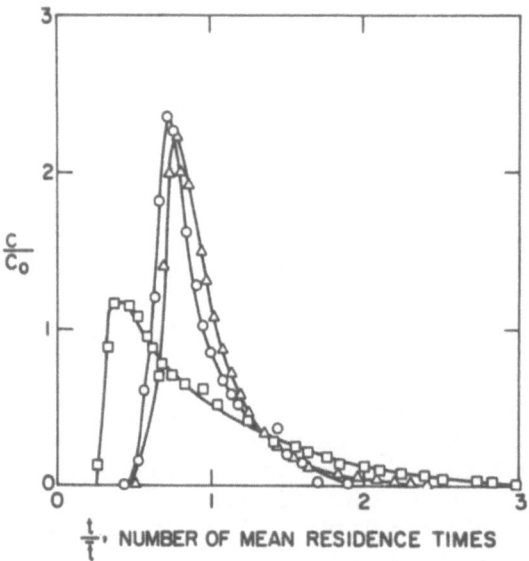

Fig. 4. Exit-age distribution curves for packed and expanded beds.

adequately by a one-step adsorption mechanism such as the Langmuir kinetic expression or a simple film transport expression. Rather, mathematic prediction of the concentration history profile requires use of a more complex model to describe the rate of adsorption (e.g., two-step film transport followed by intraparticle diffusion). Because of the physical bulk and strong adsorption energy of the sulfonated alkylbenzene molecule, intraparticle diffusion is more likely to limit the adsorption rate of this solute in the final stages of the breakthrough profile, whereas film transport likely limits the rate of adsorption of smaller and less strongly adsorbing substances such as the DNOSBP and p-toluene sulfonate throughout the entire breakthrough profile. This conclusion is substantiated by observations of adsorption rates in batch reactors.[17,18] The overall intraparticle transport coefficients for p-toluene sulfonate and DNOSBP are about 20 times that for the more bulky sulfonated alkylbenzene molecule, and about 10 times as large as the overall film transport coefficient. Hence, rates of adsorption for the former substances should be limited by only one step of the two-step mechanism over the entire breakthrough profile.

5. OPERATIONAL FLEXIBILITY

As noted previously, there are additional factors relating to practical conditions of operation of adsorbers in water and effluent treatment which cannot -- because of mathematic complexity -- be incorporated in predictive design models which rely upon analytic solutions. Three of these -- dispersion, time-variant influent concentrations, and multicomponent systems -- will be briefly discussed for purpose of illustration.

5.1 Dispersion

Using non-adsorbing tracers we have experimentally measured significant dispersion (or channeling) for both packed (N_{Pe}=0.063 - 0.165) and expanded beds (N_{Pe}=0.282 - 0.343) under normal hydraulic loading (2.5-10gpm/ft^2, or, 102-410 l/min/m^2). Figure 4 is a plot of typical exit-age curves for packed and expanded bed adsorption systems, illustrating the functional relationship between dispersion and hydraulic loading. These findings suggest that dispersive transport is responsible for a substantial amount of solute transport in an adsorption bed. In addition, a significant degree of solids mixing has been observed in expanded beds.[14] These observed features of practical operation must be incorporated in a design model to insure its validity and generality.

In the absence of dispersion, the correspnnding process

equations are relatively simple first order equations. When dispersion -- which, as noted above, is a significant performance factor -- is taken into account both the rate equations and the process equations are second order parabolic partial differential equations, and a numeric solution such as that provided by MADAM I is required.

5.2 Time variant influent concentration

Figure 5 presents the breakthrough profile prediction of MADAM I for the *p*-toluene sulfonate system with a sinusoidal variation in influent concentration, the amplitude and frequency of which are illustrated by the input curve shown in the upper left portion of the figure. The rate and equilibrium expressions employed are film transport and the Langmuir isotherm, respectively. The MADAM I prediction for attenuation of the amplitude of the influent concentration by the adsorber, even at complete break-through ($C/C_0=1$), is characteristic of repeated observations of the behavior of adsorption systems in field applications.

5.3 Multicomponent systems

Figure 6 presents the individual solute effluent and adsorp-tion profiles for a bi-solute system. The corresponding system of four simultaneous PDE's were solved by the IBFDM. The rate of adsorption is described by film transport, while the modified Langmuir expression developed by Snoeyink and Jain[10] was used to describe equilibrium conditions. Curves 3 and 4 are the effluent concentration profiles for components 1 and 2, respectively. The influent concentrations, the Langmuir energy constants, and the ultimate capacities are treated as identical for the two solutes in this system. The only difference is the overall film transfer coefficient, $\hat{k}(t^{-1})$, where $\hat{k}_1=2.5\hat{k}_2$. This difference between otherwise identical solutes can arise from differences in molecular geometry relating to substitution of functional groups on a parent molecule. (See, for example, Morris & Weber).[19] Commensurate with expectations, component 1 does not break through as soon as component 2; further, it overshoots its equilibrium uptake. Thus, the effluent concentration is greater than the influent concentration as the solute adsorbed in excess of the equilibrium uptake is eluted away.

SUMMARY

In contrast to previous attempts at comprehensive prediction and design models for dynamic adsorption systems, MADAM I is not limited by mathematic complexity to oversimplified rate and

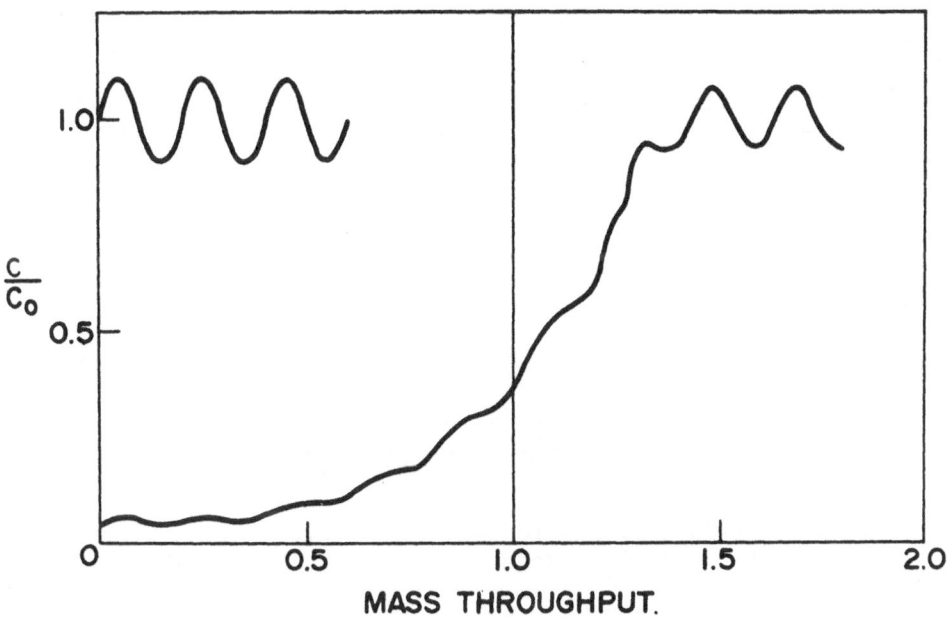

Fig. 5.   Time variant influent concentration.

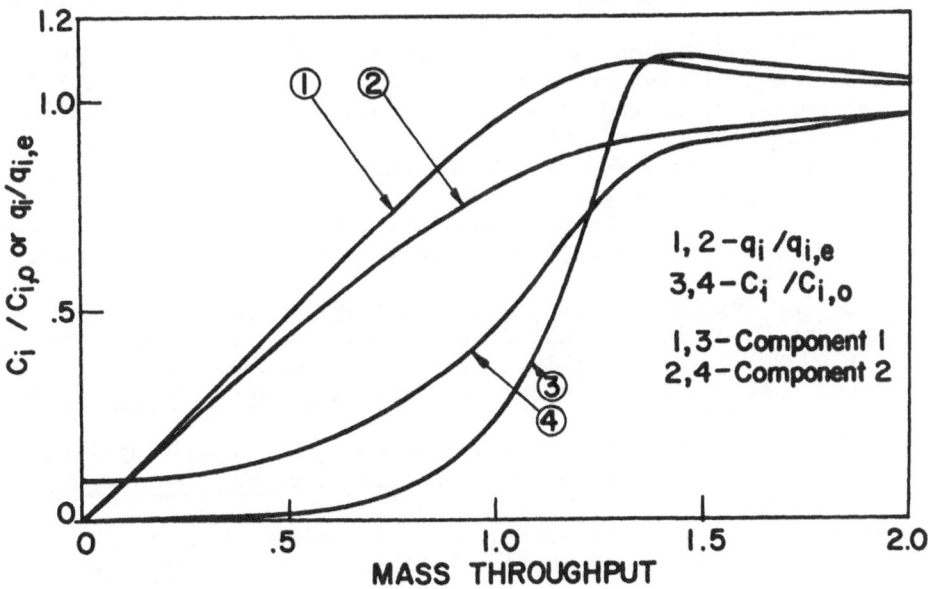

Fig. 6.   Effluent concentration and adsorption profiles for the components of a bi-solute system.

and equilibrium expressions, and is further capable of incorporating the highly practical aspects of fluid dispersion, solids mixing, multi-component interactions, multicomponent hysteresis and biological activity in adsorbers. The present lecture illustrates the general validity, accuracy, and flexibility of the IBFDM numeric approach of MADAM I for solving the appropriate PDE's describing adsorption dynamics. It remains in continuing development of the model to incorporate all of the above parameters, while more fully verifying its predictions experimentally.

REFERENCES

1. Weber, W. J., Jr., and Crittenden, J. C., MADAM I -- A numeric method for design of adsorption systems, Jour. Water Pollution Control Fed., in press, 1974.
2. Hiester, N.K. and Vermuelen, T., Saturation performance of ion-exchange and adsorption columns, Chem. Eng. Prog. (48), 10, 505, 1952.
3. Michaels, A. S., Simplified method of interpreting kinetic data in fluid-bed ion exchange, Ind. Eng. Chem., 44, 1922, 1952.
4. Keinath, T. M. and Weber, W. J., Jr., Predictive model for the design of fluid bed adsorbers, J. Wat. Pol. Con., 40, 743, 1968.
5. Rosen, J. B., General numerical solution for solid diffusion into spherical particles, J. Chem. Phys., 20, 387, 1952.
6. Thomas, H.C., Heterogeneous ion exchange in a flowing system, J. Amer. Soc., 66, 1664, 1944.
7. Tien, C. and Thodos, G., Ion exchange kinetics for systems of linear equilibrium relationships, AIChE Journal, 6, 364, 1960.
8. Usinowicz, P. J. and Weber, W. J., Jr., Mathematic simulation and prediction of adsorber performance for complex waste mixtures, Summary Report, EPA Research Project No. 17020 EPF, University of Michigan, Ann Arbor, Mich., June 1973.
9. Weber, W. J., Jr., Competitive interactions in adsorption from dilute aqueous bi-solute solutions, Jour. Applied Chem., 14, 565, 1964.
10. Snoeyink, V. L. and Jain, J. S., Adsorption from bisolute systems on active carbon, J. Wat. Pol. Con., (45), 12, 2463, 1973.
11. Weber, W. J., Jr., Hopkins, C. B. and Bloom, R., Jr., Physiochemical treatment of wastewater, J. Wat. Pol. Con., (42) 1, 83, 1970.
12. Parkhurst, J. D., Dryden, D., McDermott, G. N. and English, J., Pomona activated carbon treatment plant, J. Wat. Pol. Con., 39, 70, 1967.
13. Weber, W. J., Jr., Friedman, L. D. and Bloom, R., Jr., Biologically-extended physicochemical treatment, Proceedings Sixth International Conference on Water Pollution Research, Jerusalem, Israel, 1972; Pergamon Press, Oxford, 1973.
14. Hopkins, C. B., Weber, W. J., Jr. and Bloom, R.,Jr., Granular

carbon treatment of raw sewage, Water Pollution Control Ser., Fed. Water Qual. Admin., ORD 17050DAL05/70, May 1970.

15. Weber, W.J., Jr., Physicochemical Processes for Water Quality, Wiley-Interscience Pub., New York, N.Y., 1972.

16. Weber, W. J., Jr., Fluid-carbon columns for sorption of presistent organic pollutants, Proceedings Third International Conference on Water Pollution Research, Munich, Germany, 1966.

17. DiGiano, F. A. and Weber, W. J., Jr., Sorption kinetics in finite-bath systems, Jour. Sanit. Eng'r. Div., ASCE, SA 6, 1021, 1972.

18. DiGiano, F. A. and Weber, W. J., Jr., Sorption kinetics in infinite-bath experiments, J. Wat. Pol. Con., 45, 4, 1973.

19. Morris, J. C. and Weber, W.J., Jr., Adsorption of biochemically resistant materials from solution, Vol. AWTR-9, p. 26, 41, 1964.

CDE NON-DESTRUCTIVE TEST

F.A.P. Maggs

Chemical Defence Establishment, Porton Down,
Salisbury, Wiltshire, SP4 OJQ, England

ABSTRACT. The principles and applications of the CDE non-destructive test are outlined. Measurements of the fractional penetration of a vapour pulse through charcoal beds under several conditions of flow, concentration, bed depth and moisture are summarised and are shown to be inter-related through the residence time of the pulse in the bed.

INTRODUCTION

Three stages may be distinguished in the manufacture and use of a filter at which some form of testing is advisable. During the design stage, prototypes must be tested with respect to filter capacity and integrity of the construction; in production, samples must be examined; and in use there is a requirement to detect the point of exhaustion of the adsorbent or absorbent material. In all these situations the use of a simple non-destructive test would clearly be an advantage.

Two systems were proposed and examined at CDE. The earlier method involved the measurement of the penetration time of a 1 to 10% v/v concentration of a weakly adsorbed vapour. After the test, the test agent could be desorbed by the passage of clean air. Test vapours included propane, butane, freon and carbon dioxide. Under conditions of constant humidity, correspondence with the results of normal destructive testing with carbon tetrachloride was good. But for the more practical situation of an uncontrolled air-humidity, the test results showed a marked dependence on humidity, as might be expected.

SMALL-DOSE NON-DESTRUCTIVE TEST

Nearly all filter beds involve the measurement of a penetration time. A new approach, devised at CDE in 1966, employs an alternative principle. A well constructed filter containing an efficient sorbent, will effect a large reduction in the contaminant concentration; reduction ratios are usually better than 10^4, and values greater than 10^7 are encountered. The minimum ratio required will be a function of the toxicity, or of merely the unpleasantness of the aerial contaminant. The penetration of a filter by the known hazard, or a simulant, should therefore be less than a given value, predetermined for that filter/hazard system. There is thus no necessity for the test attack to persist any longer than is required for the estimation of the reduction ratio, and a finite <u>pulse</u> of test agent in the influent air is all that may be called for. The quantity of test agent injected can thus be small and constitute only a small fraction (less than 0.5%) of the total sorptive capacity of the filter. The test then becomes non-destructive and the filter may be tested on a number of occasions. It is implicit in this approach, that the test agent is strongly adsorbed and is used in very small quantities, in contrast with the previous method, where high concentrations of a weakly adsorbed agent are used in measuring a penetration time.

Although the injected test agent will give rise to an appreciable concentration pulse (in the mg/l range, for instance) the detector sampling the effluent air will of necessity require a high sensitivity - better than µg/l - if slightly defective filters are to be identified. Under laboratory conditions or where elaborate testing facilities are available, various forms of detecting instruments, often specific to the selected test agent, may be used. For a simple, portable system, however, the prime and sometimes limiting consideration may well be the choice of a suitably sensitive detector. The relations between the dose, the detector sensitivity and the filter parameters have been given earlier[1]; consideration has also been given to those factors governing the choice of test agent, which may even be the known hazard or contaminant, though a simulant having a similar interaction with the filter adsorbent is generally preferred. The principle would appear to be applicable to gas or liquid filtration, and to physical or chemical action by the filtering medium.

In the present paper, however, the specific system of a charcoal bed acting by physical adsorption of the contaminant will be examined to illustrate both the usefulness and the mode of action of the CDE non-destructive test.

THE CDE NON-DESTRUCTIVE TEST FOR ADSORPTIVE FILTERS

For testing charcoal filters it was desirable that the test agent should be relatively non-toxic and detectable in very low concentrations, since reduction ratios greater than 10^5 would be involved.

The selection of a suitable system was dictated to some extent by observation of the high sensitivity and simplicity inherent in the halogen (or positive ion) detector, in which a potential applied between two heated ($\sim 850°C$) platinum electrodes develops a small ionisation current which is increased markedly by the presence of a halogen, or halogen-containing vapour. For the several detector configurations we have studied the following relationship was found to hold over a wide range ($\sim 10^4$) of concentration:-

$$I = Kc^{0.6}$$

(I is the ionisation current increase resulting from a concentration c of organohalogen). If c is expressed in µg atoms halogen per 1 air, we have found that for a given detector a common curve of ln I versus ln c is obtained for many organohalogen vapours. The detector sensitivity K is dependant on the applied potential, the temperature of the electrodes and on their size. For one type of detector head examined K ranged from 20 to 800, so that ng/1 concentrations are measurable when K is high. A smaller detector head (requiring less heater wattage) gave values for K lower than 20, but by the addition of a simple amplifier, high sensitivity together with battery-operated portability has been achieved. Commercial models are now available.[2]

Although this detector exhibits specificity, the range of properties in the organohalogen group of compounds is wide. In an earlier paper[1] the use of several organohalogens in filter testing was described and the influence of adsorbed water in equilibrium with the atmosphere was referred to in relation to the choice of test agent. In the following section are summarised the experimental results, which demonstrate the influence of bed depth, air velocity, concentration and moisture content on the penetration of the pulse through the beds.

EXPERIMENTAL CONDITION AND RESULTS

Charcoal beds were held in a refillable container (cross-sectional area, 58 cm^2) into which the charcoal was filled by the Porton "snow-storm" filling machine to give reproducible close-packing. A nutshell charcoal graded 7 x 14 B.S.S. was examined. The temperature, humidity and flow rate of the main air stream through the bed could be controlled.

The test agent pulse was obtained by diverting for 2 seconds into the main air stream a 1 l/min air stream containing a known quantity of the vapour under test.

The CDE Halogen Detector, calibrated at intervals during the experiments, was used to measure the concentration of the effluent pulse. The following test agents were used: ethyl chloride, tertiary butyl chloride, carbon tetrachloride, 1,2-dibromoethane, chlorobenzene, m-chlorotoluene. The pulse concentrations in the main air stream ranged from 30 to 100 µg atom halogen/l. The majority of experiments were conducted at a flow rate of 50 l/min, with several experiments at 30 and 100 l/min. Charcoal bed depths were in the range 0.5 to 3 cm. In one series of experiments the charcoal moisture was 5% w/w and in another series, 30%.

RESULTS AND DISCUSSION

In the Introductory paper to this Conference, the basic Hinshelwood equation was given

$$-\ln c/c_o = K_3 \, d/L$$

where c is the effluent concentration through a bed of length d, when the pulse velocity through the bed is L. In the present experiments we may therefore expect log c/c_o (the fractional penetration) to be a linear function of the bed depth, and of the reciprocal of the flow rate.

The validity of this simple approach was verified for pulses of the six vapours, using both "moist" and "dry" charcoal. The penetration ratio was also found to be independent of the pulse concentration over the three-fold range examined. It will be noted, however, that the ratio d/L is the residence time in the bed; for a given vapour therefore, the results obtained by varying the bed depth, flow rate and pulse concentration may be combined in a single graph of log c/c_o versus d/L. The data for the various experiments are displayed in this form in Figs. 1 and 2 referring to the "dry" and "moist" charcoal respectively.

In both graphs it is observed that the gradient of the lines increases with increase of those parameters of the test agents which reflect the inter-molecular forces (e.g. boiling point, latent heat), as might be expected. The effect of charcoal moisture is also clear from a comparison of the two Figures. With the more weakly adsorbed test agent (ethyl chloride, butyl chloride) adsorbed moisture increases the fractional penetration. In contrast, penetration of the strongly adsorbed test agents

(such as chlorobenzene and chlorotoluene) is notably unaffected
by the presence of adsorbed moisture. An appreciable but small
effect is noted for carbon tetrachloride, whilst moisture has
little effect on the penetration of dibromoethane. Further work,
not detailed here, in which charcoal samples containing different
amounts of adsorbed moisture were examined, has confirmed these
observations, and has also shown that, where moisture is
influential, $\log c/c_o$ varies linearly with the amount of moisture.

CHOICE OF TEST AGENT

The limited value of the earlier methods using gases such as
freon and butane is emphasised by the above results, since these
gases are even more weakly adsorbed than ethyl chloride and will
be expected to give test results strongly susceptible to charcoal
moisture content.

On the other hand, the use of an apparently very suitable
substance such as chlorotoluene as a test agent is precluded -
unless a more sensitive detector is available - by the difficulty
in obtaining a sufficiently high influent pulse concentration
of this relatively involatile liquid to allow the detection of
the penetrating pulse from any but very shallow beds.

At present then, the test agent should be selected from
liquids having the highest volatility compatible with no suscepti-
bility to charcoal moisture content. For charcoal filters we
have found that bromobutane conveniently fulfils these conditions.

THE CDE TEST EQUIPMENT FOR FILTRATION UNITS

Although a description of the test developed for filtration
units, and of its use has been given elsewhere[1], it may be of
value here to provide a brief reiteration.

The influent pulse of vapour (generally of bromobutane) is
generated by use of a spray gun powered by a Sparklet carbon
dioxide cylinder. A measured quantity of test agent is dispersed
in 2 seconds into the influent air to give a pulse concentration
in the region of 10 mg/l (approximately at a dose rate of 1 ml
per 3 m^3/min air flow rate). A CDE halogen detector samples the
effluent air through a PTFE tube; the small standing current is
initially backed-off. The effluent pulse reading should be
observed within a few seconds of firing the spray gun. A simple
calibrating device for the detector is also provided.

If the test performance of the filter when new has been
determined, (or if all new filters have been required to pass a
manufacturing specification which includes the non-destructive

test) any deterioration of the filter during its life can readily be ascertained. Faults of installation may also be detected. It may be noted that the required performance of the filter in removing a given hazard should be previously related to the results of the non-destructive test.

REFERENCES

1. Maggs, F.A.P. Ann.Occup.Hyg., 351, 1972.

2. Supplier: D.A. Pitman Limited, Weybridge, Surrey.

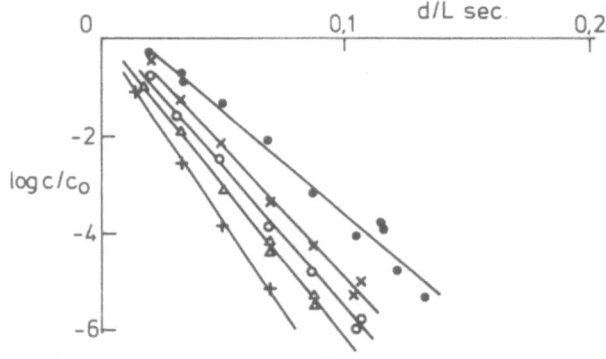

FIG.1. DRY CHARCOAL

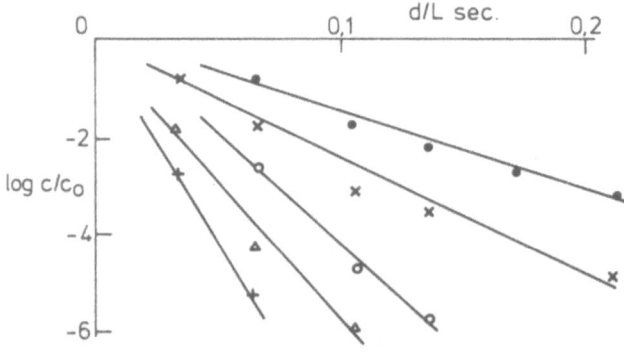

FIG.2. MOIST CHARCOAL

Figs. 1 and 2. Penetration as a function of residence time O-ethyl bromode, X-butyl chloride, O-carbon tetrachloride, Δ-bromoethane and chlorobenzene, +- chlorotoluene

TESTS FOR THE PERFORMANCE OF FILTERS USED IN PROTECTION AGAINST TOXIC GASES

M. van Zelm, R.H. van Dongen* and P.C. Stamperius

Chemical Laboratory TNO, Rijswijk, The Netherlands

ABSTRACT. To check the protective capacity of active carbon filters against toxic gases usually so-called breakthrough tests are carried out. In these tests the filter is subjected to a flow of air with a given concentration of the toxic gas concerned, under conditions which simulate the conditions of use. The performance of these tests is described. Essentially the tests are destructive and therefore a test for proper functioning of the filter which leaves as much of the protective capacity undisturbed would be very useful. To this end methods using a pulse of a weakly adsorbing gas have often been reported. This technique will be considered both from a theoretical and from an experimental point of view.

1. BREAKTHROUGH TESTS

The adsorption capacity of filters that are used to protect against toxic vapours is usually expressed in terms of breakthrough time, i.e. the time in which the maximum permissible exit concentration (breakthrough criterion) will not be exceeded. To determine the breakthrough of a given filter for a given toxic vapour or gas, air containing this gas in the desired concentration is drawn through the filter and the time noted until the appearance of the toxic gas in the effluent air at a predetermined concentration.

Each test introduces its own problems of forming the required concentration, measuring this and detecting the penetrating vapour.

*Present address: Kon. Shell Lab., Amsterdam

1.1 Forming the required concentration

Gases may be fed directly into the air stream, liquids may be injected onto a hot finger or a hot plate in the air stream or added
by bubbling an auxiliary air stream through the liquid to give the
required vapour concentration. To obtain a reasonable vapour pressure it is sometimes necessary to heat the liquid. To achieve a
constant concentration over a reasonable long period of time we
have found it very convenient to pass the air stream over the surface of the liquid, contained in a shallow vessel with a large
surface area. When the liquid is kept at a constant temperature
and the air flows are carefully controlled the rate of evaporation
is constant and consequently also the concentration.

Care should be taken to avoid the formation of aerosols when
concentrations have to be used near the saturated vapour pressure
of the liquid. Aerosols are not retained by the active carbon filter and may give rise to a spurious early penetration when the
filter is not protected by an appropriate aerosol filter.

1.2 Determination of concentrations

To obtain reproducible results it is essential that all test conditions are carefully controlled. A frequent check on the concentration of the vapour or gas in the main stream is therefore necessary and a continuous recording of the concentration would even
be better. The methods used are ordinary analytical chemistry methods and need not be discussed here.

There are several ways to detect the penetrating vapour in
the effluent air. Often the method is used of passing part of the
effluent air through a bubbler containing a suitable reagent that
will change colour as soon as a certain quantity of the penetrated vapour has passed the filter. Usually it is required that a
second bubbler with reagent changes colour within a short specified period of time. Obviously in this case there is no sharply
defined breakthrough criterion and the method is becoming obsolete
and replaced by methods in which the actual concentration is measured. As the breakthrough curves usually rise very rapidly, the
choice of the limiting concentration is not very critical and
often determined by the methods and equipment available in the
laboratory. Nevertheless, the breakthrough criterion should not
be too far from the concentration that is physiologically acceptable for a reasonable period of time, because due to improper
packing of the active carbon layer there might be a low continuous penetration right from the beginning which would otherwise
pass unnoticed. Therefore, the best way to determine the breakthrough time is to integrate the quantity penetrated to obtain
the dosage that has passed and to determine the time at which the
dosage exceeds the acceptable value for the toxic compound under

consideration.

1.3 Relative humidity

Filters that are used to protect against toxic gases, whether they are used in air conditioning installations to provide collective protection or used in respiratory protective equipment for individual protection, have to deal with air not only containing toxic compound(s) but also water vapour, being a normal, though varying constituent of the atmosphere. This aspect should be taken care of when testing filters. Up to about 50% R.H. there is little influence, above this value the influence is very large because the water vapour isotherm of activated charcoal rises very steeply there. Since in practical use any relative humidity may be encountered its value is usually fixed. It is advisable to take a high value because in that case one can be sure that in practice protection will be better, thus increasing the safety factor. We usually take 80% R.H. as this seems a reasonably high value for West-European conditions.

When dealing with very toxic substances in testing military filters, we even equilibrate the filters at this relative humidity beforehand, thus adding to the safety factor.

1.4 Flow rate

With filters applied in air conditioning installations the flow rate is constant and the tests in the laboratory are usually tube tests, using to the same height the same activated charcoal as for the filling of the filter, at a linear velocity corresponding with the actual flow rate.

With respiratory protective equipment the flow through the filter varies depending on the work rate of the wearer. For this case we judged that a flow rate should be selected corresponding with work rates that can be maintained during a period of three hours. Figure 1 shows for unmasked and healthy personnel the time a given work rate can be maintained (1). For masked personnel we assumed a somewhat lower work rate (100–125 watts) because of the extra effort imposed by wearing the gas mask. From measurements of subjects wearing different types of gas masks when performing on a bicycle ergonometer we have concluded that a reasonable minute volume for test is 30 1/min.

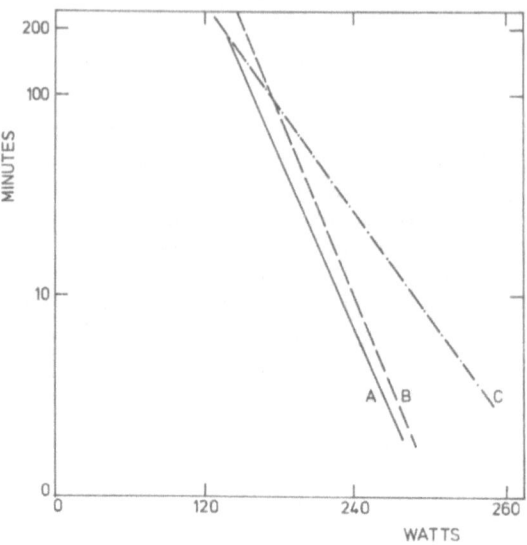

Fig. 1 Work rate vs the time this work rate can be maintained.

A comparison was made between breakthrough times obtained with constant flow and with intermittent flow, using a breathing machine which simulated human breathing. The tests were performed with charcoal layers in test tubes and with chloropicrin (CCl_3NO_2) as a test gas.

Table I Conditions of tests to compare intermittent and continuous flow		
tube diameter	30 mm	68.5 mm
chloropicrin concentration	3 mg/l	5 mg/l
relative humidity	15-20%	15-20%
flow rates	5-15 l/min	10-20 l/min
breakthrough criterion	1 mg/m^3	1 mg/m^3
height of the charcoal layer	40 mm	40 mm
type of charcoal	NORIT 1 mm impregnated charcoal	

The results of the tests are given in the Fig. 2a and 2b. The flow rates have been calculated from the test tube data for a gas mask filter with a diameter of 100 mm.

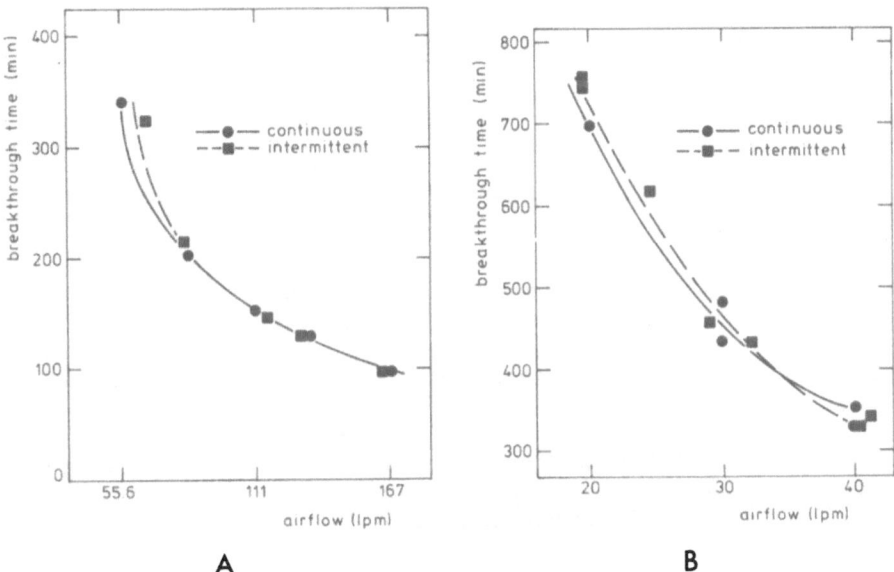

A   B

Fig. 2 Breakthrough times at several flow rates with intermittent and continuous flow. a. High flow rates, b. Medium flow rates.

The results are such close to each other that it does not warrant the trouble of testing filters of gas masks with intermittent flow.

1.5 Test temperature

Temperature has a definite effect on the results of a breakthrough test and should therefore be controlled accurately. For filters that retained toxic compunds by physical adsorption one may expect on the basis of the general theory of adsorption a decrease of protection with increase of temperature. In the case of chemical interaction between adsorbate and adsorbent an increase of temperature will in general enhance the protection. Table II may serve as an illustration.

258

	breakthrough times (min)	
temperature (^{o}C)	chloropicrin	cyanogen chloride
10	82	34
18	45	40
30	28	44

Table II
The effect of temperature on the breakthrough times
of a physically adsorbed (chloropicrin) and a
chemisorbed substance (cyanogen chloride)

1.6 Final remarks

Obviously when testing filters with highly toxic substances the
safety aspect is very important. In the construction of our
breakthrough apparatus the pressure-vacuum-balance principle is
applied. This means that by careful use of pressurised air and
vacuum we maintain an underpressure in every part of the apparatus
in which high concentrations of the toxic substance occur. An
example of a breakthrough apparatus used for tests with the highly
toxic nerve agent sarin (lethal dosage upon inhalation approximately 100 mg.min/m^3) is given in Fig. 3.

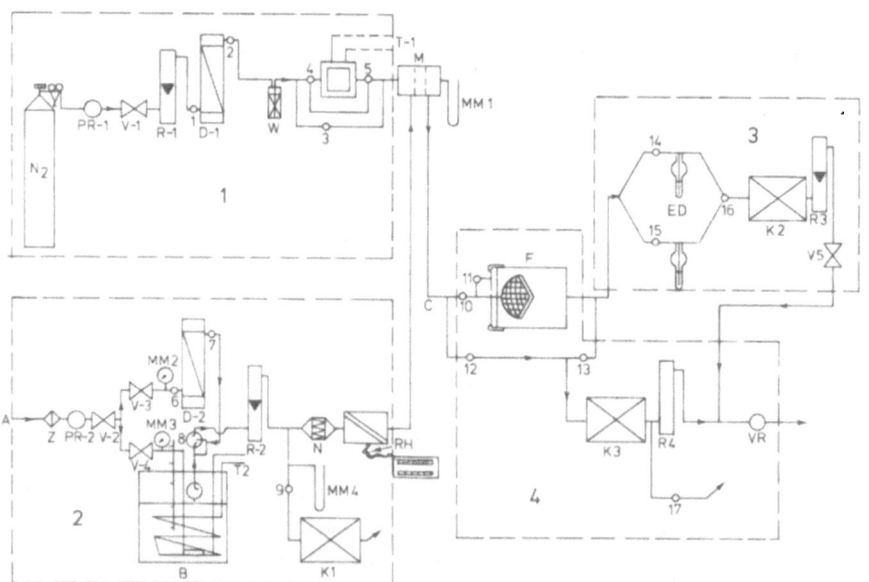

Fig. 3 Apparatus for the determination of breakthrough of filters with
sarin vapour. Main parts of apparatus: 1. air flow adjustment,
2. sarin-vapour generation, 3. detection, 4. test rig.

The apparatus has adjustable flow rates from 10 to 80 1/min. The relative humidity is adjustable from 10 to 90% by splitting the air stream, one part being passed through a drier with molecular sieves and the other through a thermostated humidifier.

2. NON-DESTRUCTIVE TESTS

Breakthrough tests, as discussed in the preceding paragraphs, are very useful to determine the protection that can be afforded by a given type of filter. It is, however, by its very nature a destructive test and the filter tested cannot be used to protect against toxic gases anymore. It does not warrant that a similar filter will protect as properly. When the lives of people depend on the protection afforded by the filters a non-destructive test, which leaves the filter as much as possible in its original condition, may be of vital importance. Also for reasons of economy it is desirable to check if a filter needs replacement, without destroying it.

Quite an amount of research on non-destructive testing of charcoal filters for air purification has been done employing the pulse technique. In this method a quantity of a test gas is injected in the influent of the adsorption column. Its concentration in the effluent is recorded as a function of time. The research in this field is aimed at finding a relationship between the retention time, height or width of the eluted peak and the residual adsorption capacity of the filter. So far, the method has been found to be applicable in such circumstances that no adsorption of water on the charcoal can take place, e.g. testing new (dry) filters (2) or testing special industrial adsorption columns (3).

However, in any practical situation where ambient air has to be freed from toxic gases also water vapour will be adsorbed on the filter. The shape of the eluted peak will not depend on the degree of surface coverage with the given adsorbate only, but also on the quantity of adsorbed water. Various attempts have been made to distinguish the effect of adsorbed water on the eluted peak from that of the contaminant (4, 5). The efforts have not been very succesful so far.

We have also been investigating the feasibility of a non-destructive test based upon the so-called pulse technique using a weakly adsorbing test gas to determine the condition of a charcoal bed under humid conditions. In the following this technique will be looked at from a theoretical and a practical point of view.

2.1 Theoretical

As to the theory of the dynamics of adsorption applied to a fixed

bed of charcoal we will confine the discussion to a brief outline
and will refer to a more detailed treatise on the subject frequently (6).

As was demonstrated before (6) the transfer of an adsorbate
through a filter is largely governed by the capacity of the filter
and by the efficiency of the transport processes taking place. A
suitable non-destructive test should cover both aspects. The capacity of the filter can be expressed in the residence time of a
pulse of adsorbate according to the equation:

$$t_m = \frac{L}{v} \cdot E_g \cdot W \tag{1}$$

in which t_m = the mean residence time in the column
L = the bed length
v = the real velocity of the gas in the bed
E_g = the porosity function
W = the adsorption capacity parameter

In fact eq. (1) is a simplified form of the first statistical
moment derived from the solution of the differential equations
covering the dynamics of the adsorption process in Laplace co-
ordinates. The identification of the first statistical moment with
the residence time is allowed when Gaussian distributions are involved. To find the effect of a strongly adsorbed contaminant on
the shape of a pulse of a weakly adsorbed gas we assumed that the
adsorption of the contaminant proceeds at such a high·rate that
there is a distinct adsorption front and that no adsorption of the
gas occurs in the contaminated part of the bed. This is equivalent
to a definition of an effective bed length according to:

$$L_{eff} = L \ (1-\theta) \tag{2}$$

in which θ is the spent fraction of the bed.

The capacity parameter W can be derived directly from the
adsorption isotherm. When water vapour is present the isotherm is
influenced because a two component adsorption process is occurring
then. As a result it is doubtful whether the effect of water can
be derived from the influence the water has on the pulse of a
weakly adsorbing gas.

A similar discussion can be built up around the efficiency
of the adsorption process, expressed in the Height of a Transfer
Unit. The H.T.U. is related to the ratio of the first and second
statistical moment:

$$HTU = \frac{\mu_2}{\mu_1^2} \cdot L \tag{3}$$

When a contaminant is preadsorbed on the bed an effective length should be used again in equation 3. As both the first and the second statistical moment contain the capacity parameter derived from the adsorption isotherm and in addition the second statistical moment contains the rate of transport in the interior of the pores, both moments are strongly influenced by water vapour. The influence of water vapour can be derived from an expression of the HTU in terms of the dynamics of the process. In this case again a simplified expression is used, the HTU is divided into an internal and an external contribution. The external contribution refers to axial dispersion and mass transfer limitations when entering the particles. The internal contribution is governed by one overall transfer coefficient. In this way one arrives at the following expression for the HTU:

$$HTU = HTU_{ext} + HTU_{int} \qquad HTU_{int} = \frac{d_p v}{6E_g K} \qquad (4)$$

in which d_p is the particle size and K the overall mass transfer coefficient in the interior. This coefficient is a complicated function of diffusion coefficients for the interior of the particles and the rate constant for adsorption. It may be imagined that K is a complicated function of the amount of water vapour in the bed.

The statistical moments can also be derived from the experimental elution peaks thus allowing a comparison between theory and experiments (10). The first statistical moment is defined by:

$$\mu_1 = \int_0^\infty C(t) t \, dt \qquad (5)$$

in which μ_1 is the first statistical moment, t the time elapsed from the moment of injection and C(t) the output signal normalised according to:

$$\int_0^\infty C(t) \, dt = 1 \qquad (6)$$

The second statistical moment is defined as:

$$\mu_2 = \int_0^\infty C(t)(t-\mu_1)^2 \, dt \qquad (7)$$

To find the statistical moments from the observed elution peaks the expression for the complete elution peak must be integrated in three different ways (Eg. 5, 6 and 7) which is rather laborious. If equations 5 and 7 are applied to a normalised Gaussian distribution it turns out that:

$$\mu_1 = t_m \qquad \text{and} \qquad \mu_2 = \sigma^2 \tag{8}$$

in which t_m is the elution time or the mean residence in the column and σ the standard deviation. Consequently μ_1 and μ_2 may be estimated from the time t_m at which the output signal acquires its maximum and from s, the width of the peak at half the peak height. In case of a Gaussian residence time distribution the latter is related to the standard deviation according to $s^2 = 0.18033\sigma^2$. The first two moments of ethane and propane have been estimated from their elution peaks and from that of an inert gas (methane) according to:

$$\mu_1 = t_m - (t_m)_{inert} \tag{9}$$

$$\mu_2 = s^2 - (s^2)_{inert} \tag{10}$$

2.2 Experimental

A flow diagram of the apparatus is given in Fig. 4. It contains an air conditioning system, an injector assembly, the charcoal bed under investigation and instruments for analysis. The humidity of the air stream can be adjusted by mixing two streams of air of which one has been saturated with water and the other dried, in the desired proportion. A seperate air stream is added carrying the required amount of chloropicrin. The injector assembly for the injection of the test gases propane and ethane and for the inert reference gas methane is shown in Fig. 5.

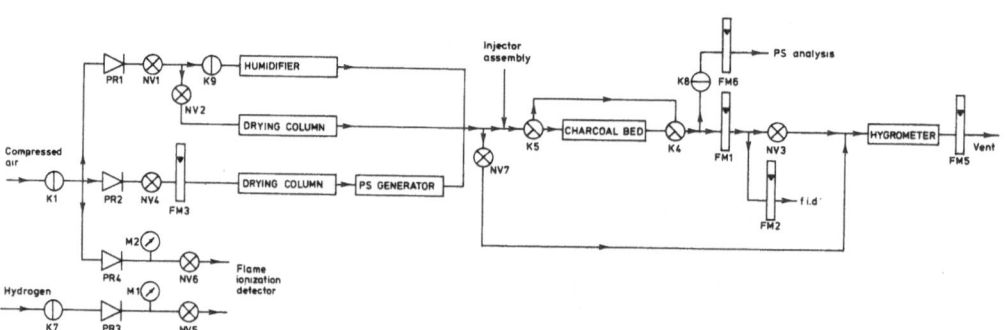

Fig. 4 Flow scheme of the adsorption apparatus.

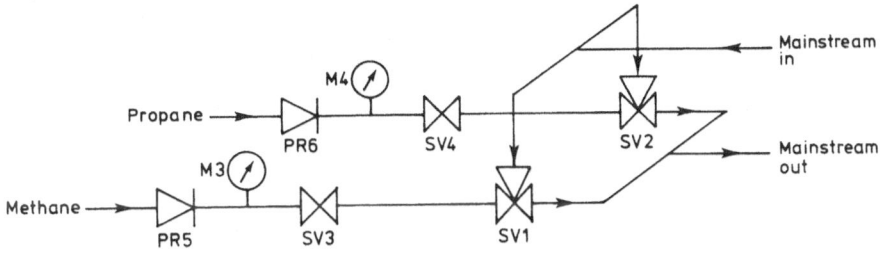

Fig. 5 The injector assembly

The various solenoid valves are powered by a switch roller which opens either SV1 or SV2 for a few tenths of a second.

The charcoal tested was Norit impregnated 1 mm charcoal pellets packed in a glass tube with an internal diameter of 2 cm. The bed height was 7 cm in all cases. Special attention was given to obtain a reproducible packing by filling and vibrating under standardised conditions.

The chloropicrin concentrations in the influent and effluent air were determined spectrophotometrically. The elution peaks of methane, ethane and propane were detected with a flame ionisation detector and registered on a recorder. A wet and dry bulb hygrometer was used to measure the relative humidity. Prior to a series of experiments the charcoal bed was equilibrated with air of the desired relative humidity.

2.3 Results and discussion

According to the modified equation (3) a plot of $\mu_1^2/\mu_2 L$ against the residual capacity $1 - \theta$ should show a straight line with a slope of 1/HTU. Such plots are given in Fig. 6 and 7. for ethane and propane respectively (7).

In most cases the individual points deviate little from the straight line. The lines do not pass through the origin indicating that some adsorption of the test gas on the contaminated part of the bed occurred. However, this might also be a consequence of the fact that the peak moments have been determined assuming a Gaussian distribution and that deviations of this become the greater the higher the relative humidity.

In all cases the slope of the line decreases as the humidity increases, which indicates that the HTU increases with the water content. Table III illustrates this for the adsorption of propane on non-contaminated charcoal at various humidities.

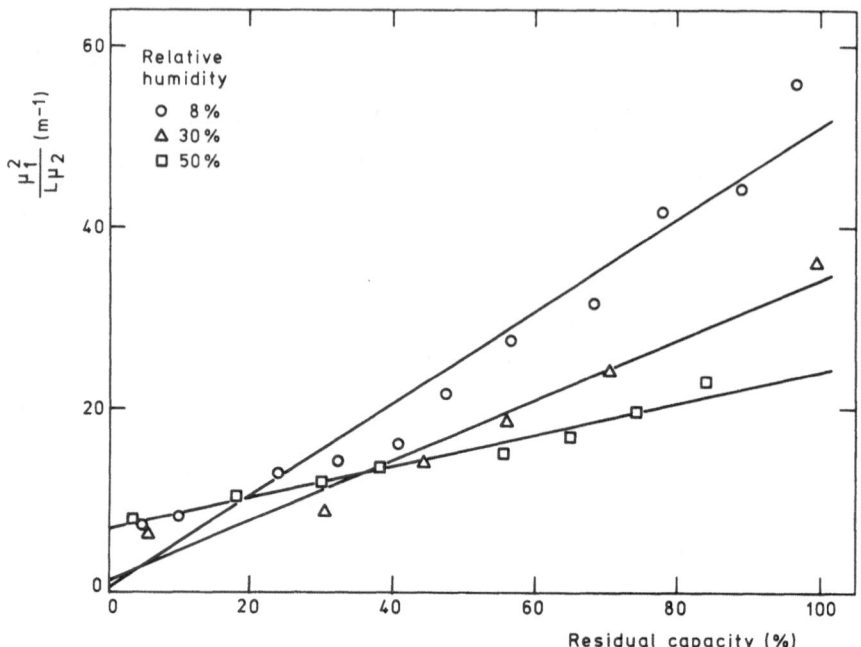

Fig. 6 $\mu_1^2/\mu_2 \cdot L$ determined from the elution peaks of ethane vs the residual capacity of the bed partially spent with chloropicrin.

Table III

The HTU_{int} of propane on charcoal at humidities (V_o = 17.2 cm/sec)

water content (weight %)	R.H. (%)	μ_1 (min)	W	HTU_{int} (mm)
2	8	97.2	20300	8.7
3	29	69.0	14400	8.2
5	48	46.2	9650	12.0
26	65	5.63	1180	23.0
34	75	0.59	123	95.6
40	84	0.17	48	207
40	90	0.10	21	439

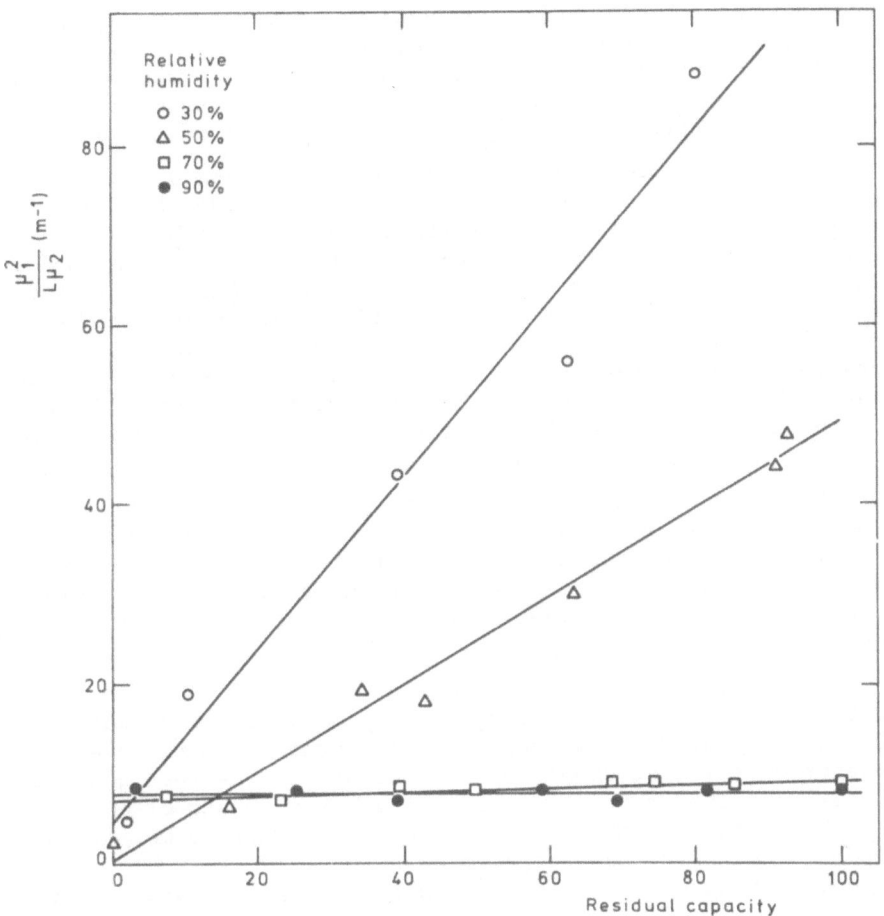

Fig. 7 Same as Fig.6, for propane.

Table III shows that as long as the relative humidity remains below 50% the water content does not exceed a few percent. At higher relative humidities the water content rises rapidly owing to capillary condensation of water vapour in the micro pores. This is attended by a sudden drop in adsorption capacity by a factor of 1000. Simultaneously the HTU_{int} increases from 1 cm to about 0.5 m!

At high relative humidities the elution peaks of ethane and methane are so much alike that it must be concluded that ethane does not adsorb on wet charcoal (in equilibrium with relative humidities over 60%). This means that ethane cannot be used to determine residual capacities in the case of moist charcoal. At the higher relative humidities propane still shows an appreciable adsorption, however, the difference with the adsorption on charcoal contaminated with chloropicrin is still much too small. This is clearly shown in Fig. 7 where the lines corresponding to high relative humidities are horizontals and no effect of the chloropicrin loading on the value of μ_1^2/μ_2 . L can be noticed.

It can be concluded that also propane is too weakly adsorbed to make it a suitable test gas. Gases that are adsorbed sufficiently well on moist charcoal are bonded that strongly that they can not be eluted within a reasonable period of time. This should be a prerequisite for a non-destructive test based upon a pulse technique.

The parameters given in equation (1) and (4) show another prerequisite for a non-destructive test. The influence of water vapour and adsorbed water on the capacity parameter W and the overall mass transfer coefficient should be equivalent for the test gas and the contaminant. Therefore, adsorption isotherms, heats of adsorption or polarisabilities must be comparable. In order to meet this requirement Maggs (8) has proposed an alternative test. It can be derived that as soon as a gas is injected into a charcoal bed a concentration gradient throughout the whole bed is built up (6). This concentration gradient is described approximately by the expression:

$$\ln C/C_o = \frac{-6E_g K.L}{vd_p} (1 - \theta) + \frac{6K.t}{d_p W}$$

From this expression it is clearly seen that even at zero time a quantity of adsorbate emerges from the bed. Maggs uses a known amount of a relatively heavy compound (bromobutane) that is injected into the influent of the filter. Most of the vapour is adsorbed in the bed but a very small amount passes the bed unadsorbed and is detected with a very sensitive method. This amount is compared with the quantity injected according to eq. (11). It has been established experimentally that the reduction ratio (output/input) increases exponentially as the residual capacity for the test compound decreases, which is in agreement with eq. (5). However, the argument still holds that the influence of water vapour on the overall mass transfer coefficient should be equivalent for test gas and gases to be adsorbed.

Although it has been demonstrated that the use of a pulse of a gas does not lead to a non-destructive test of residual capacity under all conditions of humidity, the method is valuable to determine the water content of filters in a non-destructive way (9). It was found that a unique relationship exists between the peak height of an ethane pulse and the water content as demonstrated in Fig. 8. The pulse technique is also quite useful to detect leaks or channels in a charcoal filter. A leak permits a small volume of gas to move relatively undisturbed through the bed. Consequently a double elution peak is obtained; the first peak represents the volume of gas passing through the channel and the second peak represents the volume of gas submitted to adsorption and desorption in the charcoal layer. The height of the first peak has been found to be a semi-quantitative measure for the size of the leak (9). Table IV shows the relation between the concentration of sarin (isopropyl methylphosphonofluoridate) emerging immediately from the bed and the height of the first peak.

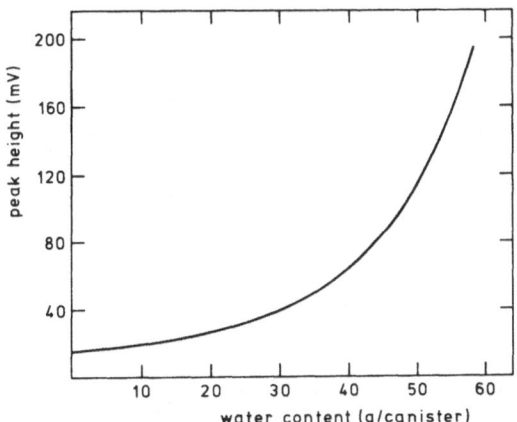

Fig. 8 Ethane peak height vs water content of charcoal in a cannister.

Table IV
Relation between effluent sarin concentration at zero time
and the height of the first peak of an eluted pulse of ethane

effluent sarin concentration (mg/m^3)	height of the first peak (mV)
0.001	0*
0.5	46
6	46
35	118

*no double elution peak present

3. CONCLUDING REMARKS

To know the protective capacity of certain types of filters as
regards certain toxic compounds under working conditions, each
time a breakthrough test has to be carried out. It would be very
useful if these protective capacities could be calculated from the
known adsorbate and adsorbent properties. As long as this is not
possible to the desired degree of accuracy it would be very helpful to have a simple non-destructive test that would allow the
determination of residual capacity without destroying the filter.

Both from a theoretical and from an experimental point of
view the so-called pulse technique is unable to provide such a
non-destructive test, although very useful information can be obtained with respect to water content and the proper functioning
of the filter. So far a semi-non-destructive test using a heavy
test agent that can be measured with great sensitivity, seems the
most promising.

REFERENCES

1. Bonjer, F.H., TNO-Nieuws 1966, 21.
2. Leger, A.E., Barnes, W.D., Development of the ethane tester,
   Defence Chemical, Biological and Radiological Laboratories,
   Report No. 495, Ottawa, 1966.
3. Muhlbauer, D.R., Standardised non-destructive test of carbon
   beds for reactor confinement application, Dupont Savannah
   River Laboratory, Report DP 1082, Aiken, 1966.
4. Bollen, L.J.M., Van der Klooster, H.W., Non-destructive test
   for charcoal filters. I. The pulse technique as a non-destructive method to determine the residual gaslife of charcoal filters, Report No. 1972-19, Chemical Laboratory TNO, Rijswijk,
   1972.

5. Van der Klooster, H.W., Non-destructive test for charcoal filters. II. Application of the ethane pulse technique. Report No. 1972-24, Chemical Laboratory TNO, Rijswijk, 1972.
6. Van Dongen, R.H., Stamperius, P.C., The dynamics of adsorption of vapours by porous adsorbents in fixed beds, Report No. 1974-4, Chemical Laboratory TNO, Rijswijk, 1974.
7. Van Dongen R.H., Non-destructive test of charcoal filters. III. The characteristics of the elution peaks of ethane and propane of partially spent charcoal filters of various moisture contents. Report No. 1974-5, Chemical Laboratory TNO, Rijswijk, 1974.
8. Maggs, F.A.P., Ann. Occ. Hyg. 15, 351, 1972.
9. Stamperius, P.C., Van der Klooster, H.W., Protection against toxic compunds, Leaks in charcoal beds and the water content of charcoal determined using the ethane pulse technique, Chemical Laboratory TNO, Rijswijk, 1973.

THERMAL ACTIVATION

A. J. Juhola

MSA Research Corporation, Evans City, Pennsylvania

ABSTRACT. Thermal processes are described for the manufacture of activated carbons. Pertinent properties of the raw materials are discussed as to their effect on the pore structure and other physical and adsorptive properties of the activated carbons produced. Each step of the processes, namely: 1. preparation of raw materials for thermal treatment; 2. carbonization; 3. activation; and 4. product preparation, is discussed in regard to the effect operating conditions have on the final product. Reference is made to types of equipment or machinery used in the performance of each step. A brief description of test procedures are included to permit characterization of product from each step.

INTRODUCTION. This lecture deals with the preparation of activated carbons from a variety of raw materials by procedures that have been used commercially. The processing can take several different directions depending on the properties of the raw material and the desired quality and type of ultimate product. Although, potentially, it is possible to take any carbonaceous raw material and convert it to any type of carbon, it is not always economical to do so.

The flow diagram in figure 1 shows the different paths a raw material may be directed through the process. A raw material, such as peach pits can be first sized; i.e., crushed to granules, and then carbonized and activated, or it can be first carbonized and then sized before activation. Some materials are easier to size after they have been carbonized. Raw materials, such as lignite, semihard coal, hard coal, can be activated directly after the sizing operation, the carbonization step being by-passed. Soft coals, because of their coking properties, must be taken through

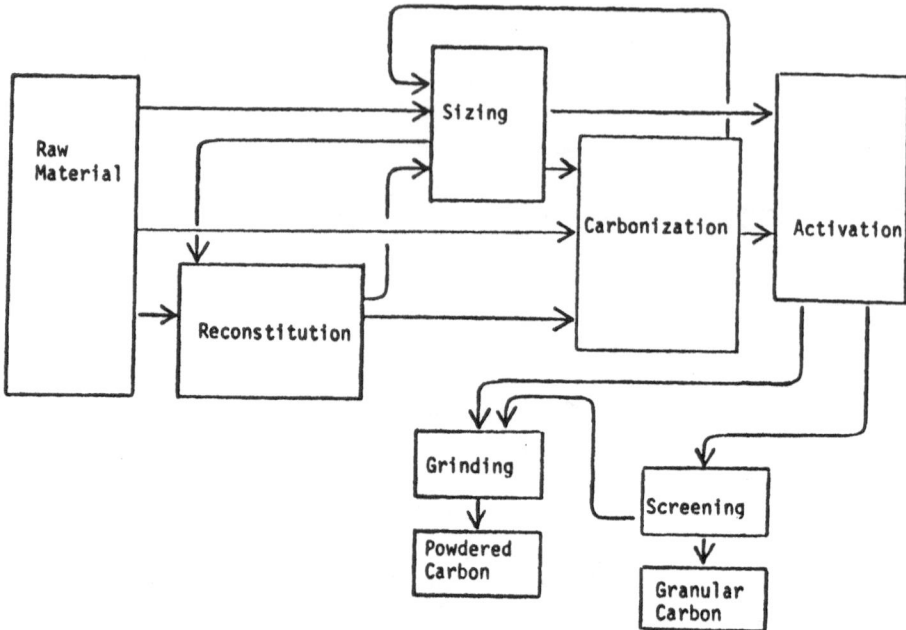

Figure 1. Flow diagram for manufacture of activated carbons

the reconstitution process which consists of grinding the coal to powder and putting it back together again by briquetting, extruding or pelletizing. Pellets and small extrusions go directly to the carbonization step while large extrusions and briquettes must be broken down to granular form first. Raw materials, other than soft coal, are also reconstituted to produce a modified product. Fines produced during the crushing operations are in some cases reconstituted and recycled through the process. The activated product, if too soft to be used in granular form, is ground to powder. Granular carbons are separated by screening to various granule size ranges. Fines and undersized granules can be converted to powder.

The processing scheme of figure 1 has a number of unit operations after the introduction of the raw material which will be topics of discussion in this lecture. In order to discuss the changes occurring to the material on passing through each unit operation, a brief description of the various tests used to characterize the product is needed. The lecture is developed around six topics or operations and will be presented in the following order: 1. product characterization, 2. raw materials, 3. preparation of raw material for thermal treatment, 4. carbonization, 5. activation, 6. product finishing.

1. PRODUCT CHARACTERIZATION

1.1 Surface properties

The two fundamental properties of activated carbons are those of
the surface and pore structure. A pure carbon surface is nonpolar.
Because of this property, it is a poor adsorbent for small highly
polar molecules, but a very good one for hydrocarbons. It is also
this property that has made activated carbons so uniquely suited
for pollution control applications.

Almost all carbons contain surface impurities in some form,
such as alkaline and alkaline earth hydroxides, as in carbons
from vegetable origin, or, such as silica, alumina and iron oxides,
as in carbons from coal. Also, depending on the final activation
temperature and post treatment, the surface may be partially
covered with chemisorbed oxygen. All of these impurities tend to
make the carbon more polar; i.e., have some of the characteristics
of polar adsorbents, such as silica gel. In some cases slight
polarity is desirable. Basicity produced by the alkaline hydroxides
is desirable in carbons used for sorption of radioactive iodine
and methyliodide.

1.2 Pore structure

The pore size distribution curve shows graphically the structure
of a carbon. The use of these curves is also a very lucid way to
show the changes to the carbon pore structure resulting from each
processing step. In figure 2 are pore size distribution curves
of commercial carbons selected to show variations in structure.

For future reference, pores smaller than 30A diameter will
be referred to as micropores and those larger than 30A as macropores. The volumes of pores smaller than 600A are measured by
adsorption techniques while those of pores larger than 30A are
measured by mercury porosimetry. Computed diameters larger than
100,000A are not true pores but irregularities in the granule
exterior surface. Mercury porosimetry measurements on powdered
carbons (10μ particle diameter) do not differentiate between
pores and intergranular space, hence this method is not suitable
for powdered carbons.[1,2,3,4]

Over 90% of the surface area of the carbon is usually in the
micropores. The micropore volume very often is the measure of the
adsorptive capacity. Macropores are also very essential because
they provide passageways through which the adsorbate molecules
can rapidly reach the micropores in the interior of the granule.
For the carbon to be effective, a balance is required between the

274

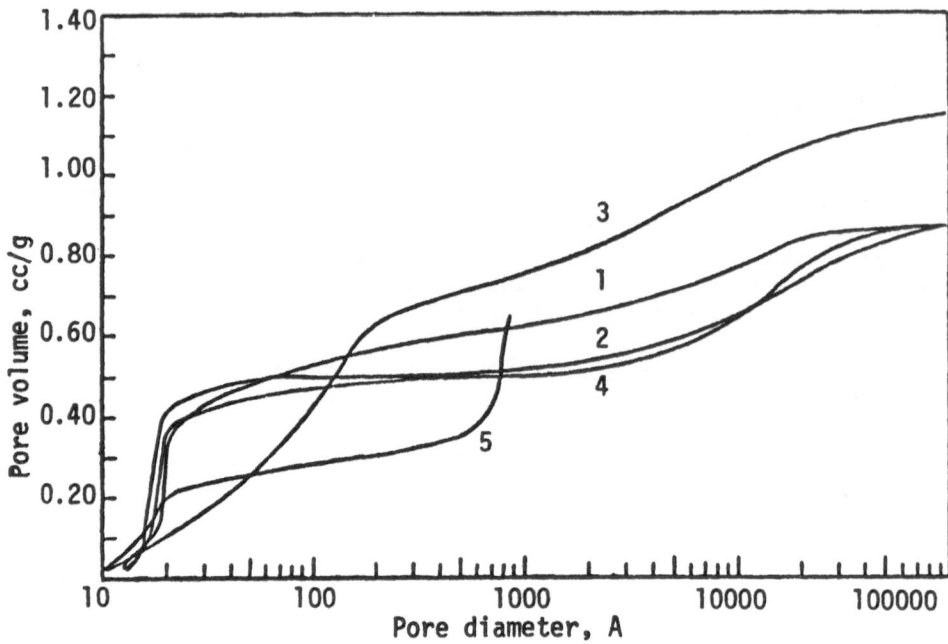

Figure 2. Pore size distribution curves of commercial activated carbons

micropore and macropore volumes. Generally, it is desirable for a carbon that is to be used in liquids to have a larger macropore volume, even if it requires a decreased micropore volume. For a gas-phase carbon a closer balance between micropore and macropore volume is desirable.

All carbons represented by the curves in figure 2 are commercially available. Carbons represented by curves 1, 3 and 5 are used primarily in liquid-phase applications while those for 2 and 4 are used in gas-phase applications. Carbons 1 and 4 are reconstituted, bituminous-coal base, granular carbon manufactured by Pittsburgh Carbon Co. Carbon 3 is a lignite base, granular carbon manufactured by Atlas Powder Co. Carbon 5 is a powdered carbon manufactured by West Virginia Pulp and Paper Co. from paper manufacturing waste materials. Carbon 2 is a granular carbon made from petroleum coke by Union Carbide Corp.

Curve 1 and other data on this carbon will be used as standards of reference when discussing other carbons. For this carbon the following data are presented for future reference: total pore volume - 0.87 cc/g, macropore volume - 0.45 cc/g, micropore volume - 0.42 cc/g, macropore surface area - 110 sq m/g, micropore surface area - 910 sq m/g.

Many of the pore size distribution curves presented are from a declassified U.S. government report.[5] Others are from in-house studies at MSA, sales literature and obtained from carbon manufacturers.

1.3 Adsorption tests

Pore size distribution curve determinations take considerable time and require special equipment and technically trained personnel. Their determinations are not suited for routine testing or quality control. For this purpose a number of adsorption tests have been developed. Those that have been related to the pore structure and will be referred to are: iodine number, carbon tetrachloride capacity, and molasses number.

Iodine number. The iodine number is the amount of iodine adsorbed in mg/g of carbon from an aqueous solution of potassium triiodide when the final iodine concentration is 0.02 mg/l. If granular, the carbon is first ground to particle sizes less than 44μ diameter and then treated with HCl. The acid treatment nullifies the effects metallic oxides might have on the iodine adsorption. This test is used to evaluate carbons intended for liquid-phase applications.

The iodine number has been correlated with the surface area of pores larger than 10A in diameter,[6] according to the equation

$$I_2 \text{ No.} = 17 + 1.07 \times (\text{s.a. of pores} > 10A) \tag{1}$$

For carbon 1, the surface area of pores larger than 10A diameter is 965 sq m/g which gives a calculated iodine number of 1040 mg/g while the actual test iodine number was 1090 mg/g.

Carbon tetrachloride capacity. The carbon tetrachloride capacity test is used to evaluate carbons intended for gas-phase application. In this test, granular carbon at 40°C is exposed to carbon tetrachloride vapors (in vacuo) at 32.9 mm pressure until equilibrium is established. The amount adsorbed, reported as weight percent, is the carbon tetrachloride capacity. The liquid volume of carbon tetrachloride adsorbed under these test conditions correlates closely with the micropore volume. For carbon 1 the volume is 0.45 cc/g, quite close to 0.42 cc/g micropore volume.

Molasses number. These tests are used to evaluate carbons for sugar decolorizing. They are performed on powdered carbons of less than 44μ particle diameter on a standard molasses solution. The molasses number correlates with the surface area of pores larger than 28A diameter.[6] The equation for the correlation is

Molasses No. = 129 (s.a. of pores >28A)   (2)

For carbon 1, the surface area of pores larger than 28A diameter is 120 sq m/g which gives a calculated molasses number of 249 while the measured number was 250.

The correlation is due to a rapid filling of the micropores with sucrose present in the molasses solution. Although the color bodies are more strongly adsorbed than sucrose, the interchange of color bodies with sucrose in the micropores is so slow that during the test the micropore surface contributes very little to the molasses decolorization. The macropore surface is accessible to the color bodies; hence, the decolorizing power correlates with the surface area of pores larger than 28A diameter.

Density. Three types of density measurements can be used in characterizing the carbon products. These are bulk, granule and real densities. Bulk density is calculated from the weight of carbon in a known volume of a container filled with the carbon. This density can vary considerably with the mesh size distribution and the way the container is filled. A reproducible method for filling the container is necessary. The void volume can vary from 0.30 to 0.45 fraction of the container volume. This density, however, can be measured quickly and is, therefore, used for control purposes during activated carbon production.

In determining the granule density (also designated as particle density), the volume of a weighed amount of carbon is measured by mercury displacement. A mercury pressure is used sufficient to fill the intergranular void and outline the granules. For this purpose a 900 mm Hg pressure is used. The granules are outlined but no mercury can penetrate pores or surface identations smaller than 100,000A diameter.

Volume measurement by helium displacement is used in determining the real density. Helium penetrates the pore structure but is negligibly adsorbed. The impervious solid volume measured by this method gives an indication of degree of activation, since the ultimate density for carbons approach that of graphite, 2.2 g/cc.

For carbon 1 the three densities are: bulk density - 0.47 g/cc, granule density - 0.78 g/cc, real density - 2.23 g/cc. The granule and real densities give the necessary data for calculating the total pore volume. For carbon 1 the total pore volume is 0.83 cc/g, but a more meaningful figure is the pore volume per cc of granules, which is 0.65 cc in this case.

Ash analysis. Total ash analysis is done by burning off the carbon and weighing the residue which now consists of the oxides of the various metals initially present. Most carbons have over

5% ash. In general, it is undesirable to have large amounts of ash. At times an analysis of the elements is desirable particularly if there is a possibility of them being leached out and contaminating the product being purified.

For carbon 1, the total ash is 5.7% of which 85% is SiO_2 and Al_2O_3.

Other tests. Granular carbons are also tested for hardness and abrasion resistance, since these properties at times determine the suitability of the carbon for various applications where frequent transport or handling could cause considerable granule breakdown if too soft. Granule softness is also the deciding factor between powdered and granular carbons.

2. RAW MATERIAL

2.1 General

Table 1 lists types of materials that are used. Included in the table are properties that affect the way the material must be treated and also determine the type of final product that it produces.

Carbon content, volatile content and density go together. Up to a point, the suitability improves with increased carbon content and decreased volatiles content. Increased density is always desirable. It is also desirable to have a low ash content, unless for some special reason, an activated carbon is wanted that has a high inorganic content.

Table 1. Types of raw materials for activated carbons and some properties calculated on dry basis.

Material	Carbon,%	Volatile,%	Density,g/cc	Ash,%
Soft wood	40	70	0.35 to 0.5	0.2 to 1.0
Hard wood	40	70	0.5 to 0.8	0.2 to 1.0
Nut shells	40	60 to 70	1.3	0.5
Lignite	50 to 70	40 to 60	1.05 to 1.35	6.0
Soft coal	60 to 80	20 to 30	1.2 to 1.5	1 to 15
Semihard coal	70 to 80	10 to 20	1.4	1 to 15
Hard coal	80 to 95	5 to 10	1.4 to 1.8	1 to 15
Lignin	40	70	---	---
Petroleum coke	70 to 80	10 to 20	1.4	0.5
Other byproducts	---	---	---	---

278

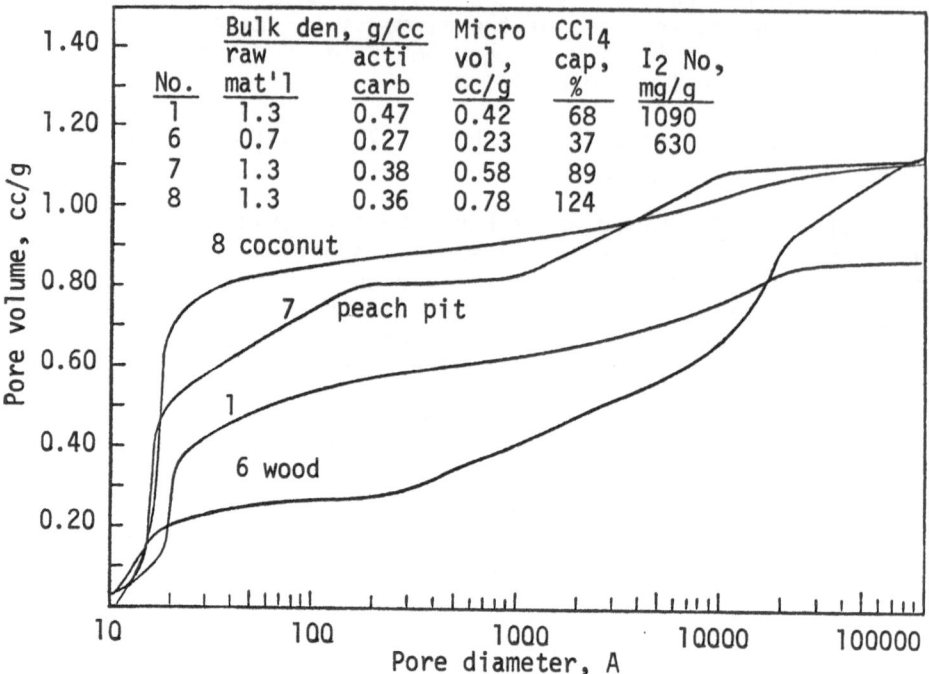

Figure 3. Effect of raw material density on the pore size distribution curve of final activated product

2.2 Vegetable base

Soft wood, hard wood, and nut shells are about 70% cellulose structure with 18% to 28% lignin. The predominant chemical group is -CHOH- which has a carbon content near 40%. On pyrolysis the hydroxyl groups and hydrogen are driven off as water leaving a skeletal carbon structure as determined by the initial cellulose fibers. The skeletal structure retains the original rigid form of the starting material throughout the carbonization and activation; hence, granular activated carbons can be made from them.

In this group of materials there is a large difference in densities, ranging from 0.35 to 1.3 g/cc. Activated carbons prepared from the low density (0.3 to 0.8) materials, by straight thermal treatment, produce activated carbons that have open pore structures, low densities, and are very soft, while the high density nut shells produce carbons of fine pore structures, higher densities and of much greater hardness. Figure 3 shows curve 6 for a wood-base carbon, curve 7 for a peach pit carbon and curve 8 for a coconut shell carbon. The micropore volumes of carbons made from the denser materials are larger as are also the iodine numbers and

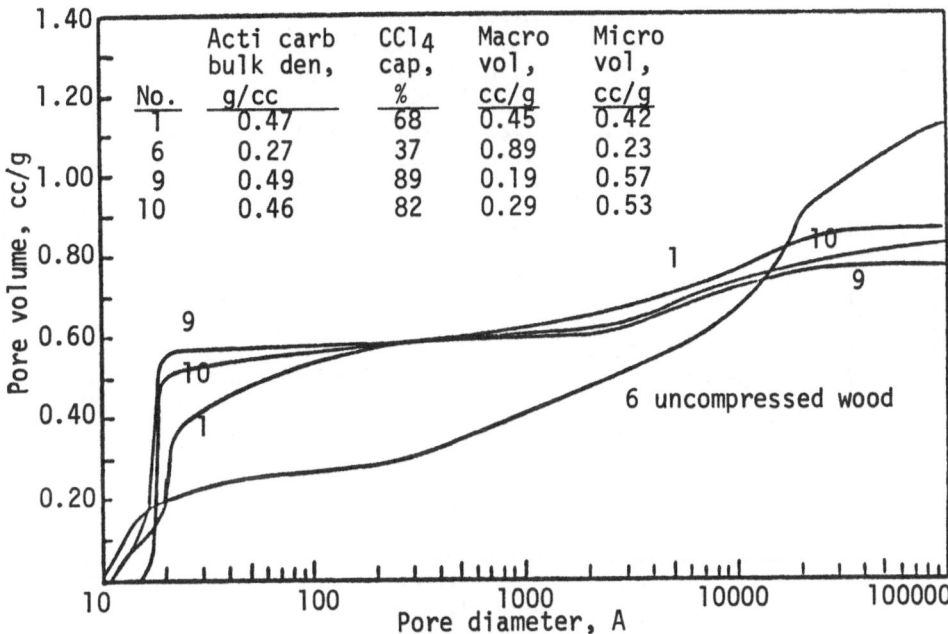

Figure 4. Effect of compression on wood char during carbonization or activation

carbon tetrachloride capacities.

High activity carbons can be made from wood by using compression during the carbonization or activation process. Figure 4 shows two pore size distribution curves, 9 and 10, that were made by such techniques. Curve 9 is for a carbon made from hardwood sawdust by chemical activation using zinc chloride as dehydrating agent. The char, while still hot from the dehydration reaction,is extruded under high pressure, which causes breakdown of the large pores into smaller ones.

Carbon for curve 10 was prepared from pressed logs commercially manufactured for fireplace use. These logs were made of sawdust and binder. During carbonization they were forced through a tube under pressure while being heated. This treatment caused a large shrinkage and, thereby, produced a much higher density and harder final product than would otherwise be obtained without compression.

On comparing curve 6 with curve 9 and 10, note the large decrease in macropore volume and correspondingly large increase in micropore volume. The densities increase in going from carbon 6

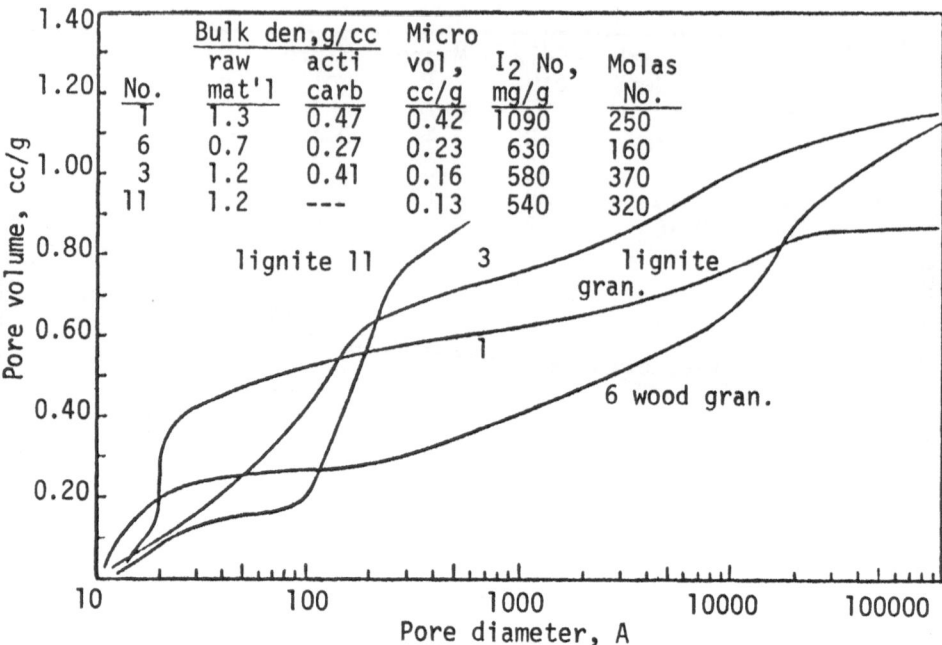

Figure 5. Pore size distribution curves for lignite-base carbons

to 9 and 10 and, also, the adsorptive capacities increase. Carbons 9 and 10 are much harder than 6. On the negative side carbons 9 and 10 are much more expensive than 6 to manufacture.

2.3 Coal base

In the transition from lignite through soft coal and semihard coal to hard coal, the material loses the cellulose wood structure by formation of free carbon. This transition is accompanied by a decrease in volatiles content.

Lignite. Activated carbon can be made from lignite by straight thermal treatment and steam activation. Granular carbons made in this way are soft and are usually ground to powder. Figure 5 shows curve 3 for a lignite-base granular carbon and curve 11 for a powdered carbon. There is some resemblance between the two distribution curves although made at different dates and, therefore, subject to differences due to changes in lignite properties. The lignite carbons, just as the wood carbons, present an open pore structure with fairly low micropore volume, low density, and low iodine number. These carbons have been used effectively in water purification.

Soft coal. They contain 60% to 80% fixed carbon and 20% to 30% volatiles. When these coals are heated, they go through a plastic state and the volatiles released during this period cause the coal to bloat and, thereby, make the char unsuitable for subsequent activation.

One way the damage from bloating can be avoided is by raising the carbonization temperature very slowly. Carbonized granules produced this way, however, crumble very easily. The crumbling appears to be promoted by the ash layers in coal. Only powdered carbons can be produced.

Another way to avoid bloating damage is to reconstitute the coal; i.e., grind the coal to a fine powder, 1 to 150μ particle diameter, and reform the granular material by briquetting or extruding. Then by a carefully controlled temperature schedule, hard granular carbonized char is obtained which can then be steam activated at normal rates. Carbonization reduces the volatiles content from near 25% to about 10% at which bloating does not readily occur. The reconstituting process develops macropore volume in the coal granules through which the gases can escape. Carbon 1 is an example in which the briquetting step was used in

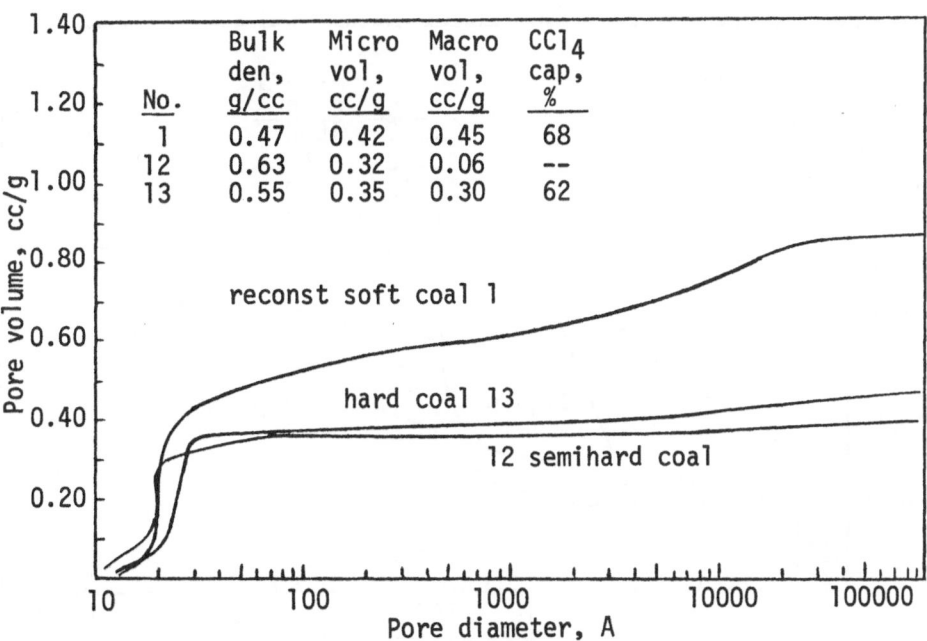

No.	Bulk den, g/cc	Micro vol, cc/g	Macro vol, cc/g	CCl_4 cap, %
1	0.47	0.42	0.45	68
12	0.63	0.32	0.06	--
13	0.55	0.35	0.30	62

reconst soft coal 1

hard coal 13

12 semihard coal

Figure 6. Pore size distribution curves for hard and semihard coal-base carbons

the manufacturing process.

Semihard coal. These coals have high densities and, in some,
the volatile content is sufficiently low so that they can be
carbonized without requiring a reconstituting step. Curve 12 in
figure 6 shows the type of pore structure possible with this type
of raw material. This was a hard high density carbon with a very
small macropore volume. By a reconstituting process, the macropore
volume could be increased and with sufficient activation, curve 12
and 1 could be quite similar. Carbon 12 would have to be activated
longer than carbon 1 because of the lower initial volatiles content.

Hard coal. In hard coal we have a material that has interested
activated carbon producers repeatedly. Most attempts at making a
granular product have not been successful. In this case the volatiles content tends to be too low. A 5% volatile content does not
appear to be sufficient to create an initial pore structure which
can then be further developed by activation. Experimental carbons
have been reported where a partially steam-activated coal was
treated with caustic solution and then dried and roasted. A carbon
was produced with a surface area of 900 sq m/g, corresponding to
an iodine number near 850 mg/g.

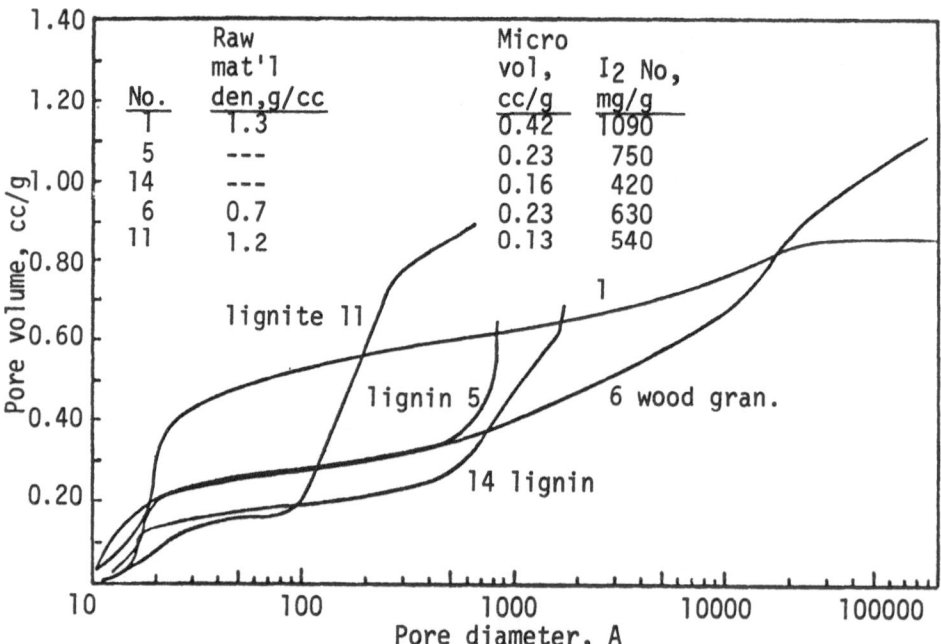

Figure 7. Pore size distribution curves for lignin-base carbons

Experimental studies at MSA Research Corp. laboratories on an anthracite coal of 4.8% volatiles and 11% ash, and on a semi-hard coal of 10.6% volatiles and 14% ash did not produce the sought for high activity granule carbon. After an extended steam activation the granules delaminated at the ash planes.

Recently, an anthracite-base carbon was received from the British Isles. Its pore size distribution curve 13 is also shown in figure 6. It bears considerable resemblance to curve 12 for the semihard coal carbon. Without the reconstitution step these coals do not develop macropore volume to any significant extent. It is quite hard, comparable to coconut carbons.

Lignin. This material is a byproduct of the paper making process. Only powdered carbons appear to be made from this source material. Curves 5 and 14 in figure 7, for lignin-base carbons, are compared with curve 6 for wood-base and 11 for lignite. They have a common feature in that the micropore volumes, and there-fore, the iodine numbers are relatively small.

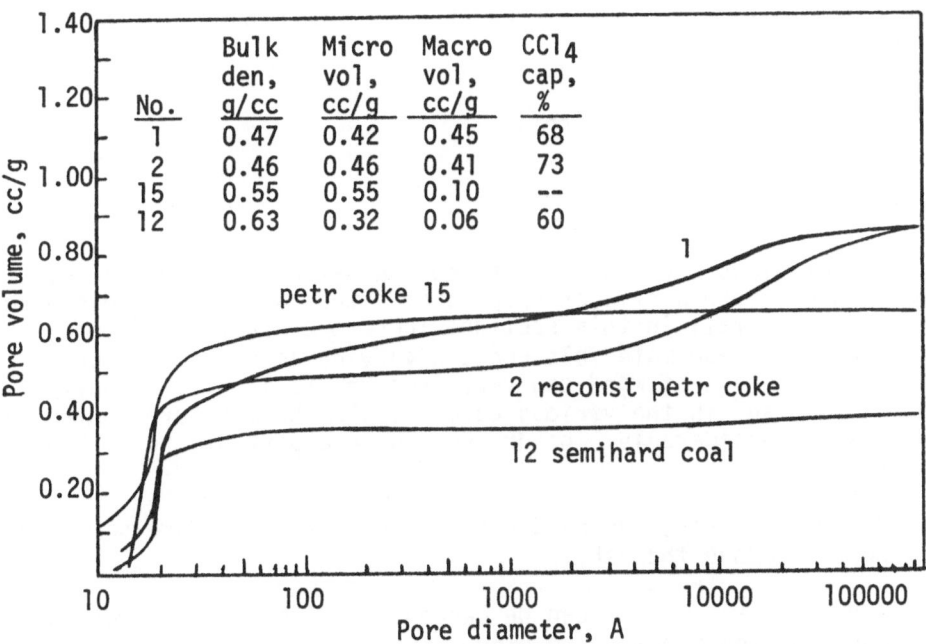

No.	Bulk den, g/cc	Micro vol, cc/g	Macro vol, cc/g	CCl$_4$ cap, %
1	0.47	0.42	0.45	68
2	0.46	0.46	0.41	73
15	0.55	0.55	0.10	--
12	0.63	0.32	0.06	60

Figure 8. Pore size distribution curves for petroleum-base carbons

2.4 Petroleum coke

This material is a byproduct from white oil manufacture. A high-grade oil is treated with sulfuric acid to remove all unsaturated hydrocarbons and on recovery of the sulfuric acid a char residue is formed. This char can be converted to activated carbon by a short carbonization and steam activation. Curve 15 in figure 8 is for one grade of carbon made from petroleum coke. This particular carbon has a relatively large micropore volume with much of it in pores less than 10A diameter. Macropore volume, however, is relatively small, comparable to that found in carbon 12 made from semihard coal.

Petroleum coke tends to produce spherical granules; the ash is low (<1.0%), and hardness is comparable to reconstituted soft coal carbons. By the reconstituting process the balance between micropore and macropore volumes is changed as in curve 2.

2.5 Other materials

Most other materials in one way or another are similar to the preceding types. Among byproduct materials in which interest has been shown, we can list: paper, sewage sludge, corn cobs, thermosetting plastics, reconstituted graphite, tree bark, cloth and blood.

3. PREPARATION OF RAW MATERIAL

3.1 Drying and sizing

In the previous section briquetting and extruding were already mentioned. These are major preparatory steps and will be discussed in greater detail in this section. There are other operations that are required. Raw materials contain varying amounts of water. Moisture content of wood can vary from 36% to 115%, of lignite about 90%, and in the various harder coals the moisture content decreases considerably. The drying can be a part of the carbonization.

Raw materials, the harder coals in particular, are sometimes washed to remove the ash.

Raw materials that can be carbonized without prior reconstituting are broken down into pieces that can be readily handled by the carbonization equipment. Peach pits and other small nuts can be carbonized whole and sized later to a granule size that will give the desired size range in the final product. Lignite is broken

285

up into lumps of about 3 cm diameter. Semihard and hard coal are
also sized to granules that will yield the desired size product.
This is also done to materials that have been reconstituted. Some
materials are byproducts in a finely divided state, such as sawdust, which can be reconstituted for thermal activation or can be
chemically activated.

The granulating operation can be done with a saw tooth crusher,
for the initial breakdown, followed by passes through a roller mill.
The product is screened, with oversized granules recirculated
through the roller mill. In granulating operations, it is difficult
to get over 70% of the desired granule size range. The fines formed,
however, can be reconstituted.

3.2 Reconstitution by briquetting

Material to be briquetted is first crushed to a fine powder by
means of a hammer mill followed by further grinding in a bowl mill.
When the powder has reached the desired particle size, it is then
pressed under high pressure in a hydraulic press to briquettes
that can be as large as 20 cm diameter. The briquettes are then

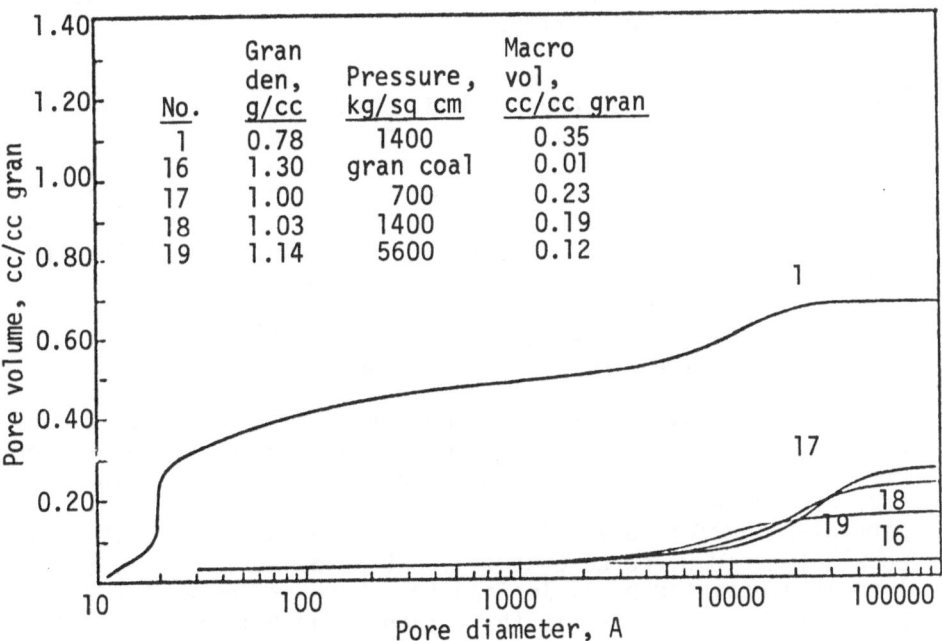

Figure 9. Effect of briquetting pressure on macropore volume with
powdered coal of 1 to 150μ particle diameter

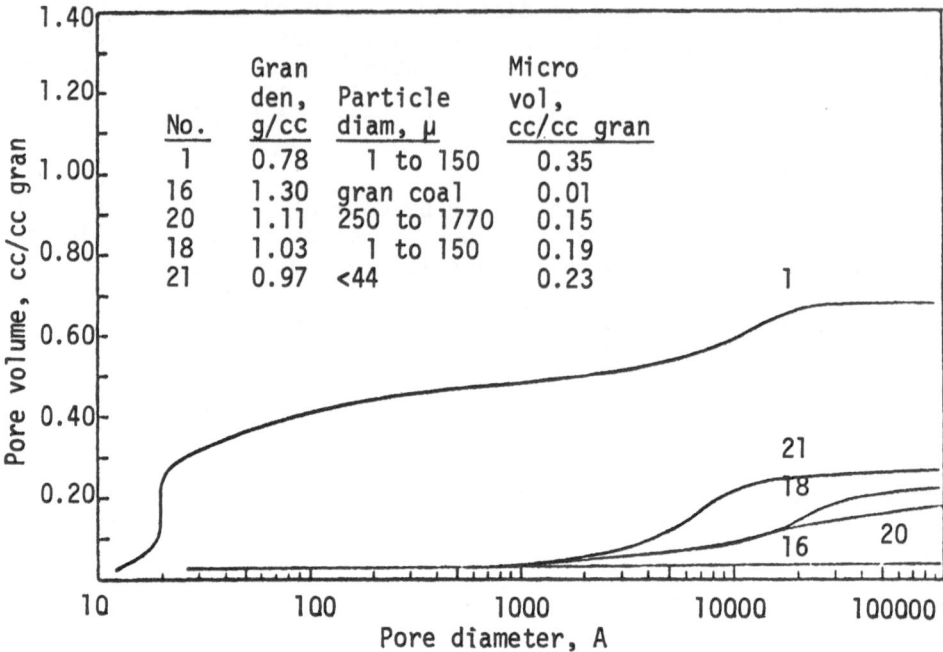

Figure 10. Effect of coal particle diameter on macropore volume
and pore diameter when briquetted at 1400 kg/sq cm

crushed in a saw tooth crusher and further reduced to precise size
in a roller mill. On screening, the oversized granules are recycled through the roller mill and undersized repulverized and
recycled through the process. On carbonization and activation the
granule sizes decrease considerably, hence, at this stage, the
granules must be considerably larger than those of the desired
final product.

To strengthen the briquettes, pitch or flour may be added to
the raw material during the grinding operations.

Experimental work has been done to determine the effect of
briquetting pressure and particle size on product hardness properties. Figure 9 shows the immediate effect of briquetting
pressure on the macropore volume for soft coal powder of 1 to 150μ
particle diameter. Granular coal, curve 16, has virtually no macropore volume (0.01 cc/cc granule). But on pulverizing the coal to
1 to 150μ diameter particles and then briquetting at the pressures given, macropores in the volume range 0.12 to 0.23 cc/cc
granule are built into the granules. Hardness of the granules
increases with briquetting pressure,but since the volume also
decreases, the briquetting process has been optimized at 1400 kg/
sq cm pressure.

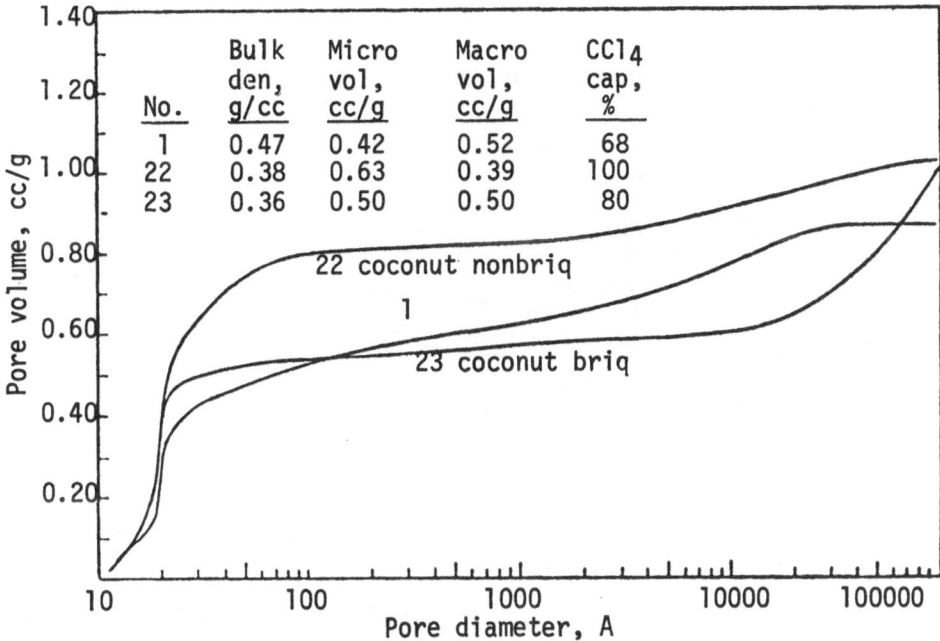

Figure 11. Steam activated carbons made from briquetted and non-briquetted coconut

Note, for these comparisons the cc granule base is used rather than the per gram base since it gives a more meaningful comparison.

Figure 10 shows the effect of particle size on macropore volume when briquetting pressure is 1400 kg/sq cm. The macropore volume increases and the mean pore diameter gets smaller as the particle size is decreased. Briquettes pressed with the 44μ particles tend to develop cracks. Apparently particles of smaller diameter and possibly of more uniform diameter distribution do not tie the briquettes together as those in the 1 to 150μ range.

The effect of briquetting carries over into the pore structure of the final activated carbon. Carbon 1 was made from a briquetted material. The curve shows a 0.36 cc/cc granule macropore volume which is approximately a two-fold increase in volume due to the thermal process.

Other raw materials can also be reconstituted although it is not essential for further processing. Sometimes briquetting does not improve the product. Figure 11 shows the effect of briquetting on coconut. In this case, briquetting reduced the micropore volume

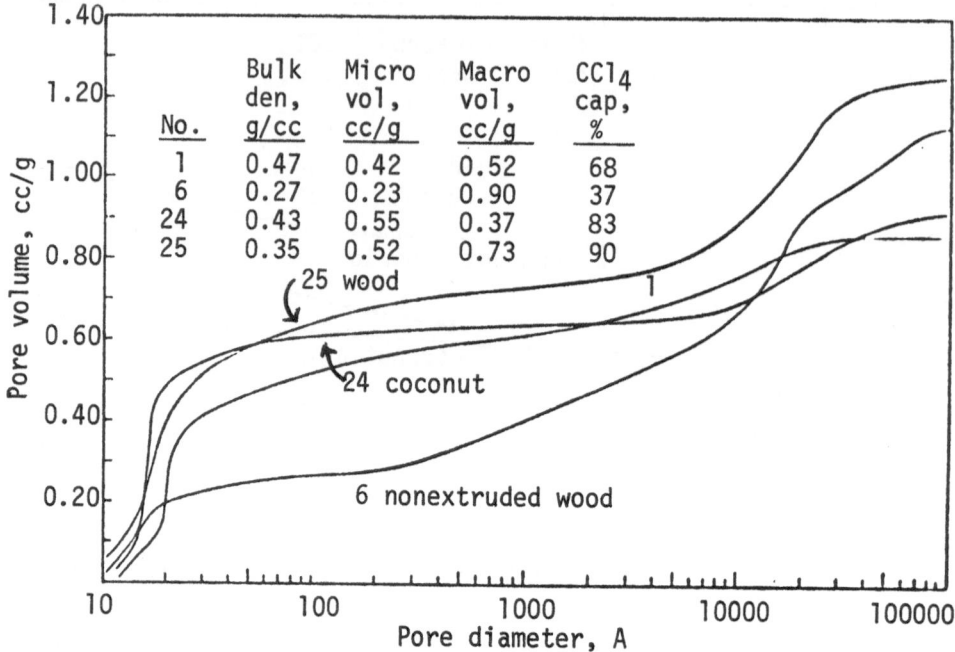

Figure 12. Steam activated carbons made from extruded powdered coconut and wood, curves 24 and 25.

and thus reduced this carbon's effectiveness for gas-phase applications.

3.3 Reconstitution by extruding

Extruded materials have been made in diameters ranging from 0.16 to 18 cm. The small diameter materials are thermally processed as is, except for cutting to length. Activated carbons made from them have some advantages over the granular variety made by crushing the large extrusions or briquettes. They tend to be harder and because the granule size can be controlled to a narrow size range, the resistance to air flow through the granule bed is less.

As in preparation for briquetting, the raw material is ground to a fine powder. It is necessary to add a binder, which also has lubricating properties, and a precisely controlled amount of water. The mixture is thoroughly worked together in mixing equipment, such as the mix-muller. This machine rolls over and scrapes the mixture until a uniformly thick paste is formed. The paste is then forced through the dies by means of an auger at a high pressure. The extrusions, as they emerge from the die, can be cut into precise lengths with a knife. Too viscous a paste can cause the dies to jam while a paste of too low viscosity will give very soft extruded material. Prior to the carbonization step the extruded material is dried at a relatively low temperature, near 100°C.

Figure 12 shows curve 24 for an activated carbon made from an extruded coconut powder. In regard to adsorptive properties, the extrusion step produces no real improvement. In product appearance, hardness and resistance to gas flow through the granule bed there are advantages.

Curve 25 is for a carbon made from extruded powdered wood. When compared to curve 6, the extrusion step produces a much better carbon. The carbon tetrachloride capacity increased from 37% to 90%. There was also a considerable increase in hardness.

4. CARBONIZATION

The primary purpose of carbonization is to prepare the raw material into a form that can be subsequently developed by steam activation. Within limits the carbonization process can be varied to influence the properties of the final activated product.

Carbonization is a thermal decomposition process usually conducted at temperatures below 850°C with an inert sweep gas, such as steam or flue gas, passed through the carbonizer. Important operating parameters are the rate of heating, final temperature, and sometimes the sweep gas if an active one is used.

The end result depends on the type of raw material being carbonized. With wood, fruit shells or pits, or other vegetable origin raw materials, which have a low carbon content (~44%), the objective is to decompose the cellulosic structure to form a rudimentary porosity which is further developed by the subsequent steam activation. The pyrolysis causes considerable shrinkage of the carbon portion. This is promoted by a slow heating rate, producing a stronger and denser carbonized product.

Soft, semihard, and hard coals contain over 70% fixed carbon. With these materials there is more than sufficient carbon present; the volatiles content now tends to be insufficient. The carbonization for these materials should be conducted in a way that will minimize shrinkage and the formation of additional free carbon, but maximize the formation of porosity. Semihard and hard coal can be heated rapidly to activation temperatures (above 800°C) where the steam can rapidly oxidize the organic matter, and also, start enlarging the pores before they can close from shrinkage. Soft coal, because of its coking properties, is treated differently. As discussed previously, it must be reconstituted and even then

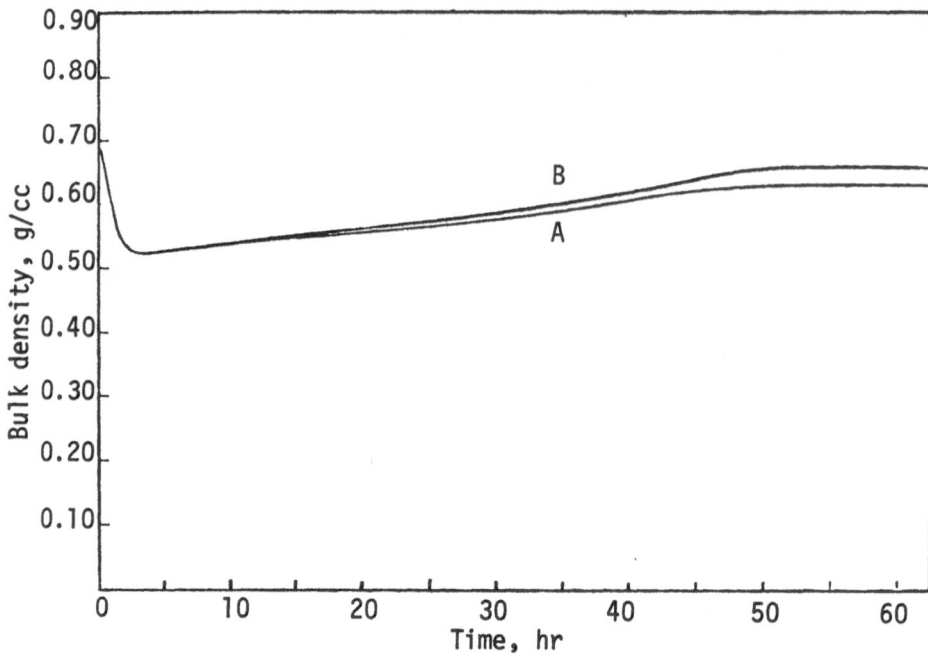

Figure 13. Effect of heating rate from 30° to 820°C on peach pit bulk density, curve A for raw peach pits sized first to 0.8 to 3.4 mm granules and carbonized, curve B for whole pits carbonized first and then sized

a carefully controlled heating schedule is required to avoid ruining the product. Lignite again can be directly activated although a slow carbonization can aid some in strengthening the final activated product.

Lignin, petroleum coke, or other byproduct raw materials require the type of carbonization treatment discussed above,based on relative fixed carbon and volatiles content.

4.1 Temperature, heating rate and shrinkage

Vegetable raw materials. The degree of shrinkage that occurs when dry wood (10% moisture) is carbonized is quite large. On heating a measured and weighed hickory block in an electrically heated oven, from ambient temperature to 615°C and holding at 615°C for one hour, the block volume reduced to a 38% yield and weight to 28% yield. Of the 72% weight yield of volatiles driven off, 68% were condensable. Of the 40% carbon present about 70%

became fixed in the char. The char bulk density was 0.55 g/cc, which is on the light side to give a strong and dense activated carbon.

Considerable data was available from studies at MSA Research Corp. on the effects of heating rates and final carbonization temperatures on peach-pit shrinkage. Figure 13 shows the effect on bulk density when whole pits and pits sized to granules first are heated at different, but steady, rates to 820°C, with nitrogen sweep. The lowest density at 0.55 g/cc and, also, the softest char is produced when the heating time is short at about 4 hr. At slow heat up rates requiring more than 50 hr heating time, the bulk density is maximized. For the carbonized whole pits (but sized to granules) the bulk density levels off at 0.66 g/cc and for the sized carbonized char, at 0.63 g/cc. The best char is obtained by slow carbonization of whole pits, assuming that a largest possible micropore volume (or largest surface area) is desired in the final activated carbon rather than an open pore structure.

Table II. Changes in physical properties and yields with change in final carbonization temperature when carbonization rate is slow for peach pits

Temp°C	Bulk den,g/cc	Granule den,g/cc	Real den,g/cc	Pore vol,cc/g	Yield % vol	wt
(30)	0.68	1.18	1.43	0.15	100	100
450	0.55	0.95	1.77	0.48	40	33
820	0.63	1.07	1.87	0.40	31	28

Table II presents density and yield data on effects of slow heating of sized raw peach pits to 450°C in 28 hr and holding at 450°C for 4 hr, and again heating the same char to 820°C in 12 hr and holding at 820°C for 4 hr. At 450°C the bulk density was 0.55 g/cc, the same as obtained for pits heated to 820°C in 4 hr or hickory heated to 615°C in 9 hr. Most of the volatiles come off at temperatures below 650°C. On further heating to 820°C, the carbon structure goes through further transition without much loss in weight, but considerable increase in bulk and granule densities. At the 450°C temperature, the volume yield was 40%, quite close to the 38% for hickory carbonization. The weight yields are near each other, 33% for pits and 28% for hickory. On further heating of the pits to 820°C, the volume and weight yields decreased to 31% and 28%, respectively.

Although the carbonization produces very similar changes in bulk density, volume and weight yields for peach pits and hickory, the two raw materials produce greatly different activated carbons, as was already shown in the section on raw materials, figure 4. Wood carbons produce carbons of large macropore volume and small micropore volumes, and are also soft. Peach pits, as also coconut shells, pecan nuts, walnuts or all nuts and shells, produce carbons with a more even balance between macropore and micropore volumes. The inherent structure of wood can be modified by applying compression during carbonization or during extrusion of $ZnCl_2$ dehydrated wood sawdust to produce carbons of structure similar to those from peach pits.

With reference to table II, the real density of the char heated to 820°C is still 1.87 g/cc, down considerably from the 2.2 g/cc (graphite density) obtained on final activation. Either not all volatiles had been driven off or the carbon had not adjusted to its maximum density.

Coals. Hard and semihard coals shrink to much less degree than the vegetable raw materials. Experimental studies were conducted at MSA Research Corp. using an electrically heated oven. Coal sized to granule was heated from ambient temperature to 930°C in 3 hr and held at 930°C for 1 hr. Nitrogen sweep was used during the carbonization. The physical properties of coals before and after the carbonization and, also, the volume and weight yields are summarized in table III. Volume and weight yields for the hard coal were 86% and 95% and for the semihard coal 76% and 88%. This change is also registered in the increases in all the densities. The pore volumes of the hard and semihard coal chars were 0.085

Table III. Changes in physical properties during carbonization and yields for hard and semihard coals

Raw materials	Coal	
	hard	semihard
Bulk density, g/cc	0.90	0.87
Granule density, g/cc	1.52	1.40
Real density, g/cc	1.62	1.46
Pore volume, cc/g	0.040	0.030
Char		
Bulk density, g/cc	0.99	1.01
Granule density, g/cc	1.67	1.61
Real density, g/cc	1.95	1.90
Pore volume, cc/g	0.085	0.095
	(0.180)	(0.280)
Volume yield, %	86	76
Weight yield, %	95	88

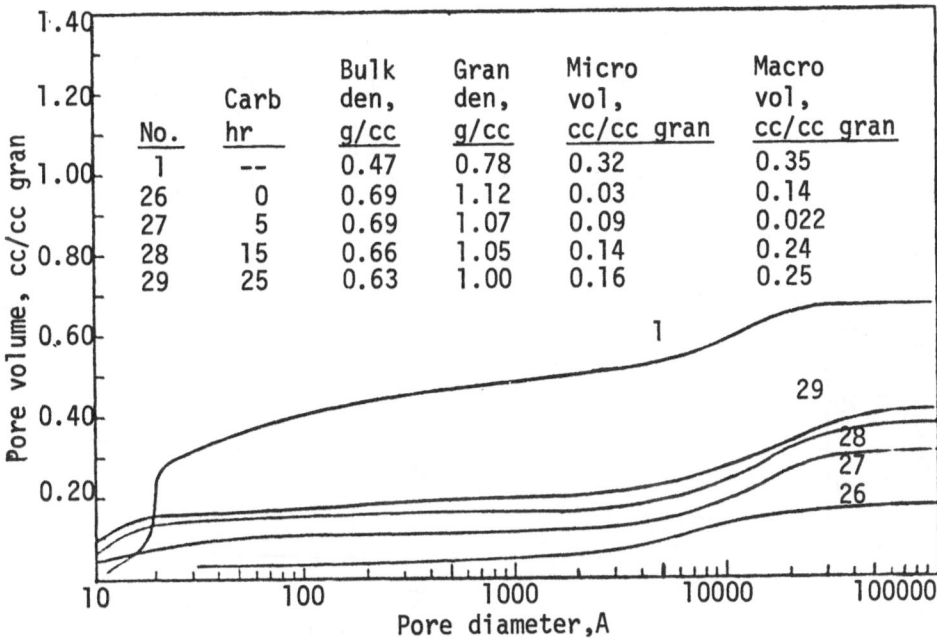

Figure 14. Reconstituted soft coal sized to 0.6 to 1.7 mm granule size and carbonized different lengths of time, 500°C final temp.

and 0.095 cc/g respectively. If no shrinkage had occurred, but other changes had occurred as measured, the pore volumes would have been 0.180 and 0.280 cc/g, respectively. A considerable number of pores closed up during the carbonization. By raising the temperatures in a steam atmosphere much of potential pore volume would have been saved. At the completion of peach pit carbonization the pore volume was 0.40 cc/g (table II) compared to 0.085 and 0.095 cc/g for the two coal chars.

During carbonization of reconstituted soft coal, the temperature must be brought up gradually over a 5 to 6 hr time period to about 500°C. During lower temperature heating schedule a small air sweep through the carbonizer improves the product. A steam sweep is used over the remaining heating schedule.

Other raw materials, depending on their fixed carbon and volatile ratios, fall in between wood and hard coal in their behavior during carbonization.

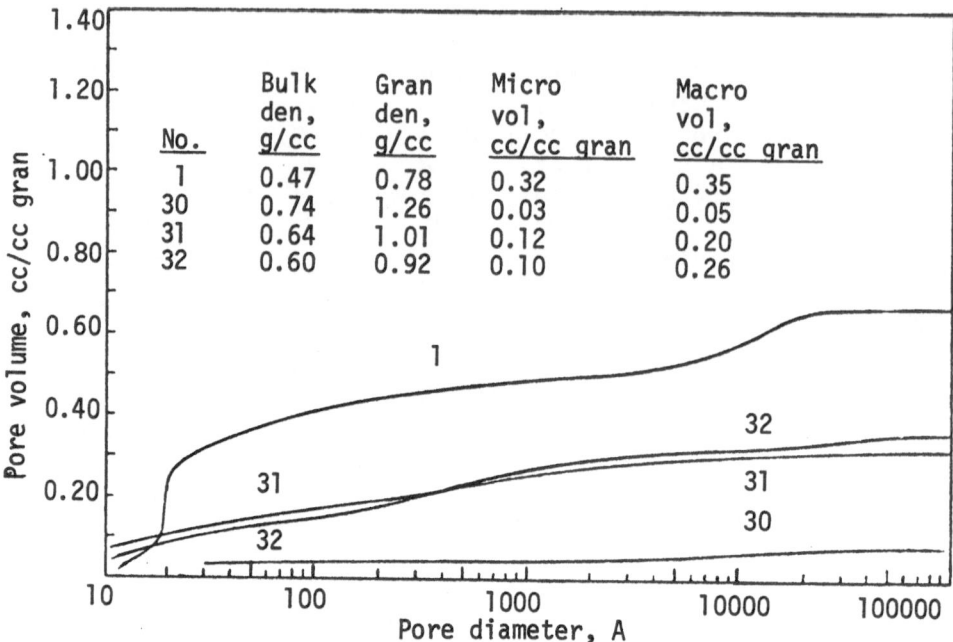

Figure 15. Peach pits carbonized, 540°C final temp, curve 30 sized pits not carbonized, curve 31 sized pits carbonized, and curve 32 whole pits carbonized and sized

4.2 Effect of carbonization on pore size distribution

Figure 14 shows the effect of continued carbonization on the pore structure of reconstituted soft coal. A comparison of the pore volume changes given in the table shows the greatest change in macropore volume occurring in the first 5 hr. Increase of micropore volume increases at a fair rate at carbonization times beyond 5 hr. However, these chars can be safely heated at a rapid rate to activation temperature after a 5 hr carbonization. It is advantageous to start activation after a 5 hr carbonization to minimize further shrinkage.

Figure 15 shows pore size distribution curves for peach pit chars when carbonized whole and then sized to granules. (These are not the same samples discussed previously and note, pore volume scale is in cc/cc granule.) A comparison of the micropore and macropore volumes, as given in the table, show a volume increase in both pore sizes on carbonization. The final temperature was 540°C. No record of time or heating rate was available.

4.3 Air pollution

For every kilogram of raw material carbonized, 0.05 to 0.7 kg of volatile matter is evolved which, if allowed to escape into the atmosphere, would pollute the air to varying degrees. The worst offender may be soft coal because many of the evolved gases are carcinogenic. Vegetable raw materials are the largest by weight percent, but much of the vapor is water.

To control emissions from carbonization some carbon plants use a thermal incinerator. Much of the vapor is combustible, hence, supplies a large share of the heat requirements.

4.4 Carbonization equipment

Where carbonization times of about 4 to 6 hr duration are required, it can be performed by one of several types of heating devices. Among those that have been used or are in use are: (1) indirect-fired-rotary tube kiln, (2) direct-fired-rotary tube kiln, (3) multiple hearth furnace, (4) fluidized-bed reactor. Each has advantages and disadvantages. The indirect-fired rotary tube kiln has an advantage over the others because of indirect firing, a large discharge of inert gases to the after-burner is avoided. On the other hand, it is the largest and most expensive to maintain. Fluidized-bed reactors produce considerably more attrition and loss of material unless recycling is possible as with processes that reconstitute the raw material. They have the smallest space requirements. The first three devices are known to be in use.

Where carbonization times of much longer duration than 6 hr are required to obtain the highest quality char, other means are used than the continuous equipment identified above. To get the necessary duration of carbonization, one manufacturer carried out the process in two steps. The first step was a low temperature operation, from 90° to 450°C. A large kiln was used through which kiln cars loaded with trays of coconut shells or the other nut shells were pushed, with retention time of six days for each car. Gas heating was used with flue gas flow countercurrent to movement of the cars. Discharged char was cooled 24 hr in air before being carbonized at a higher temperature in the second step. The furnace was a building-like structure consisting of vertical ducts or flues made of brick. The char was loaded continuously into alternate ducts at the top and discharged at the lower end. Combustion gases passed upward through adjacent ducts. Retention time was two days at temperatures between 540° and 820°C.

5. ACTIVATION

5.1 Thermodynamics

Chars can be thermally activated with CO_2, steam, or with a mixture of the two. The general commercial practice at present is to use a direct-fired activator where the combustion products of natural gas or coke-oven gas supply both the heat and CO_2 and steam. In the past, indirectly heated activators have been used where just steam or CO_2 has been used. Experimental work is also often done with electrically heated ovens where one of these two gases are used.

Steam and CO_2 are used in activation because their reactions with carbon are endothermic and are, thereby, controllable by adjusting the heat input. The reactions are as follows:

$$H_2O + C = CO + H_2 \tag{3}$$

$$CO_2 + C = 2CO \tag{4}$$

Activations are generally carried out at temperatures between 850°C and 1000°C. Over this temperature range the heat of reaction for either reaction changes very little. The average for reaction 3 is +32.4 kcal/mol and for reaction 4, +40.3 kcal/mol.

At 850°C, the free energies of reaction for both reactions is -5.3 kcal/mol. Both reactions go to a high degree of completion at this temperature but the rate is slow. At 1000°C, both reactions are very fast, hence by maintaining the temperature at any level between these two limits, the activation rate can be controlled to give the desired product.

The main objective of the activation is to enlarge the pores formed during the reconstituting process and especially those formed during the carbonization. The activation gases, however, come in contact with the exterior part of the granule first. There is, therefore, an unavoidable amount of granule exterior oxidation which is measured as a decrease in granule size. The relative rates at which the exterior surface is burned off and the pores are enlarged depends on the rate of diffusion of the activation gas into the pores compared to the rate of surface oxidation. By combining information from several sources, the indications are that CO_2 activation promotes more external surface oxidation and development of larger pores than steam. Higher temperatures also appear to have the same effect as CO_2 activation. The relative amounts of external oxidation versus internal oxidation also depends on how well developed the pore structure is in the carbonized char. Char with no porosity can only decrease in granule size.

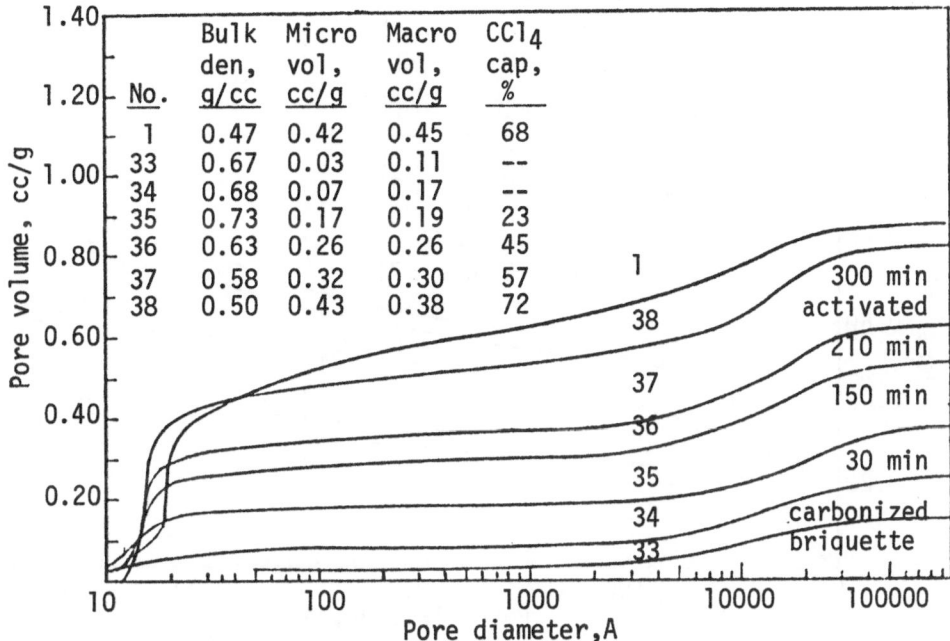

Figure 16. Steam activated series for reconstituted soft coal char

The rate of activation depends on the temperature, efficiency of contacting the char with the activation gas and concentration of activation gases. Gas utilization also depends on the same factors.

When flue gas is used, steam is usually added to moderate the gas temperature. The extra steam also suppresses the action of CO_2.

Heat requirement also vary with type of equipment and length of activation. An approximate figure is about 500 cal/g of char fed to the process.

Volume and weight yields can also vary considerably depending on the type of char and method of activation. Volume yields between 50% and 80% and weight yields between 25% and 50% are common.

5.2 Effect on pore size distribution

The effect of activation on pore size distribution is shown by the activation series in figures 16 and 17. Those shown in figure 16 are for a reconstituted soft-coal char activated in an indirectly heated rotary activator with steam at 1000°C. The activation was

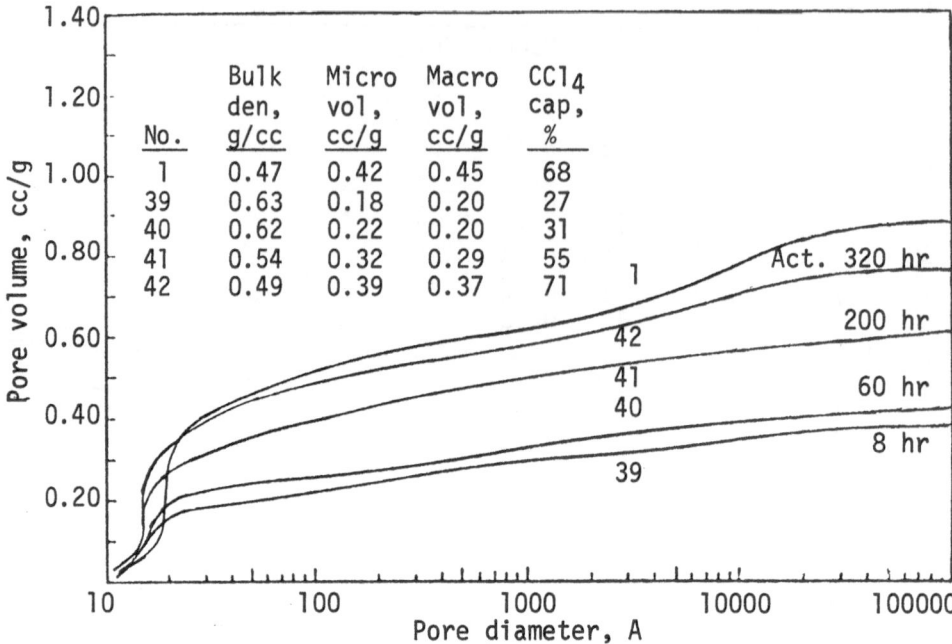

Figure 17. Steam activation series for pecan-shell base char

relatively fast, a highly activated carbon was attained in 300 min. Those in figure 17 are for a pecan shell char activated with flue gas and steam at temperatures between 850° and 900°C. Activation was done in a brick constructed furnace containing slanting brick baffles over which the carbon flowed as it was fed in from the top. Activating gases were fed upward countercurrent to the flow of carbon. Because of the lower temperature and less efficient contact of carbon and gas, the required activation time was 320 hr to produce a highly activated carbon.

Regardless of which way the char was activated there was a steady increase in both the micropore and macropore volumes. Bulk density decreased. Carbon tetrachloride capacity increased to the 70% level.

5.3 Superactivated carbons

In activating a carbon the question often arises, how long should the carbon be activated? As the activation is continued, the activity increases up to a maximum level and then begins to drop off. During this time the yield is also decreasing and the cost per unit weight is increasing. The answer to this question is that

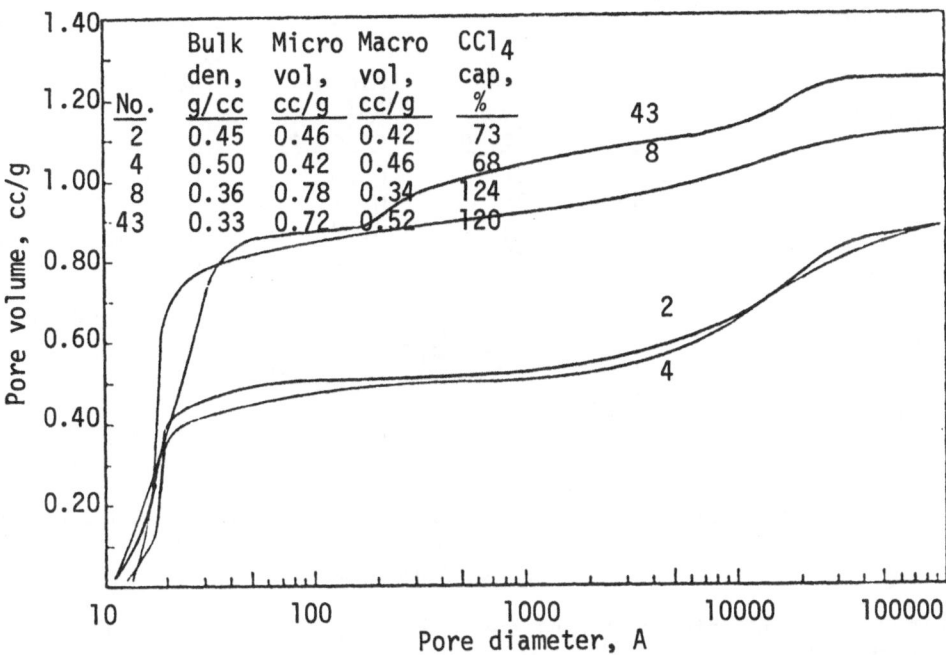

Figure 18. Pore size distribution curves for superactivated and standard carbons

activation should be carried to a level considerably below the maximum. The only exception is for carbons that are to be used in space flight where weight is the ultimate cost determining factor.

Figure 18 shows pore size distribution curves for two standard gas-phase carbons used commercially in solvent recovery (curves 2 and 4), and two superactivated carbons (curves 8 and 43). Curve 8 is for steam activated coconut. Curve 43 is for a commercial steam activated coconut which was further activated with CO_2 at 830°C for about 40 hr in an electrically heated oven. CO_2 feed rate into the oven was 57 l/min per 1000 g of charge. The weight yield was 50% which, when superimposed on the probable commercial activation yield of 10%, lowers the overall yield to 5% of the raw material.

The carbon tetrachloride adsorptive capacities of the superactivated carbons are approximately 75% greater than for the standard carbons (per gram of carbon). In space flight this is a distinct advantage.

Granular activated carbons are generally used on the volume basis as determined by the size of the adsorber. Per unit adsorber

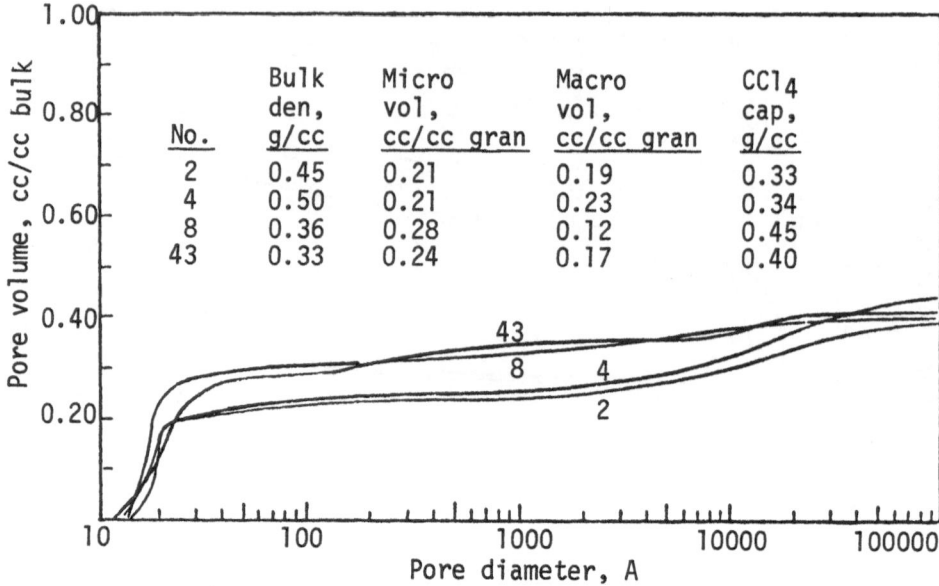

Figure 19. Pore distributions for superactivated and standard carbons

volume, the adsorptive superiority of the superactivated carbons is drastically reduced. Figure 19 shows the pore size distribution curves of the same carbons with cumulative pore volume and carbon tetrachloride adsorptive capacity now given on the cc bulk volume basis. The adsorptive capacities of the superactivated carbons are about 27% higher than the standard carbons. There are several equalizing factors here that tend to make the cost per unit of adsorptive capacity the same. The superactivated carbons cost more per unit weight, but less is needed to fill the adsorber and, per unit volume of adsorber, they adsorb more. Superactivated carbons tend to be more friable and cannot withstand regeneration as well as the standard carbons. As a consequence, carbon manufacturers have tended to optimize on the carbon properties which characterize those of carbon 1, 2 and 4.

5.4 Activation equipment

The more common type activators used commercially are the multiple hearth furnace and the direct-fired rotary tube kiln. Activation time is usually about 5 hr at temperatures between 900° and 950°C using flue gas with some steam added, usually about 1.0 kg/hr per kg carbon activated. It has been demonstrated that long slow

activations are not necessary to attain a high activity carbon
once the raw material has been properly prepared and carbonized.
Furnaces designed for a long gradual activation are being taken
out of service. Fluidized-bed activators have been investigated.
With some chars a high activity carbon can be produced in 30 min
but fluidized-bed activators do not appear to have been generally
accepted by the carbon manufacturing industry. One of the problems
is attrition loss. For the activation of hard coal, a fluidized-
bed reactor presents advantages, since the temperature of the
charge can be brought to activation temperature rapidly, and thereby, minimize the effects of shrinkage.

6. PRODUCT FINISHING

6.1 Sizing and size separation

Most often the product coming from the activator is not ready for
market. Raw materials are charged into the process in granules
and/or chunks of size larger than the desired product, since
granule reduction occurs from shrinkage during carbonization and
external oxidation during activation. Reduction of granules and
fines formation also occurs from attrition. Hard granular materials
are sieved into separate granule size ranges as required by the
market. Soft granular materials are ground to powdered carbon.
Undersized granules and fines are also ground to powder. Particle
size is usually reduced to over 98% less than 44μ diameter.

6.2 Removal of soluble ash

Carbon invariably contains at least a small percentage of ash.
When this ash is soluble, and could contaminate a product being
purified, the carbon is washed with acid and water to remove the
soluble portion of the ash. It is then necessary to dry the carbon
to meet water content specifications which are usually 2% to 3%.

6.3 Testing

Testing, although not a product finishing operation, is very
necessary to determine the quality of the activated carbon. Bulk
density, and one or more of the quality control adsorption tests,
iodine number, molasses number or carbon tetrachloride adsorptive
capacity are determined as soon as possible so that, if off-grade
product is being produced, corrective changes can be made in any
one of the steps of the process. When the carbon is intended for
a specific use, a test is conducted that relates closely to the
use.

REFERENCES

1. Juhola, A. J. and Wiig, E. O., Pore structure in activated charcoal, I Determination of micropore size distributions, J. Am. Chem. Soc., 71, 2069, 1949.

2. Ritter, H. L. and Drake, L. C., Pore size distribution in porous materials, pressure porosimeter and determination of complete macropore size distributions, Ind. Eng. Chem. Anal. Ed., 17, 782, 1945.

3. Joyner, L. C., Barrett, E. P. and Skold, R., The determination of pore volume and area distribution in porous substances, II Comparison between nitrogen isotherm and mercury porosimetry methods, J. Am. Chem. Soc., 73, 3155, 1951.

4. Juhola, A. J., Palumbo, A. J. and Smith, S. B., A comparison of pore size distribution of activated carbons calculated from nitrogen and water desorption isotherms, J. Am. Chem. Soc., 74, 61, 1952.

5. Juhola, A. J. and Blacet, F. E., Survey of pore structure in charcoal, OSRD Report No. 5500, August, 1945. (Note: Many of the pore size distribution curves came from this report.)

6. Grant, R. J., Basic concept of adsorption on activated carbon, Pittsburgh Activated Carbon Company publication.

7. Hassler, J. W., Process for producing an activated carbon from anthracite, West Virginia Pulp and Paper Co. Bulletin No. 70.

THERMAL REGENERATION

A. J. Juhola

MSA Research Corporation, Evans City, Pennsylvania

ABSTRACT. Activated carbons are being used for the control of gaseous emissions that would pollute the atmosphere, treatment of waste water and decontamination of water containing spilled chemicals. In these three areas thermal regeneration of the carbon is an economic benefit. Methods for regenerating the carbon in these three applications vary considerably. These activation procedures are discussed with consideration given to the effect adsorbate physical properties have on the selection of best regeneration route.

INTRODUCTION. Activated carbons in the past have been used primarily for air and water purifications and industrial uses, such as solvent recovery and sugar refining. In more recent years their use has expanded into cleaning waste water (sewage) to reduce pollution of rivers and lakes. Efforts are presently being made to apply the carbon-adsorption method to control of gaseous emissions to reduce air pollution. A new problem is emerging in industrialized areas and that is the accidental spillage of chemicals which may get into rivers and lakes where they can quickly spread causing widespread damage. Fast emergency action is required. Adsorption by activated carbons is under intensive study as a means for removing spilled chemicals from water.

In these applications, thermal regeneration of the spent carbon is already an integral part of systems already in use or is being investigated for those that are in the developmental stages. Economic studies generally show that it is less costly to regenerate the spent carbon, preferably at the site of the application, than to discard it and use new carbon.

The manner in which the regeneration is done depends on several factors: 1. physical and chemical properties of the pollutant, 2. the media in which the adsorption phase is carried out (gas or liquid phase), and 3. recovery value of the pollutant. The urgency of the regeneration, as would be the case in spill clean-up operations, is also another decisive factor.

Because of these factors, regeneration will vary considerably depending on whether the carbon adsorption system is used for control of air pollution, for waste water treatment or recovery of spilled chemicals from water. In this lecture, regeneration procedures applicable to these three areas are discussed. In waste water applications the discussion will center on how it is being done. In air-pollution and chemical-spill applications, the discussion will be directed more toward the way regeneration procedures may develop or are being considered.

1. POLLUTANT PROPERTIES

There are several physical and chemical properties of pollutants that determine their adsorbability by carbons and, consequently, determine the procedures by which the carbon can best be regenerated. The emphasis in this lecture is on the regeneration of the carbon for reuse, but whenever recovery of a valuable adsorbate is feasible this phase of regeneration will also be considered.

The adsorbates that are being considered here are primarily those classified as organic. Inorganic compounds, as a general rule, are not adsorbable. Some exceptions are chlorine, iodine, potassium permanganate, silver salts, gold chloride, ferric salts and molybdates. A large number of inorganic chemicals are in daily commercial transit that, by accident to the carrier, can be spilled into rivers or lakes and cause widespread pollution. Unfortunately, sorption by activated carbons is not a possible means for cleaning the water of these chemicals. Organic compounds containing halogens, metallic elements, sulfur or nitrogen are again effectively adsorbed if the hydrocarbon portion of the molecule is relatively large.

Properties of organic adsorbates that have a direct bearing on the manner in which the regeneration must be conducted are: 1. liquid or solid molar volume, 2. vapor pressure, 3. water solubility, 4. tendency to polymerize, 5. toxicity, and 6. presence of halogens, sulfur, nitrogen or metallic elements in the molecules. The effects of these properties will be discussed in detail with each of the three areas of application, since their relative importance varies with each application.

2. AIR-POLLUTION CONTROL CARBONS

2.1 Emission sources

A survey of emission sources controllable by activated carbon
adsorption was completed in U.S.A. in 1973.[1] Major sources identified are surface coating, degreasing, dry cleaning, and graphic
arts industries. Pollutants from these sources are solvents, some
of which are toxic, and some smog precursors. Process industries,
such as meat rendering plants, produce decomposition products,
primarily the organic aldehydes, acids, sulfides and amines.
Although not emitted in as large quantities as the solvents, they
compensate for this lack by being more toxic, odorous and stronger
smog precursors.

2.2 Pollutant properties

Twenty-two emission sources considered to be of major importance
were investigated in the study. Over 160 compounds were identified
as pollutants emitted from these sources. In the report these were
listed downward in decreasing order of liquid molar volume as
measured at the compound's boiling point. A partial list is presented in table I. In general, adsorbability on carbon decreases
with decreasing molar volume; those pollutants at the top of the
list are more strongly adsorbed than those at the bottom. There
is also a general trend in boiling points; they decrease with decreasing molar volume. There are also many exceptions. Those containing hydroxyl groups have higher boiling points while fluorine
and amine compounds tend to have lower boiling points. In using a
list such as table I for a guide on adsorbability, any conclusions
should be modified by also taking into consideration the compound's
volatility.

When the pollutant molar volume is in the range 80 to 190,
the pollutant can be removed from the emission gas stream with
solvent-recovery type systems where recovery of adsorbate and
regeneration of carbon is done at low temperatures (100 to 150°C).
When the molar volume is larger than 190, low temperature regeneration is not effective. These pollutants also have very low
volatilities, hence are not a threat as air pollutants unless they
are toxic or are very malodorous. In the latter case, air-purification systems are more effective.

Those pollutants of molar volume smaller than 80 tend to be
weakly adsorbed, hence difficulties are experienced in using carbon
adsorption for their control. Each member of this group should be
judged separately, taking into consideration its volatility.
Propanone and ethanol can be recovered by carbon adsorption

Table I. Properties of pollutant vapors emitted from small stationary sources

Pollutant	Chemical formula	V_m, cc/mol	Boiling Point, °C
2,6-dimethyl-4-heptanone	$[(CH_3)_2CHCH_2]_2CO$	207	168
2-ethylhexanol	$C_4H_9CH[C_2H_5]CH_2OH$	194	185
αpinene (turpentine)	$C_{10}H_{16}$	184	149
butylacetate	$CH_3COOC_4H_9$	152	129
hexanal	$CH_3[CH_2]_4CHO$	141	128
4-methyl-2-pentanone	$[CH_3]_2CHCH_2COCH_3$	141	117
1,1,2-trichloro-1,2,2-trifluoroethane	FCl_2CCClF_2	120	48
toluene	$C_6H_5CH_3$	118	111
dimethyl amine	$[C_2H_5]_2NH$	112	56
butanoic acid	$CH_3[CH_2]_2COOH$	108	164
phenol	C_6H_5OH	105	182
trichloroethene	$Cl_2C = CHCl$	98	87
benzene	C_6H_6	95	80
2-propanol	$[CH_3]_2CHOH$	84	82
propanone	$[CH_3]_2CO$	74	56
propane	$CH_3CH_2CH_3$	74	-44
propenal	$CH_2 = CHCHO$	67	94
carbon disulfide	SCS	66	81
ethanol	CH_3CH_2OH	61	142
ethanal	CH_3CHO	56	37

processes while propane and ethanal cannot.

Pollutants that can polymerize in coming into contact with carbon, such as styrene, cannot be desorbed in their molecular form.

Since low pressure steam is one of the regenerating agents used, solubility of the desorbed pollutant in the steam condensate is also an important property that can cause change in mode of regeneration.

In solvent-recovery type operations, designed for relatively high concentrations and frequent regenerations (1 to 2 hr), a deep carbon bed is used and regeneration is done in place. In air-purification type operations, where the need for regeneration is infrequent, (1 to 2 yr), shallow panel-type beds of 2.5 to 5 cm thickness are used. Regeneration is not done in place. The carbon panels are returned to the carbon manufacturer's plant where they

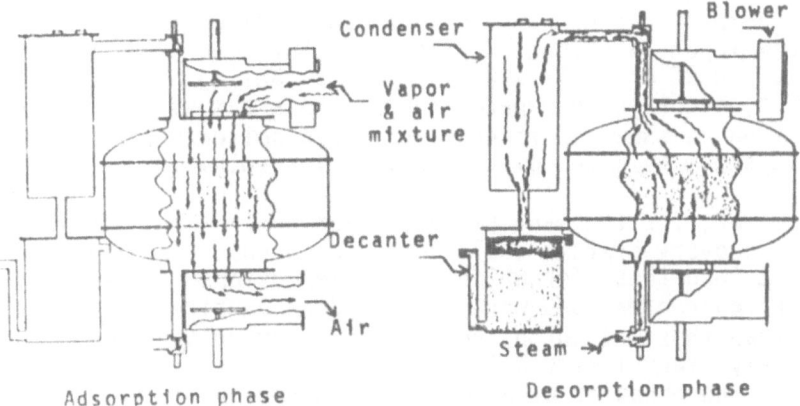

Figure 1. Cut-away diagram of solvent-recovery adsorber, showing vapor-air flow pattern during adsorption and steam flow pattern during desorption. Courtesy of Vic Manufacturing Co.

are regenerated at high temperature with steam.

These two types of systems are discussed in the following paragraphs as applied to air pollution control.

2.3 Solvent-recovery type systems

Figure 1 shows a solvent-recovery system as operated in the adsorption and regeneration phases. Recovery of solvents with these systems are economical at solvent concentrations above 700 ppm. To comply with safety regulations, the maximum concentration of combustible solvent in an airstream that may be stripped is 25% of the lower explosive limit. For most solvents of interest which are volatile enough to exceed this limit, the maximum allowable concentration range is 2,500 to 10,000 ppm. At these concentrations the adsorptive capacity is usually between 0.10 and 0.30 g/g on carbons generally used in solvent recovery.

When the carbon has become saturated with solvent, regeneration is carried out by passing low temperature steam upflow through the bed as shown in figure 1. Only a part of the solvent is desorbed. It has been found that desorption of all of the solvent would require excessively large amounts of steam. To stay within economical limits, desorption is carried to the extent where the ratio of steam to solvent is in the range 2 to 6. At these steam input ratios the operating capacities vary between 0.05 to 0.10 g/g, or 5% to 10%.

The carbon bed is usually sized so that at the design flow rate and expected concentration, the adsorption time is one hour, and steam input rate then regulated so that desorption can be completed in 45 min, allowing 15 min for cooling. The solvent, if insoluble, can then be separated from the steam by cooling and decantation as shown in figure 1. Soluble solvents are separated from the steam by means of a fractionating column.

However, for the many air pollution control applications, the solvent recovery systems would be operated under conditions greatly different than those for which they were designed. Foreseeable pollutant concentrations were estimated at considerably lower levels than those at which the solvent recovery systems operate effectively. Very little data was available that could be used to determine the behavior of a solvent-recovery system under the great variety of conditions that would be encountered in strictly air pollution control. To overcome this deficiency theoretical calculations were made based on the Polanyi potential equation for adsorption.[2]

2.4 Theoretical approach

A modified form of the Polanyi potential equation has been found very useful for estimating adsorption capacities where temperature and vapor concentration are variable. The equation has the form

$$A = [T/V_m] \log [C_0/C] \tag{1}$$

where C_0 is the vapor pressure or concentration of the pollutant at temperature T in degrees absolute, C is the equilibrium vapor pressure or concentration of the adsorbed pollutant at T, and V_m is the liquid molar volume as measured at the normal boiling point. The quantity A plotted as function of amount adsorbed gives a generalized adsorption curve. Typical curves of this type are shown in figure 2. The amount adsorbed is expressed in cc of liquid using the same density as used in V_m.

The shape of the generalized adsorption curve is determined by the pore size distribution curve and is therefore different for each carbon to the extent the distribution curves differ.

For similar compounds, as in a homologous series, a curve determined for one member is applicable to the others. Differences have been observed between greatly differing homologous series. All curves tend to meet at $A = 0$; w at this point is close to the micropore volume.

From a very limited amount of adsorption data, it is possible to construct a generalized adsorption curve and then use the curve

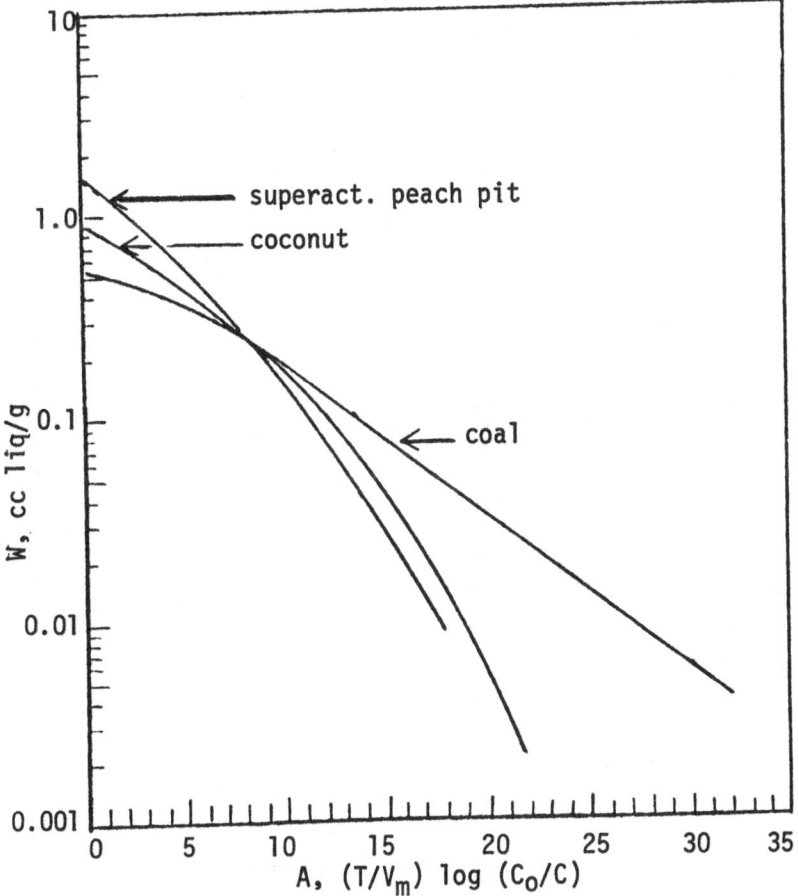

Figure 2. Generalized adsorption curves for three widely different types of activated carbons

to extrapolate to other temperatures, concentrations, and to other compounds of differing molar volumes, as found in the pollutant-emission sources. With this information it is now possible to design adsorption systems that optimized both the adsorption and desorption phases.

In regard to the desorption phase, figure 3 shows typical curves which give the amount of regenerating agent required to desorb to various adsorbate levels.

In solvent recovery operations the interest is only in higher concentrations.For the 3000 ppm curve, the optimum operating capacity is near 12%; i.e., desorption from 0.52 to 0.40 g/g. The

310

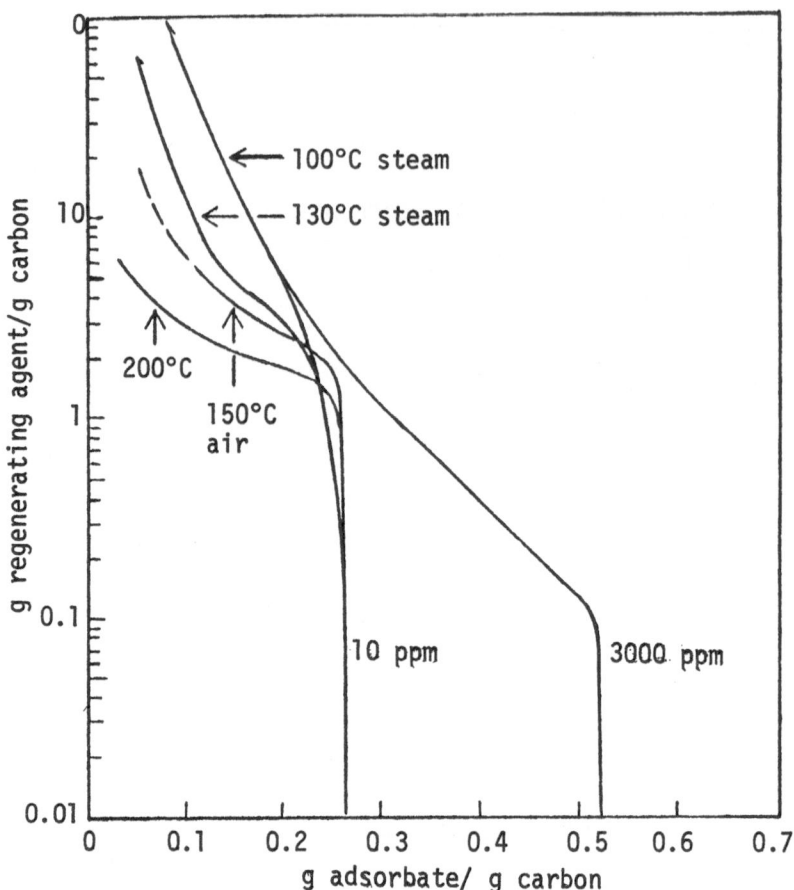

Figure 3. Amount of regenerating agent required to regenerate a coconut carbon equilibrated with 4-methyl-2-pentanone at 10 and 3000 ppm concentrations.

amount of 100°C steam required is 0.35 g/g, which gives a steam to adsorbate ratio of 2.9.

Pollutant concentration from emission sources are generally much lower, which lowers the operating capacity, and also, the adsorbate is lodged in smaller pores from which it is considerably more difficult to desorb. At the 10 ppm concentration, the optimum operating capacity is about 5%; i.e., from 0.26 to 0.21 g/g. The amount of 100°C steam required is 4 g/g, giving a steam to adsorbate ratio of 80. At these high ratios most organic compounds would dissolve completely in the steam condensate. Separation of the adsorbate from the water by distillation is costly, and further-

more, many adsorbates have no recovery value or are recovered in
such small quantities that the handling costs overbalance the value
of the recovered adsorbate. The alternatives are to discharge the
mixture into a sewer, which in effect converts air pollution to
water pollution, or modify the regenerative procedure in a way that
would also economically dispose of the adsorbate.

2.5 Carbon-resorb systems

Several systems have been proposed which by theoretical and limited
pilot scale studies show considerable promise in solving the adsorbate disposal problem. In the above referenced report these systems
were referred to as carbon-resorb systems.[1]

Two systems proposed by Mattia employ a noncondensable regenerating gas at temperatures up to 150°C.[3,4] The desorbed pollutant
is then immediately incinerated thermally or catalytically. If the
pollutant contains halogens, sulfur, or nitrogen which on incineration produce acidic gases that would repollute the air, the
desorbed pollutant is adsorbed in a secondary carbon bed from
which it can be recovered by steam regeneration.

In these systems the carbon bed acts as a concentrator. It
adsorbs a quantity of pollutant from a very large volume of air
at a low concentration and over a long period of time. During
desorption the pollutant is released from the carbon in a small
volume of regeneration gas, at a high concentration over a short
period of time.

For combustible pollutants, the concentration is high enough
over most of the desorption phase for the incinerator to be self
sustaining. Supplementary heating is required only at beginning
and end of the incineration period. The incinerator is small
relative to the size of the carbon bed and may be only one tenth
the size of the incinerator that would be required if the air
volume were incinerated directly.

Figure 4 shows a flow diagram of a carbon-resorb system which
incinerates the desorbed pollutant. Two adsorbers are used alter-
nately on adsorption and desorption. In this diagram, adsorber 1
is on-stream while adsorber 2 is being regenerated with 150°C flue
gas. The adsorber effluent gas-stream is divided into two parts.
One is recycled through the adsorber and the other part is passed
through the heat exchanger and incinerator. The incinerator effluent gas-stream, at over 720°C, is in part discharged to the
atmosphere and in part recycled through the adsorber. Its temperature is brought down to 150°C by passage through the heat exchanger,
gas cooler, and by admixture with the recycle gas-stream from the
adsorber. Enough air must be fed to the incinerator to burn the

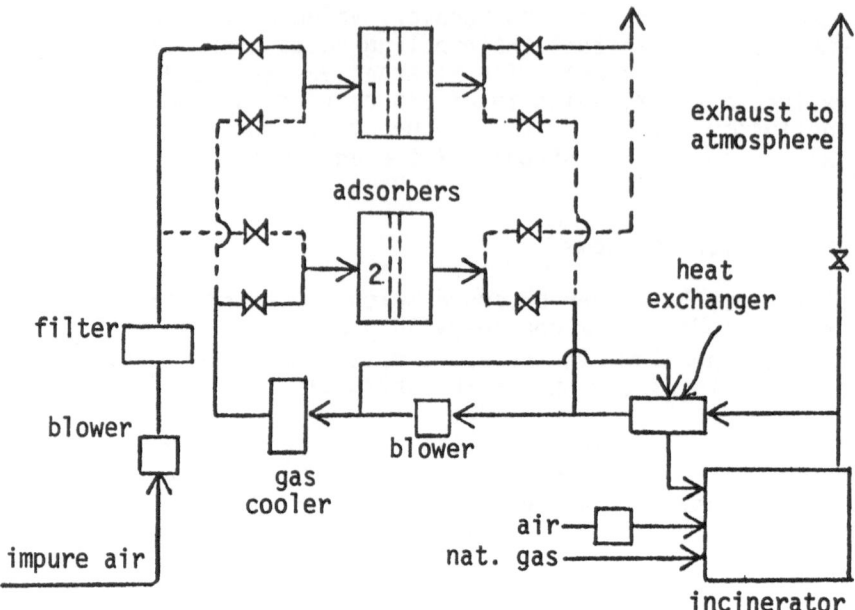

Figure 4. Carbon-resorb system with flue gas regeneration and incineration of desorbed pollutant (from Mattia[3])

pollutant but must be controlled to avoid formation of explosive mixtures in the system.

Figure 5 shows a flow diagram of a carbon-resorb system with a secondary carbon adsorber for the recovery of the pollutant. The secondary bed need be only 20% of the primary bed size. In this system an inert gas such as flue gas must be introduced into the system and is circulated through the spent primary adsorber and secondary adsorber. It is heated to 150°C for the regeneration phase and cooled down to ambient temperature for the secondary-adsorption phase. Low temperature steam is then used to recover the pollutant from the secondary adsorber.

In place of the secondary adsorber, the pollutant can also be recovered by condensation or by vacuum desorption.

2.6 Air-purification system

When the pollutant concentration is always below 10 ppm, the carbon-resorb system is not as cost-effective as the nonregenerative air-purification system. These systems consist of a blower, a particulate filter, and cells containing activated carbon. The carbon

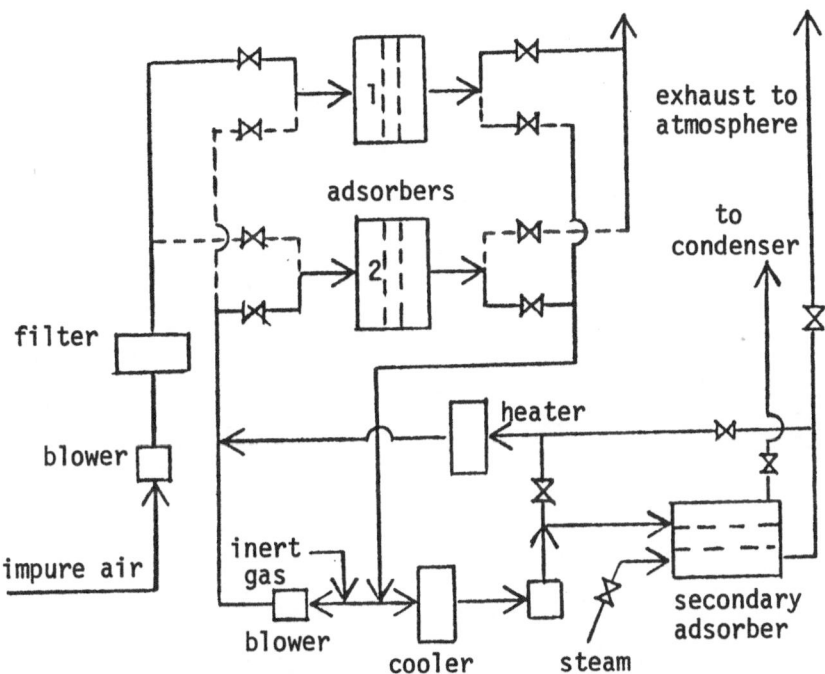

Figure 5. Carbon-resorb system with gas regeneration and secondary adsorber for recovery of pollutants (from Mattia[3])

may be contained in panel-like trays, canisters or in pleated retainers. Bed depth can range from 1.5 to 5 cm. Bed areas are large relative to air flow velocity to keep flow resistance low. Figure 6 illustrates a multiple-cell type filter. This general type is used to adsorb radioactive iodine and methyl iodide in atomic energy plants.

When regeneration is feasible, the carbon beds are removed from the cells and returned to the carbon manufacturer's plant and steam activated at high temperature. Very often the carbon has been impregnated for special applications. Reactivation is then not feasible and new carbon is used.

3. WASTE-WATER TREATMENT

3.1 Waste-water treatment plants

Pollution laws in the United States now prohibit discharge of untreated sewage or industrial waste waters into rivers, lakes,

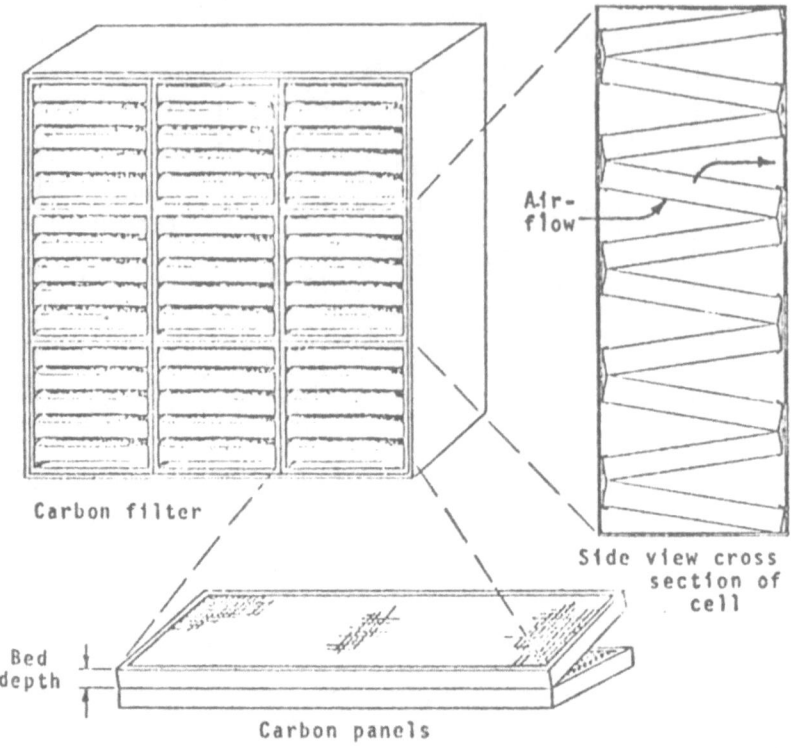

Figure 6. Multiple cell activated carbon filter for air purification

or other bodies of water that are in public use. To comply with this law new waste water treatment plants are constantly being built. Some industries are permitted to discharge into municipal sewerage systems, but most are required to treat their own waste waters. This practice is intended to limit the pollutants to biologically degradable matter and also decrease the load on treatment plants.

Waste water treatment plants vary in complexity depending on whether they are equipped to carry out the primary, secondary and tertiary treatments. The primary treatment consists of filtration and settling, usually with coagulants to clarify the water. The secondary treatment utilizes aerobic bacteria to decompose the soluble pollutants to CO_2 and H_2O. Tertiary treatment finishes the water cleaning process. Organic pollutants that the bacteria could not decompose are removed. Some turbidity and also some of the biologically degradable matter get through the secondary treatment and must be removed at this step. Both granular and powdered carbons

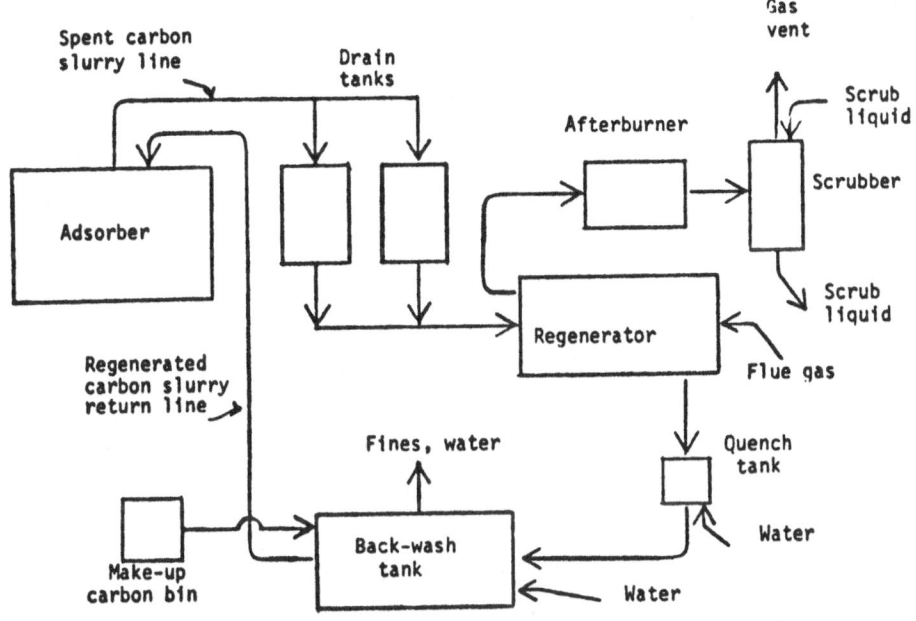

Figure 7. Regeneration of spent granular carbon from waste water treatment

are effective in removing these impurities. Both carbons are also regenerated although the methods for the powdered carbons are still under development.

There is a wide range of organic pollutants that get into the tertiary treatment. Some of the larger molecular variety have been identified as tannins, proteinaceous matter, lignins, herbicides and pesticides. They, in general, have low solubilities and are virtually nonvolatile. Molar volumes are over 200 cc/mol. They top the list given in table 1. Because of these properties they are very strongly adsorbed and cannot be desorbed without decomposition. Furnace and kiln type regenerators that can operate at temperatures up to 1000°C are required.

3.2 Regeneration of granular carbons

The overall regeneration process, as shown by the flow diagram in figure 7, involves a number of operations. First the spent carbon is transferred in a water slurry from the adsorber to a drain bin where the water content is reduced to about 50%. The partially de-watered carbon is then fed to the regenerator. The regenerated product is cooled and also reslurried by being discharged into the quench tank. During regeneration, quenching, and on each transfer,

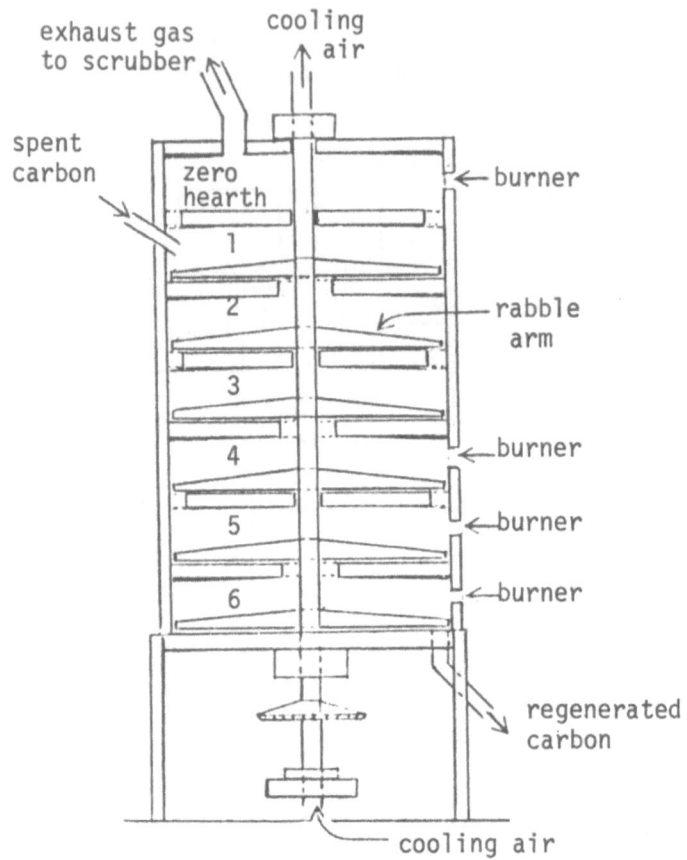

Figure 8. Six hearth furnace with zero hearth for burning released organic vapors

the granules are abraded and form fines which would increase the resistance to flow in the adsorber if not removed earlier. These fines are removed by backwashing in the tank as shown. Because of the loss of fines and also carbon loss due to oxidation during regeneration, make-up carbon must be added on each regeneration cycle to fill the adsorber to the required level.

Carbon losses, varying from 2% to 10% for each regeneration, have been reported from various waste water treatment plants. Loss of adsorptive capacity at times accompanies each regeneration. One plant reported a 26% loss over eight regenerations but then leveled off.

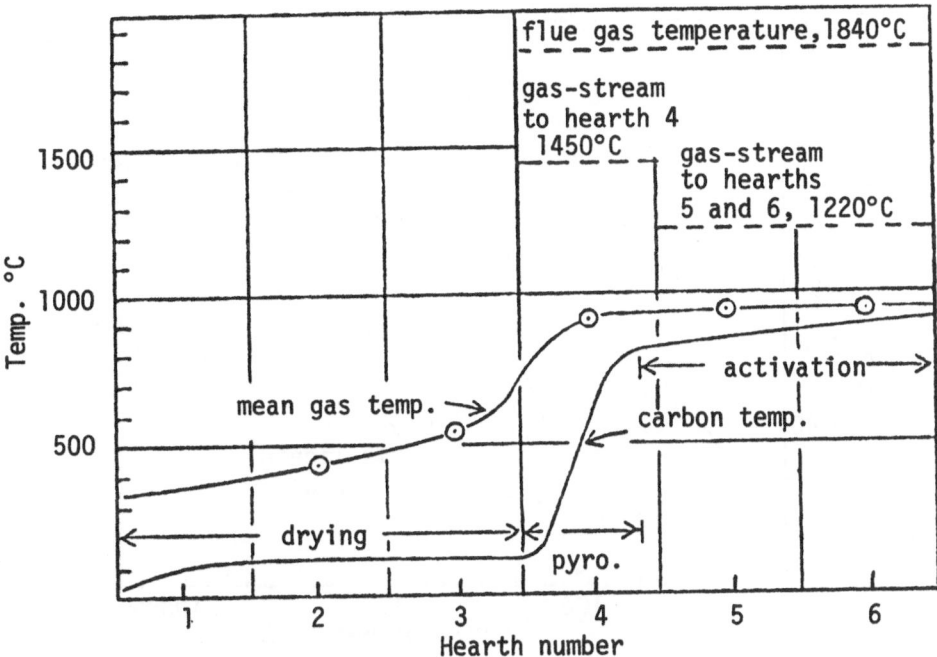

Figure 9. Temperature profiles for gas and carbon in multiple hearth furnace, regeneration of wet spent carbon

Several types of thermal equipment are suitable for regeneration, such as the multiple-hearth furnace, direct- and indirect-fired rotary tube kilns. Of these the multiple-hearth furnace is the one most used; it provides the most hearth surface area per unit floor space. Figure 8 shows the essential features of multiple hearth furnaces. This sketch shows a furnace with six hearths over which the carbon is thermally treated and one extra zero hearth at the top which serves as an afterburner to incinerate volatile decomposition products released from the carbon on the lower hearths. Rotary tube kilns employ auxiliary afterburner.

In the operation of a multiple hearth furnace, the wet spent carbon is fed to hearth 1 by means of a screw feeder. The rotating rabble arms sweep the carbon over each hearth in a spiral pattern and moves the carbon to the drop holes in each successive hearth. It is a direct-fired furnace with gas burners located at the lowest two or three hearths. The flue gas passes upward through the drop holds countercurrent to the movement of the carbon.

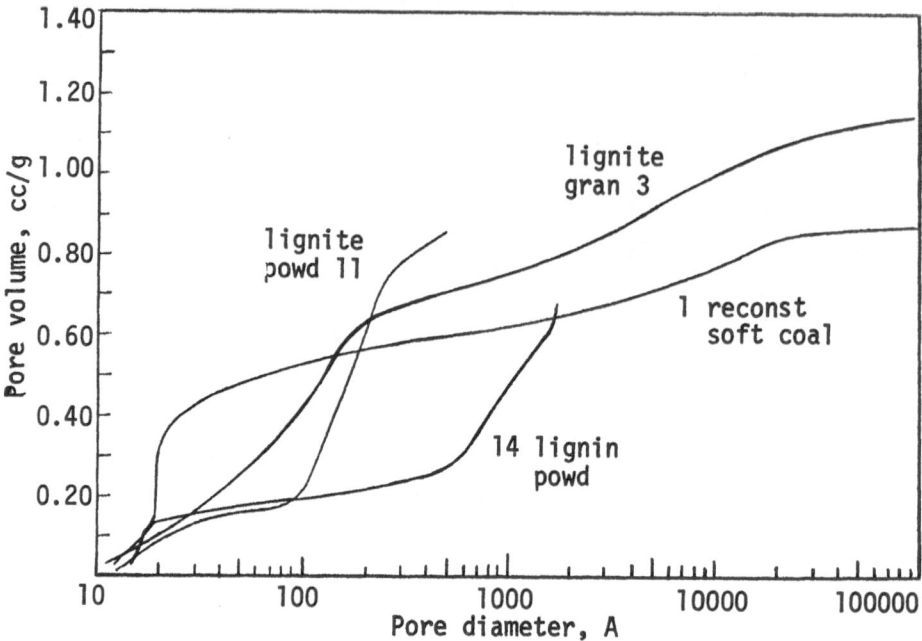

Figure 10. Pore size distribution curves for carbon 1 with close
balance between micropore and macropore volumes. Carbons 3, 11
and 14 have open pore structures; i.e., macropore volume very large

3.3 Thermodynamics of regeneration

Thermodynamics studies were done on a small six-hearth furnace used
in pilot plant scale studies at Pomona, Calif. under the sponsor-
ship of the Environmental Protection Agency.[5,6] A number of fundamental properties of the system's operation were observed which
would apply equally to other furnaces. Figure 9 shows two temperature profiles; the one for the gas was measured and the other for
the carbon bed was calculated from the operating data and thermodynamic properties of the gas-carbon system. From this study and in
conjunction with laboratory studies, it was learned that the regeneration process consisted of three steps: 1. over hearths 1, 2
and 3 the water was vaporized from the carbon, 2. over hearth 4, the
dry carbon began to heat up rapidly and the adsorbate in the carbon
was pyrolyzed, and 3. over hearths 5 and 6, the carbon was reactivated. For every gram of (dry) carbon, 0.8 to 1.0 g of water was
vaporized. Its vaporization adsorbed about 40% of the total heat
input. Also, for every gram of regenerated carbon there was
approximately 0.2 g of adsorbate. Its pyrolysis adsorbed 10% to
20% of the heat input.

The manner in which adsorbate pyrolyzes depends on the pore structure of the carbon. In a carbon, such as carbon 1 shown in figure 10, the adsorbate pyrolyzed into a volatile fraction, 70% by weight, and the rest broke down to free carbon. Carbon 1 has an even balance between micropore and macropore volumes; i.e., 0.42 and 0.45 cc/g, respectively. In an open pore carbon such as carbon 3, the pyrolysis produced very little or no free carbon. In this carbon the macropore volume, 1.00 cc/g, was very large relative to the micropore volume, 0.16 cc/g. The ease with which the gas could escape through the large pores appeared to have been the determining factor. For carbon 1, activation was required on hearths 5 and 6 to remove the free carbon.

Other results of significance from this study were: 1. impurities are preferentially adsorbed in pores of 14 to 30A diameter range; 2. the oxides, Fe_2O_3, CaO, MgO, K_2O and Na_2O, when present in the carbon, catalyze the oxidation of the carbon and lower its adsorptive capacity, and 3. steam activation recovers the original adsorptive capacity more closely than CO_2 activation.

3.4 Air pollution

The pyrolysis of impurities produces volatile matter that is very odorous and may also have limited toxicity. These are converted to CO_2 and H_2O to a large degree in the zero hearth which is heated close to 1000°C with gas burners. Many of the pollutants also contain halogens, sulfur and nitrogen which on incineration are converted to inorganic acids. These are removed from the exhaust gas-stream by water scrubbing as are also carbon fines, smoke and aerosols.

3.5 Regeneration of powdered carbon

Powdered carbons have in the past and are at present used to purify water drawn from rivers and lakes for household use. The impurities concentrations in these water sources are quite low; relatively small carbon dosages can purify large quantities of water. Because of their low cost, powdered carbons can be used economically on the once-used basis. However, in the treatment of waste water much larger dosages are required. For powdered carbons to be competitive with granular carbons, they must also be regenerated.

Powdered carbons posed handling problems because of the small particle size and they also retain a considerably larger amount of water than granular carbons. Methods used for granular carbons are not suitable for powdered carbons. An entirely new approach was necessary.

Powdered carbon treatment of water consists of a two-stage countercurrent batch process. After the second stage the carbon suspension (~0.2 g/l) is thickened to a slurry (~100 g/l or 10%) in a settling tank and pumped to the first stage. After the first stage the carbon suspension is either thickened to a slurry and pumped into the regenerator or filtered. When filtered, a cake is produced with 30% to 50% carbon which is fed into the regenerator by means of a screw feeder.

When the wet carbon becomes surface dry, at 70% carbon content, it readily becomes gas-borne due to the agitation from the steam being released and the regeneration gases. These factors have been accepted as inherent to the regeneration process. All regeneration procedures investigated employ a fluidized system with recovery of the gas-borne carbon by means of cyclones, particulate filters and water scrubbers. The last step produces a slurry which is then fed back to the water treating phase of the operation.

The U.S. Environmental Protection Agency has supported the study of several regeneration procedures. Three techniques evolved that showed promise; these were: 1. the transport-reactor system of FMC Corp., 2. the microwave-energy method of Dorr-Oliver, Inc., and 3. the fluidized-bed technique of Battelle Memorial Institute. The transport and fluidized-bed methods were further studied on a pilot plant scale at Lebanon, Ohio. Further study has recently been completed at Salt Lake City, Utah. The results are promising although all problems have not been solved.

Figure 11 shows a flow diagram of an indirectly heated fluidized-bed system based on bench scale results obtained on one of the sponsored programs.[7] In the operation of this system the carbon slurry is filtered and the filter cake dried to a moisture level below 50% by weight. The cake is partially broken up and charged into the feed bin from which it is then fed by means of a screw feeder into the fluidization gas-stream. To gain sufficient heat transfer to the carbon, the reactor is partially filled with 0.05 to 0.02 cm diameter granules of sand or 0.08 to 0.03 cm diameter flint shot. The inert granular material forms the fluidized bed while the carbon particles of diameter from 2 to 40μ pass rapidly through the bed, being entrained in the gas stream. The regenerated carbon is separated from the gas stream by means of the cyclone.

Recovery of over 90% of adsorptive capacity and over 80% weight yields were attained. Operating temperatures were maintained in the 540° to 820°C range. Carbon residence time was near 1.0 sec assuming carbon particles passed through the bed at the same velocity as the gas-stream. Various gas mixtures were tried, such as flue gas, 80% N_2 and 20% H_2O, 90% N_2 and 10% CO_2, air and 95% N_2 and 2% O_2. Oxygen control was critical; it had to be kept at a

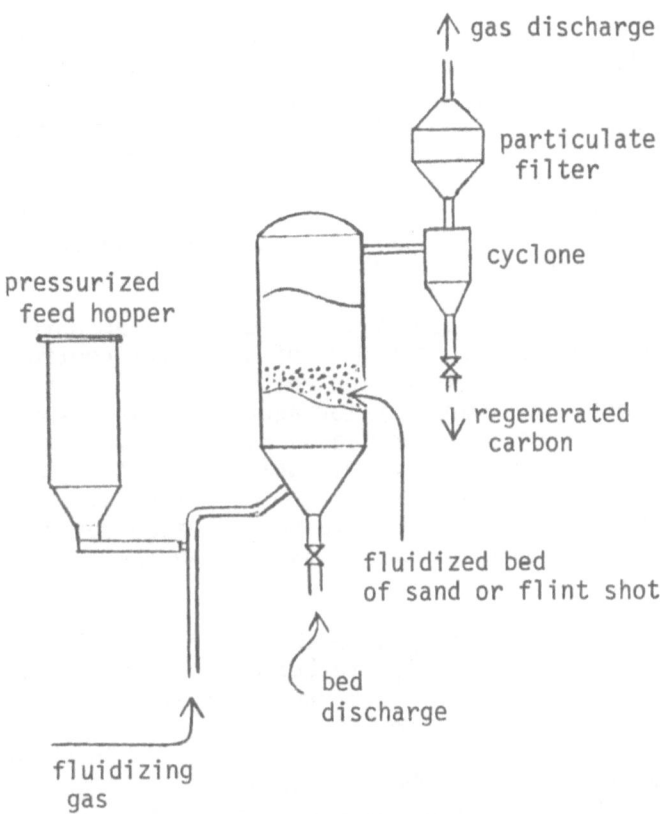

Figure 11. Fluidized inert bed system for regenerating powdered carbon spent in waste water treatment

minimum to avoid oxidation of the carbons. Aside from the loss of capacity and weight yield, the inert media abrades and the fine powder formed gets mixed with the regenerated carbon.

Powdered carbons are considerably easier to regenerate than some granular carbons because of the lack of the major portion of the macropore volume. Powdered carbons generally have an open pore structure since they are made from the softer low density chars. The open pore structure is illustrated by the pore size distribution curves 11 and 14 in figure 10. Lower regeneration temperatures are sufficient, below 820°C, while granular carbons required temperatures over 850°C.

4. SPILLED CHEMICAL CONTROL

The vast quantities of potentially hazardous materials produced and transported each year present a significant threat to the environment from accidental spills. It has been estimated that more than 3,000 major spills of hazardous polluting substances occur annually on navigable waters in the U.S. Pollution from this source is a certainty in any industrialized country. Methods for minimizing damage when a hazardous material spill occurs are of interest to everyone.

Because of the wide variability in the types, amounts and locations of chemical spills, a number of different control stratagies have been proposed and investigated. Activated carbon, because of its demonstrated capacity for aqueous phase adsorption of a large variety of organic chemicals, has been found to be an effective medium for removing the chemicals from water. Because of solubility and density differences in the potential spilled chemicals, a number of techniques have been devised for contacting carbon granular with chemicals that float, those that sink and those that dissolve.

Carbons coming in contact with essentially pure floating or sunken chemicals can become loaded to 50% or more by weight. Carbons coming in contact with dissolved chemicals adsorb from 10% to 20% by weight. As in waste water treatment, the carbon will take with itself a certain amount of water when fed into the regenerator. When the amount of chemical adsorbed is 20%, the amount of water is about 50% by weight. At higher chemical contents the water content is proportionately less.

The concern in this portion of the lecture is with the disposition of the chemical- and water-saturated carbon. Several alternatives for its disposition are practicable. It can be transported to a solid's disposal area and buried, it can be transported to an area station or to the carbon manufacturer's plant for regeneration, or it can be regenerated at the spill site with a portable regeneration system. Regeneration generally produces a savings on carbon costs and this is expected to be the case also on spill chemical cleanup operations. The spill-chemical, if it can be recovered, can to some degree offset the cost of the clean-up operation.

To carry out recovery operations a more complex, larger and heavier system is required than just for carbon regeneration. It has already been ascertained that because of size and weight limitations, it is impractical to move all the necessary equipment for chemical recovery on a mobile unit. At present it appears certain that recovery of chemical must be sacrificed with on-site mobile unit regeneration while this option is open to regeneration

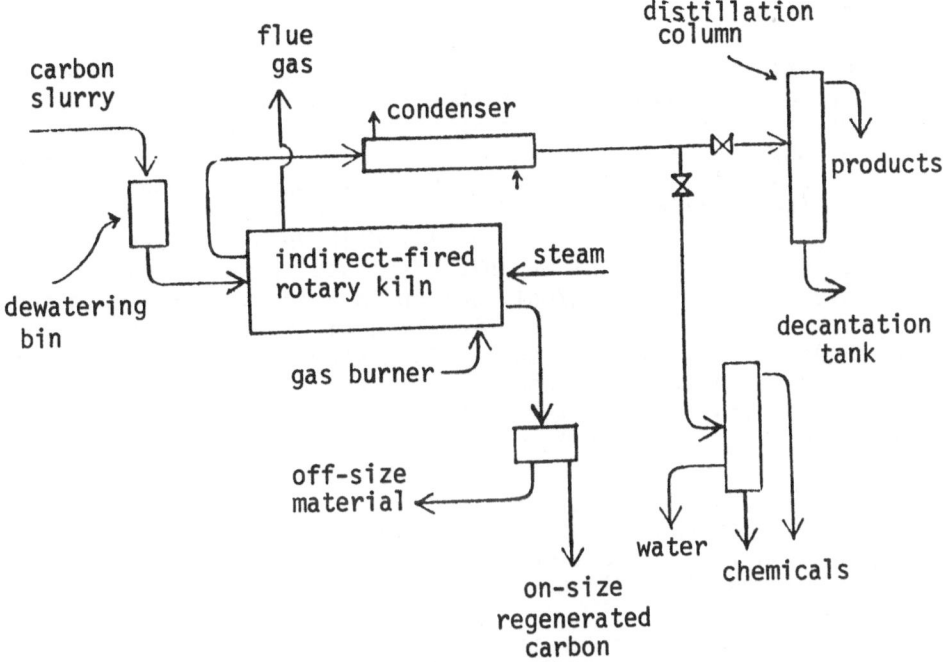

Figure 12. Flow diagram for carbon regeneration with chemical recovery

at an area station. Brief discussions of carbon regeneration without and with recovery follow.

4.1 Regeneration with chemical recovery

Volatile organic chemicals of liquid molar volume less than 190 cc/mol can be desorbed as vapors at relatively low temperatures with steam sweep, in the manner done in solvent recovery. Because of the large heat requirements to vaporize the water, more heat input is required than in desorption of solvents from a dry carbon. Regeneration of the wet carbon can be done more efficiently in an indirect-fired rotary tube kiln.

Figure 12 shows a flow diagram for the various operations required in this overall process. The carbon slurry is dewatered down to 30% to 50% water content by drainage in the dewatering bin and by means of the dewatering screw feeder. By external heating of the rotary tube a large amount of the heat can be transferred to the wet carbon without introducing noncondensable gases into the desorbed-gas stream, as would be the case with any direct-fired furnace. The efficiency of the condenser is then not impaired by passage of noncondensable gases.

Material and heat balance calculations show that at a heat input rate of 1,200 cal/hr/g carbon, the wet carbon can be desorbed of water and organic adsorbate in 45 min. The carbon discharge will be near 300°C and the steam-organic vapor mixture to the condenser will be near 150°C. About 40% of the heat input is consumed in desorbing the water and about 27% in desorbing the organic adsorbate. Over 50% of the tube length is utilized in desorbing the water and partial desorption of the organic adsorbate. Organic adsorbates of low molar volume, ~140 cc/mol, desorb to a large degree with the water. With increasing molar volume the difficulty of desorption increases and a larger percentage is desorbed from dry carbon at increased temperatures. Steam sweep is used to lower the organic vapor concentration over the carbon and, thereby, increase desorption rate. With increase in molar volume and desorption temperature, the decomposition rate also increases with the formation of noncondensable gases which greatly reduce the condenser efficiency. For organic adsorbates much over 190 cc/mol liquid molar, this system is no longer functional; a direct-fired furnace operated at gas and carbon temperatures near 950°C is now required.

The carbon is regenerated to completion by this method except where decomposition occurs and some free carbon is left in the pores. Some loss of capacity will occur until a steady state is established.

The regenerated carbon, after cooling, is sieved to remove fines and over-sized material. The carbon, as taken from the river or lake, may contain foreign matter such as silt and also other pollutants that are normally present in public water supplies. Silt becomes separated from carbon during the sieving operation. Some of the other pollutants may appear in the recovered organic chemical.

Water soluble chemicals are separated from the condensate by distillation and the insoluble ones by decantation. Many chemicals are partially miscible with water in which case the decantation is followed by distillation of one or both liquid fractions.

4.2 Regeneration without chemical recovery

Figure 13 shows a flow diagram for a carbon regeneration system where the desorbed water and organic matter are incinerated at temperatures near 1000°C and the combustion gases then scrubber with water or mild caustic solution. The regeneration kiln in this case is heated directly by gas burners at the carbon-effluent end of the tube. The rotary tube is ceramic lined to withstand temperatures up to 1000°C. The adsorbed chemicals are desorbed directly,

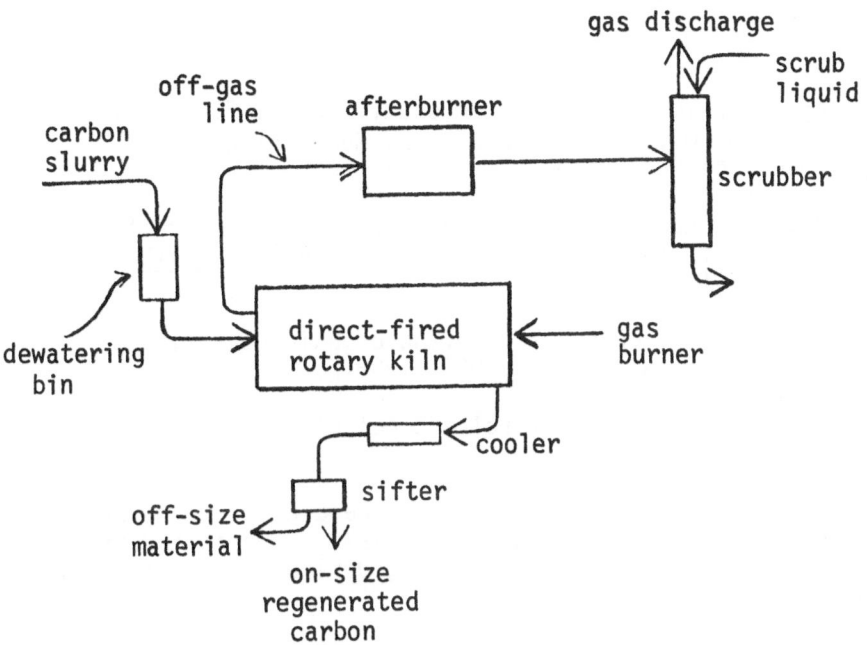

Figure 13. Flow diagram for carbon regeneration without chemical recovery

if of low molar volume, or are decomposed first and desorbed partially as volatile fragments of the adsorbed molecule and partially as CO and H_2. The reactivated carbon is discharged at temperatures near 950°C and the gases at about 300°C. In the afterburner, the off-gases are reheated to 1000°C, the retention time being at least 0.5 sec. There is sufficient steam to oxidize all combustible products to CO and H_2, but this would produce a toxic gas mixture. It is desirable to add sufficient air to oxidize all combustible products to CO_2 and H_2O. In many cases this combustion produces enough heat to make the afterburner self sustaining.

This regeneration system is the same type as used in regeneration of waste water treatment carbons but the operation varies some because of the presence of more volatile adsorbates. In an area plant, a multiple-hearth furnace could be used rather than the rotary kiln, but because of height limitations only the rotary tube kiln can be used on a portable system. Because the nature of the spill is not always known and could consist of both high and low volatile chemicals, a portable system is needed that can regenerate the carbon without regard to the type of chemical present.

REFERENCES

1. Package sorption device system study. EPA-R2-73-202,
   April, 1973, Office of Research and Monitoring, U.S.
   Environmental Protection Agency, Washington, D.C. 20460.

2. Polanyi, M., Adsorption and capillarity from standpoint
   of 2nd law of thermodynamics, Verhandl, deut. physik. Ges.,
   16, 1012, 1914.

3. Mattia, M.M., Process for removing organic contaminants
   from air, U.S. patent 3,455,089, 1969.

4. Mattia. M.M., Process for solvent pollution control,
   Eng. Progr., 66(15) 74, 1970.

5. Optimization of the regeneration procedure for granular
   activated carbon, 17020 DAO 07/70, 1970, Superintendent of
   Documents, U.S. Government Printing Office, Washington, D.C.
   20402.

6. Juhola, A.J., Tepper, F., Regeneration of spent granular
   activated carbon, Robert A. Taft Water Research Center Report
   No. TWRC-7, Superintendent of Documents, U.S. Government
   Printing Office, Washington, D.C. 20402.

7. The development of a fluidized-bed technique for the regen-
   eration of powdered activated carbon, ORD-17020FBD03/70, 1970
   Superintendent of Documents, U.S. Government Printing Office,
   Washington, D.C. 20402.

CHEMICAL REGENERATION

T. Halmø ·

The Engineering Research Foundation
at the Norwegian Institute of Technology
Trondheim, Norway

INTRODUCTION

One of the main reasons for the increase in use of activated carbon adsorption systems in water and wastewater treatment, is the improved economics that is due to improved regeneration systems.

Another reason is of course the increased demand for treatment-systems with sufficiently high performance to meet the ever sharpening standards of wastewater treatment.

With improved regeneration procedures one is also making high standards easier to approach and contributes hereby to the establishment of even stronger standards.

Whatever way we look at it, it leads to cleaner environment, but it is necessary to make strong efforts in order to reduce and provide better performance.

PRESENT SITUATION

The present situation concerning granular carbon regeneration is that thermal regeneration is still the only general applicable regeneration process that can be used for practically all regeneration purposes.

But is thermal regeneration the optimal process for carbon regeneration? - Intuition tells us "No".

The obvious advantage is that it is a general process, ready

for all sorts of jobs. Most of the adsorbed material is converted
into gases than can, if necessary, be converted further by combustion to CO_2 and water.

However, the danger for air pollution is there, and one
should be careful not to transfer a water pollution problem into
an air pollution problem.

The disadvantages of thermal regeneration is first of all the
loss of carbon in each regeneration cycle. This is reported to
be between 5 and 20 % for the different carbons, which means a
lifetime of 20 to 5 cycles. Further the investment involved in
the regeneration furnace and transportation equipment is considerable, and so are the operation costs.

This has led to a more and more intense search for a regeneration method that will have the performance and applicability of the thermal regeneration with no or low carbon loss
and low investments and costs of operation.

This indicates a process that takes place in situ where
attrition can be avoided, running on moderate temperatures to
avoid burn-off, and using a method that can remove any adsorbate
from the carbon surface without reducing the adsorption-capacity
of the carbon.

Unfortunately this method does not exist, but there are a
number of methods that come fairly close and these are all based
on some kind of chemical regeneration.

It may be useful to try to analyze the situation on hand when
regeneration of carbon becomes necessary.

THE REGENERATION PROBLEM

The carbon surface is covered by a monomolecular or multimolecular layer of the adsorbed organic molecules. These are
mainly bound to the surface by weak molecular forces - van der
Waal forces, - but to some extent the molecules will be bound to
positions on the surface by chemical bonds. This is due to irregularities on the surface, intended or unintended by the manufacturer. The molecules bound with v.d. Waal's forces are in
principle able to change positions, but the chemically bound
molecules will not easily do so.

At saturation the carbon surface has reached a situation of
dynamic equilibrium. At this point the rate of adsorption is of
the same order as the rate of desorption. The nature of the
equilibrium is dependent upon the solubility of the organic
compounds in water, the hydrofobic/hydrophilic character of the

adsorbate, but also on the organic molecules themselves; the size, rheology, symmetry etc.

To restore the adsorptive capacity of the carbon, we want the adsorbate removed from the carbon surface as quickly as possible. If the adsorbate can be removed in a nondestructive manner, this is favourable, especially when the adsorbate has some value if recovered.

So what has to be done in a liquid system is to change the equilibrium conditions in a manner that leaves us with a driving force from the solid through the liquid film and onto the bulk of the liquid. And the liquid must have a capacity to absorb this matter effectively and the new equilibrium conditions to be established must be favourable towards the liquid phase.

So when we want to remove the adsorbate from the carbon surface we have two alternatives:

1. We can introduce a force on the molecule directed away from carbon surface, and this force must then be stronger than the attraction force between the molecule and the carbon.

2. We can arrange matters in such a way that the existing attraction between the molecule and the carbon either is reduced or completely removed.

In thermal regeneration we have a combination of the two methods. When introducing higher temperature the vibration energy increases and eventually the internal agitation of the molecule breaks it away from the surface, and into solution or gas phase. If decomposition of the molecules takes place before the molecule breaks away the attraction force to the surface is reduced, being the net sum of force from each atom and therefore a function of molecular weight.

On decomposition the new molecules will easier break away also because the decomposition is exothermic, which will add to the energy of the molecules.

On chemical regneration we have the same two alternatives, only with the restraint that the heat introduced must be moderate.

The most wellknown methods for chemical regneration will follow alternative 1, and can be:

- ionization of adsorbate

    - chemical reaction changing the solubility
    - changing the solubility by changing the solvent.

Alternative 2 will be:

    - complete or partial chemical oxidation (by strong oxidants).

Ionization

Weak organic acids and bases are readily absorbed on activated carbon in undissociated form. When altering the pH-conditions organic acids and bases will ionize under basic and acid conditions, respectively.

Ionized compounds are strongly hydrophilic and will hence rapidly go into aqueous solution.

By changing the pH-conditions again the organics may be recovered.

Chemical reaction

Components that are not easily influenced by merely changing the pH-conditions may be subject to chemical reactions producing components more soluble in water. Hydrolysis and sulfonation may be such a reaction.

Extraction

Instead of water one may expose the carbon to another solvent with a higher solubility for the adsorbed components. Hereby the equilibrium is altered.

To be able to use the right technique in chemical regeneration it is therefore highly desirable to know the system involved and this is an important restraint for instance in the treatment of sewage from a mixed net. However, it is a great advantage in the treatment of industrial wastewater where the processes and wastewaters are generally well known, and where recovery of chemicals is an incentive once wastewater treatment has to be introduced or improved.

APPLICATIONS

Phenol recovery

A number of successful applications of granular activated carbon adsorption with chemical regeneration has been reported.

One of the most successful ones is recovery of phenols and phenolic compounds from regeneration with sodium hydroxide.

Reports of this have been made by among others Fox et al.[1] who has described recovery of phenols and acetic acid from briny water from a phenol plant. This makes it possible to use the brine as a feed to chlorine-caustic plants. The process is this: Hot brine containing phenol and sodium acetate at pH=7 is cooled and passed through granular adsorption columns where phenol is removed. Afterwards the brine is adjusted to pH=3 with HCl and passed through a new set of columns where acetic acid is adsorbed. Then the brine is neutralized with NaOH and passed on to chlorine cells. The regneration for both sets of columns is done with a diluted (4%) solution of sodium hydroxide. The sodium phenolate is returned to the phenolic plant and the sodium acetate is deposited.

The carbon losses are reported to be very small. The behaviour of the carbon bed with respect to restored adsorption capacity on repeated regnerations has been investigated for acetic acid. (Fig. 1). This shows that the adsorption capacity drops to about 65% of the virgin carbon at 5 regenerations but after that it drops very slowly and after 50 regenerations is still more than 55%.

The regneration strategy is also important and a later paper by Fox et al.[2] shows the regeneration cycle. Periodical introduction of stronger alkali gives improved regeneration. (Fig. 2).

The long time influence of regeneration of the carbon from phenol adsorption has not been reported and has been studied by SINTEF as part of our carbon-adsorption program. The tests were carried out in laboratory columns where saturation was obtained in less than 3 hours using a highly loaded water. The breakthrough curves were monitored. 28 cycles of regenerations with somewhat different strategy was carried out on one of the columns and the results are shown in Fig. 3. The first somewhat confusing results are mainly due to changing the regeneration procedure. 100% regeneration is based on the adsorption capacity of the virgin carbon which was 0.37 g phenol/g AC. This means that with a regeneration of 54% the adsorption capacity is still 0.2 g phenol/g AC which is rather nice. The experiments show basicly the same as Fox et al. reported for acetic acid. The capacity drops with repeated regenerations but flattens rapidly out and remains pretty good even after a considerable number of regenerations.

But it also shows that a periodic variation in the adsorption capacity must be expected. This must be due to local build-ups where the regenerant has less access for micro- or macroscopic reasons. This periodic variation was also found in pilot plant operations.

The same procedure is applicable to a good many other components and mixtures of such and should be considered whenever weak acids and bases are to be removed.

Other methods mentioned in the literature is regeneration with concentrated H_2SO_4, which has shown fairly good results. This method is, however, connected with several practical problems on the hardware side.

Sontheimer's method

Another method that claims a wide variety of applications is patented by Sontheimer et al.[3]. This method is based primarily on the application of dipolar aprotic organic solvents with Dipol Moment > 3 Debye. This group of solvents has no proton donation capacity and will therefore stay unchanged in water. Examples of such components are N-Dialkylamines like Dimethylformamide and Dimethylacetamide. Many other components are mentioned in the patent.

The regeneration can be done in batch or with circulation of the solvent and with or without elevated temperatures. Mixtures of the solvent and other solvents like polar protic solvent are also claimed to have favourable effects in some applications.

The performance and applicability of this method are, however, somewhat fogged by the chosen examples. In these, saturated carbons from practical applications are extracted with Dimethylformamine. The experiments are then carried out with a "model-water" consisting of a solution of this extract in pure water. Then hot Dimethylformamine is the basic regenerant in the regeneration. By this procedure it seems that the inventor is loading the carbon only with organic compounds that have shown ready extractability with the regenerant.

However, the method may prove to have general application, but further demonstrations will have to be made.

Degremont's method

Degremont has introduced a method based on the combined effect of alkali and solvent, as was reported by Rovel[4]. Here hot caustic soda is used to provide alkaline conditions and at the same time cause hydrolysis. A variety of solvents were tested and isopropanol was chosen. It has good solvation power over a large spectrum, displacing power at alkaline conditions, and it is easily eluted by water and vapours. Further it has low costs and is easily recovered.

The regeneration is carried out in 5 stages whereof 2 are optional.

1. **Initial acid treatment.**
   Eliminates minerals deposited on the carbon. 0.5 - 1 bedvol. of 1-2 % HCl in 1 hour.

2. **Hot soda treatment.**
   100-110°C, 1.25 bedvol. of 10 % soda soaking for 1 hour.

3. **Solvent treatment.**
   Slow percolation of solvent at 80°C
   50 % solvents sol. 0.5 bedvol. 3-4 hours.

4. **Evaporation of solvent.**
   0.75 - 1 kg vapour at 120-140°C pr. kg. carbon 1 hour.

5. **Re-acidification.**
   If needed 1-2 % HCl for 10 min.

The 90-95 % of the isoproponol is recovered by distillation. The soda can be rejected after neutralization but is preferably concentrated and recovered through alkaline fusion in a pyrolytic process. The loss of soda is 5 %.

This process has been applied to treatment of wastewater from manufacturing of organics (PTBB-Acid) and pharmaceuticals, for decolourization of beet sugar and on textile wastes, and it is claimed that 80 % of the initial adsorption capacity is restored.

This process is being investigated further and improvements should be expected. The economic comparison made by Rovel shows favourable costs based on a plant for regeneration of 10 m^3/day.

Oxidation agents

Sontheimer et al.[5] have also tested strong oxidation agents like $KMnO_4$ and ozon. By this method the adsorption capacity of the carbon is substantially reduced due to the formation of surface oxides.

This indicates that the oxidation of organics at the carbon may not be feasible when based on chemical oxidation agents.

Powdered carbon regeneration

The methods mentioned here are all concerned with granular activated carbon. However, regeneration of powdered activated carbon has proved interesting during the last years. The main methods for this is the thermal fluidized bed regeneration and

334

the "transport reactor".

Zimpro[6] has tried to regenerate powdered activated carbon
by their wet oxidation process. This process is a liquid phase
combustion at elevated temperatures (below the critical temperature of water) due to high pressures. Typical temperatures are
200-300°C and typical pressures are 15-85 atm.

The powdered carbon will have some burn-off by this process,
but after 23 regenerations the carbon content in the AC has been
reduced only from 80 to 60 % and the ash content raised from 5.4
to 14.0 %.

Biological regeneration

Biological regeneration has also been reported. This method
has obvious limitations - like many organic compounds not being
biodegradable. Further the microorganisms of which the smallest
one has a diameter of 10.000 Å will not be able to enter into the
internal pore structure of the carbon where most of the organic
matter is adsorbed. However, it may have some effects on the
adsorption/desorption equilibria, but the decomposition products
will rarely be CO_2 and H_2O, more likely small organic acids and
bases that have to be further treated.

LITERATURE

1. Fox, R., Keller, R., Pinamont, C. and Stevenson, I.
   "Purification of Waste Brine for Reuse",
   Industrial Water Engineering, March 1970.
2. Himmelstein, K.J., Fox, R.D. and Winther, T.H.,
   "Inplace Regeneration of Activated Carbon",
   Chemical Engineering Progress, 69, No. 11, 1973.
3. Sontheimer, H., Maier, D. and Rolke, D.,
   D.Pat. 2129459 "Verfahren zur chemischen Regeneration von
   erschöpften Aktivkohlen", 1973.
4. Rovel, J.M., "Chemical Regeneration of Activated Carbon",
   Applications of New Concepts of Physical Chemical Wastewater Treatment, Nashville, September 1972, Pergamon Press
   1972.
5. Sontheimer, H., and Kühn, W., "Einfluss chemischer Umsetzung
   auf die Lage der Adsorptionsgleichgewichte an Aktivkohlen",
   Vom Wasser 40, 1973, 115-123.
6. Knopp, P.V. and Glitchel, W.B., "Wastewater Treatment with
   Powdered Activated Carbon Regenerated by Wet Air Oxidation",
   25th Perdue Ind. Waste Conf., May 1970.

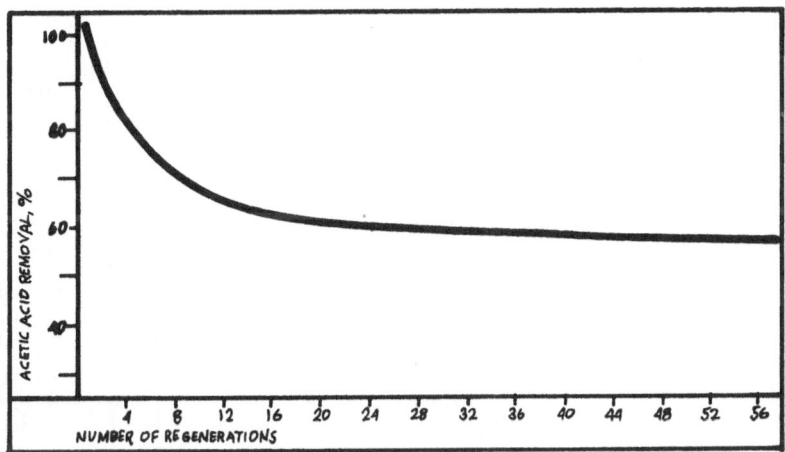

Carbon adsorption of acetic acid. After five regenerations, the carbon capacity is 65% that of the virgin carbon but levels off thereafter.

Figure 1.   Ref.: Fox (1)

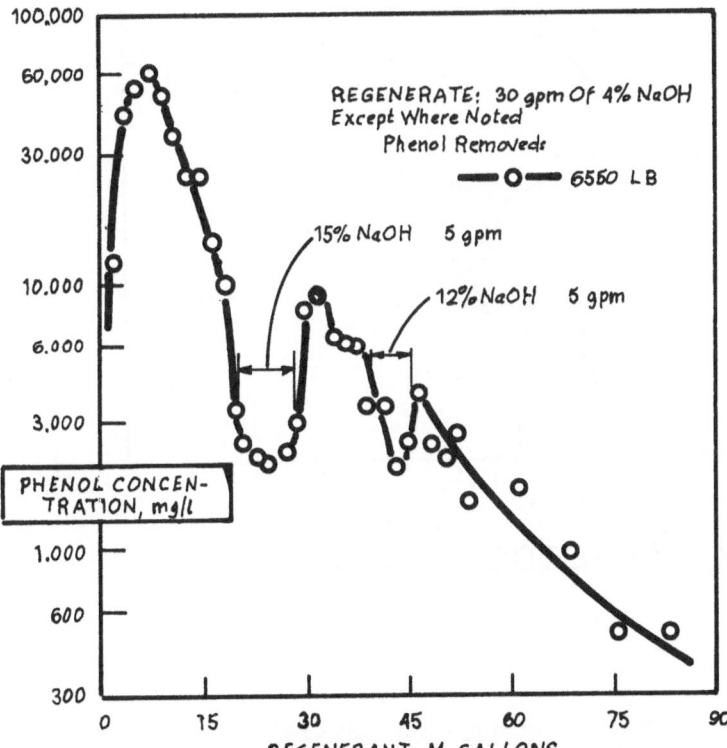

Figure 2.   Phenol desorption curves showing effect of percent
           caustic.

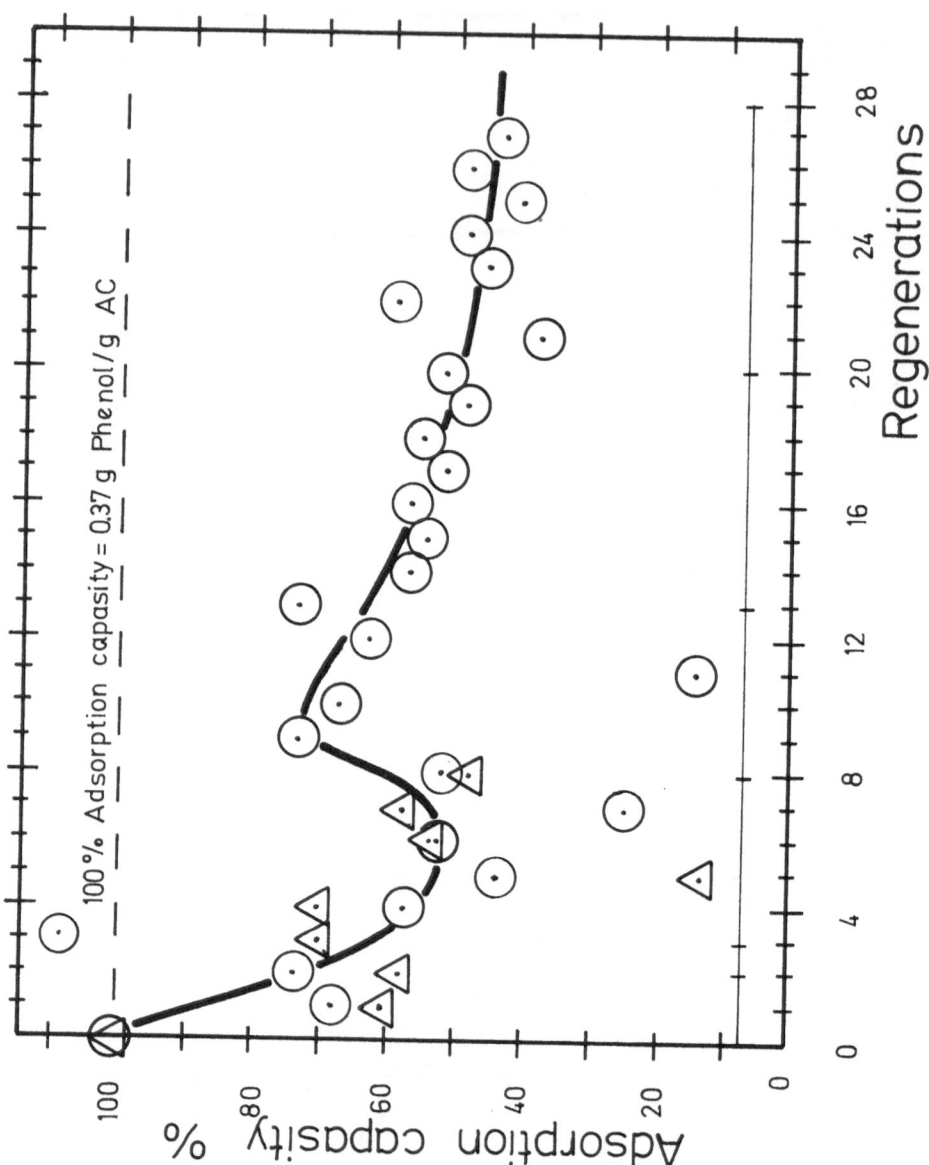

Figure 3. Development of adsorption capacity on repeated regeneration

MISCELLANEOUS REACTIVATION METHODS

Ronald Berg

Institutt for Atomenergi,
Kjeller, Norway.

INTRODUCTION

Activation in a general sense covers a large field of
physical/chemical processes by which the products (solid sorbents)
gain or regain selective sorption properties or higher physical
sorption capacities.

Sorption selectivity is of importance for separating purposes,
while the overall capacity will be highly important, for instance
for gas mask filters, where different types of toxic gases are
sorbed. By increasing the capacities of such filters a longer
service time is favoured. High capacity filters will also make it
possible to reduce the pressure drop during operation. New activation methods for increasing the capacity of such filters may be
desirable.

In many cases little is known about the mechanisms occurring
on surfaces during the activation of sorbents. This implies that
control of the activation process and consequently the reproduction
of sorption effects may be difficult.

Activation processes can be classified as:

- Physical activation
  - thermal activation
  - irradiation activation

- Chemical activation.

The present lecture deals with activation methods enabling sorbents

to gain capacity with regard to gases. The lecture is divided
into two parts. The first one deals with thermal shock treatment,
the other involves irradiation of sorbents.

Thermal shock or quenching of sorbents with water has been
used for a long time. The application of cryogenic quenching,
however, seems to have been applied to a lesser extent. Rapid
cooling as well as irradiation of sorbents are simple operational
methods and involve some advantages compared to other methods.

Pretreatment of sorbents with ionizing radiation implies the
introduction of electrical charges on the surface of the solid.
New reactive groups may also be fixed to the surface. These
methods imply that the chemisorption capacity of the sorbent may
be increased or decreased. The effects obtained are dependent
upon a large number of experimental parameters such as irradiation
dose, type of irradiation atmosphere, temperature, humidity, etc.
Some examples are presented in which the chemisorption capacity
of gas masks is increased as a result of β- and γ-irradiation.

The lecture presented is partly based on published literature
data and partly on own research efforts.

The various sorption properties discussed are illustrated by
a limited number of examples.

I. REACTIVATION OF SORBENTS BY THERMAL SHOCK TREATMENT

The properties of adsorbents are largely determined by the
pore structure and by the nature and area of the solid surface
(1 - 7). The specific area alone is not sufficient to characterize an adsorbent. The accessibility of the internal surface area
of a porous body depends on the shapes and dimensions of the pores.
Large surface areas are associated with systems consisting of net-
works of a great many micro-pores (< 20 Å), transitional pores
(20 - 1000 Å) and macro-pores (1000 - 300000 Å). A simplified
network model for active carbon is proposed by Dubinin (1).
According to the Dubinin model only a few of the micropores lead
directly to the outer surface of the carbon particle.

The pore structure is predominantly arranged in the following
pattern

- macropores open up directly to the external surface of the
  particle

- transitional pores branch off from the macropores

- micropores branch off from the transitional pores.

The term sorption capacity is understood to cover both physical sorption and chemical sorption of a fluid.

SOME ASPECTS REGARDING THERMAL ACTIVATION AND THE OBTAINED SORPTION PROPERTIES

In this lecture the term "thermal activation" comprises the steps of heating and cooling of a sorbent.

The final properties obtained from such an activation sequence will depend on the following parameters:

- the activation temperature
- the duration of the activation
- the cooling rate of the sorbent
- the activation atmosphere
- the composition (nature) of the original sorbent.

Cooling of sorbents may be carried out slowly or fast, i.e., by thermal shock (quenching). By applying a quenching technique (thermal shock) after heating a carbon, one or more of the following effects may possibly occur:

- change of pore texture (volume, size, shape, specified surface area)

- pore volume enlargements in the carbon material due to breakdown of walls between neighbouring pores

- change of the hardness of granules as a result of pore wall breakdown

- change of surface chemistry (number and types of reactive groups)

- freezing of the high temperature state, yielding thermodynamic unstable, more reactive surface groups or sites.

For impregnated carbon, the following additional effects might arise:

- creation of cracks and pores in the impregnation which results in increased (or changed) surface area of the impregnation layer
- opening of blocked pores, resulting in increased change of total surface area and pore volume.

Whether breakdown of pore walls occurs or blocked pores are opened, depends on the intensity of the thermal shock ($^{\circ}C/min$), and this is determined by:

- the temperature gradient between sorbent and cooling medium

- sorbent layer thickness

- specific heat and heat conductivity of the sorbent and void gas.

In the following some activation procedures are reviewed and emphasis is particularly laid on the effect of the cooling procedure.

SURVEYS OF SOME ACTIVATION/QUENCHING METHODS

Reactivation followed by quenching in water.

Thermal activation of exhausted carbon used in water purification is carried out in a multiple heat furnace. The activation is usually accompanied by a quenching step by which the carbon is dropped from the bottom of the hot furnace directly into water of ambient temperature. Fig. 1 illustrates such a reactivation system.

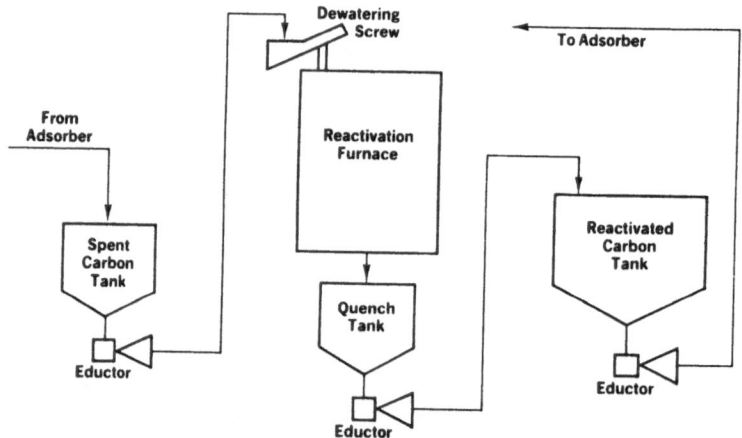

FIGURE 1. Schematic diagram of a reactivation system with water quenching.

During the quenching step some of the carbon granules will disintegrate. Quenching tests (8), however, have indicated that less than 0.4% losses will occur on water quench regenerated carbon and that essentially no change occurs in the average particle size of the carbon.

Evans (9) indicated particle breakdown on quenching, but

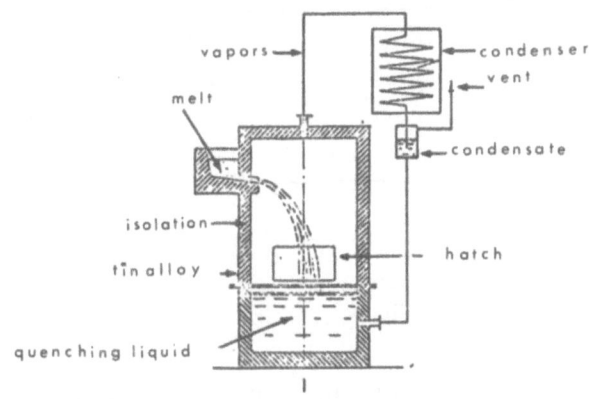

FIGURE 2. Method for quenching a melt of catalyst mass in a liquid.

actual losses during quenching could not be determined. In water purification, the carbon loss is in the range of 5%. Compared with the total carbon loss, the loss in the quenching step was rather low (acceptable). The main loss is affected by other operational parameters.

Preparation of catalysts by activation and quenching a contact mass in a liquid at ambient temperature.

A high activity catalyst is prepared from melting a contact mass of iron and ammonium-molybdate. The melt is poured slowly and directly into a suspension of ammonium-ferrocyanide and potassium-ferrocyanide in iron-carbonyl (liquid) at ambient temperature (10). In this way granules of high porosity, high specific surfaces and with a highly dispersed active layer of iron and iron cyanide, are prepared. (Fig. 2). Prior to the heating and quenching step the catalyst was reduced with hydrogen.

A modification of this method (11) involves dispersion of the mass to fine particles by means of hydrogen gas pressure before quenching the catalyst mass. In this way a much faster quenching of the mass was obtained.

The catalytic activities in the two cases regarding ammonia formation are demonstrated in Table 1.

342

TABLE 1.

Method	Reaction temp., °C	NH_3 yield, g	Ref.
Dispersed particles	450	6.05	(11)
Flowing stream	520	3.48	(10)

Parameters in both cases were:
    N_2/H_2 ratio: 1:3
    Pressure: 100 atm.
    Gas flow: 80 l/h.

Preparation of activated mixed sorbents by indirect quenching.

    Finely divided carbon is mixed with an inorganic hydro gel
such as silica gel (12). The air dried mixture is activated by
heating to 500°C, while keeping it in constant motion. The mixture is then exposed to indirect quenching, that is, rapidly cooled
in a heat exchanger surrounded by cooling water. (Fig. 3). This
technique prevents the powdered sorbents from packing in the hot
state. The activated products thus obtained are no longer subject
to aging. The carbon ensures the preservation of the system of
extremely fine pores in the gel and therefore prevents the aging
of the latter.

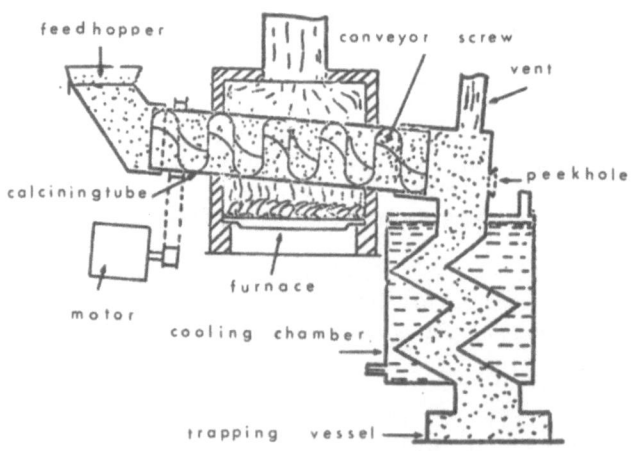

FIGURE 3. Method for activation of powdered sorbents combined
           with rapid cooling in heat exchanger.

Reduction of particle size of crystalline materials by cryogenic quenching (13).

Crystalline zeolites exhibit an internal and an external surface area, with the largest portion of the surface area being internal, and consisting of defined micropores. Blockage of the internal channels, by coke formation, poisoning of the catalyst, lattice imperfections will reduce the surface area considerably.

By reducing the particle size so as to increase the above mentioned surface ratio, the problem of channel blockage would be lessened, and in addition diffusion to the internal surface would be facilitated. The method involves a heat-quench treatment which subjects the zeolites to drastic thermal shocks in liquid nitrogen and effects a significant reduction in particle size without significantly reducing the crystallinity of the zeolite.

THERMAL ACTIVATION OF ACTIVE CARBON FOLLOWED BY CRYOGENIC QUENCHING

General description of experiments.

The activation procedure (14) involves the steps of heating the sorbent and subsequently quenching it to a temperature below - 75°C. The quenching step may be effected by submerging the sorbent directly in a cryogenic liquid such as liquid nitrogen. The sorbent can also be subjected to an indirect heat exchange process with a cryogenic liquid. The quenching step is normally followed by raising the temperature of the sorbent to ambient conditions before use. This technique has been applied on typical gas mask carbons impregnated with compounds of copper and chromium as well as silver salts. The exact composition of the impregnation has so far not been published. However, it is recognized that this type of impregnation gives good protection against toxic gases. Some properties of typical gas mask carbon of the ASC type compared to the properties of typical non-impregnated "gas" carbon of BPL type are illustrated in Table 2.

Gas mask carbons have a pore size distribution covering the range of both micropores, transitional pores and macropores. The BPL has a higher ratio of micropores than the ASC carbons.

The ASC carbons (I and II) originate from different batches; their Cu-Cr content was determined by X-ray spectrometry. Table 2 reveals that the higher loading with impregnation (ASC-II) yields a lower specific surface, probably due to blocking effects in the transitional pores, possibly also in macropores. Since a large part of the micropores are assumed to branch off from transitional pores, diffusion of gas to the micropores would thus be inhibited.

TABLE 2. Specific surface area and capacities of ASC and BPL carbon

Carbon type	Impregnation %		Specific surface	Chloro-picrin capacity	Cyanogen chloride capacity
	Cu	Cr	m^2/g	mg/ml	mg/ml
BPL	0	0	1100	130.5	4.0
ASC-I	6.2	2.0	750	105.0	13.7
ASC-II	7.0	2.5	690	52.5	21.6

Toxic gases are retained both by physical sorption and by chemisorption. Two test gases

- chloropicrin representing physical sorption (with capillary condensation in transitional pores)

- cyanogen chloride representing chemisorption (with catalytic oxidation)

were used for the capacity tests. In Tables 3 and 4 some physical data for the two gases and general test conditions are presented.

TABLE 3. Physical data of chloropicrin and cyanogen chloride.

	Chloripicrin	Cyanogen chloride
Formula	CCl_3NO_2	CNCl
Boiling point, °C	112	13.8
Vapour pressure, mm Hg at 20°C	18.1	1001
Vapour density, g/l		0.47

TABLE 4. Test conditions.

	Chloropicrin	Cyanogen chloride
Feed to test filter, mg/min	150	72
Flow, l air/min	30	30
Humidity, %	0	0
Linear velocity, m/min	10	10

As observed in Table 2, higher amounts of impregnation increase the decomposition of cyanogen chloride. This advantage is, however, accompanied by a decreasing physical sorption capacity, because chemisorption of cyanogen chloride is also dependent on the surface area of the impregnation.

The presence of water molecules is regarded as essential in the reaction between cyanogen chloride and the impregnation. Thus, some water content in the carbon (impregnation) will be beneficial to the decomposition of this gas.

In the present investigations, the influence of the activation and the cooling rate on the sorption capacities of these carbons were studied. Two cooling rates designated as "slow cooling" and "quenching" were used.

The cooling procedure was carried out by heating ASC carbon granules in a quartz tube (18 mm diam.) with nitrogen as cover gas. Slow cooling implies cooling from activation temperature to ambient temperature. These cooling curves are demonstrated in Figure 4.

For quenching, liquid nitrogen was used as cooling medium. The samples were flushed with nitrogen gas during both cooling operations.

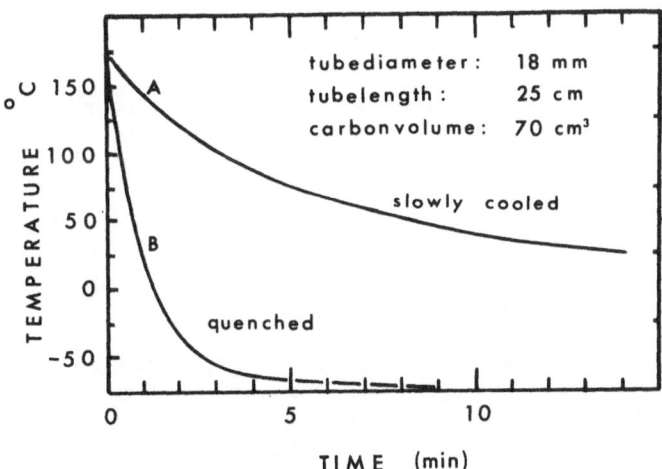

FIGURE 4. Cooling rate curves of ASC carbon activated to 170°C.

Influence of the cooling rate.

When impregnated carbon is thermally shocked, opening of blocked pores, pore enlargements, possibly takes place. This effect should be revealed by an increase in the physical sorption capacity, e.g., by increased capillary condensation of chloropicrin vapour. Fig. 5 shows the uptake of this gas as a function of the cooling rate ($^{\circ}$C/min) (15). This rate is defined in the temperature interval 170 - 0°C.

As seen in Figure 5, a plateau seems to be reached. With too rapid quenching, breakdown of the impregnation is experienced. This is the case for instance with direct contact between granules and liquid nitrogen and is also accompanied by a lower physical sorption capacity which indicates lower total pore volume.

Effect of repeated quenching.

With repeated quenching, a capacity increase (physical) should also be expected. Figure 6 shows an example of the percentage increase of the physical sorption capacity as a result of repeated quenching. Under the conditions specified, a saturation level seems to be reached. The activation temperature was as in the foregoing examples, 170°C.

The effect of activation temperature.

The dependence of the physical as well as the chemisorption capacities of ASC carbon on the activation temperature is shown in Figures 7 and 8.

The capacity decrease between 20 and 130°C is tentatively ascribed to diffusion of water from micropores to the surface of transition pores and macropores, giving rise to blocking effects in the transition pores.

At activation temperatures above 130°C a steep raise in the capacity curve is observed. The quenched carbons reach a capacity maximum for both physical and chemisorption at approx. 160°C. The decrease in sorption capacity at higher activation temperatures is probably due to decomposition of the impregnation and rearrangement of the pore textures. The figures illustrate that under the prevailing test conditions slow cooling definitely yields lower capacities than quenching.

Figures 9 and 10 show the simultaneous change in specific surface areas. For quenched carbon a remarkable coincidence of capacity maximum and surface minimum is observed.

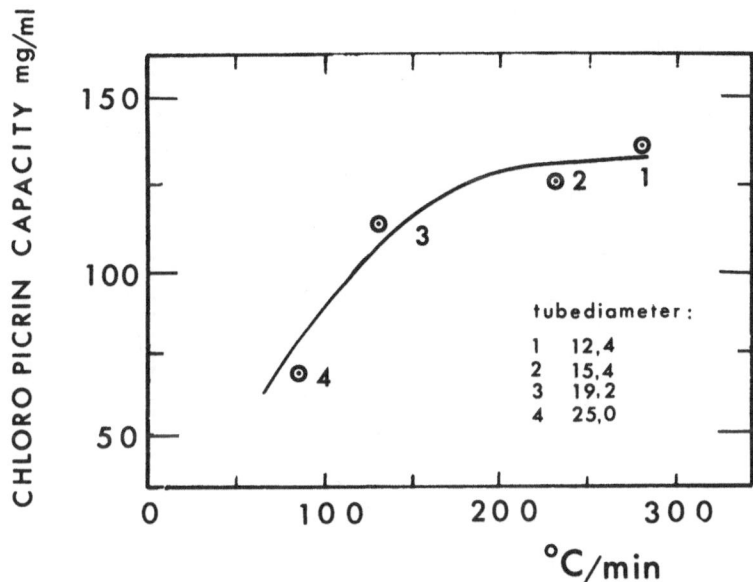

FIGURE 5. Chloropicrin capacity of activated carbon (ASC)
after quenching in liquid N_2 vs. cooling rate.

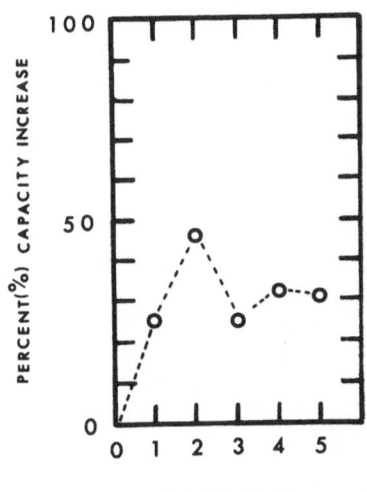

FIGURE 6. Chloropicrin capacity of ASC carbon vs. number of
heating/quenching cycles.

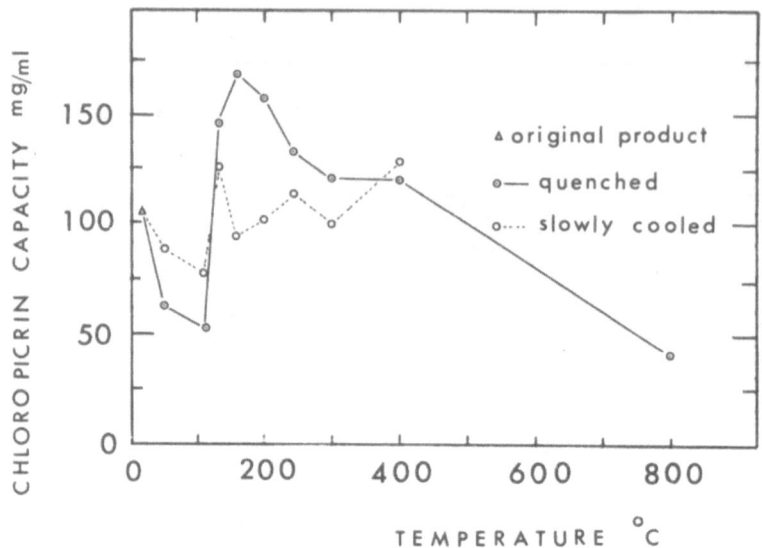

FIGURE 7. Chloropicrin capacity of activated carbon type
ASC vs. activation temperature.

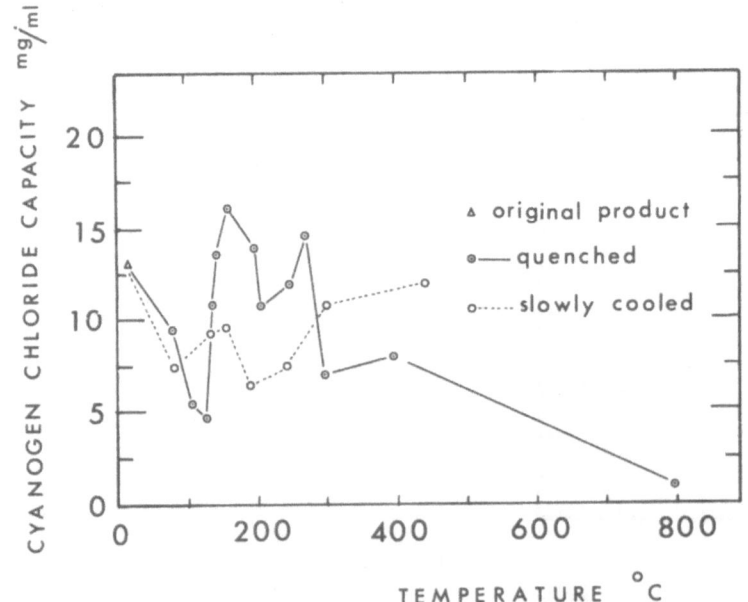

FIGURE 8. Cyanogen chloride capacity of activated carbon
type ASC vs. activation temperature.

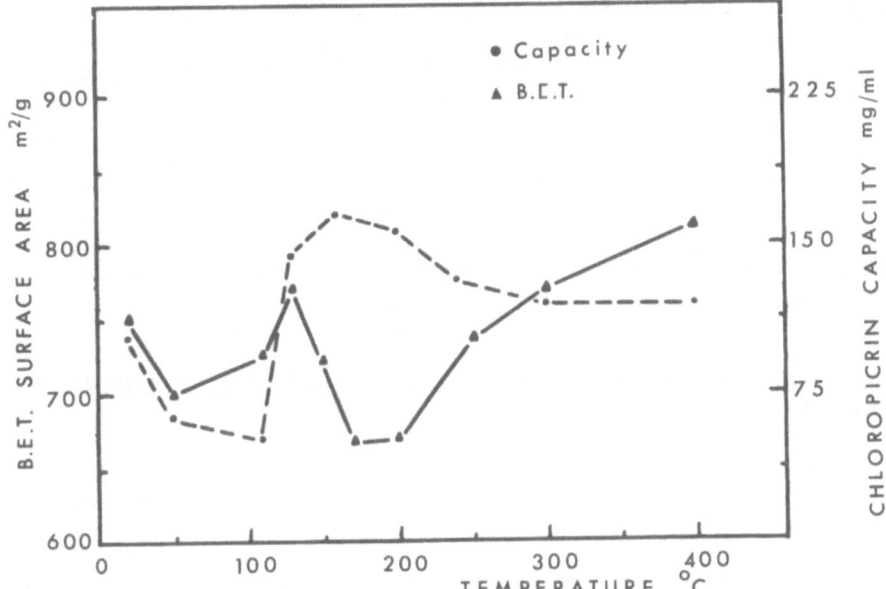

FIGURE 9. Specific surface and chloropicrin capacity of quenched ASC carbon vs. activation temperature.

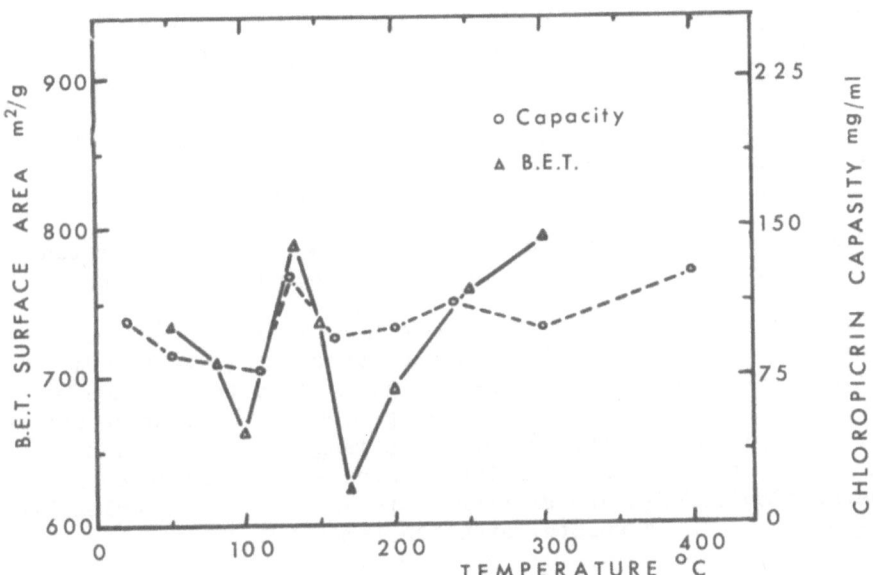

FIGURE 10. Specific surface and chloropicrin capacity of slowly cooled ASC carbon vs. activation temperature.

The increase of the chloropicrin capacity with activation temperature is ascribed to opening of pores blocked by water, organic molecules or impregnant; thus increasing the volume of pores accessible to capillary condensation of chloropicrin vapours. The increased cyanogen chloride capacity may be due to increased reactivity and/or accessible surface of the impregnant.

The favourable effect of quenching is ascribed to freezing of the more reactive high temperature state and to pore enlargement due to breakdown of walls between neighbouring pores as a result of thermal shock, which agrees with the simultaneous decrease in surface area (Fig. 9).

THERMAL ACTIVATION OF INORGANIC SORBENTS FOLLOWED BY CRYOGENIC QUENCHING

The effects obtained by application of the heating/quenching technique to inorganic sorbents involve both increased selectivity and sorption capacity (14).

Such effects are presented in two examples. Samples of molecular sieve 5 A were activated at 580°C and subjected thereafter to slow cooling and quenching, respectively. The quenching procedure was carried out as an indirect heat exchange process with liquid nitrogen.

The sorption effects were investigated by gas chromatographic techniques with the gases Ar, O_2, N_2, Kr and Xe. The results are shown in Table 5.

Alumina of the type Boehmite was treated in a similar way. The activation temperature was 600°C. Table 6 illustrates the results of the gas chromatographic investigations.

Tables 5 and 6 reveal that in both cases the longest retention times are achieved on columns being subjected to quenching.

It is further observed that the order of elution occurs according to the polarisability of the gases. This fact indicates that the polar forces (reactive groups) on the surface of alumina are influenced by the cooling procedure. Tentative interpretation implies that the number and strength of the acid sites (Lewis sites) are increased on the alumina surface by quenching. The separation effects are similar to those achieved on molecular sieves, however, the absolute retention times are shorter.

TABLE 5. Retention times of inert gases on molecular Sieve 5 A column.

Activation temp. °C	Cooling procedure	Retention times, min.				
		Ar	O_2	Kr	N_2	Xe
580	slow cooling	5.9	5.9	15.8	18.2	120
580	quenched	6.8	7.6	21.8	34.8	142
660	quenched	7.1	7.9	23.2	38.3	165

Other parameters were:

    Column temperature, °C: 22
    Carrier gas   : He
    Activation gas   : He
    Activation time, h   : 5

TABLE 6. Retention times of inert gases on activated alumina column

Activation procedure	Retention time, min.			
	O_2	N_2	Kr	Xe
None, original product	2.45	2.45	3.05	5.25
600°C and slow cooled	2.63	2.90	3.50	7.78
600°C and quenched liq. N_2	2.75	3.08	3.73	8.67

Other parameters:

    Column temperature, °C: 22
    Carrier gas   : He
    Activation gas   : Air
    Activation time, h   : 6

REFERENCES

(1) M.M. Dubinin, Uspekhi Khim (24), 3, 1955.

(2) M.M. Dubinin, J. Colloid Interface Sci. 23, 487, 1967.

(3) C. Pierce, J.W. Wiley, R.N. Smith, J. Phys. Colloid Chem. 53, 669, 1949.

(4) D.A. Everett. The Structure and Properties of Porous Materials, Ed., P.H. Everett, F.S. Stone, Butterworths London, p. 95, 1958.

(5) C. Pierce, J. Phys. Chem. 64, 1184, 1960.

(6) R.M. Barrer, N. McKenzie, J.C. Reacy, J. Colloid Sci. 11, 479, 1956.

(7a) J.C. Arnell, H.L. McDermot, Proc. 2nd Int. Congr. Surface Activity, Vol. II, p. 113, London 1957.

(7b) W.H. McAdams. Heat Transmission. McGraw Hill Book Company, 1954, 3. ed.

(8) R.A. Hutchens. Chem. Eng. Progr. Vol. 69, No. 11, 48, 1973.

(9) D.R. Evans, J.D. Wilson, R. Cl. Culp, L.G. Suhr, H.E. Moger. A Summary of Plant Scale Advanced Waste Treatment Research at Lake Tahoe, paper pres. at WPCF Mtg., San Francisco, 1971.

(10) US Pat. 1.550.805.

(11) Harter, Germ. Pat. 564.432, 1932.

(12) Brit. Pat. 492.929, 1937.

(13) U.S. Pat. 3.528.615.

(14) R. Berg. Brit. Pat. 1.263.603, U.S. Pat. 3. 769.776.

(15) R. Berg, H.P. Hjermstad, G.A. Neefjes. Work Report IFA Ch-105.

II. EFFECT OF IONIZING RADIATION ON THE "CHEMISORPTION" CAPACITY OF SORBENTS

GENERAL ASPECTS OF CHEMISORPTION

The present lecture deals with the sorption effects obtained as a result of irradiation of sorbents. Irradiation effects occurring simultaneously with irradiation as well as with post-irradiation effects are treated.

The occurrence of radiation effects on sorbents and catalysts is quite certain and has been recognized for a long time.

The mechanisms of chemisorption and catalysis under the action of ionizing radiation are complex. However, generally they are closely connected with the transfer of energy taken up by the solid to molecules adsorbed at its surface. On the surface competitive trapping of the electrons and holes created during irradiation will possibly take place between the adsorbed molecules and traps such as structural defects or impurities. Chemisorption takes place when an electronic interaction (electron transfer and/or sharing) between the adsorbate and the adsorbent occurs.

There are three possible types of bond between a chemisorbed particle (atom, molecule) and a solid surface:

- "weak" bond	an electron of the chemisorbed particle is moved close to the cation of the lattice
or	
	an electron of the anion of the lattice is moved close to the chemisorbed particle
- "strong" acceptor bond	an electron of the particle adsorbed on the cation interacts with a free electron of the semiconductor yielding a chemical bond with the lattice
- strong donor bond	an atom (molecule) is adsorbed on the anion of the lattice and interacts with a free hole of the solid.

Depending on the electrical properties of the sorbents investigated, they can be generally classified as insulators (Al_2O_3), semiconductors (copper oxides, chromium oxides, impregnated on the active carbon) or conductors (Cu-metal on active carbon).

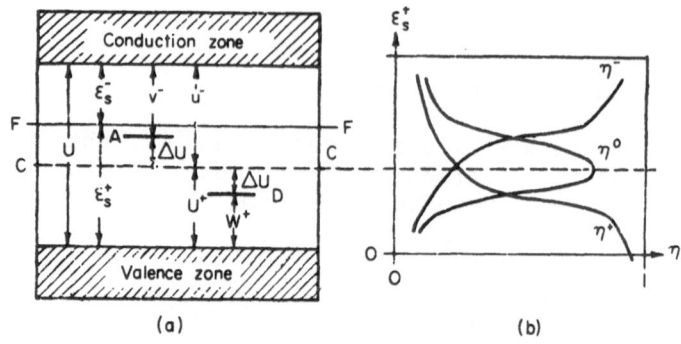

FIGURE 1   (A) Zone diagram of a semiconductor
           (B) Change of concentration of adsorbed particles
               during a change in the position of the Fermi level

ϵ_s^+ = distance from Fermi level

η = particle concentration

   The connection between the electronic properties of a solid
semiconductor and its adsorptive capacities has been proposed by
Volkenstein (1). A chemisorbed particle produces local energy
levels in the forbidden zone of a semiconductor (sorbent). Fig. 1
demonstrates a zone diagram and the change of concentration η of
the adsorbed particle during a change in position of the Fermi
level* FF in the forbidden zone of a semiconductor.

   The relative coverage of various forms of chemisorbed particles
on the surface of a semiconductor in the presence of fixed electronic equilibrium is determined by the location of the Fermi level.
The transition of an electron to acceptor level A corresponds to
the production of a "strong" acceptor bond. The production of a
strong donor bond corresponds to the withdrawal of an electron from
the donor level D, that is, the transition to it of a hole.

   An increase of the concentration of electrons in a semiconductor (i.e. electronic conductivity) displaces the Fermi level FF upwards, closer to the conduction bond. Increasing the concentration
of holes (i.e. hole conductivity) displaces the Fermi level down-
wards - closer to the valence bond. According to many investigations on the electronic theory of catalysis, a shift of the Fermi

---

*Fermi level: The critical energy level below which all energy
             levels are completely occupied by electrons, and all
             those above are empty, is termed the Fermi level at
             absolute zero (°K).

level from the valence to the conduction zone can accentuate a decrease of the catalytic activity or bring it through a maximum.

It is attractive to attribute the effect of γ and β irradiation to formation of holes or trapped electrons and heavy particle effects to formation of vacancies or interstitials.

Creation of excess charge carriers by irradiation of insulators like SiO_2, Al_2O_3, silic-alumina, MgO, KF, may result in the appearance of various adsorbed, charged and radical forms. Also in semiconductors such as ZnO, CuO, etc., charge carriers are formed. These excess charges will influence the location of the Fermi level and thus change the sorptive properties.

GENERAL ABOUT RADIATION EFFECTS

By the irradiation of solids with high energy radiation, the following primary effects take place:

- electronic excitation, ionization, radical formation
- displacement of atoms.

γ-rays, X-rays and electrons (β-radiation) produce primarily ionization and excitation of orbital electrons. Massive charged particles like α-rays lose energy through ionization and displacement of atoms. A small amount of atomic displacements also can result from fast electrons. Figure 2 illustrates schematically the distribution of deposited energy from α and β rays in matter. The heavier charged particle creates a denser track of ion pairs and excitations.

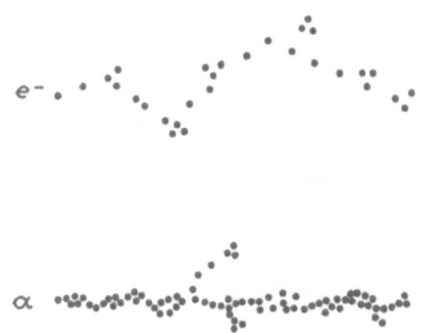

FIGURE 2. Schematic illustration of the distribution of deposited energy tracks from α and β rays in matter.

The best known quantity which is related to a particle track is the stopping power. It is defined as the energy loss suffered by the incident particle in traversing unit length in the medium along its path and is known as the linear energy transfer, LET. The LET (specific energy loss) is defined as eV/Å or KeV/μm.

Bethe (2) has developed an expression for the stopping power $-dE/dx$ of a non-relativistic charged particle in matter.

$$-\frac{dE}{dx} = \frac{4\pi\ e^4}{m_e} \cdot \frac{z^2}{v^2} \cdot N \cdot Z \cdot \ln \frac{2m_e \cdot v^2}{I}$$

where

E	= particle energy, MeV
z·e	= charge of ionizing particle
v	= velocity
N	= number of atoms per unit space of the adsorber
Z	= atomic number of adsorber atoms
z	= atomic number of ionizing particle
I	= average ionization potential of adsorber (ionized medium)
e	= electron charge
m_e	= mass of electron
x	= traversing distance of particle

For slow particles of high charge, the stopping range is shorter than for fast moving particles with lower charge (electrons).

Table 1 illustrates an example of the approximate stopping distance of α and β particles in air and aluminium depending on their energy.

The absorption of γ-rays in matter involves the interaction (collision) between orbital electrons and γ-photons. Depending on the nature (atomic number) of the absorbing substance and the energy of the γ-quantum, three effects may result (Figure 3):

- Photoelectric effect   dominating at lower energies (0.01 - 0.1 MeV)

                                               γ-energy completely transferred to the electron

- Compton effect   dominating above 0.5 MeV, elastic collision of a γ-quantum with an electron leaving scattered γ-ray with decreased energy

- Pair production   important at higher energies, formation of positive and negative electrons.

  In the actual cases the use of a Co-60 irradiation source implies that the Compton effect will dominate. The photoelectric effect, however, cannot be excluded. The energy tracks (ionization density) of the Compton electrons are usually smaller than tracks obtained by irradiation with β-rays.

  The collision between a high energy Compton electron and an atomic nuclei of solids may also result in the formation of Frenkel defects. The last effect involves the displacement of an atom from its normal lattice position into an interstitial position.

TABLE 1. Range of ionizing particles in air and aluminium

Particle type	Energy, MeV absorb medium	Distance, cm						
		0.2	0.5	1	2	3	4	5
Electrons	Air	20	120	306	710	1100		1900
Electrons	Al	0.02	0.06	0.16	0.36	0.55		0.93
α-particle	Air			0.56	1.1	1.7		3.5
α-particle	Al						0.0016	0.0023

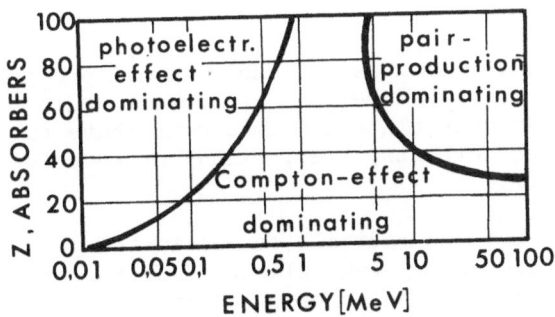

FIGURE 3. Photoeffect, Compton effect and pair formation depending on γ-energy and atomic number Z of the absorbing material.

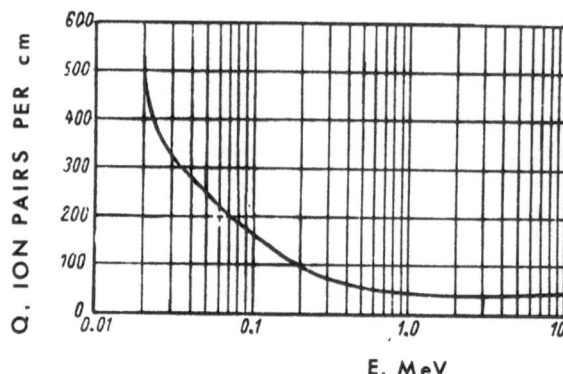

FIGURE 4.   Specific ionization of β-particles in air.

By irradiation of sorbents and catalysts, the γ-rays will penetrate both solid and gas phase (pore volume) and create both excited atoms and ion pairs in the gas phase and on the solid surface (pore volume walls). The β-rays will penetrate the gas phase, but will be absorbed by the upper layer of the pore walls (solid phase).

The concentration of excited species in the gas phase depends on the stopping power of the irradiated gas, the type, energy and intensity of the irradiation. Figure 4 illustrates an example of the specific ionization of β-particles in air as a function of their energy.

Depending on the lifetime, concentration and chemical properties of the excited species, reactions between the pore surface and these species may possibly take place and thus contribute to new chemical properties of the sorbents, i.e., fixing of reactive groups to the pore structure.

Radiochemical reactions between radicals formed in the gas phase and adsorbed water molecules may take place. Radicals and excited molecules on the surface of the sorbent possibly give rise to higher reactivity and selectivity.

Indications for such radiochemical reactions during irradiation of the sorbent would be:

- changes in selectivity and capacity with the moisture content of irradiated sorbent

- influence of the irradiation atmosphere on the sorbent properties.

Possible radiochemical reactions and the nature of the radicals formed in such heterogeneous systems may be extremely complicated and very difficult to predict.

IRRADIATION SOURCE, PRACTICAL ASPECTS

The irradiation of sorbents is carried out either by means of external or internal sources. The irradiation technique can be divided into the following categories:

External source
- Accelerators, i.e., van de Graaf, betatrons
- Nuclear reactors
- High energy γ-sources, i.e., Co-60, Cs-137, solid waste

Internal sources
- Solid α- and β-emitters, i.e., Po-210, Sr-90 deposited on the sorbents as impregnations
- α-, β- and γ-emitting gaseous sources, i.e., Rd-222, Ar-39, Kr-85, Xe-133
- Liquid sources, radioactive labelled salt diluted in solvents.

Of these sources, Co-60 is the most utilized γ-irradiator. Nuclear reactors can create induced activity in sorbents due to the neutron flux and this effect is usually not wanted.

Regarding the β-irradiation the most usual emitters with sufficiently long halflives possess β-energies in the range below 1 MeV. The reason for choosing β-emitters is the short penetration range of the β-particles and the high LET compared to γ-rays. In gas/solid systems such as sorbents and catalysts, the absorption of the particle energy is nearly 100% on the surface or close to the surface of the pores. Compared to γ-irradiation the shielding of samples and irradiation sources become a minor problem in the case of β-ray technique.

During β-irradiation the source is in immediate contact with the sorbent. A main problem is to remove the source (e.g. deposited Sr-90) without introducing unfavourable surface changes or annealing of adsorbed radiation energy. In γ-irradiation this problem is easily solved by removing the external Co-60 source.

The use of short-lived β-emitters as possible radiation sources implies that the following conditions should be satisfied. The radiation source

- should not react with the sorbents

- should not make the sorbent radioactive

- should not produce radioactive daughter nuclides with long halflives

- should not yield short-lived decay products whose daughter nuclides are reactive with the sorbents yielding negative sorption effects

- should be readily removed after the irradiation is completed.

The irradiation of sorbents with gaseous radiation emitters such as radioactive noble gases, is favourable due to their non-reactivity and the ease of removal from the sorbents.

Among the β-emitting inert gases, the following are of practical interest, partly because of their availability on the commercial market:

$$^{39}_{18}Ar, \quad ^{85}_{36}Kr \quad and \quad ^{133}_{54}Xe$$

These isotopes are produced as a result both of fission of uranium (Kr-85 and Xe-133) and of neutron activation of the atmospheric Ar-36, and Ar-38. Due to the growing number of nuclear installations in the world, the production of these isotopes, especially Kr-85, is considerable.

The removal of krypton from off-gases presents in principle no difficulties and is indeed already carried out in USA and for instance at Eurochemic to provide radio-krypton. Xenon is also recovered.

In the following some nuclear data for radiation sources referred to in the next chapter, are presented.

The complexity of parameters involved in irradiation of sorbents will be illustrated by some examples from work published by other laboratories and from own experiments. Two categories of irradiation are considered:

- simultaneous sorption and irradiation

- pre-irradiation of sorbent followed by sorption.

Solid α- and β-sources

Isotopes	Half-life	Type of emitter	Energy
$^{210}_{84}Po \rightarrow {}^{206}_{82}Pb$	138 d	α-rays	5.40 MeV (100%)
$^{210}_{83}Bi \rightarrow {}^{210}_{84}Po^x$	5 d	β-rays	1.16 MeV (ca.100%)
$^{89}_{38}Sr \rightarrow {}^{89}_{39}Y$	51 d	β-rays	1.46 MeV (99%)

Solid γ-sources

$^{60}_{27}Co \rightarrow {}^{60}_{28}Ni$	5.2 y	γ-rays	1.77 MeV 1.33 MeV
		β-rays	0.31 MeV

Gaseous β-sources

$^{39}_{18}Ar \rightarrow {}^{38}_{19}K$	265 d	β-rays	0.565 MeV, γ-rays absent
$^{85}_{36}Kr \rightarrow {}^{85}_{37}Rb$	10.3 y	γ-rays	0.695 MeV (99%) 0.15 MeV (0.65%)
$^{133}_{54}Xe \rightarrow {}^{133}_{55}Cs$	5.6 d	β-rays	0.34 MeV
		γ-rays	0.08 MeV

EXAMPLES OF SORPTION CAPACITY EFFECTS RESULTING FROM IRRADIATION

Simultaneous irradiation and sorption.

Effect of heavy particle and β-particle bombardment on oxygen sorption. The type of particle determines the kind of defects which are created during irradiation and the resulting sorptive process. Irradiation with heavy particles like neutrons and ions results in structural defects (displacements) which are responsible for the adsorption forces.

This is the case in bombardment of magnesia (MgO) with high energy positively charged particles and simultaneous sorption of oxygen.

Nelson and Duch (3) reported that the adsorption of oxygen during the bombardment was due to several damage processes, both at the surface and within the bulk of the solid magnesia. This structural defect hypothesis is supported by the effect obtained

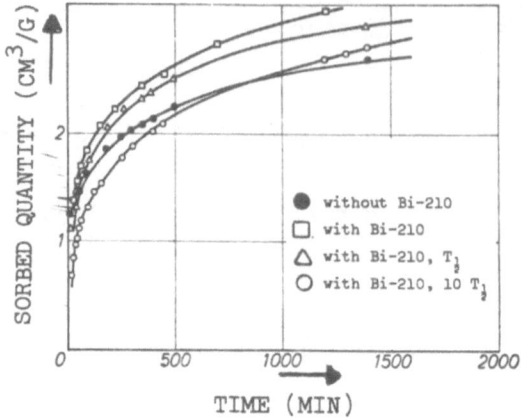

FIGURE 5. Sorption of carbon monoxide on iron catalyst during β-irradiation with Bi-210.

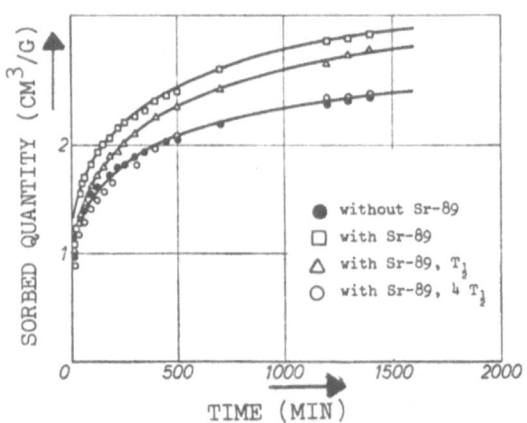

FIGURE 6. Sorption of carbon monoxide on iron catalyst during β-irradiation with Sr-89.

by Ermatov and Kusainov (4). They report that the effect of slow electron irradiation of magnesia (MgO) yields a desorptive process with respect to oxygen and carbon dioxide.

Effect of α- and β-emitters during the sorption of carbon-monoxide. The effects of the presence of the β-emitters Sr-89, Bi-210 and the α-emitter Po-210 on the adsorption of carbon-monoxide on iron catalyst were demonstrated by Kölbel et al. (5). Figures 5 and 6 illustrate the effect of two β-sources, Bi-210 and Sr-89.

In the presence of 1 mCi β-activity the amount of CO adsorbed per gram catalyst is increased as compared to radioactivity-free catalyst.

In is observed from the figures that the sorption effect increases with time but decreases with decaying activity (radiation). After 4 (Sr-89), respectively 10 (Bi-210) halflives ($T_{\frac{1}{2}}$), the obtained CO-capacities approach the original capacities.

By the application of an α-emitter (Po-210) and a starting activity of 1 Ci, higher CO-capacity is observed. Figure 7 illustrates the sorption of CO in the presence of α-irradiation.

This sorption increase is probably due to both structural and energy effects.

The higher LET of α-particles generally explains the increased sorption capacity compared with the smaller effects obtained from β-irradiation. The effects are, however, consistent only at comparable activities of the α- and β-emitters. For these cases it is assumed that energetic centers are continuously created in the presence of irradiation. This possibly results in an increased number of adsorption centers compared with the amount of centers existing without irradiation. In the case of α-irradiation, Figure 7, the adsorption capacity is less than the original capacity after two halflives. This sorption behaviour indicates structural effects (defects) on the surface of the catalyst.

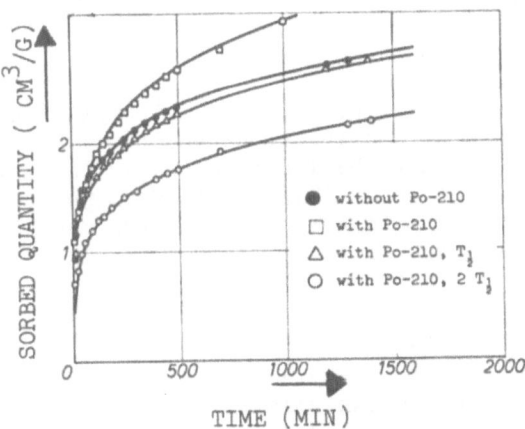

FIGURE 7. Sorption of carbon monoxide on iron catalyst during α-irradiation with Po-210.

According to the authors (5), regain of the original capacity is obtained by annealing of the irradiation effects.

   The effect of γ-irradiation on the sorption process. An example of the influence of γ-irradiation on the electron interactions during chemisorption processes is reported by Zhabrova et al. (6). They have investigated the chemisorption of oxygen and hydrogen on Al_2O_3 in the presence of γ-irradiation from a Co-60 source. The samples have different specific surfaces in the range 13 - 300 m^2/g. Chemisorption of hydrogen or oxygen was not observed for aluminium oxide at room temperature in the absence of irradiation. In the presence of irradiation, the chemisorption of these gases increased over a long period of time. No sorption plateau is observed within the time of irradiation. These effects are illustrated in Figure 8.

   One characteristic feature of the kinetics of radiation chemisorption will be noted. The chemisorption rate of oxygen per g oxide is higher than that of hydrogen. It is also interesting to note that chemisorption of these gases in the source of intermittent irradiation gave no after-effect (7). From Figure 9 it is observed that radiation chemisorption stops when irradiation is discontinued.

   Curves 1 in Figure 9A and B correspond to chemisorption of hydrogen and oxygen on aluminium oxide that was not pre-irradiated. Curves 2 refer to pre-irradiated samples.

   The results obtained by Zhabrova et al. (6) are in favour of the conclusion that shortlived radiation-induced defects, i.e., free electrons and holes, or electrons and holes in shallow traps, rather than "biographic" defects play a decisive role in radiation chemisorption.

   Comparison of the kinetic curves of oxygen chemisorption on aluminium oxide (insulator) and zinc-oxide (semiconductor) having the same specific surface shows that chemisorption on an insulator is greater than that on a semiconductor (Fig. 10). This indicates that the presence of electron traps in the semiconductor may compete with the chemisorbed molecules.

PRE-IRRADIATION FOLLOWED BY SORPTION

   The following examples refer to own efforts in this field.

Effects of γ-irradiation on impregnated active carbon.

   Some sorption capacity effects obtained on typical gas mask

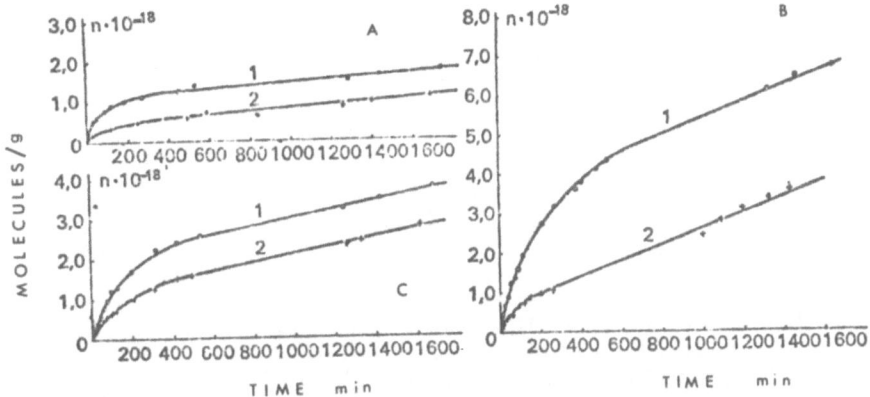

FIGURE 8. Chemisorption of O_2 and H_2 on Al_2O_3 with different sp. surfaces.
(A) 13 m^2/g (B) 150 m^2/g (C) 300 m^2/g
Curve 1 Radiation chemisorption O_2
Curve 2 Radiation chemisorption H_2

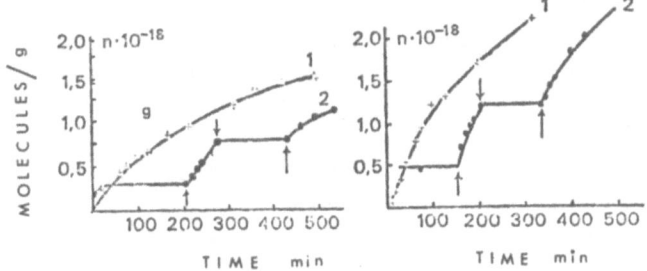

FIGURE 9. Radiation chemisorption on Al_2O_3.
(A) O_2 (B) H_2
1. Non-irradiated. 2. Pre-irradiated.

carbons are presented. Specifications for the carbons and tests gases are given in Table 2.

A Co-60 source consisting of seven cylindrical rods (each 20 cm long) mounted horizontally was used for the irradiations. The initial activity of the source was 32 300 Curie, the dose rate 725 krad/h. During irradiation the carbon granules (150 ml) were kept in polyethylene bottles at ambient temperature placed at a distance of 6 cm from the source.

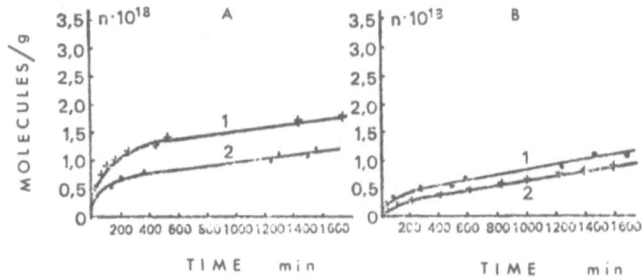

FIGURE 10. Comparison of the radiation chemisorption of (A) O_2 and (B) H_2 on Al_2O_3 and ZnO.
Curve 1. Al_2O_3 Curve 2.   ZnO.

TABLE 2.   Impregnated gas mask carbon.

Type of carbon	Carbon base	Impregnation	Reacting gas	Kind of reaction	Stoichiometric equation and reaction products
ASC Pittsburgh	Bituminous	Oxides of Cu, Cr + Ag salt	HCN	Chemisorption catalytic	$Cu_2O + 2HCN = 2CuCN + H_2O$
ASC Pittsburgh	Bituminous	Oxides of Cu, Cr + Ag salt	ClCN	Chemisorption catalytic	Details not published
Dräger	Wood	K_2CO_3	SO_2	Chemical reaction	K_2SO_4
Norit RBAA-1	Peat	Na_2CO_3	SO_2	Chemical reaction	Na_2SO_4
Dräger		$ZnSO_4$	NH_3	Complex formation	$(NH_4)_2 (ZnSO_4)_2$ $6 H_2O$

Two test gases - ClCN and SO_2 - were used in the standard procedures for control of the chemisorption capacity. The test conditions were:

Cyanogen chloride: 72 mg ClCN in 30 l dry* air/min.
Temp. 18°C. Volume of granules 75 ml.

Sulphur dioxide: 103 mg SO_2 in 15 l dry* nitrogen/min.
Temp. 20°C. Volume of granules 40 ml.
* The gases are assumed to contain some ppm of humidity.

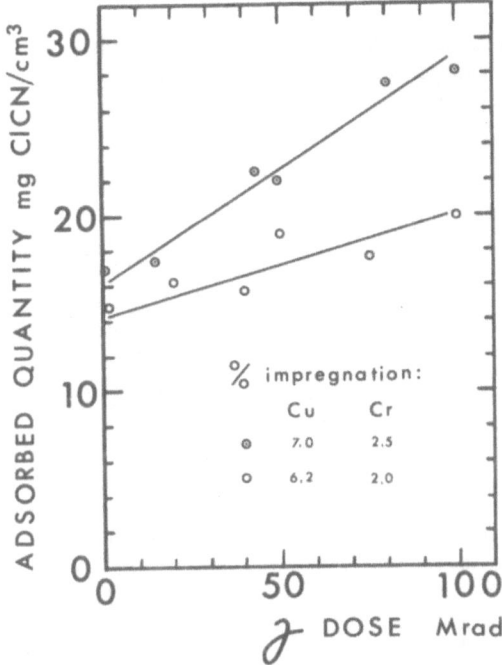

FIGURE 11.   Chemisorption of cyanogen chloride   on irradiated
              ASC carbon vs. γ-dose. Irradiation atmosphere: air.

     The following examples (8) illustrate the influence of the
irradiation dose and irradiation atmosphere on the chemisorption
capacity of impregnated carbons. Figure 11 shows the increase of
the cyanogen chloride capacity of two ASC carbons with increasing
γ-dose. The largest increase is obtained on the carbon with the
highest percentage of impregnation.

     Table 3 illustrates the influence of the irradiation atmo-
sphere at constant dose, 50 Mrad. The presence of nitrogen gas
during irradiation seems to have a remarkably positive effect on
the chemisorption (or catalytic conversion) capacity for cyanogen
chloride. Oxygen, hydrogen, carbon dioxide and inert gases have
an adversory effect. Similar effects were observed for the SO_2
capacity.

     The capacity gain could be further increased by increasing
the nitrogen pressure, as illustrated in Figure 12. The trend
shift observed at a pressure of 1 kg/cm^2 corresponds to replacement of air by pure nitrogen as irradiation atmosphere.

     In contrast to the positive effect of nitrogen on the capa-

TABLE 3. Influence of irradiation atmosphere on the cyanogen
chloride capacity of ASC carbon.

Exp. No.	γ-irrad. dosage, Mrad	Irradiation atmosphere	Capacity gain* %
1	0	–	0
2	50	O_2	– 8.8
3	50	Air	+ 30.2
4	50	N_2	+ 36.6
5	50	He	– 14.1
6	50	Ar	– 10.4
7	50	Kr	– 10.4
8	50	CO_2	– 10.0
9	50	H_2	– 69.0

* Capacity gain compared with the capacity of the original product 14.2 mg/ml bulk, exp. No.1.

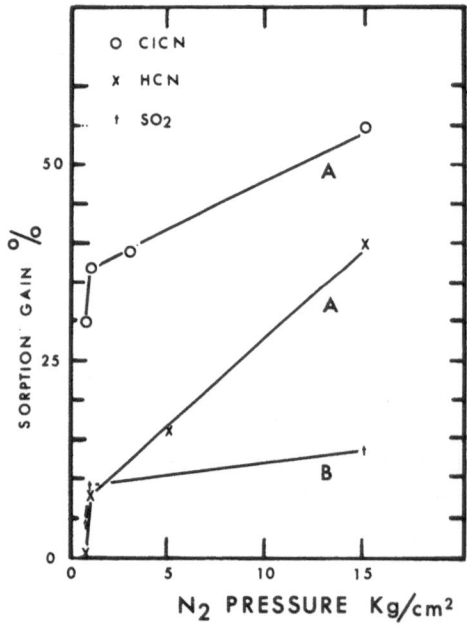

FIGURE 12. Chemisorption of gases on impregnated and irradiated
active carbon vs. nitrogen pressure. Impregnation:
(A) Cu,Cr,Ag (B) K_2CO_3. γ-dose: 50 Mrad.

city for acid gases (ClCN, SO$_2$) an adversary effect was observed for the uptake of ammonia - i.e. of a basic gas - on zincsulphate impregnated carbon. Figure 13 shows how γ-irradiation in a nitrogen atmosphere results in decreasing sorption of NH$_3$ with increasing γ-dose, while hydrogen has the opposite effect, e.g., results in a positive capacity trend.

As illustrated by these examples the effect of γ-irradiation on impregnated carbons depend on several experimental variables such as:

- total γ-dose

- composition of the impregnation

- irradiation atmosphere

- traces of water and impurities in sorbents and gases

- storage conditions before and after irradiation.

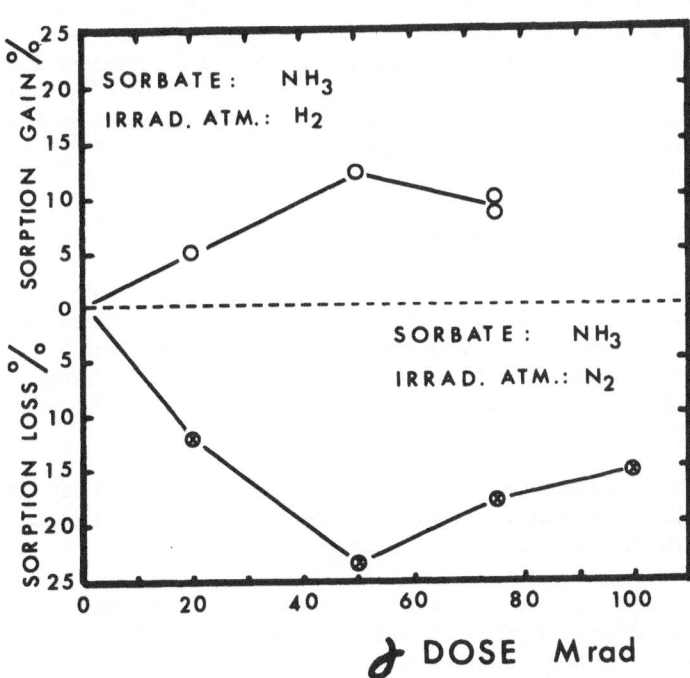

FIGURE 13.   Chemisorption of ammonia on irradiated carbon impregnated with ZnSO$_4$.

Due to the complexity of the problems involved and the many unknown reactions the actual activation mechanisms are difficult to define.

The influence of the irradiation on the electronic properties of semiconductors (e.g. on a Cu-Cr oxide impregnation), for instance change of the Fermi level, may contribute to changes of sorptive (catalytic) properties.

The interaction of irradiation (e.g. Compton electrons) with gas molecules and surface compounds will result in the formation of excited molecules, radicals and ions. The presence of water molecules or ammonia on the sorbent surface may imply formation of hydrogen and hydroxyl radicals or fixation of reactive aminogroups on the surface during irradiation.

Excited species formed in the gas phase will partly diffuse to the pore surface and possibly stick to it as reactive sites, or neutralize other reactive sites. The sticking probability will depend on their concentration, lifetime and chemical nature, which again depend on radiation flux, composition and density of the gas.

The effects obtained with nitrogen in the irradiation atmosphere seem to imply a formation of basic groups on the sorbent surface. The nitrogen molecules may be involved in charge transfer mechanisms resulting in the formation of hydronium ions and hydroxyl radicals (9), they may form charged N_2^+ species or basic groups like amines or amino compounds.

Effects of β-irradiation on impregnated carbons.

Sorption effects obtained on impregnated active carbon after irradiation with the β-emitters Kr-85 and Xe-133 have been investigated. Both isotopes emit a γ-quantum, but the energies of these quanta are only a small fraction of the total emitted energy. Thus the received γ-doses are much too low to induce essential changes of sorbent properties.

The sorbents investigated are the same as those listed in Table 4. The irradiation by means of Kr-85 was generally carried out as described in the following example.

Granulated carbons were placed in a 50 cm^3 container. The container was evacuated before 20 mCi Kr-85 diluted in Kr-carrier were added. The krypton gas was thereafter mixed with pure nitrogen to atmospheric pressure. Under this condition the carbon was stored for approx. 100 days at room temperature which corresponds to an absorbed β-dose of 1 Mrad.

TABLE 4. Test conditions of irradiated gas mask carbon

Carbon type	Test gas	Dilution gas	Test gas concentr. vol %	Test gas concentr. mg/l	Total flow cm³/min	Linear velocity m/min
ASC	ClCN	Air		2.4	$3 \cdot 10^4$	10
Norit RBAA-1	SO$_2$	He	5	146	425	3.7
Dräger	SO$_2$	He	5	146	425	3.7

TABLE 5. Cyanogen chloride capacity of β-irradiated active carbon ASC

Exp. No.	β-source	β-dose Mrad	Irrad. atm.	Irrad. atm. press. kg/cm²	Capacity mg/ml	Capacity change %	γ-dose Mrad	Capacity change %
0	-	-	-	-	8.06	-	-	-
1	Kr-85	1	Air	1	8.76	8.7	1	4
2	Kr-85	1	N$_2$	1	9.00	10.4	1	6
3	Kr-85	1.1	NH$_3$	1	11.5	42.7	1	9
4	Kr-85	1	He	1	5.1	-36.6	1	0
5	-	-	-	-	-	-	50	-13.9
6*	Kr-85	1	N$_2$	1	12.2	26.9	-	-

* This ASC carbon had some higher Cu, Cr concent.

The β-dose was measured by means of a TL-dosimeter (thermo-luminescence) of the LiF-type. Three dose meters were placed along the center axis and three along the wall of the irradiated carbon bed. The test conditions in these cases were as shown in Table 4.

As in the case of γ-irradiation, the sorption capacity for these gases seems to be extremely dependent on the type of the irradiation atmosphere.

Table 5 illustrates the obtained effects. The table reveals that the presence of ammonia has a positive effect , even stronger than pure nitrogen. The effects obtained with β-irradiation are higher compared to the results from γ-irradiation.

372

By β-irradiation in nitrogen atmosphere of Norit RBAA-1 and
Dräger carbons, the following results with respect to the sulphur
dioxide capacities were obtained:

TABLE 6. Sulphur dioxide capacity of β- and γ-irradiated active
carbon

Exp. No.	Carbon type	β source	β-dose Mrad	γ-dose Mrad	Capacity mg SO_2/ml	Capacity change, %
0	Dräger	-	0	0	10.6	-
1	Dräger	-	-	2	11.5	+ 7.8[*]
2	Dräger	Kr-85	0.71	0	14.5	+ 36.4[*]
3	Norit RBAA-1	-	0	0	23.5	-
4	Norit RBAA-1	-	0	2	22.1	- 6.7[**]
5	Norit RBAA-1	Kr-85	1.61	0	25.3	+ 7.5[**]

The table demonstrates that β-irradiation yields greater SO_2
capacity increase than γ-irradiation at comparable doses.

The number of experiments which was carried out with β-irradiation are few. However, the experiments indicate an increase
with higher β-doses with respect to the cyanogen chloride capacity.
Table 7 illustrates the capacity increase of ASC carbon β-irradiated in an ammonia atmosphere.

The presence of basic groups or electron donating groups
seems to be essential to the increased sorption of cyanogen chloride.

TABLE 7. Cyanogen chloride capacity of β-irradiated ASC active
carbon

Exp. No.	β-dose Mrad	Irrad. atmos.	Irrad. atm. pressure kg/cm²	Capacity mg/ml	Capacity increase %
0	0	-	-	8.06	-
1	1.1	NH_3	1	11.5	42.7
2	5.6	NH_3	1	12.2	51.0

The effect of increasing gas pressure during irradiation on the sorption of cyanogen chloride indicates diminishing capacity values. In the case of γ-irradiation higher nitrogen pressure during irradiation results in increasing cyanogen chloride capacity, Fig. 11.

A tentative explanation of this fact is connected to the sorption behaviour of the gaseous irradiation source Kr-85. Gas chromatographic investigation of the behaviour of Kr in nitrogen carrier gas on ASC carbon column indicates a decreasing sorption (distribution) coefficient K (cm^3/cm^2) on increasing absolute pressure in the column. This sorption behaviour of krypton suggests that the effective irradiation takes place when the source is sorbed on the solid phase.

Compared to krypton, xenon has a higher sorption coefficient, Φ (mol gas/g carbon/mol gas/cm^3 gas) with respect to ASC carbon. By using Xe-133 as radiation source, this implies that the concentration of activity on the solid surface is higher than for Kr-85 irradiation.

ASC carbon was irradiated with 0.75 Ci of Xe-133 diluted in 0.02% Xe carrier and 99.98% nitrogen. The activity was removed after the carbon had received a β-dose of 1 Mrad. The sample was tested regarding its cyanogen chloride capacity. Experimental data and test results are tabulated below together with results obtained on the same carbon irradiated with Kr-85.

The results reveal that the cyanogen chloride capacity increases more after irradiation with softer β-rays from Xe-133 than after irradiation with harder β-rays from Kr-85.

TABLE 8.   Experimental data concerning irradiation of ASC carbon in nitrogen

Irrad. source	Max β-energy MeV	Source activity mCi	Irrad. time days	β-dose Mrad	Amount carbon g	Total surface m^2	Sorption coeff. Φ
-	-	-	-	-			
Kr-85	0.695	20	100	1	25	18750	12.7
Xe-133	0.34	180	20	1	20	15000	261

TABLE 9.   Cyanogen chloride capacity of ASC carbon β-irradiated
            with Kr-85 and Xe-133

Carbon treatment	Irrad. source	β-dose Mrad	Energy emitted to surface cal/m^2	ClCN capacity mg/ml	Capacity increase %
Original	-	-	-	8.06	-
Irradiated	Kr-85	1	0.003	9.0	10.4
Irradiated	Xe-133	1	0.003	16.2	102

REFERENCES

(1) F.F. Volkenstein, Electrical Theory of Catalysis on Semi-
    Conductors, Fizmatgiz, 1960. Oleg V. Krylov Catalysis by
    Nonmetals, 1970.

(2) H. Bethe, Handbook ed. Physics 2 Auf. Springer, Bd. 24 (1)
    S 273, 1933.

(3) R.L. Nelson, M.J. Duch, Uses Cyclotron Chem. Met. Biol. Proc.
    Conf., p. 88-100, 1969. Publ. 1970.

(4) S.E. Ermatov, S.K. Kusainov, Izv. Akad. Nauk. Kaz. SSR, Ser.
    Fiz-Mat., No. 2, 56-57, 1971.

(5) H. Kölbel, I. Witte, H. Hammer. Zeitschr. für Phys. Chemie
    (Frankfurt), 73, (1-3), 97, 1970.

(6) G.M. Zhabrova, V.I. Vladimirova, A.A. Gesalov, B.M. Kaden-
    satsi, Symposium on Electronic Phenomena in Chemisorption
    and Catalysis on Semiconductors, p. 231-243, Moscow 2-4 July,
    1968, ed. K. Hauffe.

(7) A.A. Gezalov, G.M. Zhabrova, K.N. Spiridonov, Khimiya Vyso-
    kikh Energii, 2, No. 1, 1968.

(8) R. Berg, H.P. Hjermstad, G.A. Neefjes, N. Pat. Appl. 2482/72.
    "Auslegeschrift" 129722, U.S. Pat. Appl. 378161 (1973).

ACKNOWLEDGEMENT
     Part of the own work referred to in this lecture was carried
out in cooperation with the Department of Toxiology, Norwegian Defence Research Establishment (FFI).

     The author is indebted to H.P. Hjermstad, A. Gulbrandsen and
G.A. Neefjes (Institutt for Atomenergi).

THE DESIGN OF A SELF DECONTAMINATING ADSORBENT

J. Medema, J.J.G.M. van Bokhoven and A.E.T. Kuiper

Chemical Laboratory TNO, Rijswijk, The Netherlands

ABSTRACT. Nerve gases, organophosphorus anti-cholinesterase compounds, adsorb onto charcoal physically. In this study an evaluation of an impregnant that will promote the decomposition of the so-called G-agents is made. By means of infrared spectroscopy the adsorption structure and the mechanism of the decomposition are elucidated. From decomposition activity measurements with various solids it is derived what sites on the surface are involved in the decomposition. The kinetics of the decomposition reaction have been followed microcalorimetrically. On the bases of the results obtained a self decontaminating impregnant for charcoal is proposed.

1. INTRODUCTION

Respiratory protection against chemical warfare agents is usually achieved using charcoal as adsorbent to filter contaminated air (1). The physical adsorption capacity of charcoal for particular warfare agents such as Arsine, Hydrogen Cyanide and Cyanogen Chloride is not sufficient. During World War II an inorganic impregnant was applied to charcoal in order to increase the protection. This impregnant consisted of copper, silver and chromium salts and was able to decompose the agents mentioned (2).

G-agents, being volatile organophosphorus anti-cholinesterase compounds (3) are adsorbed very well onto the surface of charcoal, but the decomposition capacity of impregnated charcoal for these agents is limited. For respiratory protection this is not a problem because desorption occurs only in very rare cases. However, G-agents possess the capability to penetrate through the skin (4). Appropriate skin protection can be obtained by wearing impermeable clo-

thing; a serious drawback of this clothing is the high physiologi-
cal stress. Permeable clothing is preferred therefore. The protective layer in this clothing should consist of a highly efficient
adsorbent. To increase the service life of this clothing and to
minimize contamination hazards from adsorbed chemical warfare
agents it is advantageous to apply a protective adsorbent with self
decontaminating properties.

There are two possible methods to obtain a self decontaminating impregnant:
one can either screen as many impregnated charcoals as possible
until an active product is found; or one can investigate how the
adsorbed compound might be decomposed and then select an impregnant
that promotes this reaction. The second route was chosen because
of the advantage that basic knowledge about adsorption mechanisms
might be obtained.

It is recognized that charcoal is the best adsorbent known (2).
Therefore it was decided to develop an impregnant for charcoal to
give it the desired decontaminating properties. In order to simplify
the study the impregnant was investigated as such. Since a broad
variety of agents had to be covered, a general type of decomposition
reaction had to be chosen. Two possibilities present themselves
hydrolysis and oxidation. For both types of reaction the additional
reactant, water or oxygen, is profusely present in the air. Generally hydrolysis proceeds faster than oxidation at ambient temperatures. Therefore hydrolysis seemed to be the more promising type of
reaction.

The primary intention of the present study is to gain insight
into the activity of the impregnant c.q. adsorbent towards hydrolysis of adsorbed organophosphorus compounds. Attention is directed
towards one definite adsorbate-adsorbent system. Active alumina is
taken as an adsorbent, because its chemical and physical properties
have been studied extensively (5). Isopropyl methylphosphonofluoridate (sarin) is chosen as a representative organophosphorus warfare
agent.

$$
\begin{array}{ccc}
CH_3 & & CH_3 \\
| & & | \\
HC - O - & P & - F \\
| & & || \\
CH_3 & & O
\end{array}
$$

Other adsorbents, like magnesia and impregnated carbon as well as
other organophosphorus compounds are included in this study to
allow for some conclusions about a more general applicability of
the results obtained with the alumina-sarin system.

The first step in the process to be clarified is the adsorption of the agent onto the adsorbent. This has been studied by

infrared spectroscopy. Next the possible routes of decomposition
of the adsorbed agents must be investigated. Infrared spectroscopy
together with gas chromatography and chemical analysis has been
used to obtain this information. The same techniques have been used
for the evaluation of the functions of the adsorbent responsible
for a certain decomposition route. The kinetics of the decomposition have been established from microcalorimetric investigations;
the rate of reaction can be measured directly under various conditions. Finally, with the knowledge thus obtained it is possible to
select an active impregnant.

2. STRUCTURE OF ADSORBED SARIN

Succesful application of infrared spectroscopy to the study of the
adsorption of sarin on alumina may be expected when three conditions are fulfilled (6,7):
- The sorbent must be sufficient transparent to infrared radiation.
  This was gained by using a special type of alumina, transparent
  from 4000-1000 cm^{-1}.
- The surface area of the adsorbent must be such large that a per-
  ceptible concentration of adsorbed sarin molecules can be built
  up. The surface area of the alumina used amounted to 120 m^2/g.
- The adsorption complex must be stable to such an extent that the
  residence time of the adsorbate on the surface is large enough
  for spectroscopical investigations.

For the spectroscopic investigations of adsorbed sarin it is
necessary to be acquainted with the spectrum of the pure liquid.
Sarin has no elements of symmetry other than E, so the molecule
belongs to point group C_1. With 18 atoms there are 48 fundamental
vibrations, all of them are allowed in the Raman and infrared. In
Table I the vibrational spectrum of sarin is given. The assignment
have been made on the basis of the literature in the field of organophosphorus compounds (8) and comparison with the vibrational
spectrum of deuterated sarin.

As a consequence of the low symmetry no new vibrations of the
sarin molecule itself are to be expected upon adsorption. Even if
the symmetry would be increased by the adsorption process all vibrations are still Raman and infrared active. On the other hand,
adsorption will affect the bonds in the sarin molecule, which will
be reflected in the spectrum by frequency shifts and/or changes in
intensity. In addition bond(s) formed by adsorption between sarin
and alumina, might be observed as well as alterations in the surface layer of the adsorbent. For these reasons the obvious procedure is to compare the spectrum of the adsorbed species with that
of the pure liquid.

Table I

Vibrational spectrum of sarin

Infrared		Raman		dep. ratio	assignment
		3006	sh	0.7	ν_{as} CH$_3$ (me)
2985	s	2992	s	0.7	ν_{as} CH$_3$ (ipr)
2932	m	2935	vs	0.0	ν_s CH$_3$ (me)+ ν_s CH$_3$ (ipr)
2878	w	2880	m	0.0	ν ⟩C-H
		2835	w	0.4	2 x 1426 = 2852
1724	w				721 + 1014 = 1735
1468 ⎫ 1461 ⎭	m	1460	m	0.5	δ_{as} CH$_3$ (ipr)
1419	w	1426	m	0.6	δ_{as} CH$_3$ (me)
1390 ⎫ 1380 ⎭	m	1394	w	0.5	δ_s CH$_3$ (ipr)
1351	w	1360	m	0.6	δ ⟩C-H
1320	s	1328	w	0.1	δ_s CH$_3$ (me)
1277	vs	1279	m	0.1	ν P=O
1180	m	1185	m	0.1	⎫
1145	m	1147	m	0.7	CH$_3$ (ipr) rock
1106	m	1105	m	0.2	⎭
1014	vs	1018	mw	0.4	ν C-O-(P)
		936	w	~0.7	ν_{as} C-C-C
921	s	926	w	~0.5	⎫
905	s	912	w	~0.7	CH$_3$ (me) rock
884	sh	888	m	0.2	ν_s C-C-C
835	s	840	w	0.7	ν P-F
790	sh				ν P-C isomer 2
778	ms	780	m	0.2	ν P-C isomer 1
721	ms	726	vs	0.0	ν P-O-(C) isomer 1
		700	sh	~0.0	ν P-O-(C) isomer 2
		685	w	~0.2	278 + 410 = 688
					154 + 258 + 278 = 690
504	ms	506	m	0.2	P-O-C bend ?
450	w	455	w	~0.7	
		410	m	0.3	
		316	m	0.4	
		278	m	0.6	
		258	m	0.5	
		154	m	0.4	

m, s, sh, v, w denote medium, strong, shoulder, very, weak, respectively.

2.1 Experimental

Most infrared experiments have been performed with a commercial alumina, known as type C from Degussa (Frankfurt, Germany). The properties of this alumina as well as the procedure to prepare infrared transparent disks from this material have been described before (9). All organophosphorus compounds were synthetized by the department of organic chemistry according to standard procedures (10). Infrared spectra of adsorbed molecules were recorded on a Cary-White Model 90 double beam spectrometer. Two identical pyrex or quartz i.r. cells were used. Both cells contained an alumina disk; adsorption was carried out on the sample disk only. The cells were connected to a vacuum line and could be heated thus permitting spectra of adsorbed molecules to be recorded in situ.

2.2 Results and discussion

A rough survey of the changes in the infrared spectrum of sarin caused by adsorption on γ-alumina, may be obtained from a comparison of Fig. 1 and 2.

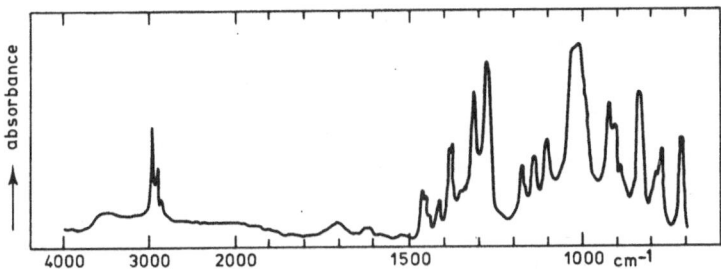

Fig. 1 Infrared spectrum of sarin.

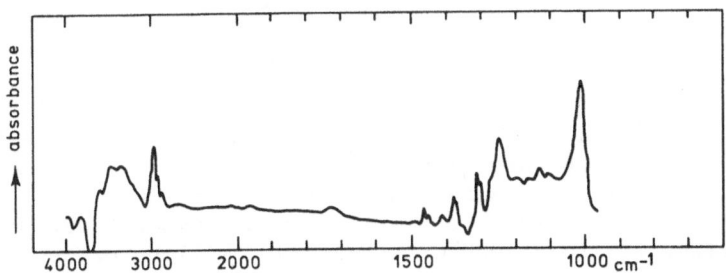

Fig. 2 Infrared spectrum of sarin adsorbed on γ-alumina.

If we look at the spectrum, obviously adsorption has not introduced appreciable changes with regards to carbon-hydrogen stretching and deformation vibrations. The C-O-P stretching absorption has not shifted and only the P=O seems to be affected. However, the P-F vibration is not visible due to the absorption of the alumina. The shift of the P=O band upon adsorption can be explained in two ways.
- sarin adsorbs on γ-alumina via the P=O group;
- sarin adsorbs on γ-alumina via the P-F group, causing a decrease in the electro negativity of the fluorine substituent and consequently a shift of the phosphoryl frequency to a smaller value.

In an attempt to ascertain which substituent of the sarin molecule forms linkages with the surface the adsorption of a series of sarin derivatives has been investigated. From this it appeared that the P=O frequency is always shifted to a lower value whether the compound that was adsorbed contained a fluorine or not. Therefore it is very likely that sarin adsorbs via the phosphoryl oxygen.

The phosphoryl oxygen is charged negatively and adsorption on the surface will therefore occur on a positive centre. Alumina contains two types of positive (electro accepting acid) sites (11). One is related to protons from the surface hydroxyls and the other to incompletely co-ordinated aluminium atoms in the surface layer. Two adsorption structures can be drafted.

$$
\begin{array}{ccc}
\begin{array}{c}
\quad \text{B} \\
\quad | \\
\text{A} \diagdown \! | \! \diagup \text{C} \\
\quad \text{P} \\
\quad \Updownarrow \\
\quad \text{O} \\
\quad \downarrow \\
\text{---Al---}
\end{array}
&
\begin{array}{c}
\quad \text{B} \\
\quad | \\
\text{A} \diagdown \! | \! \diagup \text{C} \\
\quad \text{P} \\
\quad \Updownarrow \\
\quad \text{O} \\
\quad \downarrow \\
\quad \text{H} \\
\quad | \\
\quad \text{O} \\
\quad | \\
\text{---Al---}
\end{array}
&
\begin{array}{l}
\text{possible adsorption} \\
\text{structures}
\end{array}
\end{array}
$$

Although it is difficult to discriminate between the two possibilities the structure with the incompletely co-ordinated aluminium is preferred because alumina heated at 900°C adsorbs sarin just as well as alumina pretreated at room temperature. On an alumina surface heated at 900°C no hydroxyls will be present anymore (12). Moreover, other compounds like benzaldehyde adsorb on the surface of alumina by expelling hydroxyl groups from acid sites (13). In this context it is interesting to note that the changes upon adsorption in the infrared spectrum in the range of hydroxyl stretchings (3800-3300 cm^{-1}) are the same for benzaldehyde and sarin.

3. ROUTES OF DECOMPOSITION

3.1 Infrared spectroscopy

When a spectrum of adsorbed sarin is recorded after a certain time
(24 hours) additional changes are observed (Fig. 3).

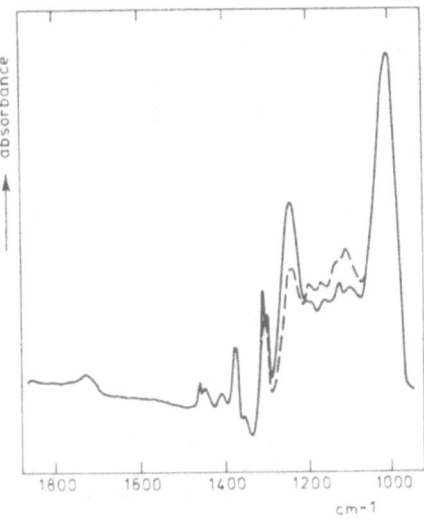

Fig. 3 Infrared spectra of adsorbed sarin
———— 1 hour after adsorption,
----- 24 hours after adsorption

All bands remain unchanged except for the phosphoryl absorption
whose intensity has decreased considerably. In the same period a
rather broad absorption has appeared in the region between 1050
and 1200 cm^{-1}. This change was not observed if the sarin derivative
did not contain fluorine. Moreover, the addition of water to adsorbed sarin accelerates the alterations in the infrared spectra.
It took a lot of experiments to be able to assign these absorptions
properly, but finally Kuiper (14) succeeded by ascribing these
bands to adsorbed fluorine and a phosphonato adsorption complex,
as shown below. Obviously hydrolysis of the P-F group has taken
place

$$\begin{array}{ccc}
\text{CH}_3 & & \text{CH}_3 \\
| & \text{CH}_3 & | & \text{CH}_3 \\
\text{HC} - \text{O} - \text{P} - \text{F} + \text{H}_2\text{O} \rightarrow & \text{HC} - \text{O} - \text{P} - \text{OH} + \text{HF} \\
| & || & | & || \\
\text{CH}_3 & \text{O} & \text{CH}_3 & \text{O}
\end{array}$$

The ratio in which adsorbed sarin and its conversion product
show up in the infrared spectrum after a certain time is a measure
for the rate of the hydrolysis reaction.

Table II
Hydrolysis of some phosphorus compounds on
alumina, compared with alkaline hydrolysis

compound	$\dfrac{I\ (1260\text{-}1200\ cm^{-1})}{I\ (1200\text{-}1050\ cm^{-1})}$	k_{OH}^- ($1.mole.^{-1}min^{-1}$)
DMPF	- / s	430×10^2
allylsarin	w / s	50×10^2
ethylsarin	w / s	29.5×10^2
n-propylsarin	m / m	25.3×10^2
sarin	s / w	16.3×10^2
DFP	s / w	0.5×10^2

In table II the ratios for various sarin derivatives are compared with the rate constant for alkaline hydrolysis. This comparison suggests that the reaction is based catalyzed. In order to verify this supposition some additional experiments have been carried out. When the basic surface sites of alumina are exchanged beforehand by fluorine the sarin is still adsorbed but no reaction can be observed; even when water is added the infrared spectrum remains unchanged. When a spectrum of sarin adsorbed on a more basic solid (magnesia) is recorded the spectrum changes much faster. When magnesia is acidified with hydrogen fluoride the rate of conversion drops again to a very small value.

In conclusion one may say that sarin is adsorbed on alumina via its phosphoryl oxygen and hydrolyzed under the influence of basic surface sites. This hydrolysis is accelerated by the addition of water. Spectroscopic investigations of the desorbed reaction products revealed a small formation of propene. This must be due to C-O bond fission of the sarin molecule. The exact nature of this reaction will be dealt with in section 3.3.

3.2 Hydrolysis studied by chemical analysis

The hydrolysis reaction was studied further by analyzing the adsorbed compound after definite time intervals on various solids. For this purpose sarin was dosed to the adsorbent and after a certain time the adsorbent was extracted and the amount of sarin in the solution was determined by means of a routine test (15). The reproducibility of this method was poor and we therefore carried out some experiments with radio-active labelled sarin. It appeared again that the hydrolysis was strongly accelerated by the addition of water to adsorbed sarin. Moreover, the impregnation of alumina with sodium hydroxide caused an increased rate of hydrolysis. Impregnation with chromium tri-oxide did not increase the rate of conversion. Doping of alumina with hydrogen fluoride stopped the hydrolysis completely. With regards to the kinetics it was found that the reaction did not follow simple first or second order rate laws. In addition the reaction rate was only very slightly dependent on temperature. In Fig. 4 the results of various experiments plotted as C/C_o (the ratio between the concentration extracted and the concentration adsorbed) is plotted versus log t. In some cases straight lines are obtained indicating a particular type of kinetics (see below).

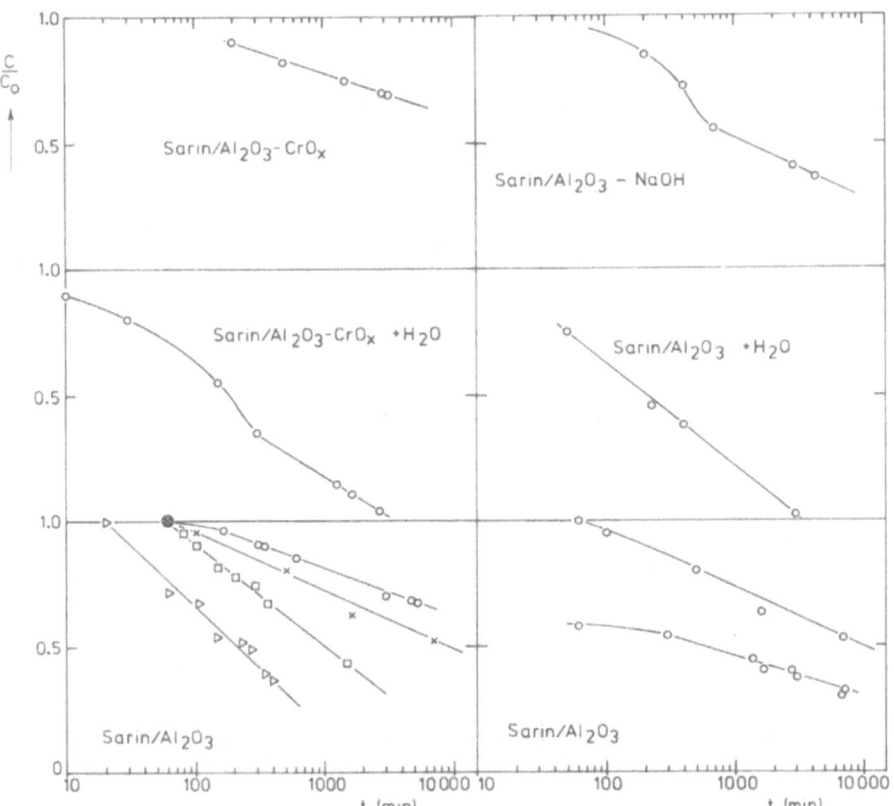

Fig. 4 Results of hydrolysis experiments plotted vs log t
o = 0°C, x = 22°C, □ = 32.5°C, Δ = 40°C.

3.3 Gaseous decomposition products measured by gas chromatography

In these experiments a tube filled with alumina to which sarin was
dosed was incorporated into the carrier gas stream of a chromatograph. In every twenty minutes interval the flow passed through
the tube for two minutes. During this the gaseous products were
fed into the chromatograph for analysis. It appeared that propene
and acetone were the only volatile reaction products. Acetone was
produced only when chromium impregnated alumina was used. In order
to check whether the propene was formed directly from adsorbed
sarin or via the dehydration of isopropyl alcohol the latter compound was adsorbed on alumina and attempts were made to measure
decomposition. It appeared that the alumina did not convert the
alcohol with a measurable rate at room temperature.

From measurements with varying surface coverage with sarin it
appeared that no matter how much is adsorbed the same quantity
propene is formed in a definite time. This indicates that the alumina has a limited capacity for dealkylation. Addition of water
eliminates the activity for dealkylation completely. As shown in
Fig. 5 the dealkylation activity increases with pretreatment temperature.

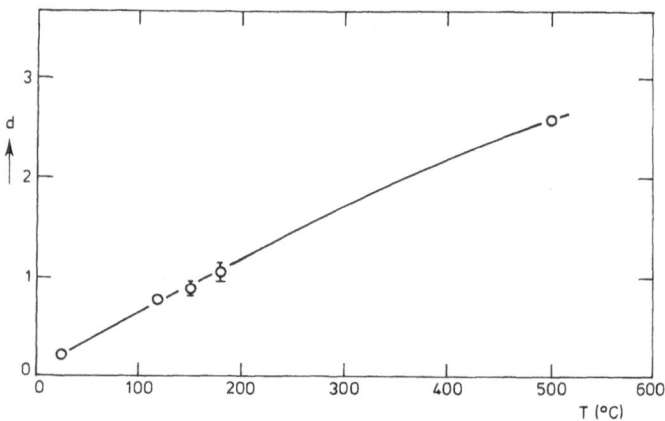

Fig. 5 mmols sarin/gram adsorbent decomposed through dealkylation
    in 1250 minutes (d) as a function of pretreatment temperature.

In table III some results with various solids are given. From this
table it appears that alkaline impregnation reduces the activity
whereas it is increased by acidic impregnation. More basic solids
like magnesia have a considerable lower activity. A chromium tri-
oxide impregnation increases the activity, producing also comparable quantities of acetone.

Table III

Dealkylation experiments performed at room temperature with different adsorbents

exp. no.	adsorbent	pretreatment temperature °C	degree of coverage	% decomposition at 1250 min	overall decomposition at 1250 min
2	Ketjen Al_2O_3	25	0.30	0.7	0.25
23	Ketjen Al_2O_3 + HF	25	0.40	2.5	1.25
17	Degussa Al_2O_3	25	0.30	0.6	0.10
24	Degussa Al_2O_3 + HF	25	0.30	10.7	1.60
5	Ketjen Al_2O_3	150	0.45	1.8	0.95
25	Ketjen Al_2O_3 + NaOH	150	0.50	0.7	0.40
26	MgO	25	1.0	0.01	0.01
27	MgO	150	1.1	0.07	0.06
13	Ketjen Al_2O_3	180	1.0	1.0	1.20
28	Ketjen Al_2O_3 + 2.5% Cr	180	0.80	1.8 + 2.5	1.70 + 2.30*
29	Ketjen Al_2O_3 + 5.9% Cr	180	0.95	5.3 + 7.6	5.95 + 8.45*

*based on the production of acetone.

These findings lead to the following conclusions. The activity of
alumina for dealkylation is much smaller than the activity for
hydrolysis. The dealkylation is accelarated by acid sites as appeared from the impregnation experiments as well as the experiments
with increasing pretreatment temperature; the acidity of alumina is
increased by dehydration of the surface (16). The experiments demonstrate that the C-O band in the sarin molecule undergoes fission.
In the case of chromium tri-oxide impregnation the P-O may be rup-
tured.

4. KINETICS OF THE DECOMPOSITION

As the dealkylation reaction on alumina is of minor importance the
decomposition must be due mainly to the hydrolysis of the phospho-
rus-fluorine bond. The measurement of the kinetics of this reaction
presents a difficult problem. The results of the procedure described before were nor accurate enough to establish the kinetics.
For several reasons it seems favourable to use a calorimetric technique. The heat development is directly related to the rate of
reaction, the reaction can be followed continuously in situ and the
technique is not dependent on the nature of the compound used.
However, the amount of heat developed by sarin adsorbed on alumina
is small. Therefore a sensitive calorimeter of the heat flow type
with a high stability had to be constructed (17). A second problem
appeared to be the thermal inertia of the system, the response time
being in the order of ten minutes. As every decomposition experiment starts with the adsorption of sarin, giving rise to a huge
amount of adsorption energy, the signal of the calorimeter during
the first 100 minutes is mainly due to the heat of adsorption and
no information about the kinetics of the decomposition process is
obtained. This problem was overcome by developping a correction
procedure with which the heat effects originating from adsorption
and reaction could be seperated. Having available an accurate method for the determination of the kinetics, several experiments
were performed.

In order to find a rate law for the reaction a number of relations between reaction rate and time were tested (zero, first, second order and poisoning effects). None of these equations fitted
the experimental results. Some experiments could be described reasonably by the sum of a series of exponentials corresponding to
first order reactions. However, when the time of duration of an
experiment increased the number of exponentials increased. Therefore, it was tried to develop a kinetic equation based on an extended series of first order reactions with decreasing rate constants or increasing activation energy. Van Bokhoven (17) showed
that with a homogeneous distribution in activation energy the reaction rate law assumes the form:

$$r = \frac{A.T}{t} (1-e^{-k_1 t})$$

After the initial period the exponential containing the rate constant of the fastest reaction k_1 becomes zero and the reaction
rate is inversely proportional to time.

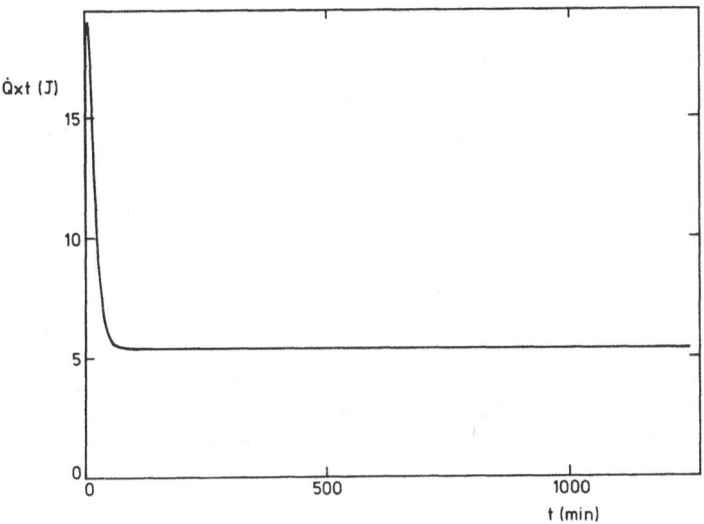

Fig. 6 Product of heatflow (Q) and time (t) versus time.
The heatflow is linear proportional to the reaction rate.

A plot of the results of a typical experiment, shown in Fig. 6,
reveals that the rate equation describes the observed kinetics
very well. A remarkable feature of the rate equation is that the
reaction rate after the initial period is proportional to the absolute temperature contradictory to the well-known Arrhenius temperature dependence. Experiments at different temperatures proved
that the rate law was correct. From the temperature dependence in
the initial period of the reaction the activation energy of the
fastest reaction can be found. This activation energy appeared to
be 4 kcal/mol. A very interesting finding is that the entropy
change in the formation of the transition state is much larger
than is expected on the basis of the theory of absolute reaction
rates. This large entropy change causes a relative low reaction
rate despite of the fact that the activation energy amounts to
4 kcal/mol only. A favourable effect of this temperature dependence
is that the reaction proceeds at comparable rates in the range of
$-20/+40°C$. As to the cause of the distribution in activation energy

388

no firm conclusion can be drawn but it is believed that it is related to the basicity of the different sites on the alumina surface. Having stronger basic sites on its surface magnesia shows another distribution in activation energy.

SELECTION OF THE IMPREGNANT

From the foregoing it is clear that a basic impregnant of charcoal must be applied for relatively fast hydrolysis of the G-agents. The kinetics of the hydrolysis are such that the temperature dependence of the reaction rate is of no importance. The stronger the basicity of the impregnant the higher the reaction rate will be. However, very strong basic solids might be poisoned by carbon dioxide from the air. Therefore it will be tried to incorporate a solid of medium basicity into charcoal. The favourable results with magnesium oxide induce the choice of magnesium hydroxide. Preliminary experiments with charcoal impregnated with magnesium hydroxide show a considerable increase in the decomposition activity of the charcoal for G-agents. At the moment we are investigating whether there are even better impregnants than magnesium hydroxide. Simultaneously we are trying to elucidate the adsorption and decomposition mechanism of so-called V-agents in order to develop an impregnant for the efficient decomposition of these agents too.

REFERENCES

1. The problem of chemical and biological warfare, Vol.I, "The rise of GB weapons", Edited by SIPRI, Almqvist & Wiksell, Stockholm (1971), Chapter I.
2. Noyes, W.A., Jr., Science in World War II, Chemistry, Boston (1948), p. 296.
3. Handbuch der Experimentellen Pharmakologie, Cholinesterases and Anticholinesterases Agents, Band XV (Ed. G.B. Koelle), Springer Verlag, Berlin (1963).
4. Wilhelm, Katja, Acta Biol. Jugoslav., Ser. C 3, 471 (1967).
5. Lippens, B.C., Thesis, Delft (1961).
6. Little, L.H., Infrared Spectra of Adsorbed Species, Academic Press, London (1966).
7. Hair, M.L., Infrared Spectroscopy in Surface Chemistry, Dekker, New York (1967).
8. Thomas, L.C., Interpretation of the Infrared Spectra of Organophosphorus compounds, Heyden Ltd., London (1974).
9. Medema, J., van Bokhoven, J.J.G.M., Kuiper A.E.T., J. Catal., 25, 238 (1972).
10. Houben, J., Weyl, Th., Methoden der Organischen Chemie, I.: Organische Phosphorverbindungen, George Thieme Verlag, Stuttgart (1963).

11. Peri, J.B., J. Phys. Chem., $\underline{72}$, 2917 (1968).
12. Peri, J.B., J. Phys. Chem., $\underline{69}$, 211 (1965).
13. Kuiper, A.E.T., Medema, J., van Bokhoven, J.J.G.M., J. Catal., $\underline{29}$, 40 (1973).
14. Kuiper, A.E.T., Thesis, Eindhoven (1974).
15. Gelauf, B., Epstein, J., Wilson, G.B., Witten, B., Sass, S., Bauer, V.E., Rueggeberg, M.H.C., Anal. Chem., $\underline{29}$, 278 (1957).
16. Peri, J.B., J. PHys. Chem., $\underline{69}$, 220 (1965).
17. van Bokhoven, J.J.G.M., Thesis, Eindhoven (1974).

RESPIRATORY PROTECTION AGAINST HYDROGEN CYANIDE

F.A.P. Maggs

Chemical Defence Establishment, Porton Down, Salisbury, Wiltshire, England

Whilst hydrogen cyanide is not a very toxic substance in terms of a lethal dose, its high vapour pressure (boiling point $26^{\circ}C$) and rapid action make it a dangerous chemical to deal with, and respiratory protection against this vapour must therefore be effective. The physical properties of the substance make it unlikely that physical adsorption on an adsorbent will be satisfactory, and indeed, the amount adsorbed by a canister filled with charcoal is insufficient to afford much protection. Desorption also occurs.

Chemical interaction is therefore necessary, and a canister filled with alkali granules or with an impregnated charcoal may be employed. Various forms of alkali granules, such as soda-lime, cement lime, have been used, with the object of providing a reactive solid which is also resistant to rough usage leading to granule breakage and attrition. Durability and reactivity tend, however, to be mutually incompatible, and impregnated charcoals are displacing the denser alkali granules as a canister filling.

The charcoal impregnations which have proved successful for a number of years are based on copper. In one form copper oxide is incorporated in the raw coal before carbonisation and activation, whilst in another preparation an ammoniacal solution of copper and silver salts is neutralised with chromic acid and applied to the charcoal itself.

It is of some interest at this point to refer to unpublished work from Professor Stone's laboratory at Bath University, which I quote with his permission. He found that reduced copper adsorbed hydrogen cyanide, and that hydrogen cyanide only was present in the equilibrium gas phase; on desorbing at 100°C, all the hydrogen cyanide was recovered unchanged. The extent of the coverage of the available copper surface suggested that each hydrogen cyanide molecule was attached without dissociation to two adjacent copper atoms. However, after exposing the reduced copper surface to air and re-evacuating much greater quantities of hydrogen cyanide were taken up on admitting this gas; carbon dioxide, nitrogen and water were formed and the reaction continued for some time without reaching equilibrium. After evacuation, these three gases were desorbed only on raising the temperature to above 270°C.

The reactions between hydrogen cyanide and solid cupric oxide were also instructive. Carbon dioxide, water and cyanogen were formed, and were also evolved on heating. It is known, moreover, that cyanogen is a product of the interaction between hydrogen cyanide and a cupric salt; a charcoal impregnated with copper sulphate, for instance, provides a most undesirable respirator filling for protection against hydrogen cyanide, for this reason. It may be concluded, therefore, that the copper surface exposed to air contains no reactive cupric ion but may be regarded as a cuprous oxide surface.

Aspects of this work may be applied to copper-impregnated charcoals. A cupric salt on the charcoal, for instance, may be reduced by heating in hydrogen, and the charcoal produced is effective in removing hydrogen cyanide without forming cyanogen. However, we have found that heating the cupric impregnated charcoal in nitrogen yields a product which is also as effective – presumably through the formation of a cuprous oxide impregnant.

Of practical interest is the observation that deterioration with time may result if chlorine or sulphur are present in the raw material (e.g. coal) in a form which can lead to the formation of cupric salts. The effluent air in dynamic testing against hydrogen cyanide will then contain cyanogen. It is thus prudent to ensure that the detecting system in use in such tests will respond to cyanogen as well as to hydrogen cyanide.

It has been suggested that charcoal may catalyse the polymerisation of hydrogen cyanide, and thus provide protection. Part of the hydrogen cyanide adsorbed on an unimpregnated nutshell charcoal, for instance, is desorbed as such only on heating to ca. 100°C. The influence of inorganic impurities has, however, not been studied. It should be added that the quantity adsorbed is, in any case, small, as was noted earlier.

A number of laboratories have conducted extensive research, generally unpublished, on charcoals impregnated by treating the active charcoal with ammoniacal solutions of copper, chromium and silver. The charcoal is then given a mild heat treatment (150° to 200°C) to decompose the complexes and remove ammonia, care being taken to avoid ignition. The dry product has a high capacity for the removal of hydrogen cyanide (and also of cyanogen chloride)..

The mode of action of such adsorbents is still not clear. With respect to hydrogen cyanide protection, it is clear that copper is the necessary constituent; the role of chromium, and of its valency state (only chromates and dichromates are effective) are still obscure, and silver probably adds little to the hydrogen cyanide interaction. Storage under humid conditions leads to deterioration, the more so when access to air is restricted.

We have found that the presence of copper dichromate alone in a charcoal can also lead to much enhanced protection against hydrogen cyanide; in this case, humid conditions do not lead to deterioration, and consequently the respirator canister can, when not in use, be left open to the atmosphere.

Many random observations have been made with these types of charcoal. For example they adsorb oxygen very slowly; hot, humid conditions can lead to deterioration; deterioration does not occur with dry charcoal; mild heat treatment, even in nitrogen, restores the hydrogen cyanide activity. It is clear that whilst the studies undertaken by Professor Stone will further understanding of the reactions involved, some modification of the final picture will undoubtedly be required when the reactions take place on an impregnated charcoal surface.

PURIFICATION OF INDUSTRIAL EFFLUENTS WITH ACTIVATED CARBON*

Walter J. Weber, Jr.

Professor of Environmental and Water Resources
Engineering and Chairman, Water Resources Program,
College of Engineering, The University of Michigan,
Ann Arbor, Michigan, U.S.A.

1. INTRODUCTION

Industrial effluent purification is one of the most effective and widespread application areas for adsorption processes in the water and effluent treatment field. Extensive experience with a broad range of different types of industrial effluents indicates that treatment by activated carbon is technically and economically suitable for many industrial applications. A few examples of full scale installations of activated carbon treatment for industrial effluent purification are presented in Table 1.

The list in Table 1 is not inclusive with respect either to type of industry, to organic parameter of concern, or to range of plant size, but does give typical examples of industries that have had success with the use of activated carbon for effluent purification. Activated carbon adsorption technology is being employed with increasing frequency in the United States, in a number of European countries, in Japan, and in Africa.

The suitability of activated carbon adsorption systems for industrial effluent purification is attributable to a number of different properties of the process, primary among which are:
 a. wide applicability to removal of many different types
     of organic impurities and some heavy metals;
 b. superior effluent quality and the possibility for water

---

* Lecture 1. Session V. NATO Advanced Study Institute: Scientific
  Aspects of Sorption and Filtration Methods for Gas and Water
  Purification, Fauske, Norway, 23-29 June 1974

Table 1. Examples of industrial effluent treatment by activated carbon

Type of industry	Typical organic parameters of concern	Range of plant sizes, MGD*
Textiles and dyestuffs	TOC, color, dyes	.05 - 1.5
Oil refinery and petro-chemical	COD, BOD	2.2 - 4.2
Detergents, resins and chemicals	TOC, COD, xylene, alcohols, phenolics, resin intermediates, resorcinol, nitrated aromatics, polyols	0.015 - 3.0
Herbicides and insecticides	Chlorophenols, cresol	0.15 - 0.50
Pharmaceuticals	Phenol	0.025 - 0.050
Explosives	Nitrated phenol	0.005 - 0.020

*MGD (million gallons per day) 3,785 = cu.m./day

       reclamation and reuse;
- c. no generation of additional waste products;
- d. no effects of toxic wastes;
- e. capability to accomodate fluctuations in hydraulic load and/or waste composition and strength;
- f. compactness and relatively low land area requirements;
- g. ease of construction and operation; and,
- h. favorable economics.

Additional impetus for the use of adsorption for industrial effluents in the United States has been provided by recent legislation calling for application of "best practicable" effluent treatment by 1977 and "best available" technology by 1983. Sharply increased use of activated carbon adsorption treatment in the years immediately ahead may therefore be anticipated.

The principles of adsorption and the properties of activated carbon have been discussed in previous lectures. This lecture will focus on specific aspects of the process relating to industrial effluent applications. In this regard, the great majority of industrial applications utilize granular activated carbon rather than powdered activated carbon, because of advantages discussed in previous lectures. In addition to its use as a singular purification process, granular carbon treatment has been applied in some instances to the effluents from biological treatment plants, and in other instances to the influents to biological treatment plants. In the former case adsorption usually is intended for removal of biologically resistant chemicals, such as nitrated aromatics in some chemicals industries. In the latter case, the purpose of adsorption is to remove substances, such as chlorophenol in the pesticide industries, which may be toxic or otherwise capable of inhibiting biological treatment processes.

2. PRETREATMENT REQUIREMENTS

As noted above, granular carbon is most commonly utilized for industrial effluent treatment; this normally involves the use of either fixed-bed or moving-bed contact systems. Pretreatment in these cases is primarily for removal of excess suspended solids, oils, and greases; materials which, although they can be removed effectively by the adsorption bed, will likely cause problems of head loss, fouling, and plugging of the adsorber.

Suspended solids in amounts exceeding approximately 50 mg/l should be removed prior to adsorption. Oils and greases in concentrations above about 10 mg/l should not be applied directly to adsorption beds. In addition to causing head loss problems, these materials, particularly oils and greases, can coat the carbon particles and reduce adsorption effectiveness dramatically.

For concentrations of suspended solids, oils, and greases below the values given above, carbon beds are frequently used for the dual role of filtration and adsorption. Carbon bed engineering incorporates well established filtration experience regarding backwash requirements, surface wash or air scour features, and appropriate flow rates conducive to filtration. In applying established filtration experience to carbon beds, which are normally deeper than ordinary filters, it has been demonstrated that the top few feet of the carbon bed function in the same manner as the equivalent depth of a conventional single media filter. Backwash bed expansion should be based on this effective filtration depth rather than the total adsorption bed depth.

Chemical clarification, air flotation and filtration are common pretreatment processes. It is not unusual, however, to find that pretreatment is not required when adsorption is applied at the point of origin of the contaminant of concern. Treatment of combined waste streams, on the other hand, invariably requires pretreatment for removal of suspended solids, oil, and grease.

Adjustment of pH is sometimes employed to enhance adsorption efficiency. Dissolved organic compounds are normally adsorbed best at the pH of minimum polarity. For example, weak acids, such as phenol, are better adsorbed at lower pH values, while amines can be expected to adsorb best at higher pH values.

Potential advantages of pH adjustment can be quickly determined by laboratory tests.

Flow equalization is desirable for many industrial effluent treatment processes. While adsorber systems can be designed to meet widely fluctuating influent hydraulic loads and organic concentrations, the treatment system can be more economically designed and operated when these fluctuations are minimized.

3. ADSORBER SYSTEMS

Common adsorber systems are illustrated in Figure 1. For any system, a particular combination of flow rate and bed depth is used to give an effective design "contact time," the time required to reduce the organic contaminants from the influent level to the desired effluent level. Flow rates under 10 gpm/ft^2 (0.4 cu.m./day/sq.m.) are usually employed to minimize pumping costs associated with high head loss. Bed depths of less than 10 ft (3.05 m.) are rare in industrial effluent applications.

Municipal sewage treatment experience has indicated good treatment at contact times between 20 and 50 minutes. Significantly longer contact times are normally required for industrial waste streams, consistent with generally higher organic concentrations. Adsorption theory and practice indicate that treatment efficiency and economics are favored by higher concentration. Thus, at some installations, concentrated waste streams are treated individually at their respective sources to optimize overall treatment system design and economics.

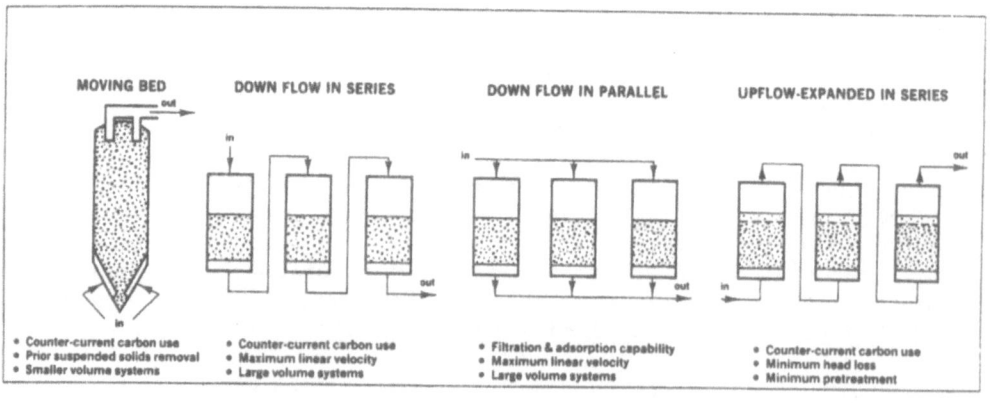

Fig. 1. Adsorber configurations.

4. TREATABILITY EVALUATION

Adsorption isotherm tests are standard first-step procedures for determining the feasibility of adsorption for a specific effluent purification. Isotherm procedures and theories have been discussed in previous lectures. The isotherm indicates the degree of purification that might be achieved and the approximate amount of carbon required to reach a treatment objective. It also indicates the dependence of the amount of adsorption on contaminant concentration.

The Calgon Corporation (Pittsburgh, Pa., U.S.A.) has recently tested 222 samples of different industrial effluents, representing 68 different manufacturing operations, to evaluate the removal of organic contaminants by activated carbon.[1]

In the work reported by Calgon, samples were membrane filtered prior to the adsorption tests to remove suspended material which otherwise could be incorrectly associated with adsorption treatment. This is generally a good experimental procedure, and the type and concentration of suspended material removed in this step gives a preliminary indication of the desirability of pretreatment.

The adsorption isotherm results are summarized in Table 2, grouped according to SIC (Standard Industrial Classification, U.S. Dept. of Commerce) classification of the wastewater treated. This grouping allows comparison of adsorption performance on like organic contaminants and provides a reference list for indication of the feasibility of adsorption as a treatment for each class of effluent.

Table 2 is a useful reference when considering treatment alternatives available for a specific industrial effluent problem, with respect to a preliminary indication of the feasibility of adsorption treatment. The data presented in Table 2 can be further categorized according to organic concentration. This categorization is given in Table 3.

Because of the generally high levels of TOC evidenced in most untreated industrial effluents, it is likely that a combination of treatment processes will be required to meet rigorous TOC reduction objectives at lowest cost. The selectivity of adsorption on activated carbon for color and phenol suggests the possibility of using adsorption as a pretreatment to biological systems to remove toxic, inhibitory and bio-resistant substances. Adsorption treatment of industrial effluents for selective removal of specific substances prior to discharge to municipal systems appears to be technically and economically feasible in many instances.

Table 2. Results of adsorption isotherm studies on different industrial effluents.

SIC number	Type of industry	Number tested	Initial TOC (or phenol), mg/1	Initial color, O.D.	Average % reduction	Carbon exhaustion rate, lb/ 1000 gal
2000	Food and kindered products	16	25-5,300	--	90	0.8-345
2100	Tobacco manufacturers	1	1,030	--	97	58
2200	Textile mill products	33 28	9-4,670 --	-- 0.1-5.4	93 97	1-246 0.1-83
2300	Apparels and allied products	2	390-875	--	75	12-43
2600	Paper and allied products	9 1	100-3,500 --	-- 1.4	90 94	3.2-156 3.7
2700	Printing, publishing and allied industries	2	34-170	--	98	4.3-4.6
2800	Chemicals and allied products	137 13 18	19-75,500 (0.1-5,325) --	-- -- 0.7-275	85 99 98	0.7-2,905 1.7-185 1.2-1,328
2900	Petroleum refining and related industries	17 3	36-4,400 (7-270)	-- --	92 99	1.1-141 6-24
3000	Rubber and miscellaneous plastic products	8	120-8,375	--	95	5.2-164

(continued)

Table 2. Results of adsorption isotherm studies on different
industrial effluents (continued).

SIC number	Type of industry	Number tested	Initial TOC (or phenol), mg/l	Initial color, O.D.	Average % reduction	Carbon exhaustion rate, lb/ 1000 gal
3100	Leather and leather products	2	115-9,000	--	95	3-315
3200	Stone, clay and glass products	7	12-8,300	--	87	2.8-300
3300	Primary metal indus- tries	8	11-23,000	--	90	0.5-1,857
3400	Fab- ricated metal products	1	73,000	--	25	606
3700	Trans- portation equipment	2	190-2,850	--	91	12-361
4200	Motor freight trans- portation and ware- housing	4	320-3,480	--	87	20-72

Table 3. Summary of adsorption data on industrial effluents.

Category	Number of samples
Initial TOC < 100 mg/l	24
Initial TOC = 100-1000 mg/l	100
Initial TOC = 1,000-10,000 mg/l	86
Initial TOC > 10,000 mg/l	12
TOC reduction > 90%	140
TOC reduction = 85-90%	29
TOC reduction < 85%	53
Color reduction > 95%	36
Color reduction = 90-95%	5
Color reduction < 90%	1
Phenol reduction > 99%	12
Phenol reduction < 99%	1

Carbon exhaustion rates for industrial effluents are clearly in excess of those associated with municipal effluent treatment. This is expected, considering the relative levels of initial TOC. Municipal effluents can be successfully treated at costs between 10-30 U.S. cents per 1,000 gallons (per 3.8 cu.m.). For the most part, treatment of industrial effluents costs in the U.S. dollars per 1,000 gallons (per 3.8 cu.m.) category for effective TOC reduction. For selective removal of color or phenol, treatment costs might well be in the U.S. cents per 1,000 gallons (per 3.8 cu.m.) range.

For certain types of industrial effluents, such as those from textile mill products operations, treatment by adsorption alone appears suitable to meet organic removal objectives. Experience with textile effluents indicates the possibility of water reuse following adsorption treatment. Such reuse can off-set pollution abatement costs. Cost reduction opportunities exist for water reuse in many other industrial areas as well. Point source treatment of specific wastes is also a substantial factor in the use of adsorption technology for industrial effluent purification.

5. REACTIVATION

Thermal reactivation of granular activated carbon has been a successful practice for several decades. Rotary kilns or multiple-hearth type furnaces operated at temperatures between 1600-1800 °F are normally used. The activation atmosphere in these furnaces is maintained at low oxygen levels to effect selective oxidation of adsorbed organic contaminants rather than of the activated carbon. Carbon losses range between 2-10% per cycle, with larger systems generally experiencing smaller losses. Large systems are normally designed for continuous operation, thereby facilitating control of the process.

Afterburners and scrubbers are used to strip the furnace exhaust gases of air pollutants. Industrial wastes containing halogens or other corrosive substances require special materials of construction in the kiln or furnace to avoid corrosion problems. It is important that each industrial effluent be evaluated from this standpoint.

The process and functional features of a thermal reactivation system are illustrated in Figure 2. Detailed engineering design features of granular activated carbon reactivation systems are readily available in the literature.[2,3]

Experience with chemical reactivation of granular carbon has been largely unsuccessful to date. Recovery of specific

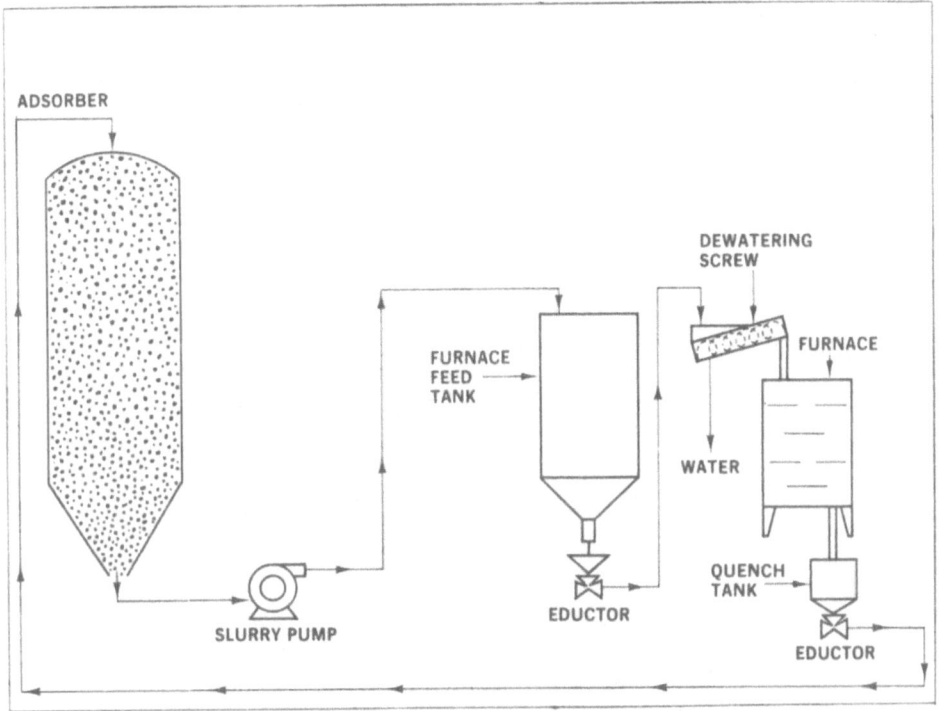

Figure 2. Granular carbon reactivation cycle

by-product adsorbates by extraction has been demonstrated in some instances, but generally the carbon does not sufficiently recover its adsorptive properties. Additionally, by-product recovery frequently suffers from the fact that the material of interest is in a waste mixture and a pure product is difficult to recover.

REFERENCES

1. Hager, D. G., Industrial wastewater treatment by granular activated carbon, Indust. Water Engin., 11, 1, 1974.
2. Smith, C. E., Principles and practice of granular carbon reactivation, in: Applications of New Concepts of Physical-Chemical Wastewater Treatment, Pergamon Press, Inc., (September 18-22, 1972), pp. 179-184.
3. Swindell-Dressler Company, A Division of Pullman Incorporated Process Design Manual for Carbon Adsorption, for the Environmental Protection Agency Technology Transfer, October, 1971.

URIFICATION OF INDUSTRIAL EFFLUENTS WITH ACTIVATED CARBON.

R.A.DAVIES

CHEMVIRON S.A.°

1. INTRODUCTION

In recent years increasing emphasis on waste water treatment throughout the world has seen more stringent effluent limits imposed, and fines for polluting discharges rise drastically. Many industrial waste waters can be satisfactorily treated and reused in the factory of origin as a cost saving measure. Most industrial effluents contain a complex mixture of organic compounds and a number of unit processes may be required in purification of the effluent. It is therefore necessary to equip the waste water treatment plant designer with as many tools as possible.

One possible treatment method is adsorption of organic contaminants on activated carbon. The activated carbon stage may be the sole treatment used, or adsorption can be used in conjunction with other treatment methods. For example the adsorption stage may be used after a chemical or biological treatment process for polishing the effluent down to the required discharge standard. Alternatively the adsorption stage can sometimes be usefully employed as a pretreatment before a biological treatment plant, thereby optimizing the performance of the biological plant by reducing toxic organic compounds, or inhibitors.

° Chemviron S.A.
  P.O.Box 17 - Ixelles 1
  1050 Brussels, Belgium

2. BASIC FEASIBILITY

When faced with an industrial effluent which would appear to be amenable to carbon treatment, the basic feasibility of carbon must be established. It is necessary to know if activated carbon treatment can remove the organic compounds present to below the permitted discharge level, or level required for reuse. At this time such a basic feasibility must be established empirically. The waste water which can be specified precisely in terms of each organic compound and its concentration is rare.

The traditional tool used for this empirical feasibility testwork is the adsorption isotherm. The isotherm test permits the calculation of the maximum amount of organic compounds that can be loaded onto a given activated carbon at equilibrium with the effluent at its initial concentration. This equilibrium will be attained or closely approached on the carbon when used in granular carbon columns which have been correctly designed.

The data for the isotherm plot is obtained by treating fixed volumes of the effluent with different weights of carbon. The carbon is usually powdered to decrease the test duration by speeding the attainment of the equilibrium condition. The carbon liquid mixture is shaken for fixed time at constant temperature. In waste water treatment the normal temperature of large scale treatment will be ambient, so the apparatus is relatively simple.

A typical example of an isotherm calculation is shown in Table I and Figure 1. Organic concentration in this case is expressed in mg/l COD.

TABLE I.

M Weight of carbon g/l of solution	C Residual organic concentr. mg/l COD	X mg COD adsorbed	x/m mg COD loaded per gram carbon
0	120		
0.06	99	21	350
0.25	70	50	200
0.40	56	64	160
0.70	36	84	120

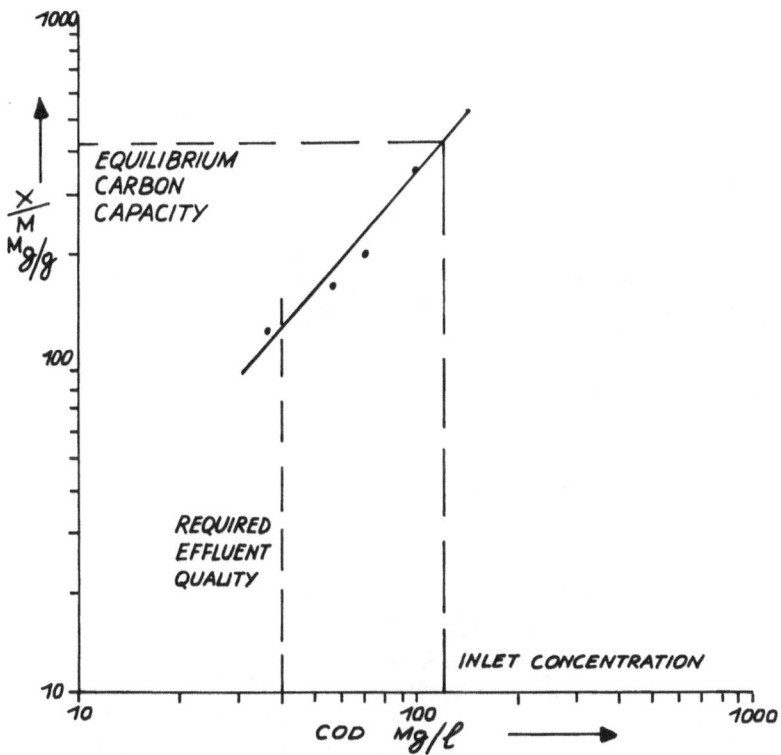

Fig. 1 Typical waste water isotherm
COD, F 400, 20°C

This isotherm indicates a good adsorption capacity of 430 mg COD per gram carbon when extrapolated to the influent concentration. It also indicated that the plot down to the indicated required limit is linear, and no difficulty would be expected to attain the required effluent limit.

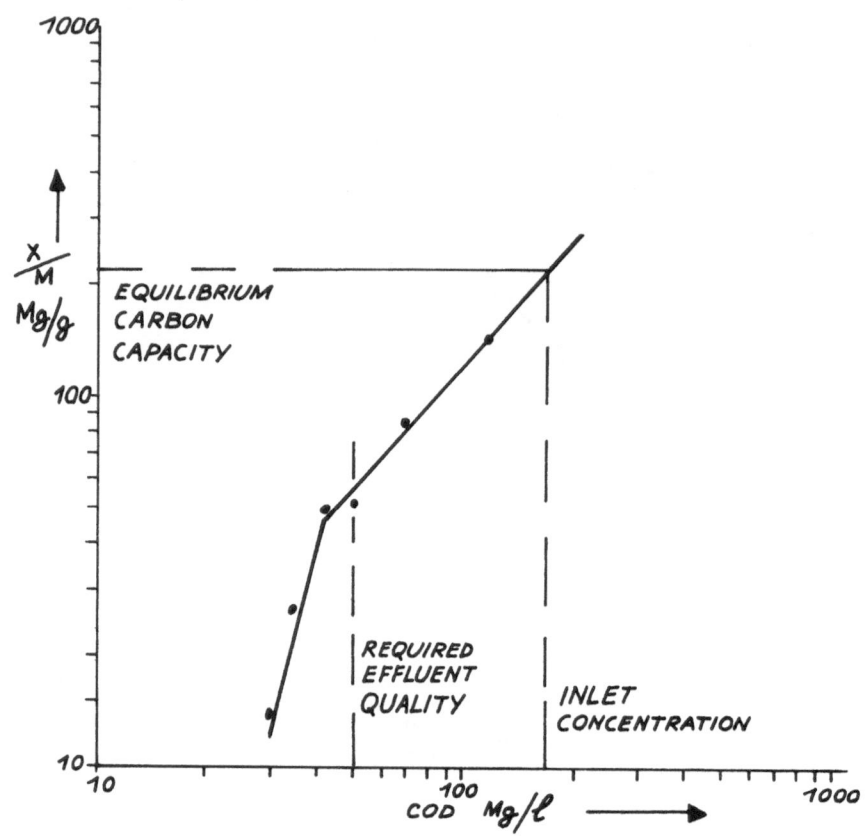

Fig. 2 Slope change isotherm
F 400, 20°C, COD

Figure 2 shows a case where the isotherm changes slope. This is normally presumed to be an indication of the presence of two groups of compounds. The adsorption capacity at equilibrium with the influent concentration is 220 mg COD per gram carbon, a rather lower than average figure. The curve breaks at about 45 mg/l COD, and since in this case required effluent concentration was 50 mg/l, the basic feasibility of application of carbon appears proven. However, had the required effluent limit been 20 or 30 mg/l, the isotherm would have been regarded as inconclusive, and a column test would have been necessary to demonstrate feasibility.

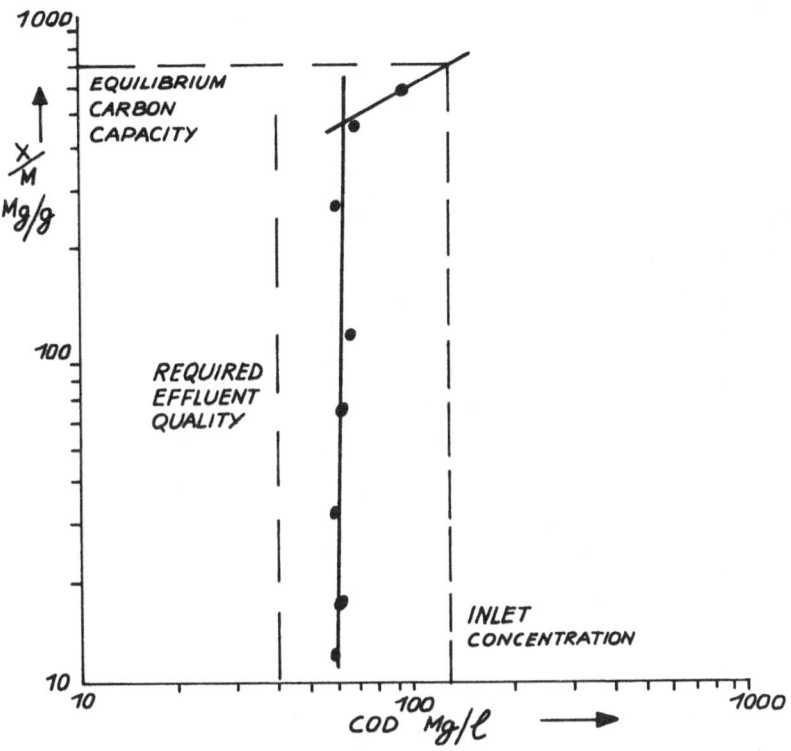

Fig. 3 Slope change isotherm
F 300, 40°C, COD

Figure 3 shows, for the sake of completeness, a waste water which is not amenable to carbon treatment alone. The curve indicates a minimal reduction in organic content at whatever dosage level applied, after the level of 60 mg/l COD has been attained. However, the high capacity of 700 mg/l for the adsorbable content (which would have to be verified with another low dosage isotherm since it depends only upon one point) might lead the designer to investigate the residual waste water at 60 mg/l COD to find out whether an additional treatment stage such as air stripping could remove what is likely to be a very light molecular weight residual contaminant.

The effect of preliminary treatment, such as pH adjustment can also be evaluated by means of the isotherm technique.

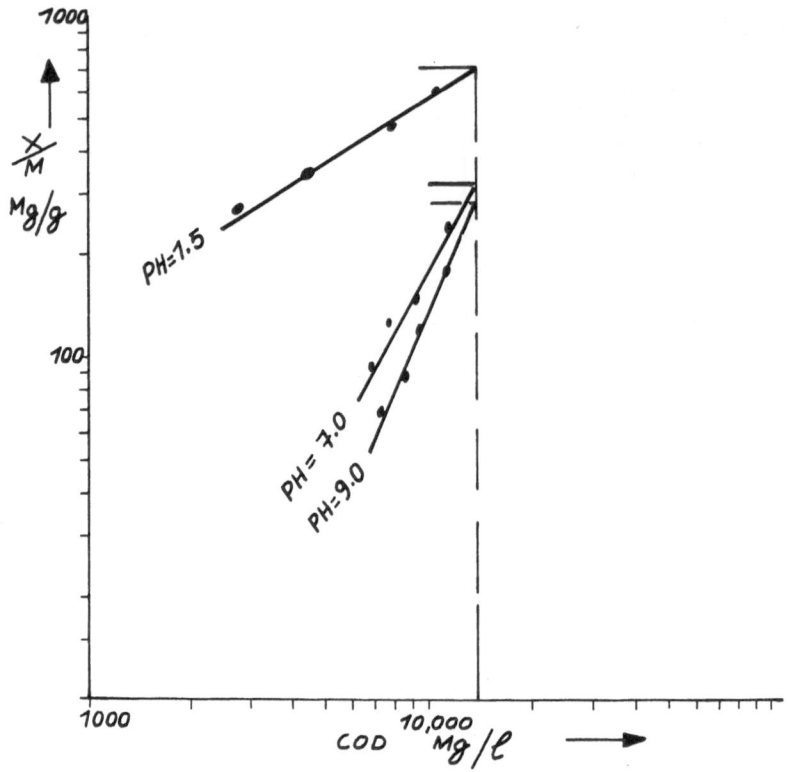

Fig. 4 pH effect on paper mill effluent
F 300, 30°C, COD

Figure 4 shows the effect of pH on the slope and the capacity of a specialist paper mill effluent.

The efficacity of various commercial carbons can be determined at this stage, although it is important to note that because the isotherm is a measure of e- quilibrium loading a final choice of carbon is impossi- ble at this stage. Contact time, which varies in func- tion of the carbon pore structure and type, must be evaluated to determine the overall system economics in conjunction with the capacity data.

3. DESIGN CRITERIA

A column test is used to define the major design necessity of contact time and to confirm the capacity data of the isotherm. On a typical waste water it will be necessary to run the column test for several weeks continuously sampling at regular intervals. For ease of sampling at intermediate depths in the total carbon column, a set of 4 columns through which the liquid is passed in series, is typically used. The total bed depth can be 4 to 5 metres in a set of 4 columns of 2 metres height, thereby permitting backwashing of the columns to remove retained suspended solids. This configuration of test apparatus is a very close approximation to the bed depths and hence velocities found in industrial systems, and the data generated can be directly used for plant design purposes. Column diameters should be as large as possible and a 10 cm diameter column has been found to be a good compromise between compactness and the certainty of eliminating wall effect. It is rather difficult however to duplicate the stratification of the carbon bed after backwash with such a diameter column, as the bed has a tendency to turn en mass. It may be necessary to air scour the bed to break up mud balls.

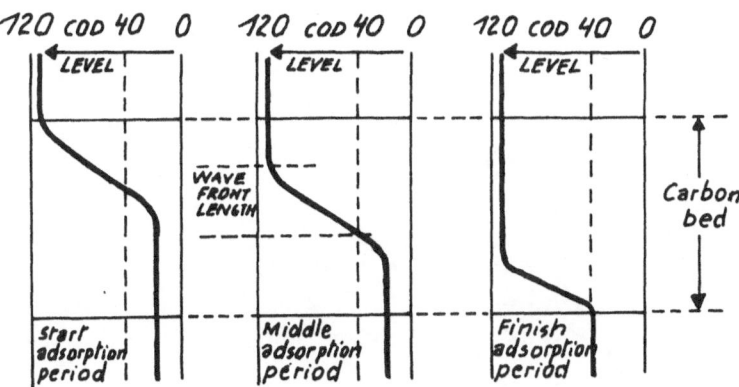

Fig. 5 Wave front diagram (schematic)

One may visualise the dynamic adsorption picture through the total carbon bed as represented in Figure 5. Ref. 1. The wavefront length (mass transfer zone or MTZ) can be expressed as a time of contact (normally considering an empty bed).

Restating the two questions to be answered in design of a carbon system, these are :

1. the necessary time of contact (which must also be always greater than the MTZ expressed as time) ;

2. the equivalent dosage rate or the carbon consumed per unit volume of liquid treated.

If one runs an evaluation as described above with a series of columns on the water to be treated and obtains samples after each column, one may plot COD (or any other measure of organic concentration) against volume throughput, at a number of different contact times, as shown in Figure 6. These curves enable one to cross-check the isotherm capacity by measuring the area A_S, which, being concentration multiplies by volume, is the mass of material removed. Thus, knowing the mass of carbon employed in column 1, one may evaluate the capacity of the carbon for the organic species being removed and hence the dosage rate for a given removal requirement.

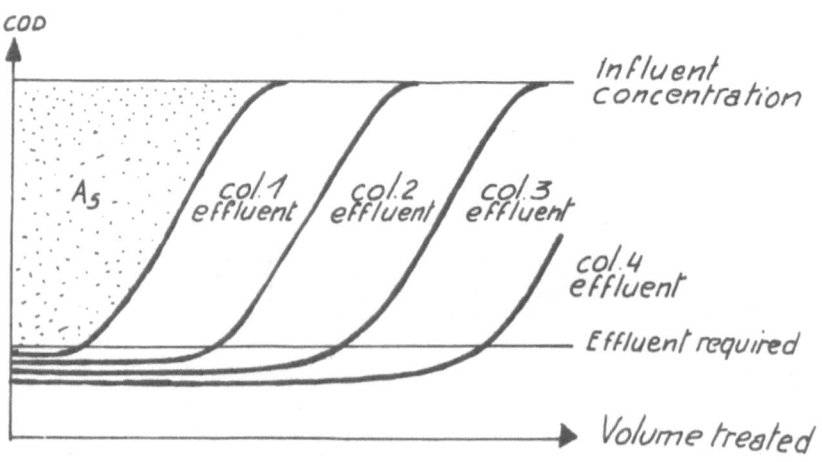

Fig. 6A Column test results plotted as COD vs. volume treated

It is advisable to run the column test until the first two columns are saturated as a crosscheck on the dosage rate.

This dosage rate is the theoretical minimum dosage rate, assuming carbon is discharged from the system only in a fully spent condition. Assuming that one will operate in full scale only one column, then one may also evaluate dosage rates at each contact time, by evaluating areas under the graph at the point where the treatment objective is exceeded, for each contact time plotted, as shown in Figure 7.

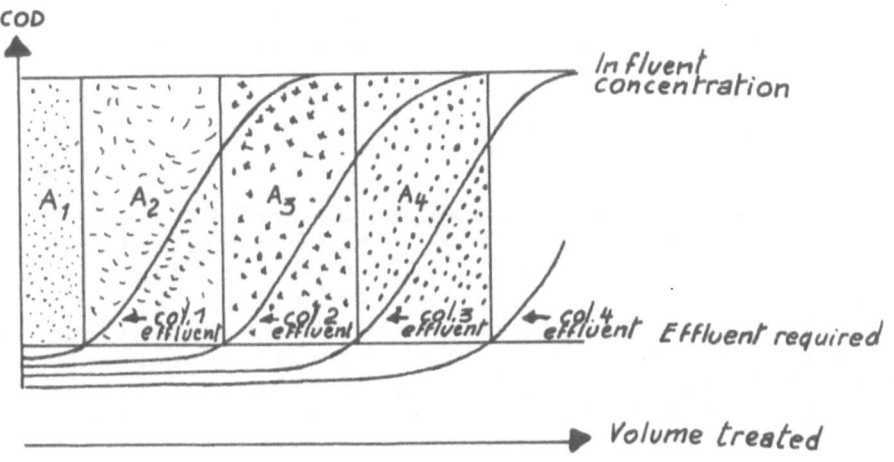

Fig. 7B Column test results plotted as COD vs. volume treated.

One finds, clearly, that at short contact times one has a high dosage rate and at long contact times one has a dosage rate approaching the theoretical minimum, since a greater and greater proportion of the carbon is fully saturated. It is then possible to plot the contact time and dosage rate, for the influent concentration and effluent concentration required, for a single column case. Such a plot is termed the operating line (Fig. 8) and can be considered to contain all of the basic data needed for system design. Ref. 2.

414

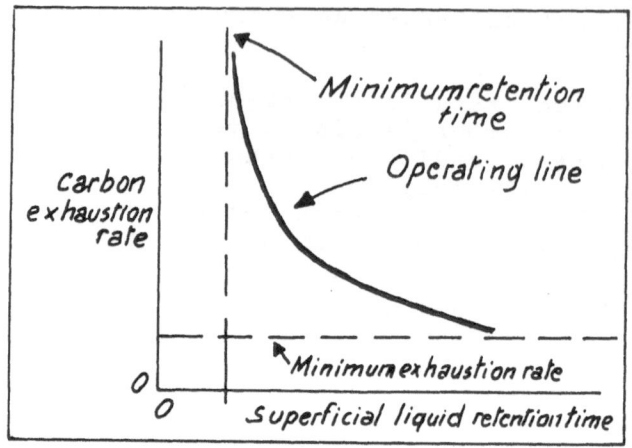

Fig. 8 Operating line

The line approaches on one axis the minimum exhaustion rate, corresponding to the highest theoretical equilibrium carbon capacity as found from column tests or isotherms. The line approaches the minimum retention time on the other axis. The minimum retention time is an indication of the carbon bed volume needed to treat the liquid, at an infinite carbon dosage rate. It is to be noted that the curves shown above are smooth, and serve to demonstrate the principles behind the relationship between contact time and carbon dosage rate. Figure 9 shows an actual set of column test data which is far from the ideal curves shown above. These breakthrough curves were obtained from tests on an aqueous effluent stream containing chlorinated hydrocarbons. A good deal of judgement is needed in interpretation of this type of data. It can be seen that the curves initially break sharply, but then tend to plateau. Column 1 for example plateaus at about 500 ppm TOC.

The situation if further complicated by the peaks in influent concentration at 18 and 26 hours. Since these peaks are based on one analytical result it is difficult to draw the likely curve over that time period.

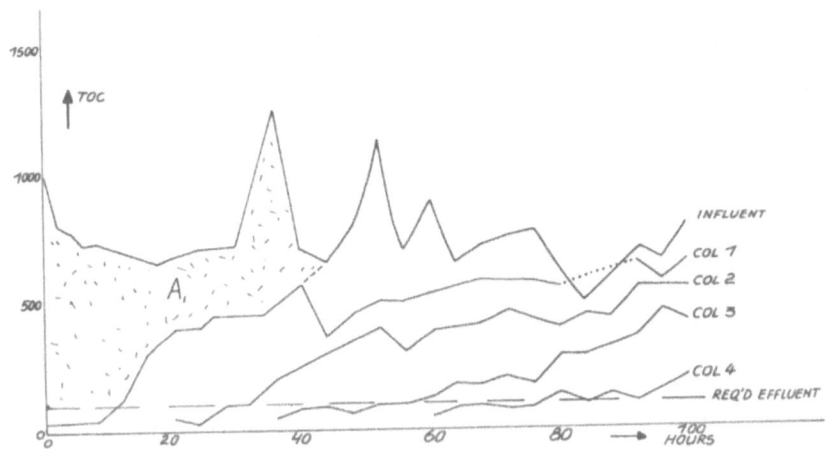

Fig. 9 Column test data
       Flow 100 l/hr, Carbon CAL
       Temp. ± 18°C, Carbon weight per col. = 10 Kg
       Contact time per col. = 15 min.

     It was felt that the plateau effect denoted a re-
lease of weakly adsorbed organic materials, and replacement by more strongly adsorbed materials.In the long
term these weakly adsorbed materials will need to be
removed from the system in the adsorbed state on carbon.
One cannot therefore calculate the capacity ignoring
this type of effect. Thus judgement should be exerci-
sed when dealing with this type of column data. In this
case it was decided not to calculate the capacity, by
measuring the total area between influent and column 1
effluent, but to measure the area A only. A corresponded to a pick up of

   (45 AREA) x (100 mg/l COD) x (4 HOURS) x (100 l/hr) =

       1,800,000 mg TOC.

Column 1 contained 10 Kg of carbon. Hence the maximum
loading was estimated to be 18% by weight of TOC. This
was crosschecked by reference to column 2 loading and
a figure of 15% was obtained for column 2 loading.
Since column 2 was not quite saturated at the end of
the test run it was decided that 18% was a good loading
capacity to use in design calculations as the maximum.

The measurement of areas at breakthrough gave the operating line shown in Figure 10.

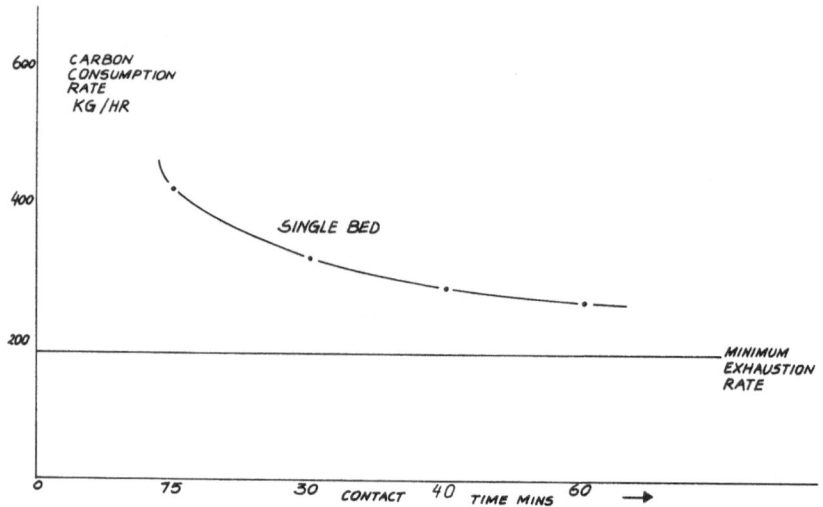

Fig. 10 Operating line

The carbon exhaustion rate has been calculated for a flow of 50 m3/hr at an average influent of 750 mg/l TOC and an average effluent of 80 TOC, thereby giving a required removal of 33.5 Kg/hr of TOC.

It should be noted that because of the fluctuations of concentrations and nature of the impurities in most waste streams safety factors should be applied if a severe control or guarantee must be given on system performance.

4. PLANT DESIGN

When an operating line is produced the designer can choose the type of system he will employ in practice. It is rare that one adsorber alone is used for industrial waste water treatment, and Figure 11 shows a choice of systems that can be considered.

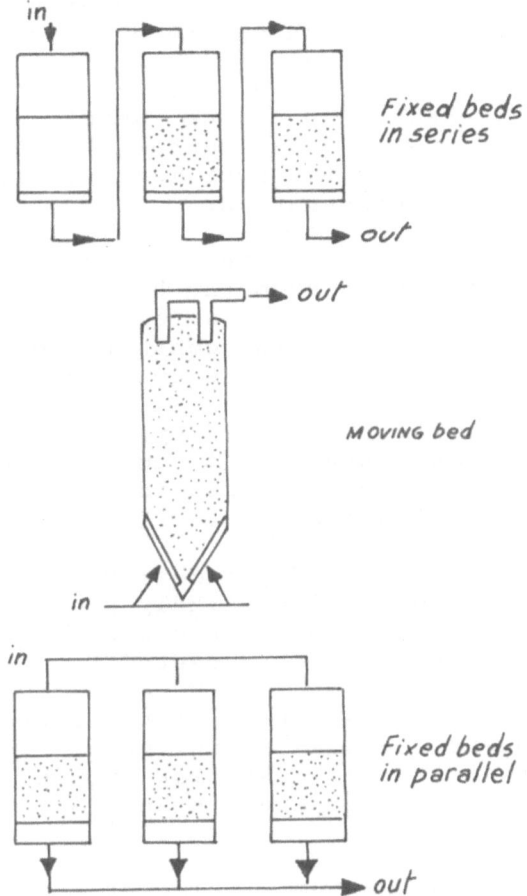

Fig. 11 Adsorber configuration for
granular carbon treatment

A number of beds may be operated in series, so that the
first bed is completely saturated while the second or
subsequent beds still produce effluent of suitable quality. The first column is then taken out of use, the
carbon replaced and put back in line as the last column.
This system will give low operating cost at the expense
of a high capital cost.

A number of beds may be operated in parallel, using the fact that for a major part of the service time of a bed, it is producing effluent better than the desired quality, which can be blended with effluent which is worse than the desired quality to provide a mean effluent of the desired quality. In other words one staggers the start-up of the vessels at equal intervals and can run a vessel longer than the true service time. A lower capital cost is given with this type of adsorber configuration.

A pulse bed system may be used, where some carbon is removed at intervals from the bottom of the column and replaced at the top by fresh carbon, enabling one to design near to the minimum contact time and for full saturation of the carbon before discharge.

For each of these possible system types an operating line may be constructed. A number of points on the operating line may then be taken for evaluation of system capital and operating cost.

In fact this phase of the design must include a judgement of factors such as likely government grants for pollution control installations and the cost of money to the purchaser of the system.

Ideally a full evaluation of the costs brought down to a common basis, such as annual cost, should be made. An example of discounted cash flow applied to the choice of system designs is given in Ref. 3.

Generally, the system with lowest capital cost is fixed beds in parallel, and as such this configuration is often optimum for low concentration waste streams where the higher operating cost implicit from the discharge of partly spent carbon is insignificant.

The pulse bed system is normally the lowest operating cost configuration, so that this system finds best application in high concentration waste waters where the operating cost is a significant part of the annualized system cost.

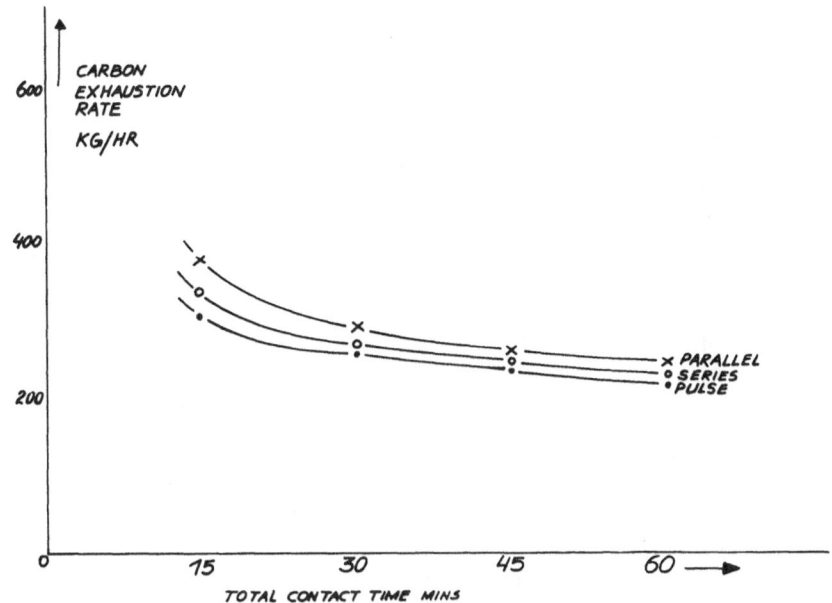

Fig. 12 Operating lines (estimated)

Figure 12 shows operating lines for 2 beds in parallel, 2 beds in series, and two pulse beds. These curves have been constructed using the data given in Figures 9 and 10. The systems postulated assume a holdup reservoir for the waste water during changeover of carbon in the fixed beds or slugging in the moving bed.

Because of the magnitude of the carbon dosage rate a regeneration furnace is indicated for this plant.

Spot capital and operating cost estimates showed that the pulse bed solution was the cheapest in annualized cost at all contact times considered for this case.

The moving bed optimum contact time was then evaluated as shown in Table II.

T A B L E   II

Annualized Costs of Moving Beds + Furnace Loop at various Contact Times

Case	1	2	3	4	5
Contact Time in Minutes	15	30	45	60	90
Exhaustion Rate in Kg/hr	310	249	238	218	210
Capital Cost – Pounds Sterling x 1000	243	232	200	230	270
Operating Cost – $*$ Pounds Sterling x 1000/yr	125	105	101	95	91
Total Annualized $*$ – Cost at 10% Annual Value of Capital – Pounds Sterling x 1000	163	141	132	131	133

$*$ Note : 1. No tax benefit due to depreciation is assumed.

2. Subsidies on Capital or Operating Cost, accelerated depreciation allowances, etc. may radically change these figures.

It can be seen in the cases 3 and 4, at the assumed straight
10% annual value of capital,that there is little to
choose between each solution. In this case it was felt
that the rather variable influent quality would dictate an overdesign, and in fact a further column test
showed that at 90 minutes contact time only a slightly
higher annual cost resulted. Hence a design figure of 80
minutes contact time was chosen, resulting in the choice
of 2 moving bed adsorbers of 35 m^3 useful volume each
being chosen.

An average carbon consumption of 200 Kg/hr was assumed.
In fact the reactivation loop was overdesigned in order
to permit containment of sustained influent levels up
to 1000 ppm, and to guard against the possibility of an
incomplete recovery of the carbon activity.

5. PULSE BED SYSTEM DESCRIPTION (Figure 13)

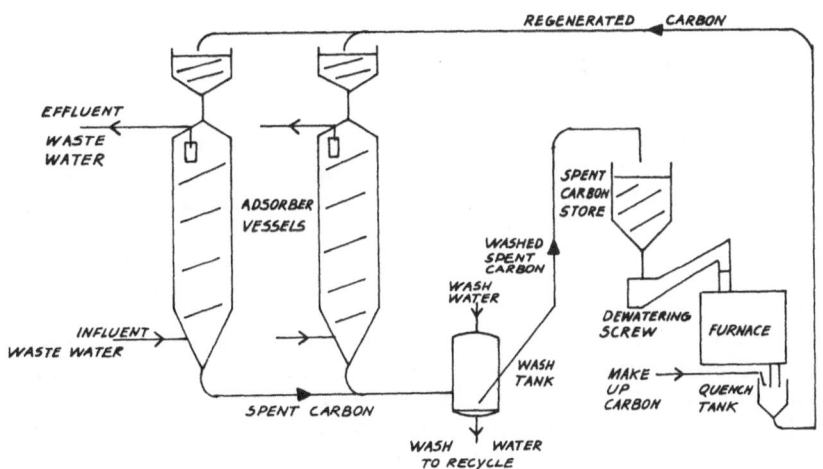

Fig. 13 Pulse bed carbon system

In a pulse bed liquids upflow through the adsorber vessels. Periodically about 10% of the adsorber volume of spent carbon is discharged from the conical bottom of these vessels. Simultaneously fresh carbon is allowed to enter at the top of the vessels from charge tanks. The carbon thus pulses down the vessel in a counter-current fashion to the liquid flow. The vessel design is such as to allow the carbon to move in plug flow. The vessels must always be maintained full, and tightly packed with carbon. Depending on the influent suspended solids level, a prefiltration may be necessary. The rejected carbon is washed in certain circumstances (for example to remove reactivation inhibiting inorganic salts) before being passed to a storage tank. The spent carbon is metered from the storage tank to a regeneration furnace. The regeneration furnace may be a multi-hearth furnace, or rotary kiln, and a new development is the fluidized bed regeneration furnace. The regenerated carbon is quenched with water and slurried back to the carbon charge tank on top of the adsorber columns. There is a necessity to add a percentage of fresh carbon to make up the losses due to attrition in the system and burning in the furnace.

6. OPERATING INSTALLATIONS

a) A plant installed in Switzerland treats a red coloured waste water arising from the production of explosives. Ref. 4. The waste water contains a number of nitrated aromatic hydrocarbons such as Dinitrotoluene and Trinitobenzoic acid. Four fixed bed columns of 1.3 metre diameter and 5.6 metre straight side are operated in series to treat this waste water. Overall contact time is an average of 6 hours. COD is reduced from 2300 mg/l to less than 120 mg/l and toxic materials and colour are completely eliminated. The columns are operated upflow to readily allow disengagement of gas formed in decomposition of some organic materials. At this time the activated carbon is not regenerated, because of concern over the explosive nature of the contaminating organic materials.

b) Water reuse is practised at a carpet mill in the United States. Ref. 6. A limited sewer capacity, and general water shortage led to this design which provides recycle of 80% of the water from the dyeing and rinsing operations.

This 80% recycle is primarily rinsing water and pastel dye solutions. The deep colours are discharged directly to the sewer. The inlet water quality to the unit contains up to 200 mg/l of dye in solution, and a coarse prefiltration is provided for lint removal. Contact time in the column is in the order of 45 minutes giving column dimensions 3 metre diameter and 8.3 metre straight side for a flow of 80 m3/hr. The spent carbon is reactivated in a multiple hearth furnace installed on site. Depending on the exact mix of dyes passing to the system, the carbon consumption rate fluctuates around 500 Kg/day.

c) A herbicide manufacturing plant in the United States generates a waste water with phenols, chlorophenols and heavy alcohols with a COD of 3,600 mg/l. Ref. 5 and 6. The waste water is discharged to a municipal treatment plant. Tests showed that a concentration of phenolics higher than 1 mg/l could affect the operation of the treatment plant. Thus the system is designed to remove phenolics from an influent of 50 - 200 mg/l, to below 1 mg/l, at a flow of 23 m³/hr average. The system chosen was 2 columns in series, operated upflow, at a design contact time of 53 minutes per adsorber. The columns are constructed of wood, the most economical local construction material for the highly acid pH of the liquor. Reactivation is run on site with a multiple hearth furnace having 6 hearths and 135 cm internal diameter. The furnace can reactivate up to about 4,000 Kg/day. The dosage rate in fact averages about 2,500 mg/l.

424

REFERENCES

1. Clemens, M.M., Davies, R.A. & Kaempf, H.J.,
   Removal of Organic Material by Adsorption on
   Activated Carbon, Chemistry and Industry, Sept. 1,
   1973.

2. Erskine, D.B. & Schuliger, W.G., Chem. Engng. Prog.,
   67, 11, 41, 1971.

3. Grandjacques, B.L. & Waller, G., Design of Carbon
   Beds, in Activated Carbon in Water Treatment,
   Papers and Proceedings of Water Res. Ass. Conference
   Univ. Reading April 1973, Water Res. Ass., Marlow,
   Feb. 1974, 159.

4. Cahen, J.P., Traitement des Eaux résiduaires de
   Nitration avec Charbon Actif en grains Pittsburgh,
   La Technique de l'Eau et de l'Assainissement, 273.

5. Henshaw, T.B., Adsorption/Filtration Plant cuts
   Phenols from Effluent, Chem. Engng., May 31, 1971.

6. Rizzo, J.L., Adsorption/Filtration...A new unit
   process for the Treatment of Industrial Waste Waters,
   1970

THE ELIMINATION OF TOXIC ORGANIC COMPOUNDS FROM WATER BY
ABSORPTION ON ACTIVATED CARBON

A. van Vliet and M. van Zelm

Chemical Laboratory TNO, Rijswijk, The Netherlands

ABSTRACT. A procedure for the preparation of drinking water from
surface water, whether or not contaminated with chemical warfare
agents, has been developed using active carbon. The most suitable
type of active carbon has been chosen on account of a comparative
study of the adsorption of lindane on five types of powdered active carbon. A calculation procedure for the quantity of activated
carbon necessary to eliminate a known concentration of pollutant
is proposed.

1. INTRODUCTION

For the preparation of drinking water from surface water in emergencies the Chemical Laboratory of the National Defence Research
Organisation TNO has developed a reliable purification procedure,
to be carried out in the so-called Paterson installation in use
by the Netherlands Army (1). The procedure involves chlorination,
dechlorination, coagulation, filtration and post-chlorination,
successively. Thus drinking water of good quality is obtained,
even when the surface water is contaminated with harmful chemicals
like warfare agents, pesticides and heavy metals.

In the dechlorination/adsorption step activated carbon is
used to account for dissolved harmful chemicals that have not been
decomposed during the chlorination as well as for the excess active chlorine and for decomposition products from the chlorination
step.

A type of activated carbon different from the one mentioned
in reference (1) has been used here because of the results of an
investigation into the adsorptive properties of various carbons as

regards toxic organic compounds dissolved in water. In this paper it will be discussed how the most suitable type of carbon to eliminate a specific pollutant, can be selected from rate and capacity determinations. A calculation procedure as to the quantity of activated carbon necessary to eliminate a known concentration of pollutant is proposed.

2. EXPERIMENTAL

Lindane (γ-BHC) was chosen as the pollutant, as this compound is very stable and no decomposition effects (e.g. hydrolysis) will interfere the measurements. A solution of lindane in water was prepared by adding an excess of lindane to 25 litres of water and mixing with compressed air during several days. Before use the non-dissolved lindane was removed by filtration. The solution was stored in a glass container. The concentration of lindane in the solution thus obtained was about 5 mg/ℓ (maximum solubility 7.52 $\pm$ 0.04 mg/ℓ at 25°C) (2). Lower concentrations were obtained by dilution with water. The concentration of the lindane was determined gaschromatographically.*

Five types of activated carbon have been selected for the study, viz. SA-1, SX-1, FNW, TNW and W, all produced by Norit N.V. (Amsterdam, The Netherlands).

For every carbon the minimum quantity was determined necessary to decrease a given initial concentration (C_o) of lindane to a final concentration of 0.056 mg/ℓ, being the U.S. Public Health Service drinking water standard (3).

For the capacity determinations the adsorption time was fixed at 24 hours, since longer contact times proved not to affect the lindane concentration any further.

In a second series of experiments the effect of contact time upon the removal of lindane was determined. All experiments were carried out at room temperature.

*Apparatus   : Hewlet Packard, Model 5750
 Column   : glass, 1.80 m
 Liquid Phase   : 3.8% SE-30 on Diatoport S, 80-100 mesh
 Gas phase   : Nitrogen, 30 mℓ/min
 Purge Gas   : 95% Argon/5% methane, about 60 mℓ/min
 Detector   : ^{3}H electron capture detector
 Temperature   : Injection port: 215°C; column: 200°C; EC-
                   detector: 205°C
 Injection   : 1 $\mu\ell$
 Retention time : about 210 seconds
 Detection limit : 0.1 ng

3. RESULTS AND DISCUSSION

According to the method described by Cohen et al (4) the minimum quantity of carbon has been determined using the Freundlich equation:

$$\frac{X}{M} = K.C_f^{1/n} \tag{1}$$

In this formula
  X = the adsorbed quantity of lindane in mg/ℓ
  C_f = concentration in mg/ℓ
  K and 1/n = constants depending on the system
As $X = C_o - C_f$ equation (1) can be written:

$$\frac{C_o - C_f}{M} = K . C_f^{1/n} \tag{2}$$

When the Freundlich isotherm equation is applied to the removal of lindane to a concentration of 0.056 mg/ℓ, equation (2) can be transformed in:

$$C_o - 0.056 = a.M_{0.056} \tag{3}$$

In this formula $M_{0.056}$ is the quantity of carbon necessary to reach a final concentration of 0.056 mg/ℓ of lindane, and a is a constant.
  If the Freundlich equation is obeyed a plot of $M_{0.056}$ versus C_o will yield a straight line through the point $C_o = 0.056$.

  To test the validity of the Freundlich equation we used it in the form of equation (4).

$$\log \frac{C_o - C_f}{M} = \log K + 1/n \log C_f \tag{4}$$

A plot of $\log C_o-C_f/M$ versus C_f did not give a straight line. However, it is still possible to determine the value of $M_{0.056}$ by interpolation of the line of equation (4).

  The fact that the Freundlich isotherm equation is not obeyed implies that more experimental work is required to determine the isotherm reliably.

  In this respect it is interesting to note that also other isotherm equations could not fit the experiments properly. The Freundlich equation probably gives the best results for these adsorption experiments.

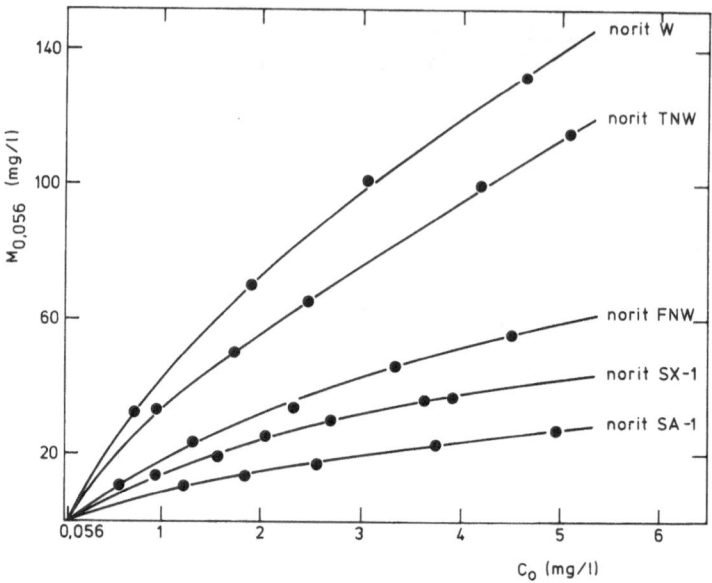

Fig. 1 Initial concentration of lindane in tap water versus minimum
quantity of activated carbon (5 types) to reach a final
concentration of 0.056 mg/l of lindane at a contact time
of 24 hours (equilibrium conditions).

Values of $M_{0.056}$ for several activated carbons and various
concentrations of lindane are given in Fig. 1, where C_o is plotted
versus $M_{0.056}$ for an adsorption time of 24 hours.

Although the relation between C_o and $M_{0.056}$ is not proportional, for the different carbons the ratio of the values of $M_{0.056}$
at one and the same C_o is within 5% constant over the whole range
of concentrations, viz.:
SA-1 : SX-1 : FNW : TNW : W = 1.0 : 1.6 : 2.2 : 4.1 : 5.1.
When also the different costs of the various carbons are taken
into account the ratio becomes:
SA-1 : FNW : TNW : SX-1 : W = 1.0 : 1.2 : 2.0 : 2.2 : 2.7.

However, the presented data refer to equilibrium conditions.
In practice shorter contact times will be applied and adsorption
velocities will play an important part then.

Figure 2 gives an impression of the relative adsorption velocities for the different types of activated carbon when the minimum quantity of carbon (determined from figure 1) is applied at a
concentration of about 4 mg/ℓ of lindane.

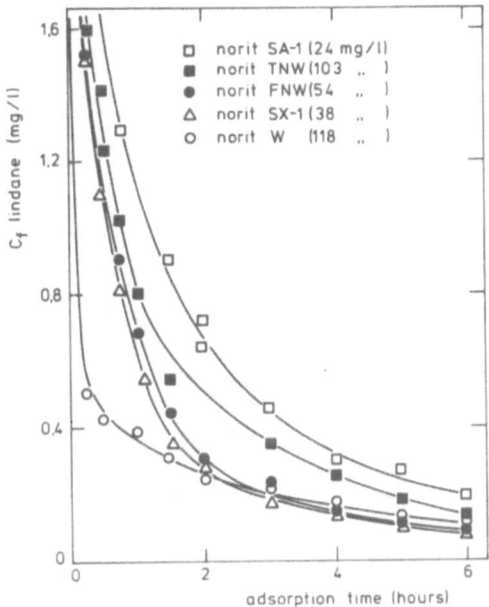

Fig. 2 Decrease of lindane concentration (C_o = 4 mg/l) with adsorption time for five different types of activated carbon. The quantity of carbon is the minimum quantity to reach a lindane concentration of 0.056 mg/l.

Figure 2 shows that SA-1, which has the highest adsorption capacity according to Fig. 1, has a low adsorption velocity. So the sequence of the suitability of the carbons as determined from equilibrium conditions may change. Therefore, values of $M_{0.056}$ for various initial concentrations and contact times were determined. The results for a contact time of 1 hour are shown in Fig. 3.

Just as in the equilibrium experiments there appeared to be a constant ratio (within 5%) between the amounts of each type of carbon at a given C_o of lindane over the whole concentration range. For 1 hour contact time and including the costs of activated carbons the ratio is:
SA-1 : FNW : SX-1 : W = 1.0 : 1.2 : 2.1 : 2.2.

Analogous experiments have been carried out for parathion in distilled water. For the removal to a final concentration of 0.1 mg/ℓ and a contact time of one hour (+ 30 minutes of flocculation prior to the filtration of the charcoal) we have found the following ratio (costs included):
SA-1 : FWN : TNW : SX-1 : W = 1.0 : 1.0 : 1.6 : 1.8 : 1.9.

430

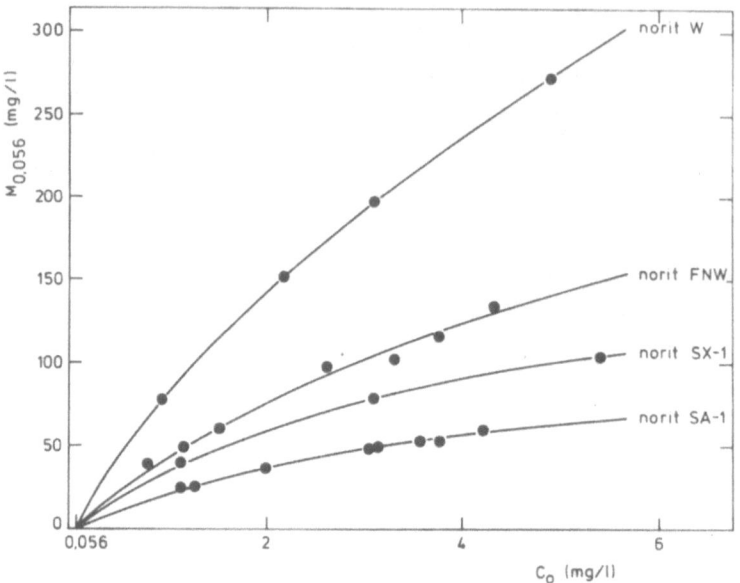

Fig. 3 Initial concentration of lindane in tap water versus quantity of activated carbon (4 types) necessary to reduce C_o to 0.056 mg/l within 1 hour.

In comparison to the lindane experiments some levelling has occurred, however, SA-1 activated carbon appears to be the best choice for both pollutants.

As the ratio of the amount of Norit SA-1 to reach a final concentration of 0.056 mg/ℓ of lindane for various contact times appears to be constant over the whole range of concentrations measured (see figure 4) the adsorption system lindane/Norit SA-1 can be described by means of:

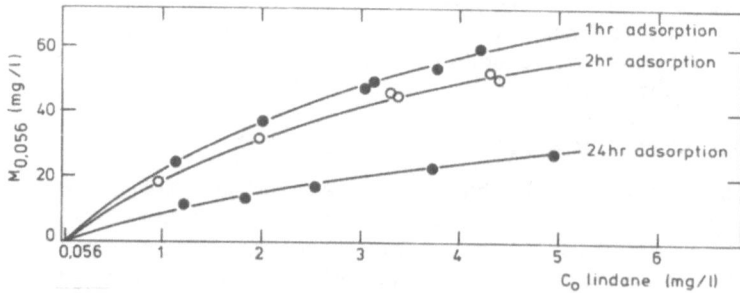

Fig. 4 Initial concentration of lindane versus quantity of Norit SA-1 to reach a concentration of 0.056 mg/l at different contact times.

- the relation between $M_{0.056}$ and the initial concentration of lindane under equilibrium conditions, and
- the relation between $M_{0.056}$ and the adsorption time at one concentration of lindane (cross-section of figure 4).

From these two relations a factor $M_{0.056}/M_{0.056\ minimum}$, depending on the adsorption time, can be determined. The amount of Norit SA-1 necessary for the elimination of lindane from a given initial concentration C_o to a final concentration of 0.056 mg/ℓ at a given adsorption time can now be calculated by multiplying $M_{0.056\ minimum}$ and the factor mentioned above.

To check the validity of this we establish the relation between $M_{0.056}$ and the adsorption time for lindane concentrations of 1 and 4 mg/ℓ with Norit SA-1 (see figure 5).

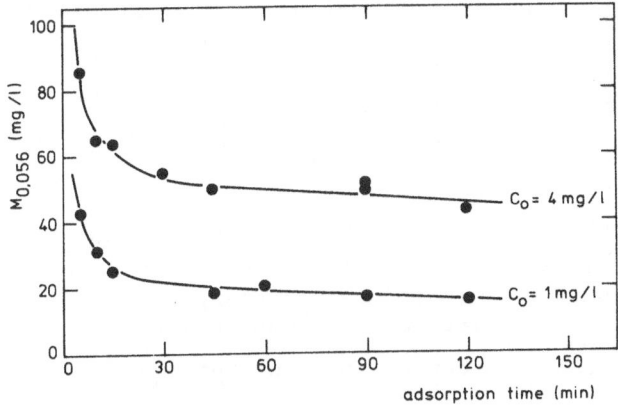

Fig. 5 The relation between the quantity of Norit SA-1 needed to reach a lindane concentration of 0.056 mg/l and the time.

Using this curve and the curve for SA-1 in figure 1 we calculated the factor $M_{0.056}/M_{0.056\ minimum}$ for different adsorption times. The results are given in table I.

Table I
The ratio between $M_{0.056}$ at an adsorption time t and at an adsorption time of 24 hours (0.056 minimum)

C_o lindane (mg/ℓ)	$M_{0.056}/M_{0.056\ minimum}$ for given adsorption times (min)							
	5	10	15	30	45	60	90	120
1	5.0	3,7	3,2	2.6	2.4	2.2	2.0	1.9
4	3.6	2.9	2.6	2.3	2.2	2.1	2.0	1.9

Table II
Comparison of the calculated and experimental values of
$M_{0.056}$ at various lindane concentrations and adsorption times

lindane concentration (mg/ℓ)	adsorption time (min)	$M_{0.056}$ (calc) (mg/ℓ)	$M_{0.056}$ (exp) (mg/ℓ)	deviation (%)
2.45	90	34.6	35.0	1.1
2.28	60	35.3	35.2	0.3
1.16	30	24.7	24.3	1.7
2.27	102	31.8	30.6	3.9
3.41	76	43.1	43.8	1.6
2.08	53	33.4	31.0	7.5
4.35	38	56.3	56.0	0.5

It is seen from Table I that the factor $M_{0.056}/M_{0.056 \text{ minimum}}$ depends not only on the adsorption time but also on the concentration of lindane. It is therefore necessary to determine the factor $M_{0.056}/M_{0.056 \text{ minimum}}$ for various concentrations of lindane, so that the factor needed can be found by interpolation or extrapolation.

In Table II the values of $M_{0.056}$ (calculated) and $M_{0.056}$ (experimental) are given for various values of the lindane concentration and of the adsorption time. $M_{0.056}$ was calculated from Fig. 5 and Table I.

The data of Table II show that the calculated values agree very well with the experimental values.

Analogous experiments have been carried out with parathion in distilled water. In this case the amount of carbon necessary to reach a final concentration of 0.1 mg/ℓ of parathion ($M_{0.1}$) has been determined. Just as in the case of lindane the value of $M_{0.1}$ has been calculated and compared with the experimental value of $M_{0.1}$. The results given in Table III allow to conclude that the calculation method applies also in this case.

Table III
Comparison of the calculated and experimental values of $M_{0.1}$
at various parathion concentrations and adsorption times

parathion concentration (mg/ℓ)	adsorption time (min)	$M_{0.1}$ (calc) (mg/ℓ)	$M_{0.1}$ (exp) (mg/ℓ)	deviation (%)
1.25	70	11.1	11.0	0.9
4.85	105	36.7	34.0	7.4
4.88	120	36.7	36.7	0
2.90	40	26.7	25.3	5.0
6.85	80	48.3	46.7	3.4

4. CONCLUSIONS

From a plot of the initial concentration (C_o) of a water pollutant versus the amount of activated carbon (M_{Cf}) necessary to eliminate this pollutant to a given final concentration (Cf), at a desired adsorption time, the most suitable activated carbon can be found by comparing the curves for the various carbons.

Once the values of M_{Cf} at 24 hours adsorption time and the relation between M_{Cf} and the adsorption time for a few concentrations of the pollutant are known, the value of M_{Cf} for all concentrations and adsorption times can be calculated within about 5% accuracy.

REFERENCES

1. van Vliet, A., The Military Engineer, no. 421, 319 (1972)
2. Masterton, W.L. and Tei Pei Lee, Env. Sci. & Techn. 6 (10), 919 (1972)
3. National Technical Advisory Committee Report, Raw-water Quality criteria for public Supplies, J. Am. Waterw. Ass. 61, 133 (1969)
4. Cohen, J.M., Kamphake, L.J., Lemke, A.E., Henderson,C. and Woodward, R.L., J. Am. Waterw. Ass. 52, 1551 (1960).

THE APPLICATION OF REVERSE OSMOSIS AND ULTRAFILTRATION TO THE PURIFICATION AND TREATMENT OF NATURAL WATERS AND EFFLUENTS

W.H. Hardwick

Process Technology Division, AERE Harwell, Didcot, Oxfordshire, England

ABSTRACT.   The several designs of reverse osmosis equipment that are now available commercially are reviewed.   These include the systems that use membranes in sheet form supported on porous substrates, membrane coated onto the exterior and interior sur- faces of porous tubes, and membrane formed as very fine hollow fibres which are self supporting.

   Results are given to illustrate the performance of commercial equipment and some account is given of operational problems encountered with feeds which range from saline water and sewage to industrial effluents.

   A similar review is given of ultrafiltration equipment and its application.

SYMBOLS

B	solute permeation constant
C_i	concentration of component i mass/unit volume
D_i	diffusion coefficient of component i
F_i	total flux of component i
J_i	diffusive flux of component i
k	mass transfer coefficient

P	pressure
Rg	hydraulic resistance due to gel layer
Rm	hydraulic resistance due to membrane
x	distance perpendicular to membrane surface
π	osmotic pressure

Subscripts

1	water
2	solute
s	value in solution

Superscripts

'	variable evaluated in solution at interface between concentrated solution and membrane
b	variable evaluated in the bulk of the feed solution.

1. REVERSE OSMOSIS PLANTS

1.1 Engineering Design of Reverse Osmosis Plants

In the first lecture, the basic requirements for a reverse osmosis plant were described. Several different engineering solutions to the problems of meeting those requirements are embodied in the plants offered by different manufacturers. The principal types are shown in Figures 1 and 2 and it is instructive to examine and compare them, particularly in the way in which the membrane which may be in sheet or tubular form is used.

An area of membrane must be provided which allows the specified product flow to pass through it at the design pressure. The greater the membrane permeation constant, $A(= \frac{F}{P})$, the smaller is the area required, or alternatively, the operating pressure may be lowered providing of course that it is always greater than the opposing osmotic pressure difference across the membrane. The membrane has to be supported against this applied pressure on a porous support, and there are obvious advantages in keeping the volume of equipment as small as is consistent with satisfactory

hydrodynamics. In other words, high membrane packing density is
desirable.

1.2 Sheet Membrane – Plate and Frame

The first and perhaps most obvious way to use sheet membrane
is in the plate and frame concept which is widely used for industrial filtration operations. Aerojet in the USA spent a great
deal of effort developing what turned out to be very unwieldy and
unsatisfactory equipment in which discs of membrane were bonded
to support plates with filter paper between to conduct the permeate water to grooves or slots on the plate surface which led the
water to a convenient collection point. The feed would flow from
the periphery towards the centre through the space between two
facing membrane discs to an axial collection point which conducted
the reject away. The plates were compressed together and sealed
at the edges with O-rings, the whole assembly being mounted inside
a pressure shell.

A much more elegant version of the plate and frame concept
was devised by the De Danske Sukkerfabrikker (DDS). By clever
design of the circular plastic membrane support discs the need
for a separate pressure containment vessel has been avoided.
The feed is introduced and contained between successive pairs of
discs which are compressed to form a pressure-tight seal at their
rims. The Aerojet system is defunct, but the DDS system is
commercially available.

Both of these plate-and-frame designs employ discs of
membrane sheet and both require the assembly to be completely
dismantled to replace membranes, although the DDS design does
permit the isolation of defective membrane pairs with only
marginal decrease in the output of a working unit.

1.3 Sheet Membrane – Spiral Wrap

The system was largely pioneered by Gulf General Atomics
under contract to the Office of Saline Water (OSW) in the USA,
and is undoubtedly one of the most successful. It is illustrated
in Figure 3 and consists in essence of a long narrow rectangular
envelope of membrane with the active surface outside and with a
porous backing sheet inserted in it. The open end of the envelope is sealed to an axial tube and water passing through the
membrane envelope will pass to the tube for discharge as
product. This water is derived from feed brine which flows
axially across the membrane surface. The envelope is rolled
in a spiral round the off-take tube including a plastic mesh
spacer which keeps the membrane surfaces apart and allows passage

for the feed to flow. The spiral assembly is inserted in a
cylindrical pressure vessel with feed inlet and concentrate outlet as illustrated in Figure 4, several units or modules being
coupled (axially) together to give the required total membrane
area for that pass. The standard module is four inches (100 mm)
in diameter and three feet (~1 m) long although eight inch
(200 mm) diameter and even twelve inch (300 mm) diameter modules
are mooted for large plant assemblies.

1.4 Tubular Systems - Internal Membrane

This is an obvious and attractive system in which the membrane is inserted in, or formed as the lining of, a strong porous
tube (Figure 2). Tubes ranging 2.5 to 50 mm in diameter have
been proposed or used but 12 mm appears to offer the best compromise between the advantages and disadvantages that are related
to tube diameter. In the pioneer plant at Coalinga, designed and
operated by UCLA, tubular membranes were cast, wrapped with nylon
cloth and inserted in perforated metal tubes which provided pressure resistance, while Havens developed a porous fibre-glass tube
in which the membrane was cast directly on to the inner surface.
Other manufacturers have braided glass or plastic fibres round a
pre-cast membrane; others have made paper tubes which, after
coating with internal membrane, are carried in perforated metal
tubes. Paterson Candy International Limited have developed a
method of simultaneously forming a paper tube and depositing an
internal membrane by a continuous process.

Most tubular module assemblies are made by sealing the tubes
into turn-round headers so that in effect a long tubular membrane
is folded up into a convenient length for handling. Such a module
unit may contain 18 × ½ inch (12 mm) tubes and measure 2.5 or 4 m
in length.

1.5 Tubular System with External Membrane

The alternative use of tubes is in compressive stress where
the membrane is deposited as an external coating. For example a
perforated stainless steel tube can be wrapped with a water
permeable sleeve and a membrane then coated on this. Such a
design was used by the French in a successful, if expensive,
plant for producing potable water from sea water on the Isle of
Houat. A simpler approach has been developed at Harwell. A
grooved flexible polypropylene rod, approximately ⅛ inch (3 mm)
diameter, is braided with terylene and a membrane coated on the
braid by a continuous process; the membraned element which looks
like spaghetti is cut into lengths, sealed with a plastic cap at
one end, the other end being inserted in a "header" which may take

37 or 109 elements. The bundle is introduced into a cylindrical pressure vessel through which the brine feed is pumped. Product water passing through the membrane is conducted to and along the grooves in the rod for collection at the header. The element and assembly are shown in Figures 5 and 6.

1.6 Hollow Fibre Membrane

If a tubular membrane with an external active layer is made of sufficiently small diameter and adequate wall thickness it will be able to withstand the pressures necessary to effect reverse osmosis without collapse. With the sort of polymers that can be made into membranes the tubes become fibres of 80 μ external and 40 μ internal diameter. Although the individual fibres have a tiny surface area, so many fibres can be packed into a module that the flux per unit volume is very high for such devices even when the membrane permeation constant is low compared with cellulose acetate sheet membrane. Two companies in particular are associated with this approach. Dow, who have worked for many years on cellulose triacetate hollow fibres, and DuPont, who have achieved considerable commercial success with their nylon hollow fibres (Figure 7). Flow is in a radial direction across the fibres from the feed supply dispenser which is an axial tube.

2.   OPERATIONAL FACTORS

2.1 Concentration Polarisation

The designs described above fall into two broad categories in their approach to minimising the effects of concentration polarisation. While the polarisation effect cannot be completely eliminated, the effects can be significantly reduced:

(a) by operating at low water fluxes with very narrow feed channels in order to reduce the concentration gradient that develops as product is removed, or

(b) by continuous mixing of the feed by turbulence.

The following treatment is taken from H.K. Lonsdale's chapter in "Industrial Processing with Membranes".[2]

The mass transport equation expressing the steady-state condition within the polarization region at any point along the feed flow channel is

$$\frac{J_1 c_{2s}}{c_{1s}} - D_{2s}\frac{dc_{2s}}{dx_s} = B\Delta c_{2s} \qquad (1)$$

where D_{2s} = the solute diffusion coefficient in the solution

x_s = the distance parameter in solution perpendicular to the membrane surface.

The first term is the rate of solute transport to the membrane surface resulting from bulk flow toward the membrane; the second term represents the diffusive flux back into the bulk of the solution, assumed to obey Fick's law; the right-hand side expresses the flow of solute through the membrane. When the membrane is solute-impermeable, the solute permeation constant, B, = 0 and the equation can be integrated to give

$$\frac{c'_{2s}}{c_{2s}^b} = \exp\frac{J_1 \Delta x_2}{c_{1s} D_{2s}} \qquad (2)$$

where c_{2s}^b = the "bulk" solute concentration far from the membrane surface

c'_{2s} = the interfacial concentration

Δx_2 = a characteristic polarization distance in the solution.

In the laminar flow regime the polarization distance increases with distance down the feed channel (or with time in the absence of feed flow), until eventually the polarization region extends over the entire feed channel height. When the flow is turbulent, most of the feed channel is well mixed and, according to the film theory model of heat and mass transfer, the polarization region extends only over a boundary layer adjacent to the membrane surface, the thickness of which is determined by the linear velocity, channel geometry, and solution properties.

The solution to equation (1) is made difficult by the fact that water flux through the membrane decreases down the channel because π steadily increases as water is removed and the boundary layer is developed. In addition P will be reduced by friction, and solute flux increases because of increased salt concentration. Assuming a membrane flux of 100 US gal/d/ft^2 at the operating pressure and a 50% removal of water from the inlet brine, Brian[3] has calculated the required inlet brine velocity as a function of tube diameter with a polarisation ratio of 2 and computed the frictional pressure drop involved. The results are given in Table 1. They illustrate how a laminar flow approach tends to result in a multiplicity of short small diameter tubes, while a turbulent flow plant tends to be long and relatively thin because of the high feed velocities required. Thus the hollow fibre DuPont system with the fibres in close proximity operates virtually in the laminar flow region. Table 1 illustrates how it is possible to achieve 50% water recovery in very small laminar plants, but for a one inch diameter tube plant under these conditions the minimum size would be $10^6/30 = 300,000$ gpd. Lower membrane fluxes which prevail in practice reduce this minimum and there are other devices such as 'stepping' which help but the fact that it is not possible to make small plants to produce only a few hundreds of gallons of water a day at high product recovery with turbulent flow geometry without resource to recycle. Incidentally economic plant design demands a polarisation ratio of 1.2 rather than 2.

The Table also illustrates the importance of frictional pressure drop which can be very low in laminar flow and this is a particular advantage (see also Ref. 2 p.165). The requirement of laminar flow for very narrow and well defined passages implies high membrane packing density but such passages are very easily blocked and fouled which means that feed must be very well filtered.

The 'Spiral Wrap' system is not truly laminar. The passages are narrow but are formed by the membrane surfaces closing on a mesh which must act as a turbulence promoter. The pressure drop through the equipment is low for the fairly long path lengths – perhaps three successive 4" diameter 20 ft long pressure vessels. On the other hand the open tubular systems are subject to much higher pressure losses much of which occurs on the turn-round at the ends of modules. The tubes are open and accessible to cleaning by the passage of a foam ball swab and are immediately attractive for feeds carrying a high burden of solids. The "R" or 'Spaghetti' system is a turbulent flow system and not so susceptible to fouling.

In practice the performance of equipment is not readily deduced from theoretical analysis. There are several computer programmes which optimise performance in terms of the several design variables but practical and economic constraints usually reduce the design of plant to the most effective assembly of standard modules containing the best membrane available with adequate power and control features. Thus the tapering of the cross section of a turbulent flow plant to maintain a constant or optimum feed velocity across the membrane surface is ideally

continuous but in practice is a stepped process where the flow
from several modules in parallel is led to a smaller number also
in parallel. Allowance has to be made for the change of membrane
characteristics with time and this is probably best achieved by
steadily increased feed pressure. However, variation of membrane
properties to suit the change of conditions along an operating
plant is not usually practicable. For many practical situations
temperature variations will have an appreciable effect on plant
performance.

There are fairly simple relationships which for example
enable the desalination factor averaged over a plant to be
related to the membrane desalination factor, but there is little
opportunity or need for very sophisticated design techniques.

2.2 Performance

As far as the desalination of brackish water is concerned
there is little to choose between the different systems in terms
of technical performance. The choice will depend upon factors
such as cost and reliability and so if I describe the sort of
results we have observed with the 'Spaghetti' system, we may
take these as indicative of what the modern reverse osmosis plant
can achieve.

In Table 2 results are given for field trials on cellulose
membranes on high chloride-containing water. This water was
present in a disused gravel pit near the sea. The high rejection
of divalent ions is observed with both membranes, while the monovalent ion rejection is much lower. Not all inorganics are
rejected as well as those listed here. Boric acid is among the
more permeable together with ammonia, fluoride and cyanide. The
second illustration of salt rejection is given in Table 3. This
shows how sulphated waters can be concentrated well beyond the
saturation limit of $CaSO_4$ by small (ppm) additions of polyphosphate. In this instance the solubility product of $CaSO_4$ has been
exceeded by a factor of 5.

2.3 Organic Pollutants - Sewage Effluent

We may now consider the treatment of water polluted by
organic as well as inorganic material. As already stated a
desalting membrane is a most effective barrier to all organic
molecules other than certain low MW species, particularly alcohols and phenols, and therefore reverse osmosis should, and
does, produce water of very high quality from polluted sources.
The principal difficulty is that the larger organic molecules
and colloidal material present in a feed is concentrated at the

membrane surface as product water is removed and this surface film inhibits the water flow and reduces performance as will be discussed under Ultrafiltration. The film, which in the treatment of such feeds as dirty river waters and sewage effluents, can take the form of an adherent slime and can usually be removed by physical or chemical means. Thus for open tubular systems, foam ball cleansing can be used, but flushing with an enzyme-containing detergent solution - "BIZ" - is very effective for several systems. To illustrate this the graph in Figure 8 shows the decline and restoration of flux that occurred during one of many tests with "Spaghetti" modules with secondary sewage effluent, that is, settled effluent from a percolating filter.

There are several instances in the literature reported from the USA of the beneficial effects of treating secondary sewage effluents with activated carbon before feeding to a 'Spiral Wrap' RO plant. It is interesting that Table 4 shows that there was no apparent improvement in the performance of reverse osmosis modules when treating effluent from Rye Meads sewage works which had been given activated carbon treatment. (Bailey[4]). Whether the difference of experience stems from different sewage, active carbon, or module design is not clear.

The rejection of sewage effluent constituents is shown in Table 5 where effluent treated by reverse osmosis is seen to compare very well with the original tap water from which it was derived. The reverse osmosis treatment should remove bacteria and virus but it would be prudent to chlorinate such a product before putting it into public supply. Its low salinity makes it an excellent feed to an ion exchange polisher, to give highly pure water for high pressure boiler feed or for the electronics industry - a quality not achievable by ion exchange alone.

2.4 Industrial Wastes

In Table 6 a list of applications and possible treatments is given with the status as of some months ago. The treatment of secondary sewage effluent has been described above.

2.5 Pulp and Paper Mill Effluent

These have been processed in a tubular reverse osmosis pilot plant and a description is given by Wiley et al.[5]. Because of the fibres and suspended solids an open geometry is essential and a 'tight' membrane is needed to reduce the BOD and COD to acceptable limits. The problems encountered were:

(a) Membrane scaling due to $CaSO_4$ deposition which could be alleviated by polyphosphate addition.

(b) Organic fouling which was removed or reduced by detergent washing and bactericide addition and increased feed velocities.

(c) Permeation of low MW acids (contributors to BOD) which was reduced by neutralisation to give salts.

Fluxes were about 7.5 US gallons/day/ft^2, (3 m^3/m^2/d).

2.6 Acid Mine Drainage Waters

The water from disused mine works can contain sulphuric acid and iron which is converted to ferric hydroxide in rivers with most unpleasant polluting effects. This problem has received a lot of attention in the US where it has been shown to be possible to concentrate the iron sulphate by reverse osmosis and release the bulk of the water in clean condition.[6].

2.7 Effluent from Metal Finishing and Plating

The wastes arise from the rinsing of plated articles after removal from the plating tank and while the concentration of metal in the plating tank may be 0.5 to 3%, the waste stream concentration usually varies from 25 to 1000 mg/l. The flow rate of most rinse streams is from 400 to 2000 gph (2-9 m^3). Traditionally, wastes are treated chemically by precipitating metal hydroxides which are disposed of as sludges. But there are advantages in using reverse osmosis which can recover the metal from the rinse baths for re-use, or treat the effluent from a flocculation process to give water suitable for recycle.

2.8 Food Processing

The concentration of wastes from the Food Industry to give a water effluent of high quality for recycle, while at the same time producing a valuable concentrate, is an increasingly attractive proposition. In some circumstances it is necessary to use a tight membrane (reverse osmosis) rather than ultra-filtration, and the treatment of whey illustrates this. Whey is the by-product of cheese making, and since it contains 5% lactose it has a very high BOD, and is a serious pollutant when discharged to water courses. The protein content can be largely recovered by ultrafiltration although a percentage of the lactose is also retained. To remove the rest of the lactose from the permeate, reverse osmosis must be used.

3.   ULTRAFILTRATION

The use of membranes that impede the passage of large molecules but not dissolved salts is practised particularly for the
recovery of valuable substances from dilute wastes.   Production
of clean water for recycle or discharge is usually a secondary
consideration but it is likely to become increasingly important
as water scarcity and concern for the Environment grows.

3.1 Equipment

The equipment developed for reverse osmosis with the exception of the hollow fibre type is useful for ultrafiltration,
particularly tubular geometry.   In addition there are designs
specifically for ultrafiltration such as that by Dorr-Oliver
where membrane sheet is mounted on plastic supports in a compact
geometry.   This equipment does not stand reverse osmosis pressures and such strength is a redundant feature in reverse osmosis
equipment used for ultrafiltration but is offset by the savings
which result from standardisation.

3.2 Operation

The principal difference between reverse osmosis, as exemplified by the desalination of clean water, and ultrafiltration is
that membrane fouling and concentration polarisation effects are
much more severe in ultrafiltration.   The use of asymmetric
membranes is advantageous in that the tendency to block pores as
in "straining" is much reduced, but even so, the rejected species
must tend to concentrate at the membrane interface.   It is argued
that with the low velocities used in laminar flow, the concentrated species are much less likely to be impacted into the
membrane surface as may happen under turbulent flow, and fouling
rates are significantly lower.   But the situation in ultrafiltration is that a 'gel' layer accumulates at the membrane surface
which in many cases offers a substantial hydraulic resistance in
addition to that presented by the membrane itself.   (Porter[7]).

The solvent flux is given by

$$F_1 = \frac{\Delta P}{Rg + Rm}$$

where F_1 = the transmembrane solvent flux

ΔP = the transmembrane pressure drop

Rm = the hydraulic resistance due to membrane

Rg = the hydraulic resistance due to gel layer.

This expression explains why F_1 often tends to a constant value in spite of increasing ΔP. F_1' will increase with pressure until the concentration at the membrane surface reaches some critical solute concentration (typically between 20 - 70 volume percentage) which represents a close packing of colloidal particles or the point of gel formation. This layer must then increase in thickness, thus increasing Rg to values that can be much greater than Rm until the convective transport of solvents F_1C_b towards the membrane is reduced by virtue of the increased flow resistance to a value just equal to the diffusive back transport of solute away from the membrane into the bulk solution.

$$F_1C_b = D \frac{dc}{dx} = 0$$

C_b = Average bulk concentration of solute

$\frac{dc}{dx}$ = the concentration gradient of solute with distance x

D = the diffusion coefficient of the solute.

Further increase of ΔP causes a thickening of the gel layer which is balanced by the increase in Rg.

For the "gel-polarised" region of Figure 9 (Ref. 7) the equation may be integrated to yield

$$F_1 = k \ln C_g/C_b$$

where k is a mass transfer coefficient and C_g is the solute gel concentration.

Decreasing C_b or increasing k, for example by stirring or turbulence, will increase diffusive back transport and, in turn, the flux through the membrane. By using thin channels (25 - 75×10^{-2} mm in depth) high levels of hydrodynamic shear are created at the membrane surface which cause an increase in k without excessive energy consumption and Amicon equipment is designed to achieve this objective. In tubular geometry the resistance of the gel layer is reduced by the turbulence resulting from high velocity flow, ie Reynolds Numbers of the order of 10,000.

3.3 Applications

3.3.1 Sewage. The treatment of sewage effluent by ultrafiltration is a good example of the usefulness of the process with limited objectives. An activated sludge tank is coupled to an ultrafiltration unit to enable a proportion of the liquor to be passed continuously over the membrane. The suspended solids and a large percentage of BOD and COD are removed allowing the permeate to be recycled for toilet use. A periodic purge is needed to reduce the salt build-up in the cycle.

3.3.2 Electrophoretic Paint Residues. This was perhaps the first and most important application of ultrafiltration. The effluent from the rinse bath which formerly represented a serious problem both as pollution and waste of valuable paint material 'dragged out' of the paint bath is now treated by ultrafiltration. The filtrate is used as rinse spray and the paint in the concentrate returned to the plating bath. (Fig. 10).

3.3.3 Others. There are many other potential applications of ultrafiltration in the treatment of industrial wastes, particularly in the food processing industries for the recovery of protein and starches, but it is difficult to say to what extent the applications described in the literature are actually practised industrially. The technology is well established and there is now a wealth of experience. The degree of adoption of the process must depend on having, firstly, the right membrane and, secondly, equipment which satisfies both technical and sanitary requirements and is cheaper or more efficient than competing processes.

REFERENCES

1. Treille, P. and Rovel, J.M., Sea Water Desalination by Reverse Osmosis. Recent Experience, in Proceedings of the 4th International Symposium on fresh water from the sea, Vol 4, Heidelberg, September 1973.

448

2. Lonsdale, H.K., Theory and Practice of Reverse Osmosis and Ultrafiltration, in <u>Industrial Processing with Membranes</u>, Lacey, R.E. and Loeb, S., Eds, Wiley–Interscience, New York, 1972.

3. Brian, P.L.T., Mass Transport in Reverse Osmosis in <u>Desalination by Reverse Osmosis</u>, Merten, U., Ed., MIT Press, Cambridge, Mass., 1966.

4. Bailey, D.A. et al, <u>The Reclamation of Water from Sewage Effluents by Reverse Osmosis</u>, Water Pollution Research Laboratory, Department of the Environment, UK, 1973.

5. Wiley, A.J., Ammerlaan, A.C.F. and Dubey, G.A., Applications for Reverse Osmosis in the Pulp and Paper Industry, In <u>Industrial Processing with Membranes</u>, Lacey, R.E. and Loeb, S., Eds, Wiley–Interscience, New York, 1972.

6. Gulf Environmental Systems Company, <u>Acid Mine Treatment using Reverse Osmosis</u>, EPA Water Pollution Research Series, Programme Number 14010 DYG 08/71.

7. Porter, M.C., Ultrafiltration in Food and Beverage Processing, presented at <u>Symposium on Less Common Methods of Separation</u>, held at the University of Birmingham, UK, April 1971.

TABLE 1

Operating Conditions for Polarisation Ratio = 2

with Flux of 100 US gpd/ft^2

Tube diameter (in)	Required $\bar{u}^i$ (ft/sec)		Frictional pressure drop (psi)	Approx. length	Approx. number of tubes for 10^6 gpd
0.001	Any value	(0.18 (1.8	1 100	0.15" 1.5"	3×10^9 3×10^8
0.0025	Any value	(0.18 (0.8	0.4 40	0.36" 3.6"	5×10^8 5×10^7
0.01	50		19,700	33'	120,000
0.1	9		84	61'	6,000
1	19		385	1,200'	30
10	31		893	21,000'	$^1/5$

$\bar{u}^i$ = average feed velocity at channel inlet.

TABLE 2

Field Trials Performance of C.A. Membranes on High

Chloride Water

	Undosed Feed	Product Module I	Product Module II
pH	8.3	5.1	5.0
Electrical conductivity µmho at 20°C	8500	130	325
TDS mg/l	5750	70	175
Carbonate hardness as $CaCO_3$	230	0	
Non carbonate hardness as $CaCO_3$	935	4.6	
Total hardness as $CaCO_3$	1165	4.6	
Alkalinity as $CaCO_3$	230	0	
Ca mg/l	120	0.5	0.6
Mg mg/l	220	0.6	0.9
Na mg/l	1690	24	62
SO_4 mg/l	460	1	1
Cl mg/l	3000	38	81
NO_3 mg/l	3	0	<0.5
Silica mg/l	6	<0.1	
Conductivity D.F. (main due to NaCl)		65	25
DF (for divalent ions)		250–500	200–500

TABLE 3

Analyses of Feed, Product and

Concentrate from Recycle Operation

on High Sulphate Water at 65% Recovery

	Feed	Product	Concentrate
pH	6.7	6.15	6.95
Electrical conductivity μmho	2150	249	4600
TDS mg/l	2533	120	6131
Carbonate hardness as $CaCO_3$	271	30.6	686
Non carbonate hardness as $CaCO_3$	1282	49	3650
Ca mg/l	562	30	1600
Mg mg/l	37.6	1.2	85
Na mg/l	15.2	2.5	50.7
SO_4 mg/l	1251	43.6	3479
HCO_3 mg/l	332	35.4	845
Cl mg/l	35.5	7.8	71
K_{CaSO_4} $K_{CaSO_4(sat)} = 30 \times 10^{-5}$	18.3×10^{-5}		145×10^{-5}

TABLE 4

Influence of Pretreatment with Activated Carbon on the Quality of Reverse-Osmosis Permeates

(average performance over 30 weeks)

Sample	BOD	COD	Org. C	NO$_2^-$	Total oxidized N	Soluble phosphate (as P)	Anionic detergent	Total solids	pH value	Conductivity (μmho/cm)	Colour (Hazen units)	Turbidity (ATU)
					(mg/l)							
Feed: Sand filtered eff.	3	32.9	9.9	0.08	29.4	9.3	0.6	754	7.3*	1252	24.4	5.6
R.O. product	1	9.8	2.7	0.01	4.1	0.3	0.2	55	6.7	83	0	0
Feed: Activated carbon eff.	2	12.5	4.4	0.06	30.5	9.5	0.3	720	7.4	1124	1.8	1.7
Activated carbon + R.O. treatment	1	8.3	2.7	0.01	9.1	0.33	0.2	77	6.5	121	0	0

* Reduced to pH 6.1 with sulphuric acid

TABLE 5

Comparison of Composition of Sewage Works Effluent

with that of Reverse Osmosis Product and Original

Tap Water

	Filtered sewage effluent	Reverse osmosis product	Tap water
Gross properties (mg/1 except where stated)			
Total solids	660	58	390
Suspended solids	6	0	0
Colour (Hazen units)	22	0	0
Turbidity (A.T.U.)	9.4	3.6	0
Conductivity (μmhos/cm)	990	81	620
Inorganic constituents			
Ammonia (as N)	<0.4	<0.4	0
Nitrate (as N)	31.4	6.0	5.0
Nitrite (as N)	0.03	0.02	0
Chloride	75.3	5.3	17.0
Sulphate	66.8	2.0	14.8
Total phosphorus	2.75	0.07	0.03
Total hardness (as $CaCO_3$)	262	13	280
Organic constituents			
Permanganate value	4	0.6	<0.3
COD	52	~1	~3
Organic carbon	8.5	<0.5	<0.5
Anionic detergent (as Manoxol OT)	0.9	~0	0

454

TABLE 6

Application of Membrane Processes in Waste Treatment

Waste	Processes considered	Current status
Pulp and paper mill effluents	ED, TD, RO, UF	RO 200 m^3 per day pilot (5)
Sewage	RO, UF	UF 120 m^3 per day production (6) RO 70 m^3 per day pilot (7)
Rinses from electro-phoretic painting	UF	UF production (8)
Effluent from metal-finishing and plating	ED, RO, UF	ED laboratory tests (9) UF laboratory tests (10) RO 1000 m^3 per day (11)
Acid mine drainage waters		RO 50 m^3 per day pilot (12)
Whey treatment	ED, TD RO+UF	ED, TD 10 m^3 per day production (13) RO+UF 10 m^3 per day production (14)

ED   Electrodialysis
TD   Transport depletion
RO   Reverse osmosis
UF   Ultrafiltration

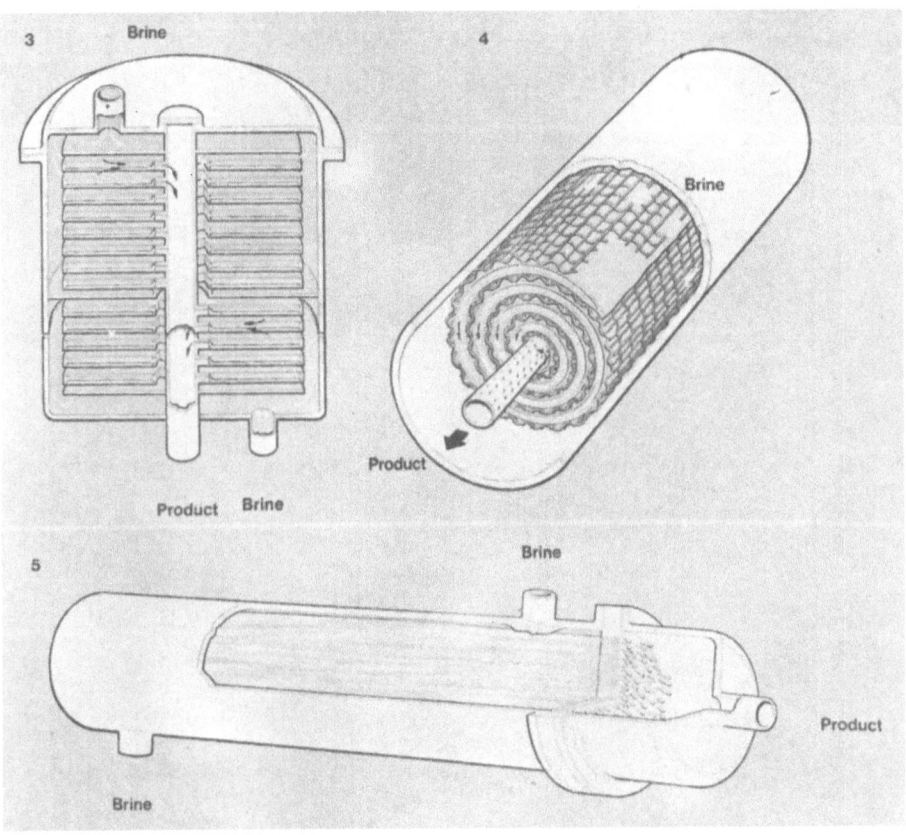

FIGURE 1. REVERSE OSMOSIS MODULES - 3 PLATE AND FRAME
                                     4 SPIRAL WRAP
                                     5 HOLLOW FIBRE

456

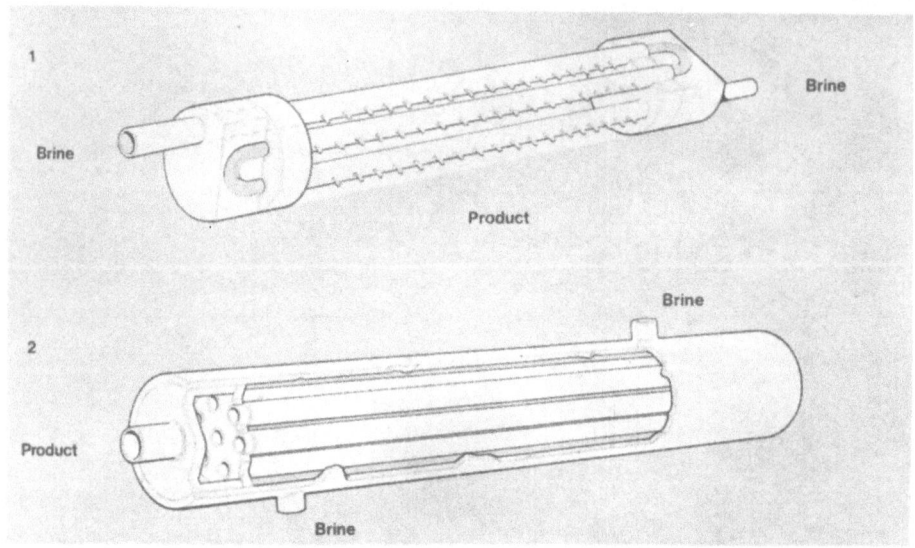

FIGURE 2. TUBULAR MODULES

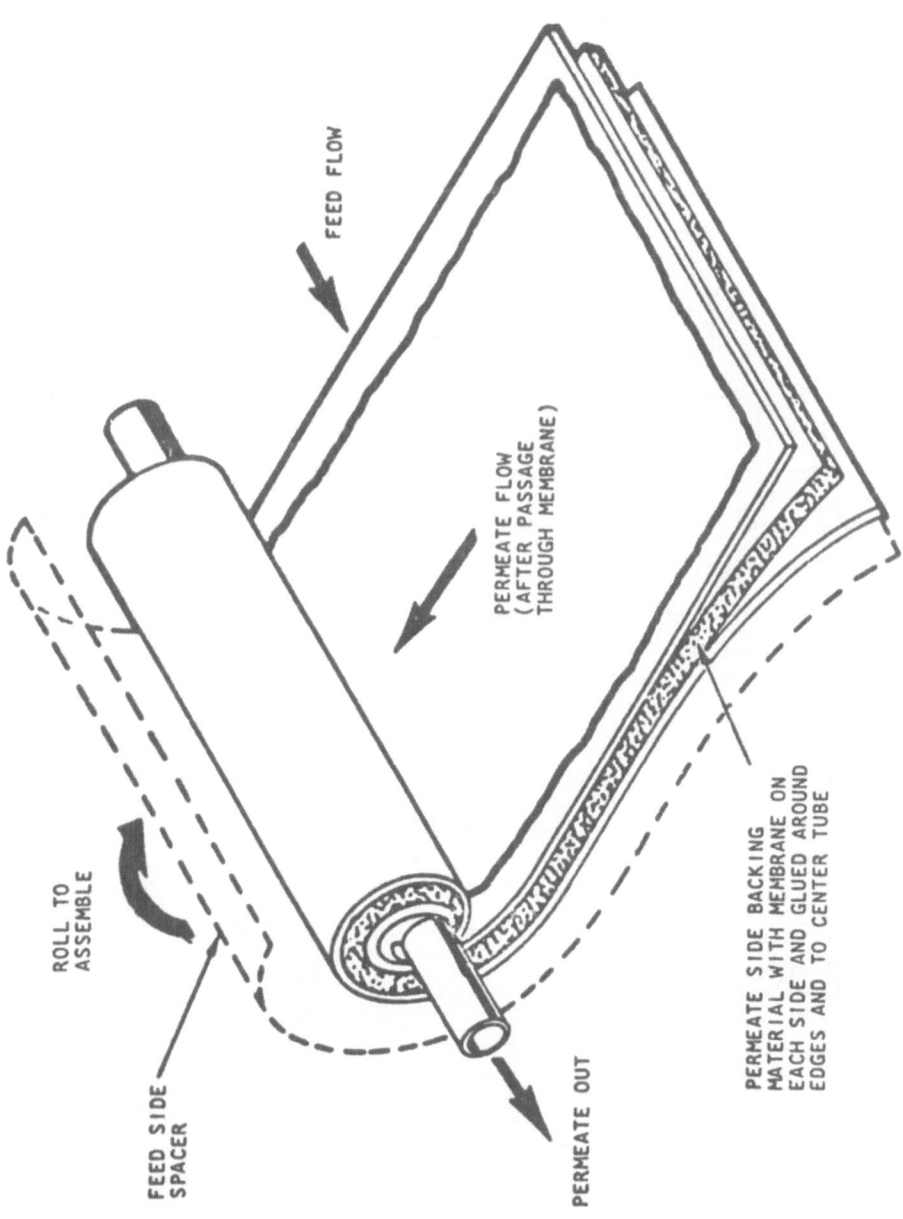

FEED FLOW

PERMEATE FLOW
(AFTER PASSAGE
THROUGH MEMBRANE)

ROLL TO
ASSEMBLE

FEED SIDE
SPACER

PERMEATE OUT

PERMEATE SIDE BACKING
MATERIAL WITH MEMBRANE ON
EACH SIDE AND GLUED AROUND
EDGES AND TO CENTER TUBE

FIGURE 3. SPIRAL MEMBRANE MODULE

458

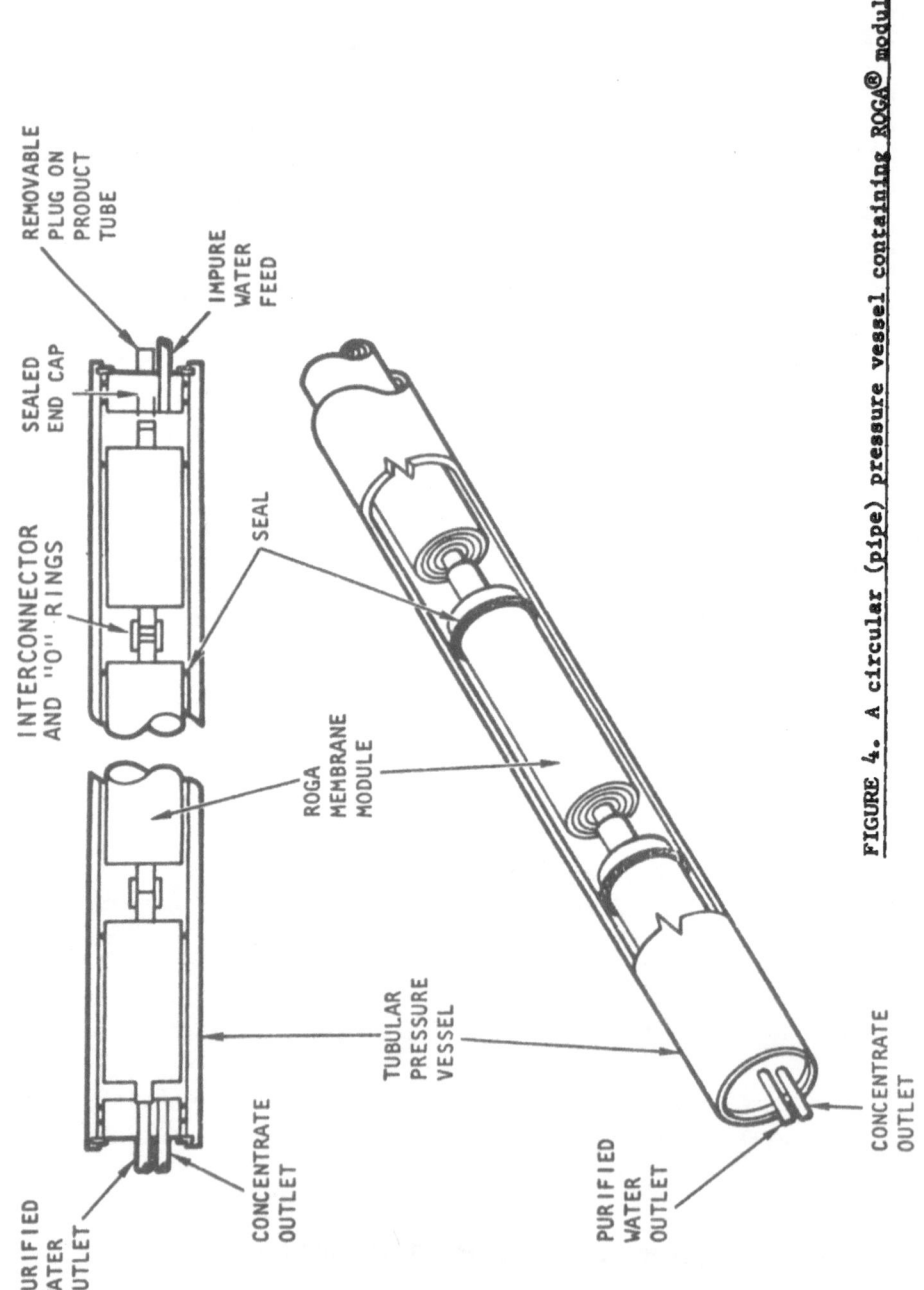

PURIFIED WATER OUTLET

CONCENTRATE OUTLET

TUBULAR PRESSURE VESSEL

ROGA MEMBRANE MODULE

SEAL

INTERCONNECTOR AND "O" RINGS

SEALED END CAP

REMOVABLE PLUG ON PRODUCT TUBE

IMPURE WATER FEED

PURIFIED WATER OUTLET

CONCENTRATE OUTLET

FIGURE 4. A circular (pipe) pressure vessel containing ROGA® modules

459

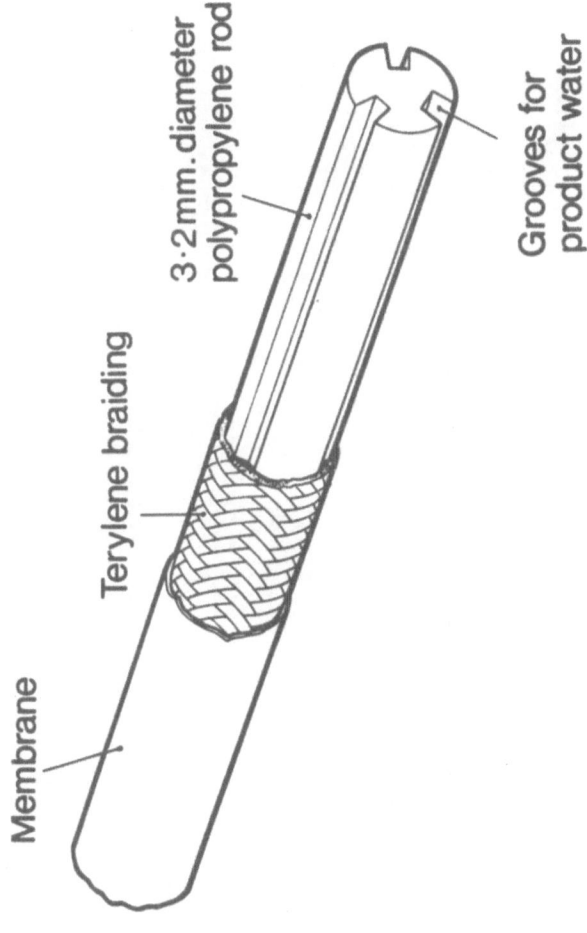

Membrane

Terylene braiding

3·2mm. diameter polypropylene rod

Grooves for product water

Illustration of Membrane Support

Figure 5

460

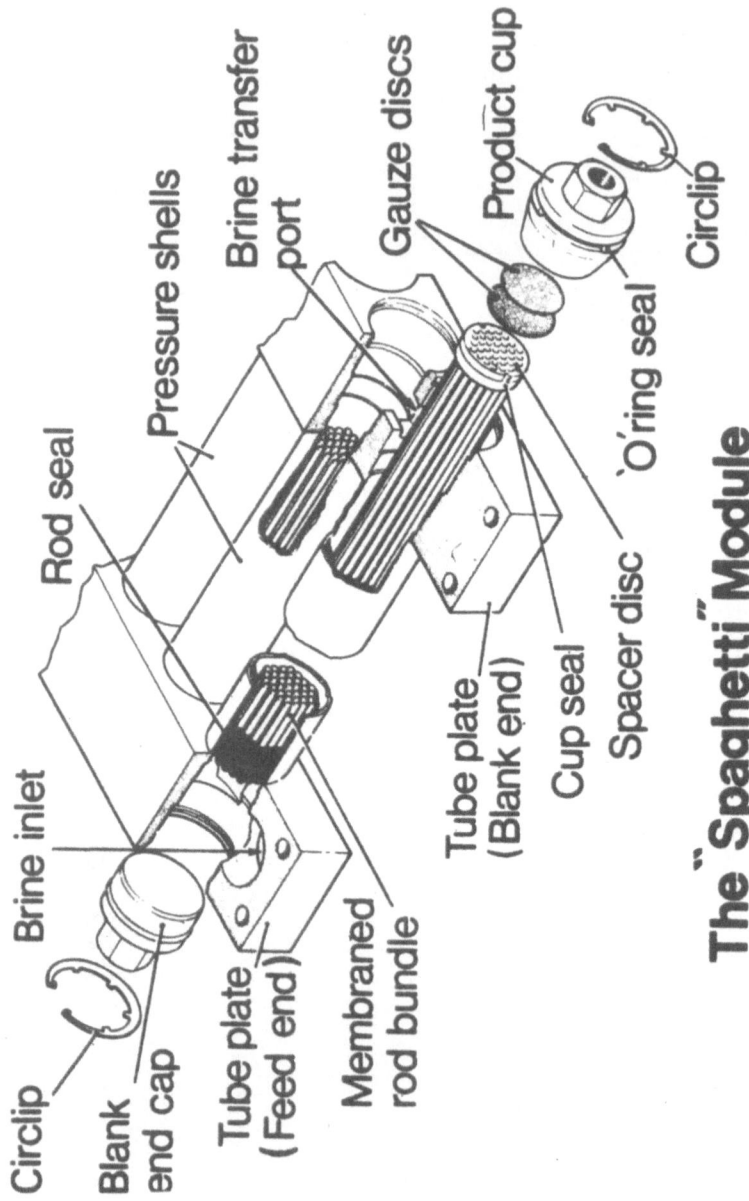

Circlip

Brine inlet

Rod seal

Pressure shells

Brine transfer port

Gauze discs

Product cup

Circlip

Blank end cap

Tube plate (Feed end)

Membraned rod bundle

Tube plate (Blank end)

Cup seal

Spacer disc

'O'ring seal

The "Spaghetti" Module

Figure 6

461

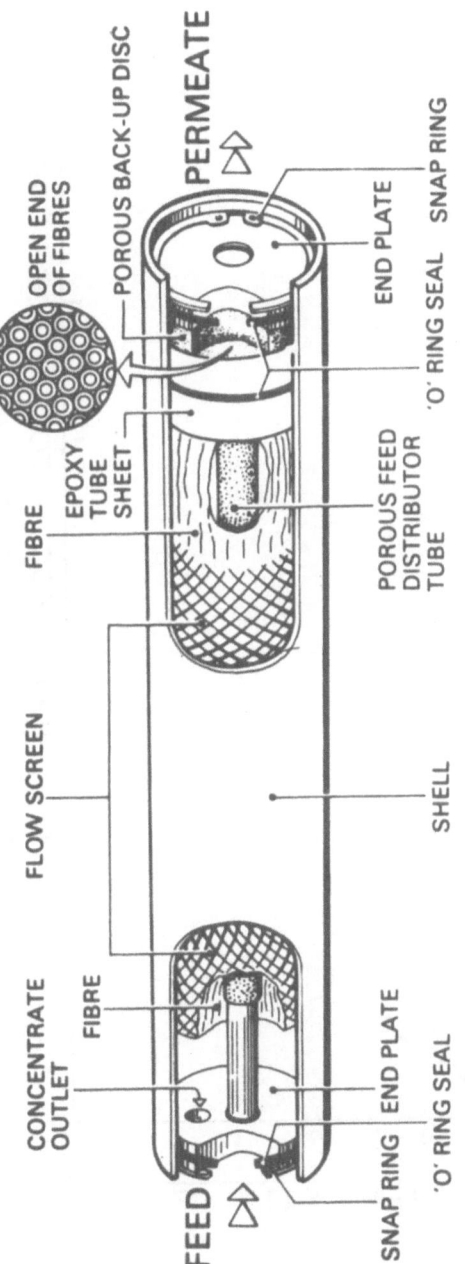

PERMEATE

OPEN END OF FIBRES

POROUS BACK-UP DISC

END PLATE

SNAP RING

'O' RING SEAL

FIBRE

EPOXY TUBE SHEET

POROUS FEED DISTRIBUTOR TUBE

FLOW SCREEN

SHELL

CONCENTRATE OUTLET

FIBRE

SNAP RING

END PLATE

'O' RING SEAL

FEED

CUT AWAY DRAWING OF PERMASEP® PERMEATOR

Figure 7

462

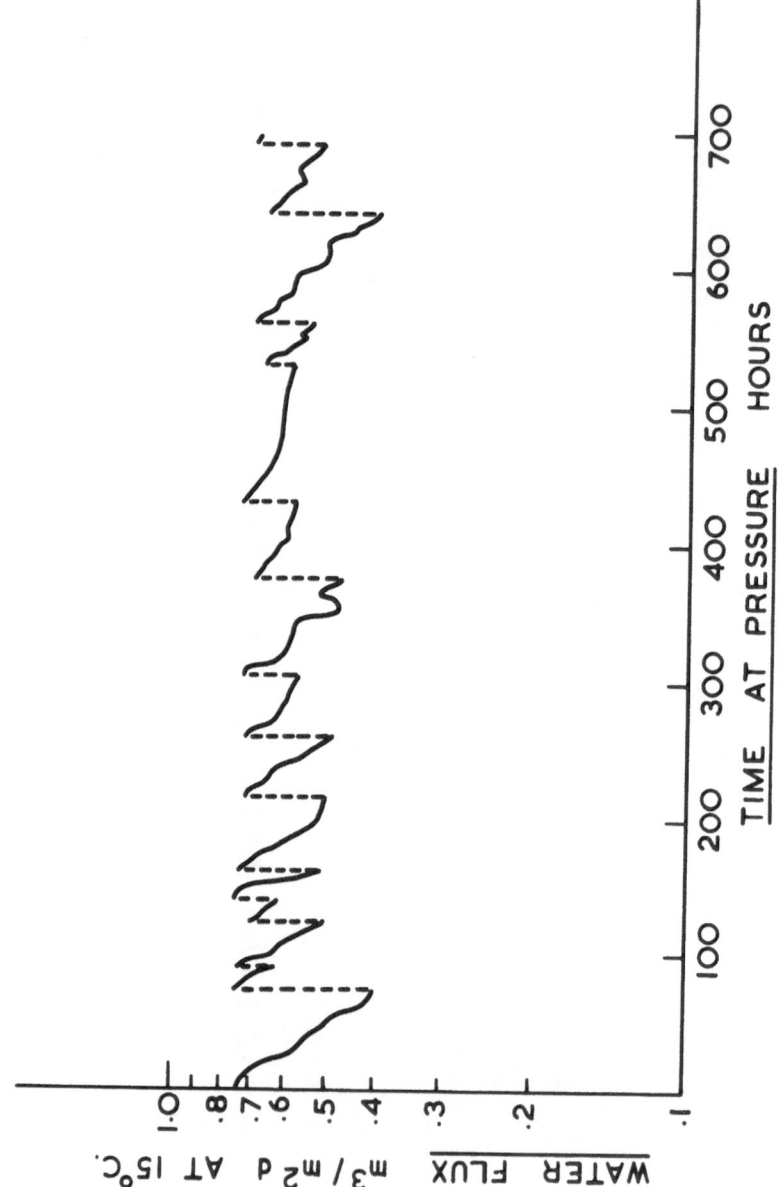

FIGURE 8. HARWELL SECONDARY SEWAGE EFFLUENT, EFFECT OF DETERGENT FLUSH ON WATER
FLUX, OPERATING PRESSURE 4.14 MN/m²

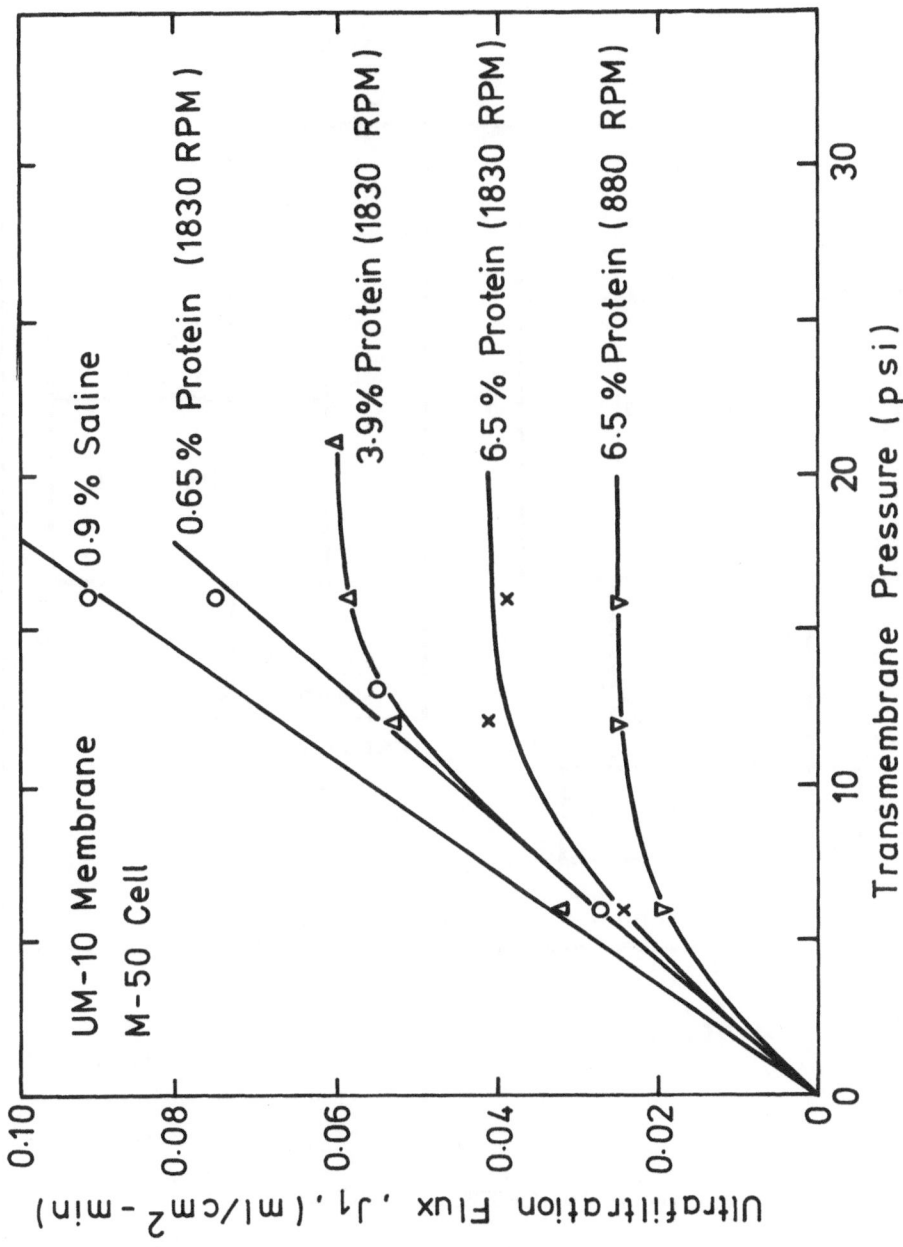

FIGURE 9.   FLUX PRESSURE RELATIONSSHIPS FOR BOVINE SERUM
ALBUMIN SOLUTIONS IN A STIRRED BATCH CELL

464

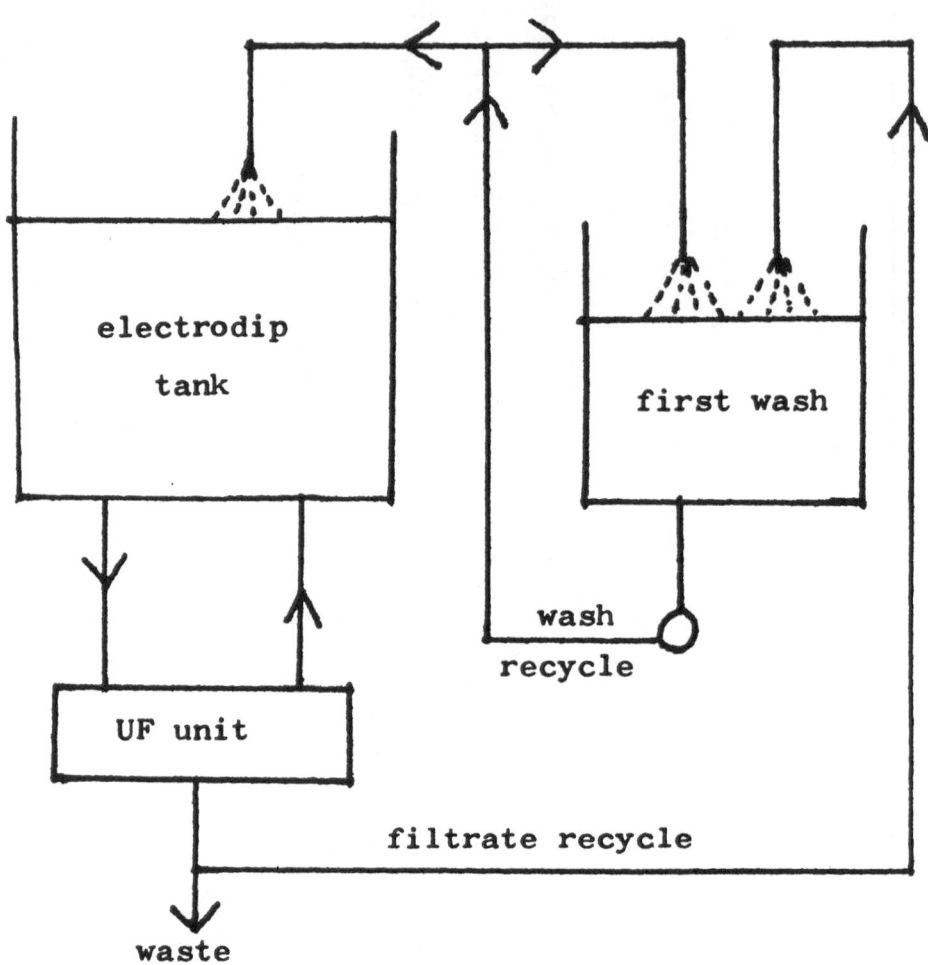

FIGURE 10.   USE OF UF IN ELECTROPHORETIC PAINTING

TREATMENT OF RADIOACTIVE WASTE SOLUTIONS

K.H. Lieser

Department of Inorganic and Nuclear Chemistry
Technische Hochschule Darmstadt

SUMMARY. A survey on the following subjects are given:
1.Sources of radioactive wastes (nuclear fuel cycle,
application of radionuclides, nuclear explosions)
2. Waste solutions of high activity (storage problems,
solidification, possible separation procedures) 3. Waste
solutions of medium and low activity (precipitation,
ion exchange, adsorption, other methods).

1. SOURCES OF RADIOACTIVE WASTE

1.1 Radioactive wastes from the nuclear fuel cycle

Fig. 1 gives a survey on the main steps involved in the
production of nuclear energy. The central part is the
nuclear reactor which contains the nuclear fuel in the
form of fuel elements, the moderator and construction
materials. The nuclear fuel is produced from uranium
ores. The ores are leached and concentrates with high
uranium content are obtained. This is generally done
in uranium mills near the sites where the ores are
found. The uranium concentrates obtained may then be
shipped for further chemical treatment as in the case
of most uranium ores found in Africa. The waste produced
in the uranium mills is of two kinds, the residue obtained after leaching and the liquid waste in the
effluents or the off-water of the uranium mills. The decay
products of uranium present in uranium ores are listed
in Table 1. Those having longer half-lives and being

present in larger amounts are more important than others. Proctactinium hydrolyses very easily even in strong acid and therefore remains in the solid residue after leaching. Thorium is leached to a great extent and found either in the effluents or it is precipitated together with the uranium, according to the procedure applied. Radium and lead follow the solid residue in form of sulfates if sulfuric acid leaching is used. If not they are found in the effluents. Thus the solid residue as well as the effluents contain appreciable amounts of long lived radionuclides. The liquid waste solutions are decontaminated in most cases by precipitation or coprecipitation.

The next step is the chemical treatment of the uranium compounds obtained in the uranium mills in refineries in order to produce uranium of nuclear grade purity. The chemical procedures applied most frequently are extraction or ion exchange separation of the uranium. These procedures again lead to effluents containing thorium and other radioelements. For decontamination these solutions have to be treated, too.

The chemical procedures involved in the production of nuclear fuel from the compounds obtained in the refineries are shown in Fig. 2. The compounds ammonium diuranate, $(NH_4)_2U_2O_7$, or uranyle nitrate hexahydrate, $UO_2(NO_3)_2 \cdot 6H_2O$, which are produced in the refineries represent the state of the highest purity of uranium. They are transferred into nuclear fuels such as UO_2 or into UF_6 for enrichment of ^{235}U by isotope separation. In these steps the waste production is generally rather small and may be neglected. The same holds for the production of fuel elements and for the isotope separation plants.

Another important source of radioactive waste are the nuclear reactors themselves. The fuel is introduced in the form of fuel elements which consist of the fuel and a cladding material or other combinations such as cermets. The heat created by nuclear fission in the fuel is transferred by a circulating coolant to a secondary cycle which comprises the electricity generator. A very small percentage of the fission products produced in the nuclear fuel are able to escape from the fuel elements and enter the coolant. In order to keep the radioactivity in the coolant as low as possible, in water cooled reactors a part of the water is circulated continuously through an ion exchange column where all ions present including those originating from fission products are sorbed. From time to time the ion exchange resins are

taken out and new resins are installed. The resins contain
multicurie quantities of radioactivity and represent
appreciable amounts of waste. In addition there are
other secondary sources of radioactivity in reactors due
to leakages in the system and off-water from the reactor.
The waste solutions arising from these sources have to
be decontaminated which is done primarily by precipitation.

Another source of radioactivity in nuclear reactors
is the off-gas which contains the volatile components
such as the noble gases krypton and xenon as well as
other radionuclides like ^{131}I and tritium. These radionuclides make off-gas control or treatment necessary.

After burn-up the fuel elements are taken out from
the reactor and stored under water, generally for a
period of some months, before they are transported to a
reprocessing plant. Some of the fission products are
released into the water tanks. Therefore this water has
to be decontaminated, too, giving rise to certain
amounts of radioactive waste.

By far the greatest amount of radioactivity, however, is released in the reprocessing plants where the
fuel elements are cut into pieces and the fuel is processed chemically. The chemical procedures involved
are illustrated in Fig. 3. Wet processes are generally
adopted and used, beginning with dissolution of the fuel.
Uranium, plutonium and the fission products are separated
by extraction and reextraction. Several streams of radioactive waste solutions are coming out from the reprocessing plants. Those containing the major part of the
fission products are of very high activity, other solutions of medium activity level. The composition of the
high active fission product waste solutions is given in
Table 2. Radionuclides of shorter half-live have decayed
during storage. Beside the waste solutions solid waste
is produced in appreciable quantities from the dismant-
ling of the fuel elements. The off-gas originating from
the dissolution process contains the volatile fission
products which make special treatment necessary.

The uranium recovered in the reprocessing plants
may be used again for fuel production thus closing the
fuel cycle. The plutonium may be taken for fuel production as well.

For thorium breeders additional steps are necessary.
Thorium ores have to be leached and the thorium obtained

has to be transferred into suitable thorium compounds,
such as ThO_2, giving rise to liquid and solid wastes
containing the radioactive daughter products of thorium
which are also listed in Table 1. The reprocessing of
thorium breeder elements involves the chemical separation
of thorium, uranium and fission products, furthermore
the separation of plutonium if the fuel elements contain ^{238}U as well. ^{233}U and ^{239}Pu are produced by the
nuclear reactions

$$^{232}Th \ (n,\gamma) \ ^{233}Th \xrightarrow{\beta^-} \ ^{233}Pa \xrightarrow{\beta^-} \ ^{233}U \qquad (1)$$

$$\text{and} \ ^{238}U \ (n,\gamma) \ ^{239}U \xrightarrow{\beta^-} \ ^{239}Np \xrightarrow{\beta^-} \ ^{239}Pu \qquad (2)$$

These reactions are essential for breeding, that is for
the production of fissile material. If the thorium is
reused for the production of breeder elements the cycle
of breeding is closed, too.

The amount of radioactivity produced throughout the
world by fission in nuclear reactors is estimated to be
of the order of 10^{12} Ci per year at the end of this
century or about 1 million cubic meters of highly radioactive waste solutions per year. The handling of these
large quantities of radioactivity makes intense technolo-
gical development, research and international collabora-
tion necessary.

1.2 Radioactive wastes from the application of radionuclides

The total amounts of radioactive wastes from the application of radionuclides is by many orders of magnitude
smaller than the amount of radioactivity which is produced in nuclear power reactors. On the other hand radionuclides are applied with great success in many fields,
mainly in nuclear medicine, and on this way more equally
distributed.

The fields of application of radioactivity may be
divided into three sections : Application in medicine,
in industry and in other disciplines of science. On the
other hand the radionuclides applied may be divided into
two classes, radionuclides in closed form and in open
form. Closed forms are the radiation sources such as
^{60}Co or ^{137}Cs. If handled with care they do not give
rise to contamination and therefore they need not to be
considered here.

By far the most extensive user of radioactivity in

open form is medicine. Radionuclides are applied in medicine for diagnosis in smaller amounts of the order of
0.1 Ci and in therapy in higher amounts of the order of
many curies. With respect to the radiation burden of the
patients preferably radionuclides of short half-lives are
used. These do not present themselves appreciable waste
problems if the half-lives are shorter than 1 day. Radionuclides of longer half-lives, such as ^{131}I, are applied
in still increasing quantities. The waste of these radionuclides which is produced in hospitals as well as the
possible contamination of the off-water from the patients
have to be considered carefully. The aqueous waste
solutions may be stored in tanks in order to allow for
decay of the radionuclides.

The major long term problem, however, are the long-
lived radionuclide impurities which are produced as by-
products and which are present in smaller or higher
amounts in the different radionuclides or radiopharmaceuticals applied in medicine and nuclear medicine. The
long-lived radionuclide impurities which are present in
the samples used for medical purpose are widespread in
the environment after application. International colla-
boration and better control of radionuclide purity in
radiopharmaceuticals are therefore necessary and need
to be considered more carefully.

The amount of radioactivity applied in open form
in industry is small compared with the quantities which
are used in medicine. The same holds for the application
in other fields of science. If the radionuclides are
short-lived and of high radionuclide purity and if they
are handled with care they do not represent any danger
and they do not lead to appreciable amounts of waste.

1.3 Radioactive wastes from nuclear explosions

Nuclear explosions of the kind of nuclear weapon tests
release huge amounts of radioactive fission products
into the environment. The surrounding area is contaminated with large quantities of radioactivity. If this
area is leached by rain water or ground water it represents a long-lived source of radioactive waste
solutions for the environment. According to the kind
of explosion a smaller or greater part of the radioactivity is immediately transferred to the upper layers
of the atmosphere where it is circulating like a cloud
for many years. A small percentage of that radioactivity
is found in the rain water and in this way it is spread

continuously in form of radioactive fall-out over that
part of the surface of the earth where the radioactivity cloud is circulating. It represents a long-lived
source of waste solutions.

If the nuclear explosions are performed deep enough
underneath the surface of the earth the release of radioactivity into the atmosphere is avoided. Only the surrounding layers of the earth are contaminated for a
longer period of time.

2.WASTE SOLUTIONS OF HIGH ACTIVITY

2.1 Storage problems

In the reprocessing plants waste solutions of high activity (of the order of several kCi per liter) are produced and stored in large tanks. These solutions contain
the fission products mainly in nitric acid. They have
to be cooled because of the heat created from radioactive decay. Off-gas is obtained by radiolysis which
leads to the release of hydrogen, oxygen and nitric oxides,
a mixture which has to be handled carefully. Furthermore
the waste solutions are highly corrosive and therefore
special attention has to be paid to leakage. It is
obvious from these reasons that the high activity waste
solutions may be kept in these tanks for some time, for
instance for several years, but not for decades or
centuries. They have to be transferred into a suitable
solid form for storage. In this connection the following
questions arise: (1) What is the optimal chemical form
for storage ? (2) What are the most suitable places for
storage ? Both questions have been discussed in much
detail but they have not yet finally been solved.

Fig. 4 gives the radioactivity of the fission products as a function of time after the end of the use
of a fuel element in a nuclear reactor. It can be seen
from this figure that the activity is decreasing very
rapidly during the first days. During the following
months the activity is decreasing more and more slowly.
Therefore a storage time of some months seems reasonable
before reprocessing in order to reduce the amount of
radioactivity which is to be handled.After some years
the radioactivity decreases approximately with the
half-live of the fission products ^{90}Sr and^{137}Cs which
is 28 and 30 years respectively. After a period of
about 1 000 years long-lived radionuclides such as ^{129}I

are still present and furthermore the long-lived actinides which have been produced from uranium or plutonium by neutron capture. Their activity is still several orders of magnitude to high for release to the environment and besides they are highly toxic. They have to be kept under safe conditions for a time of the order of 10^6 years. These considerations clearly show the main problem of storage: safety for very long periods of time.

The solidification of the high activity waste will be discussed in more detail in the next section. As to the safest places for storage geological and hydrological aspects are involved. Most suitable places are those in which leaching by rain water or ground water is strictly excluded for periods of time which are longer than the history of man. In other words, the storage places must be safe with respect to the future development of the earth crust. In the moment salt mines are considered to be the safest places available. They are impermeable for water if cracks are absent. Therefore many investigations have been performed in salt mines which are not used any more. As the number of those salt mines is limited geological prospection is necessary in order to find more places of this kind suitable for deposition. The places have to be selected and tested carefully. They must not be influenced by future movements of the earth crust or volcanic activity.

2.2 Solidification

As already mentioned in the previous section the solid state is the only form suitable for the storage of radioactive waste.The solids must exhibit the following properties: (1) Ability to take up high quantities of radioactive waste in order to keep the volume as small as possible. (2) Physical and chemical stability at high temperatures because the heat production due to the radioactive decay of the fission products will lead to temperatures of several hundreds or even more than thousand degrees centigrade depending on the size and the heat conductivity of the solid. (3) This stability must be garantueed for long periods of time. (4) Stability to high temperature gradients is also necessary because the difference in temperature between the interior and the surface will be appreciable (several hundred degrees centigrade at the beginning of the storage). (5) Stability against radiation decomposition. (6) Resistance to leaching by water and salt solutions.

Two kinds of solids have been investigated in more detail for the purpose of solidification of radioactive waste solutions, glasses and ceramics. Glasses have the following advantages: They can easily be produced in form of melts from the radioactive waste solutions and are able to take up relatively high amounts of fission products. Some kinds of glasses are also rather well resistant to leaching. The main problems of the glasses are the physical and chemical stability at high temperatures, in particular on the long time scale which has to be considered, and the stability to high temperature gradients. Both influences, temperature and temperature gradient, may lead to crystallization of parts of the glass and to migration of the fission products.

Glasses may be of different kinds. Borosilicate glasses, for instance, are more resistant to leaching than others. Phosphate glasses are able to take up high amounts of fission product oxides, but on the other hand they are highly corrosive, a property which is important if metal containers are used for the glass.

Ceramics have many advantages with respect to stability at high temperatures, stability to high temperature gradients and radiation damage. They are real crystalline solids and they are found in nature, too, where they exhibited stability at high temperature for long periods of time. Glasses, on the other hand, may be considered to be undercooled melts and they are thermodynamically unstable with respect to the transition into the crystalline state. This transition is a very slow process, but it is promoted by high temperature and may also be promoted by radiation on the long time scale.

The disadvantages of ceramics are the relatively low percentage of fission products which can be incorporated if ceramics with suitable properties are to be produced, and furthermore the more complicated way of production including the higher temperatures necessary for production.

In summarizing it can be said that the production of ceramics from radioactive waste solutions seems to be safer, but it may be more expensive.

2.3 Possible separation procedures

The fission products present in the high radioactive waste solutions may be divided into three classes according to their half-lives as already indicated in section 2.1.

First class: The short-lived fission products with half-lives up to about 1 year. They will decay within a period of about 10 years. Second class: The longer-lived fission products with half-lives up to about 30 years, mainly ^{90}Sr and ^{137}Cs. These nuclides have high fission yields and represent the predominant sources of radio-activity in the fission products after 10 years of storage. They will decay within a period of about 1 000 years. Third class: Radionuclides with very long half-lives. Some of these radionuclides are fission products such as ^{129}I which is contributing, however, only to a very small extent to the radioactivity because of its very long half-live. The most important members of the third class are the actinides, such as neptunium, plutonium and americium.

On the background of these considerations the question may be raised whether it is of interest to think on the following separation: (a) Separation and isolation of the actinides (and perhaps of iodine and others) from the fission product solutions. The radio-activity in the remaining fraction would then be negli-gible after about 1 000 years. (b) Separation and isolation of ^{90}Sr and ^{137}Cs. After having performed separations (a) and (b) quantitatively the storage problem for the rest of the fission products would be very simple. It could be discarded after about 10 years and the transformation of this remaining part of the fission products into a suitable solid form would not be necessary.

When we are accepting this argument the following questions arise: Are there simple procedures available for quantitative separation and isolation of the radio-nuclides mentioned above, the actinides, strontium and cesium? What is or may be the use of these radionuclides after isolation? If there is no use at all the advantage of the separation and isolation could still be an appreci-able reduction of the storage volume and an increase in safety by preparing certain suitable compounds of these radionuclides, such as strontium titanate. The main ob-jections in this case, however, are economical considera-tions.

The answer to the first question raised above is that there are not yet simple procedures available for the quantitative separation and isolation of the acti-nides, strontium and cesium. It should be pointed out that the term "quantitative" in this connection means separation factors of the order of 10^{13}. Some work has been done. In any case, in my opinion, research should be initiated and sponsored to a reasonable extent because the separation and isolation of the radionuclides mentioned will certainly be of practical importance some day either in the course of normal development or in the case of an accident.

The answer to the question of the possible use of the radionuclides after isolation is the following: ^{90}Sr may be used in isotope batteries for energy production. ^{137}Cs may be used as radiation source, for instance for large scale sterilization of the sludge obtained in off-water treatment. The actinides may be used for the production of radionuclides such as ^{238}Pu and others which again are of interest for energy production in isotope batteries. It should be clearly stated, however, that in the moment the demand for these radionuclides is much smaller than the production rate in nuclear reactors.

In addition to the nuclides mentioned so far there are others among the fission products which are of practical interest. ^{147}Pm, for instance, may also be used in isotope batteries for production of energy. ^{129}I may be transferred into other useful isotopes of iodine by irradiation. Rhodium is present among the fission products in form of stable isotopes in appreciable amounts due to the rather high fission yield of 2.9 %. Its isolation would be of interest because this element is rather rare in nature. The inactive isotopes of xenon which are also of practical interest may be isolated from the off-gas during dissolution of the fuel.

3. WASTE SOLUTIONS OF MEDIUM AND LOW ACTIVITY

3.1 Precipitation

Waste solutions of medium and low activity are mainly produced in nuclear reactors and reprocessing plants. Their radioactivity may amount up to many millicuries per liter. In many cases, however, the radioactivity is rather low and exceeds the maximum permissible level for water only from one to several orders of magnitude. Waste solutions of that kind are often present in rather large volumes, for instance in form of the off-water from uranium mills. Therefore the question of volume reduction is very important for low activity waste solutions. Waste solutions of low activity are also produced in laboratories which are engaged in research, production or application of radionuclides and in nuclear medicine.

With respect to reduction in volume and simplicity of the method precipitation reactions are preferred whenever possible. Some ions, however, are not easily precipitated and need special treatment. Calcium phosphate precipitation is very often used for decontamination of radioactive waste solutions in nuclear centers. As the phosphates of most elements are insoluble in water they are coprecipitated. This holds for thorium, the lanthanides, the alcaline earth, iron and many others. The ions of the alkali metals, such as Cs, are not precipitated by phosphate. But as cesium ions are sorbed very strongly on hexacyanoferrates, the addition of small amounts of iron and of equivalent quantities of hexacyanoferrate ions, $[Fe(CN)_6]^{4-}$, during precipitation bring the cesium into the precipitate as well.

If anions such as I^- or molecules such as I_2 are to be precipitated special procedures are necessary. In the case of I^- and I_2 the addition of small amounts of silver nitrate and of iodine as carrier brings the iodine also into the precipitate. Carriers are always necessary if the solubility product is not reached otherwise. Higher decontamination factors are often obtained by a sequence of precipitations of the same kind. Special precipitation procedures may be worked out for different mixtures of radionuclides in waste solutions. It should be pointed out, however, that precipitation is sometimes not effective when colloids are present or formed.

3. 2 Ion exchange

By ion exchange reactions radionuclides in ionic form may be separated. Ion exchangers are subdivided into two classes, those reacting with cations and those reacting with anions. Organic resins for cation and anion exchange are on the market. They have rather high exchange capacities, 4 to 5 milliequivalents per gram. The selectivity, however, is low. Therefore they have to be provided in quantities large enough that all ions present can be fixed. If the salt concentration is high relatively large amounts of resins are necessary. This may give rise to economical considerations. By cation exchange resins in acid form neutral solutions are transferred into the corresponding acids, for instance

$$\overline{H} + NaCl \rightleftharpoons \overline{Na} + HCl \tag{3}$$

The bar indicates the ion which is fixed. In a similar way the effluent from an anion exchange resin which is applied in the OH-form and fed   with a neutral solution shows an alcaline reaction

$$\overline{OH} + NaCl \rightleftharpoons \overline{Cl} + NaOH \tag{4}$$

If cation exchange resins and anion exchange resins are applied in form of a mixture or in form of two columns working in series desalinated neutral water is obtained. If the resins are loaded with higher amounts of activity appreciable radiation damage may occur because the resins are rather sensitive to radiation decomposition. As radiolysis of the water takes place simultaneously, giving rise to the production of hydrogen and oxygen, the columns may not work properly if standing for some time due to the formation of gas bubbles.

The resins can be regenerated by treatment with an acid and a hydroxide or by treatment by an appropriate salt solution. The following reactions take place:

$$\overline{Na} + H^+ \rightleftharpoons \overline{H} + Na^+ \tag{5}$$

$$\overline{Cl} + OH^- \rightleftharpoons \overline{OH} + Cl^- \tag{6}$$

or $\overline{Ca} + 2Na^+ \rightleftharpoons 2\overline{Na} + Ca^{++}$ $\tag{7}$

$$\overline{SO_4} + 2Cl^- \rightleftharpoons 2\overline{Cl} + SO_4^{--} \tag{8}$$

Inorganic ion exchangers are of very different kinds. Natural or synthetic silicates may be used, such as clay minerals, zeolites or molecular sieves. Other compounds suitable for special separations include various oxides, such as Al_2O_3, SiO_2, TiO_2, ZrO_2. These products are used as powders of small grain size (Al_2O_3) or as gels consisting of a matrix with a high content of water (silica gel, $TiO_2 \cdot xH_2O$, $ZrO_2 \cdot xH_2O$). Finally there are some phosphates and hexacyanoferrates, for instance those of Ti and Zr, all of non-stoichiometric composition. The sorption mechanism is in some cases an ion exchange reaction. The clay minerals, for instance, exhibit ion exchange properties, the zeolites and also the phosphates and the ferrocyanides. In other cases chemisorption takes place, as on the oxides mentioned above or in the molecular sieves. Finally chemisorption and ion exchange

properties overlap in some cases, such as the clay
minerals, zeolites or molecular sieves (which may also
exhibit ion exchange properties). Therefore the inorganic compounds have here been listed together.

A common feature of all inorganic ion exchangers
is that they are more or less selective, that means they
are able to sorbe some kinds of ions to a much greater
extent than others. Furthermore they all have lower
exchange capacities than the resins. Very often they
also show small rates of exchange as compared with the
resins. On the other hand the inorganic compounds are
more stable to radiation decomposition than the resins.

The possibilities for the application of inorganic
ion exchangers cannot be discussed here in detail. Some
examples of selective separations are the following:
Aluminium oxide for molybdate and tellurate, silica gel
for protactinium, clay minerals at pH 3 for yttrium
and the rare earth elements. Titanium phosphate and
even better titanium hexacyanoferrate and other hexacyanoferrates for cesium, antimony pentoxide for sodium,
antimony hexacyanoferrate for strontium, silver chloride
or silver nitrate or zeolites loaded with silver ions
for iodine.

3.3 Adsorption

Another method which can be used for the treatment of
medium and low activity waste solutions is adsorption.
Several adsorbents have already been mentioned in
section 3.2 in connection with some classes of inorganic
compounds, such as clay minerals, molecular sieves,
Al_2O_3, SiO_2, ZrO_2. Charcoal or powdered graphite are also
often used as adsorbents.

Although adsorption generally is not very specific
it may be applied in many practical cases, especially
for the treatment of dilute solutions. If the salt
concentration of the solution is high the adsorption
of the salt competes with the adsorption of the species
which are to be eliminated. For the treatment of solutions
with high salt concentrations rather large amounts of
sorbent may therefore be necessary.

In contrary to ion exchange adsorption methods are
also applicable for neutral species such as organic
compounds containing radionuclides or products of hydrolysis of inorganic substances which may in some cases be

present. Colloidal solutions which are not affected by
ion exchange may also be decontaminated by adsorption.

3.4 Other methods

The remaining methods which may be applied to the treatment of medium or low activity waste solutions are not
of great practical importance.

Extraction by organic solvents is applicable for
the separation of many radionuclides. Appropriate
laboratory methods exist which may be used. But because
of economical reasons solvent extraction is not suitable
for the treatment of large volumes of waste solutions.
Solvent extraction may have advantages in some cases
for relatively small volumes of aqueous waste containing
high amounts of solutes.

Electrolytic separation of radionuclides will also
rarely find application for the treatment of waste
solutions.

Evaporation of waste solutions may sometimes be
applied if no other suitable method is available. It is
a relatively expensive procedure, but it may have some
advantages if the concentration in the solution is already very high. Volatility of radionuclides has to be
taken into account, however, if evaporation is applied.
Volatile species of radionuclides may be formed during
evaporation by redox processes, for instance iodine may
be produced from iodide or volatile ruthenium tetroxide
from ruthenium salts in nitric acid solutions.

Literature

Radioactive Waste Disposal into the Ground,
IAEA Safety Series No. 15, Vienna, 1965

Nuclear Materials Management,
IAEA Vienna, 1966

Practices in the Treatment of Low and Intermediate
Level Radioactive Wastes,
IAEA, Vienna, 1966

Disposal of Radioactive Wastes into Seas, Oceans and
Surface Waters,
IAEA Vienna, 1966

The Management of Radioactive Wastes Producted by
Radioisotope Users,
IAEA Safety Series No. 19, Vienna, 1966

Disposal of Radioactive Wastes into the Ground,
IAEA Vienna, 1967

Environmental Contamination by Radioactive Materials,
IAEA Vienna, 1969

Management of Low and Intermediate Level Radioactive
Wastes,
IAEA Vienna, 1970

Environmental Aspects of Nuclear Power Stations,
IAEA Vienna, 1971

Radioactive Contamination of the Marine Environment,
IAEA Vienna, 1972

Management of Radioactive Waste from Fuel Reprocessing,
NEA / IAEA Symposium Paris, 1972

Colloque International l'Energie nucléaire et
l'environnement,
Institut Electrotechnique Montefiore (A.I.M.)
Liège, 1973

Radionuclides	Half-live	Radioactivity per ton U and Th resp. [Ci]
Uranium		
U-238	$4.51 \cdot 10^9$ a	0.331
U-235	$7.10 \cdot 10^8$ a	0.00240
U-234	$2.47 \cdot 10^5$ a	0.331
Pa-231	$3.25 \cdot 10^4$ a	0.00240
Th-234	24.1 d	0.331
Th-231	1.06 d	0.00240
Th-230	$8.0 \cdot 10^4$ a	0.331
Th-227	18.2 d	0.00240
Ac-227	22 a	0.00240
Ra-226	1602 a	0.331
Ra-223	11.4 d	0.00240
Rn-222	3.82 d	0.331
Po-210	138.4 d	0.331
Bi-210	5.0 d	0.331
Pb-210	21.0 a	0.331
Thorium		
Th-232	$1.41 \cdot 10^{10}$ a	0.109
Th-228	1.91 a	0.109
Ra-228	6.7 a	0.109
Ra-224	3.64 d	0.109

Table 1. Decay products of uranium and thorium (only nuclides with half-lives longer than 1 day are listed).

Fission product	Fission yield (%)	Half-live
Sr-89	4.8	52 d
Sr-90	5.8	28.1 a
Tc-99	6.3	$2.1 \cdot 10^5$ a
Ru-103	2.9	40 d
J-129	1.0	$1.7 \cdot 10^7$ a
Cs-137	6.0	30 a
Ba-140	6.4	12.8 d
Pr-143	5.9	13.6 d
Ce-144	5.6	284 d
Sm-147	2.6	$1.05 \cdot 10^{11}$ a

Element	Relative abundance (%)	Relative neutron absorption (%)
Noble gases	7	72
Sm	} 70	14
Other Lanthanides		11
Tc	10	1
Cs	4	0.5
Mo	1	0.2
Others	8	1.3

Table 2. The most important fission products in waste solutions.

482

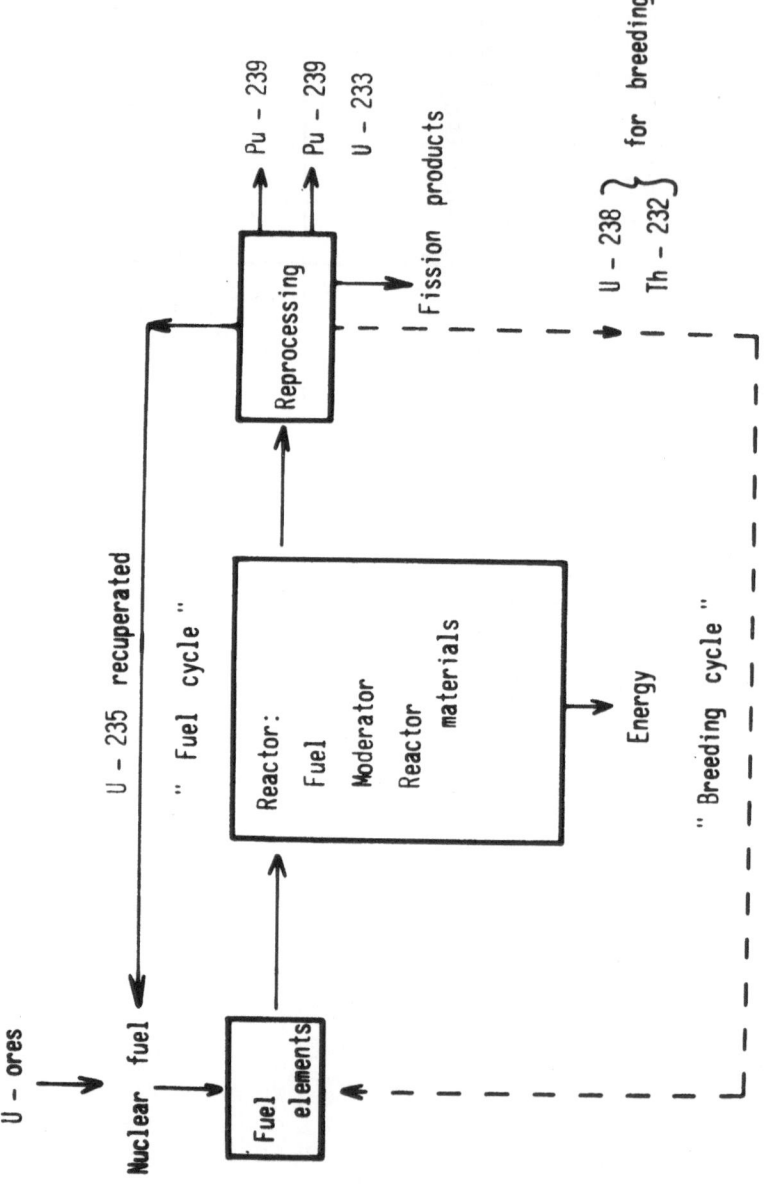

Fig. 1. The nuclear fuel cycle

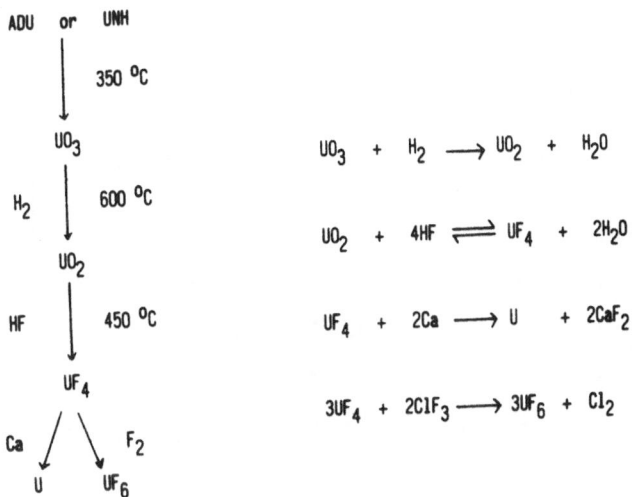

Fig. 2 Chemical procedures for uranium

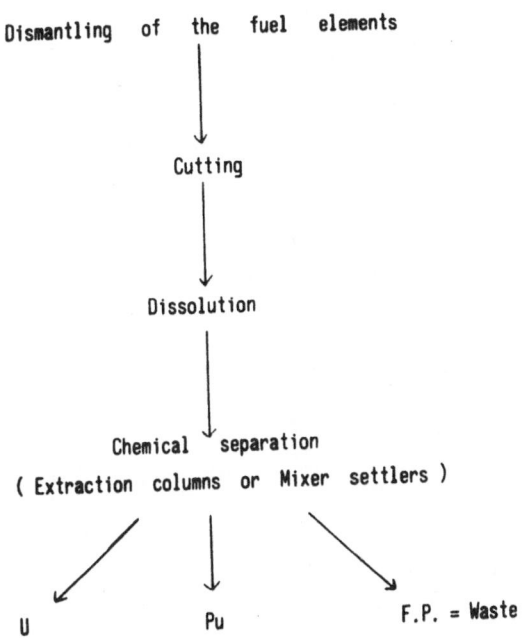

Fig. 3 Reprocessing procedures

484

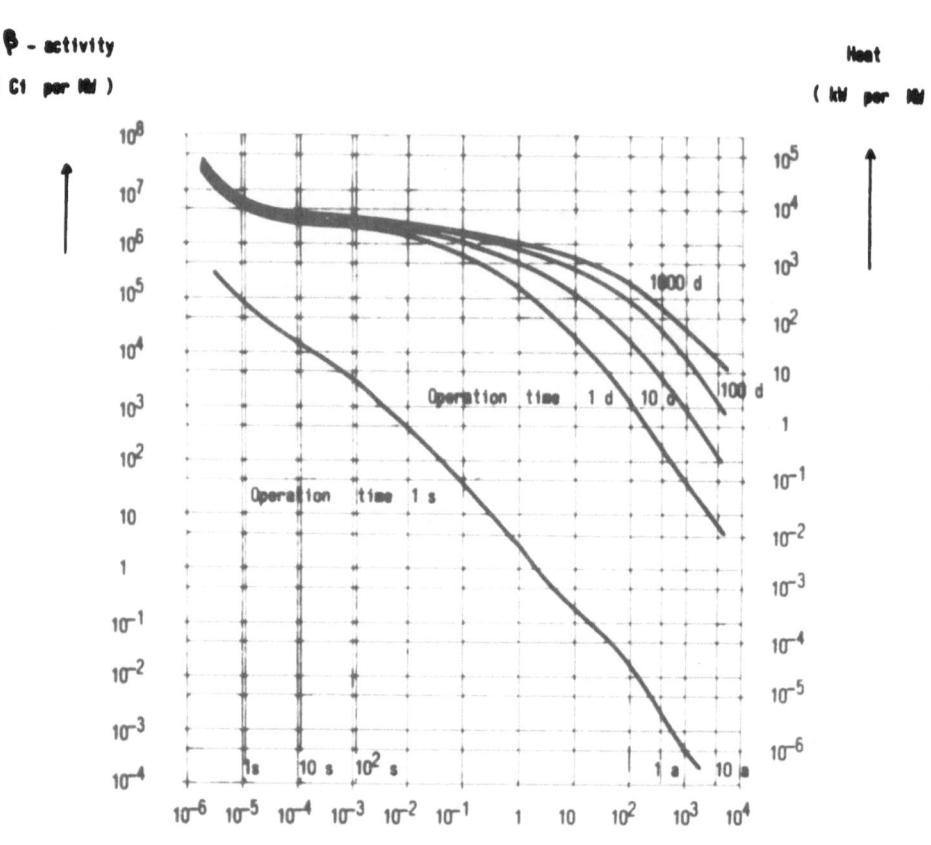

Fig. 4. Activity of fission products as
a function of time

INDUSTRIAL OFF-GAS PURIFICATION WITH ACTIVATED CARBON AND
INORGANIC SORBENTS

Dr. H. Krill

Lurgi Apparate-Technik GmbH
Frankfurt (Main), Germany

The growth of world population, the increase in consumer demand and the related expansion of production facilities to meet
that demand have, in heavily industrialized countries, induced
both a waste water problem and an increasing degree of air pollution. Objectionable odours, the health hazard caused by many substances and the growing and irreversible damage to the environment
have led many countries to enact more or less stringent regulations.
However, compliance with these laws must be technically feasible.

Technology has provided various processes which contribute to
waste gas purification, which essentially means purification of
exhaust air. Such processes are thermal and catalytic combustion,
physical and physico-chemical scrubbing as well as adsorption processes.

Adsorption processes require the use of adsorptive media,
substances with extensive open pore structures and above all large
inner surfaces. Examples are alumina, silica gel, molecular sieves
and activated carbon.

Figure 1 shows in tabular form a comparison of the physical
properties of a number of important adsorbents.

Adsorption processes are used in separating techniques mainly
designed to separate adsorbable substances of relatively low concentration from a voluminous exhaust air or waste gas stream which
is not or is only negligibly adsorbable. Of the indicated adsorption media or so-called adsorbents, activated carbon has the widest
range of application.

486

Adsorption agent	Specific surface m²/g	Micro-pore vol. cm³/g	Macro-pore vol cm³/g	Bulk density g/l	True density g/cm³	Apparent density g/cm³
Granular carbon for gas and vapour adsorption	1000-1500	0,6-0,8	0,5-0,8	300-450	appr.2,0	appr.0,6
Granular carbon for water purification	500-800	0,3-0,6	0,3-0,4	300-550	appr.2,0	appr.0,6
Powdery carbon for de-colorization	700-1400	0,45-1,2	0,5-1,9	250-500	appr.2,0	-
Fine-pore silica gel	600-850	0,35-0,45	0,1	700-800	appr.2,2	appr.1,1
Wide-pore silica gel	250-350	0,3-0,45	0,05-0,1	400-800	appr.2,2	appr.1,1
Activated alumina	300-350	0,4	appr.0,1	700-800	appr.3	appr.1,2
Molecular sieves	500-1000	0,25-0,30	0,35-0,4	600-900	appr.2,6	appr.1,1-1,5

FIGURE 1. Physical data of adsorption agents.

Adsorption technology is relatively old. Charcoal (not activated carbon) with a small surface area and porous structure was used for medical purposes already in ancient Egypt. Scheele has extended the application to gases but failed to achieve technical significance. Towards the end of the last century patents were granted on an increasing scale, showing a marked trend in direction to higher activated carbon. There was, however, no indication of any application at least in the way of gas technology until the first world war, when the use of adsorbents on carbon basis became necessary as protection against gas warfare. This significantly influenced the development of adsorption technology. The originally empiric application made rapid progress by systematic scientific-theoretical study of adsorption dynamics parameters. But the great breakthrough for the use of adsorption technology came when the dynamic and continuously operating adsorption process could be used economically on a commercial scale.

This applies to all adsorbents now employed in the adsorption technology, but above all to the activated carbon processes used for solvent recovery. In addition the adsorbents have found a wide range of applications in the field of treatment of gases for synthesis, purification of waste gas or exhaust air, different methods of the separation of gas mixtures, for protection against toxic gases in gas masks and, recently, for the solution of the many problems in air pollution control.

With very few exceptions adsorbents on inorganic basis are not widely used in the applications mentioned due to the structure of their surfaces or the influence of moisture has a negative effect. These adsorbents serve mainly as efficient dessicators, as catalysts or catalyst carriers and they are commercially available as granular or broken material, as balls, or as extrusion press products.

Nomenclature of the molecular sieves

Type	Summationformula	Diameter of pores
3 A	$0,75\ K_2O\cdot0,25\ Na_2O\cdot Al_2O_3\cdot2SiO_2$	3
4 A	$Na_2O\cdot Al_2O_3\cdot2SiO_2$	4
5 A	$CaO\cdot Al_2O_3\cdot2SiO_2$	5
10 X	$CaO\cdot Al_2O_3\cdot2-3SiO_2$	ca. 8
13 X	$Na_2O\cdot Al_2O_3\cdot2-3SiO_2$	ca. 9
Y	$Na_2O\cdot Al_2O_3\cdot3-6SiO_2$	ca. 9

FIGURE 2. Nomenclature of the molecular sieves.

Silica gel is colloidal silicic acid, nearly free from water. There are a number of production processes which all have in common the fact that they are based on alkali silicates which are hydrolized by addition of mineral acid. In one of the usual processes the water is almost completely removed from the gel under pressure so that washing out is facilitated and the gel can subsequently be dried. For formation of the various structures therefore, the washing or drying process is influential, rather than the pH-value necessary for precipitation. Depending on whether or not the wash process can be effected in acid, neutral or alkaline medium, fine-pore, medium-pore or wide-pore gels are obtained, which are distinguished by different properties, bulk densities and fields of application. The value of the bulk density is also influenced by the type of drying. The stated differences are shown in Fig. 1.

A great variety of production methods are available for alumina, depending on the field of application. In addition to synthetically produced activated alumina, other natural or artificially active products are in use, such as bauxite. In addition to dry conversion, above all wet conversion according to the Bayer process has significance for the production of these adsorbents on a large scale. A large-scale process which has gained significance in the field of the Claus-Tail-Gas desulphurization will be mentioned later.

Molecular sieves (Figure 2) are alkali and alkaline earth aluminium silicates with hollow spaces of defined size determined by the crystalline structure. Crystallographically they belong to the group of the zeolites and this in turn determines the mode of production. Technically, almost only artificial units are used which have the general chemical composition

488

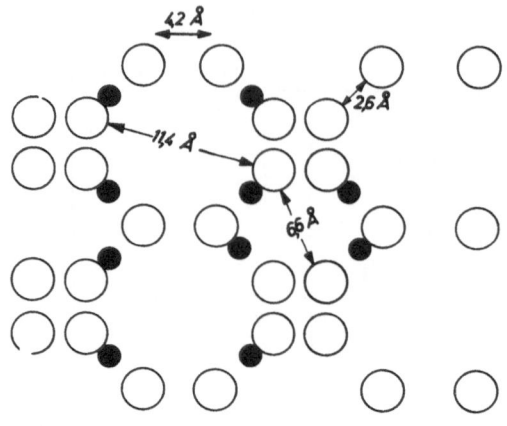

FIGURE 3. Crystal lattice of the zeolite 4 A.

$$M_{2/n}O \cdot Al_2O_3 \cdot xSiO_2 \cdot yH_2O$$

M means a metal cation of the valence n (usually Na, K, Ca); x
and y are ratios which depend on the type of molecular sieve. In
accordance with the given composition, production will be based
on raw materials which already have the groupings typical for molecular sieves, i.e., based on aluminates and silicates of the alkali
or alkaline earth, the corresponding zeolites will be made synthetically by a hydrothermal crystallization process. By hot gas activation the existing crystal water is expelled, and the pores are
freed. Different from the other more or less amorphous adsorbents,
it is their crystalline structure which determines their essential
properties:

1.   Polarity of the adsorption medium and therefore the property
      to separate polarized or polarizable molecules from non-
      polarized ones.

2.   Define and uniform pore size in the range of molecular or
      atomic diameters.

      Three different crystal lattice types are available today,
identified by the letters A, X, and Y (Figure 2). The indicated
pore diameters are consistent only with the A-types. They refer
to pore openings which are decisive for the adsorbability and the
so-called sieving effect (Figure 3). The A and X-types are typical

adsorption agents and the Y-zeolites are exclusively used as catalysts.

Molecular sieves are very versatile in use. In addition to the above mentioned gas drying (to a residual water content of 1 ppm), there are a great number of separation processes in which molecular sieves are used.

These processes on the other hand provide separation into the individual components such as oxygen and argon or carbon dioxide and air and other gases, or the removal of impurities, such as oil vapour from compressed gases, or sulphur compounds from any type of gas. However, in principle this application has nothing to do with the subject of this lecture. In the field of waste gas or exhaust air purification few, but important, cases of application have been dealt with, where it is always a prerequisite that no, or only very little, moisture is present, possibly by appropriate predrying in order to provide the conditions necessary for the use of molecular sieves.

Some processes have become known recently, in particular those of Union Carbide who has pioneered in the field of molecular sieves. They developed processes which have great significance, such as for example the fixed bed adsorption system for concentrating and removing SO_2 from H_2SO_4-plant vent gas. The desorbed SO_2 is recycled back to the front end of the acid plant for recovery and conversion to H_2SO_4. The recovered SO_2 contributes a 2 - 3% increase in H_2SO_4 production rate. SO_2 vapour and mist are also removed and recovered. The SO_2 tail gas emission level is reduced to less than 25 ppm. On the other hand there was developed a process for the removal and recovery of mercury from waste gas streams, above all in the chlorine industry. From the publications available it appears that a waste gas purity better than 60 ppb has been reached. Another problem which can only with difficulty be solved otherwise is the removal and recovery of NO_x from nitric acid plant tail gas. The NO is oxidized to NO_2 by the oxygen contained in the waste gas. Slight traces of water will not impair the efficiency of the process. The desorbed NO_2 is recycled back to the nitric acid plant's absorber tower.

In this connection it might be pointed out that recently the molecular sieve effect could be observed also in activated carbon. Such carbons are not only produced by thermal decomposition of synthetic polymers such as polyvinyl chloride but also by meeting definite conditions during the usual activated carbon production processes, above all when wood is used as raw material. The time has not yet come for a statement on the technical significance, especially in the field of gas technology based on tests known so far.

Fields of Application:	Activated carbon	Alumina gel	Silica gel	Molecular sieves
Gas- and air conditioning	+	+	+	-
De-sulphurization of petroleum vapour or natural gas	+	-	-	+
Oil vapour adsorption (air and gas filtration)	+	+	+	-
Solvent recovery	+	-	+	-
Elimination of odours and tastes from carbon dioxide	+	-	-	-
Elimination of odours and de-toxification of respiration air	+	+	-	-
Hydrogen sulphide adsorption	+	+	-	+
Operating conditions:				
Common technological grain sizes (diameter in mm)	1,7-4	2-6	2-8	3-6
Technically common gas velocities (m3/kg · h) (cm/sec)	- 10-60	1,5 12,5-50	1,5-4,5 12,5-50	1,5-4,5 15-25
Pressure drop (mm water column/m bed height) at 10 cm/sec) at 20 cm/sec } air velocity at 30 cm/sec)	30 70 130	45 80 145	30 75 130	30 80 145
Effective adsorption temperature (° C)	5-50	0-25	5-35	15-40
Reactivation temperature (° C) in specific cases up to	105-115 300	175-320	155-175	200-300 550
Reactivation Processes:				
Flowing steam (in direct contact)	+	-	-	-
Superheated gas or air	+	+	+	+
Waste gas (of gases or oils)	-	+	+	-
With Chemicals	+	-	-	-

FIGURE 4. Fields of application for granular adsorption agents
in gas technology, pertinent operating conditions.

Figure 4 shows in a comparison of important data on the use
of adsorption technology (based on some examples already mentioned)
that activated carbon has by far the widest range of application.

A more detailed description of the application of activated
carbon is now necessary. Some of the practical requirements for
its use in most cases also apply to the above mentioned adsorption media. Thus, as a matter of principle. the adsorbents discussed here are suited for separation of impurities from air or
gas. Dust, airborne particles, mist, aerosoles must be previously removed by other conventional methods. The fields of application of activated carbon processes are solvent recovery, gas and
air purification, removing or reducing odour and gas protection.

The solution of an exhaust air problem begins with the selection of a suitable activated carbon type. It must have properties - and this must be taken into account in production - as
follows:

- High selectivity for the component to be separated

- Stability against water, gases or steam, acids, alkali

and the substances to be separated

- Mechanical stability

- Thermal stability

- High adsorption capacity with lowest concentrations of
  the compounds to be separated (particularly important
  for air pollution control).

Two activation processes are most widely used:

Gas activation where steam, carbon dioxide, fuel gases or
mixtures of these gases are used as oxidation agents. This process is based on precarbonized substances, for example charcoal,
peatcoal, coconut coal. The desired pore structure is obtained
either by the oxygen contained in the gases or by free oxygen
which burns carbon at temperatures between 700 - 1000°C. Another
possibility is carbonization followed by treatment with steam at
700 - 900°C.

Chemical activation (here a solution of zinc chloride, mixture of phosphoric acid and sulphuric acid is used) is generally
based on non-precarbonized substances, for example sawdust, peat.
On heating, the added chemicals absorb water from the raw materials
so that a porous carbon structure is obtained.

Activated carbon can contribute to air pollution control in
a great variety of ways. A few examples will show the versatility
of its application.

In the field of gas and air technology, i.e., exhaust air
purification, including recovery of valuable substances, the activated carbon is used as formed carbon (granular carbon) in the
form of small cylinders with a grain diameter of approx. 0.5 - 4
mm or as broken carbon (lump carbon). At a bed height of 0.25 - 1
m the pressure loss is between 60 and 400 mm WG. The activated
carbon is used almost exclusively in fixed beds. The inner surface of the activated carbon is between 1000 and 1500 m^2/g. The
pores are classified as micro, transitional and macro pores. For
adsorption of the compounds carried in waste gas and exhaust air
most of the pores should have a radius in the micro pore range of
4 - 10 Å. The macro pores with the greatest pore radius mainly
serve as transport channels. They also have significance for the
separation of large molecules and for use of activated carbon as
catalyst or as carrier for impregnation.

A condition for the production of suitable activated carbon
or its use for a definite application is a knowledge of the physical and chemical properties of the substances to be adsorbed.

Influential factors are concentration, temperature, relative humidity, partial pressure of other accompanying matter that can be adsorbed and which would result in mixed or also displacement adsorption, molecular weight, boiling point, electrical properties, chemical reactivity, chemical stability, and many other properties. Tests, experience and studies in this field have shown that straight molecules are adsorbed relatively fast, but that molecules that are branched or exhibit steric hindrances present some difficulties. Significant differences are reported regarding the adsorption of straight chain compounds and of those with branched molecules. Also the different arrangement of a substituted group is important, and different adsorption is mentioned also in the aromatic and aliphatic compounds with the more favourable adsorbability being found in the aromatics. These factors have practical significance when the properties of the molecules show great differences, for example in boiling point and molecular weight. In some cases, only an appropriate experiment in the laboratory or in the bypass of a plant will furnish proof of suitability of a carbon grade.

By preparation of adsorption isotherms, interactions are shown in practical terms which are characteristic for a definite adsorbate and for the selected carbon type. Usually the adsorption isotherm is plotted graphically on a double-logarithmic coordinate system where the adsorbed quantity is shown as a function of the substance concentration in the carrier gas at constant temperature. The isotherm at constant conditions corresponds to the equilibrium loading.

Loading of activated carbon with adsorbable substances is a function of time. The lower the partial pressure, i.e., the concentration of the substance to be separated, the more time is required for reaching the equilibrium condition. The governing factor is the speed of migration, by diffusion in the inner carbon spaces through the macropores to the adsorbing surfaces of the micropores.

If a gas mixture carrying an adsorbable substance flows through an activated carbon filter, an equilibrium loading is reached only when flow speed and diffusion speed are identical. In practice, however, a much higher flow velocity is used than would correspond to the diffusion velocity, in most cases 0.1 - 0.6 m/sec. Here equilibrium loading is not aimed at but the much lower breakthrough point, which practically means the moment when the first traces of the adsorbate to be separated can be detected downstream of the activated carbon bed. In addition the application of a definite activated carbon type for a specific requirement depends on the properties of the adsorption agent:

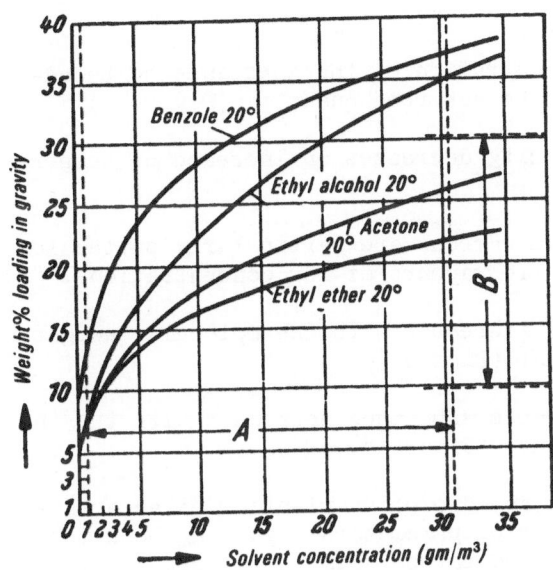

FIGURE 5. Function of substance concentration and loading capacity.

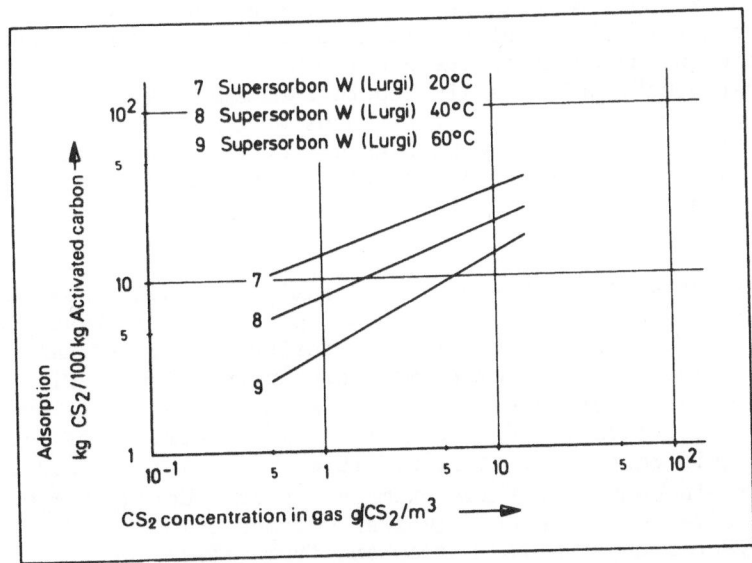

FIGURE 6. Function of adsorption capacity and increase of temperature.

- The larger the inner surface area, the higher the adsorbed quantity.

- The higher the concentration of the substance to be adsorbed, the higher the adsorbed quantity (Figure 5).

- The adsorption capacity decreases on increase of temperature (Figure 6).

- Fine-pored activated carbons are particularly suited for adsorption of volatile vapours of low concentrations.

- Adsorption capacity increases with the molecular weight and the boiling point (Figure 7).

- Loading with adsorbable substance decreases with the increase of air moisture (Figure 8).

For optimum design of an adsorption plant or of an activated carbon filter quite a number of parameters regarding both the adsorbent, in our case activated carbon, and the adsorbed materials or airborne substances must be known and related to each other.

Whereas in recovery plants the emissions are in most cases known as regards concentration and chemical properties, the situation is much more complex in the field of exhaust air purification. The removal of individual substances is necessary in many more cases than the removal of a large number of compounds - which in most cases have very different constituents and properties. Often they are not even compounds in the usual meaning of the definition "odorous substances", which is limited to substances with disagreeable odour, but they may be substances which are odourless but more toxic for the environment even at low concentrations (for example dimethyl sulphate, mercury). It appears reasonable therefore to modify this definition and to define it to mean airborne gaseous substances.

Difficulties will arise when large waste air amounts must be purified from low concentrations, such as for example in bakeries, sweet factories, breweries, chocolate factories or factory halls. Other inconveniences to the environment are known which result from the waste air of dry cleaning shops, fritting installations and carcass utilization plants etc. Of the many carriers of odour contained in this waste air, it is possible by suitable measures to remove certain compounds - for example typical odorous compounds and thus to abate the nuisance. Typical odorous compounds, often termed as key-compounds, are: Mercaptane from oil refineries, amines from fish factories, either fatty acid or amines from carcass utilization plants. Considering the waste air of a coffee roasting plant which among others contains furane, furfurol, fur-

Gas	Molecular weight	Boiling point (°C)	Volume adsorbed at 15°C by 1g of activated carbon (cm³)
$COCl_2$	99	+ 8	440
SO_2	64	− 10	380
H_2S	34	− 62	99
CH_4	18	−162	16
H_2	2	−252	5

FIGURE 7. Adsorption quantity as function of boiling point and molecular weight.

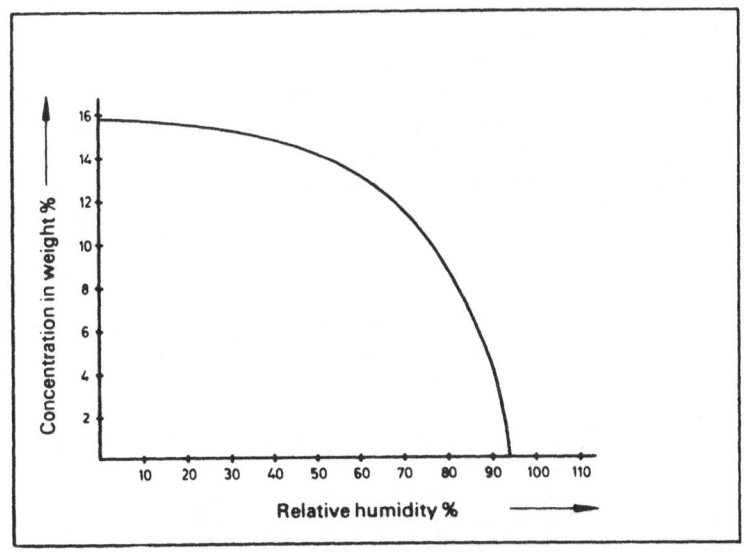

FIGURE 8. Influence of air humidity on the adsorption of benzene (concentration: 200 mg).

furylalcohol, acetaldehyde, pyridine, diacetyl, methylethylcarbi-
nol and hydrogen sulphide the problems become apparent with which
process engineers are faced in design of filters or other suitable
equipment. Even more difficult is the solution of the waste air
problems which may occur in chemical plants: the waste air rates
with reasonably constant contents of substances to be removed will

not always be the same, but in many processes air will require treatment discontinuously and more often than not with different concentrations of pollutants. Many production processes are operated in batches, and leakages, refillings, depressurizing, cleaning and operational faults result in waste air of extremely different compositions. For obvious reasons this air cannot be emitted through stacks.

In many cases the efficiency (i.e., enrichment on the surface of the activated carbon of the substances to be removed) can be increased by coupling with a chemical reaction.

By impregnating a suitable activated carbon with small quantities of specific salts, which for certain related contaminating materials invoke almost quantitatively the required chemical reactions, various other substances may also be removed. In addition the running time of the filters can thereby be greatly increased.

Ammonia, mercaptane, phosgene, ethylene, hydrochloric acid, hydrogen sulphide, mercury and hydrocyanic acid and amines - to name only a few of the more common compounds - can be removed by pure adsorption only to a minor extent, whereas a corresponding impregnation with organic or inorganic salts - depending on the substances to be removed - results in a more effective and efficient separation of the contaminating materials concerned. The most common substances for impregnation are compounds of copper, zinc, silver, chromium, cobalt, manganese, vanadium, molybdenum or a combination of these.

Alkaline and acid impregnations also have significance. Impregnations of this type have also gained importance for industrial gas protection and for use in gas mask filters.

The impregnation effect can be purely chemical - following the stoichiometric laws - or catalytic. By purely adsorptive methods hydrogen sulphide can be loaded on activated carbon up to one weight-percent only. Using a specially impregnated activated carbon loading up to above 100 weight-percent is possible. The reaction is thereby at the same time an example of catalysis. H_2S is oxidized into elemental sulphur with formation of water. Also known is for example the catalytic reaction on conversion of hydrocyanic acid on activated carbon impregnated with zinc or silver salts.

The properties of the activated carbon in many cases may influence catalytic reactions, for example decomposition of peroxides, decomposition of ozone, decomposition of hydrazine, but most of the impregnations effect chemical reactions. Thus hydrochloric acid, sulphur dioxide, COS are converted on alkaline impregnated carbon, but amine, ammonia, putrescine are converted on

Application*	Example	Type of pollution	Impregnation
Compressed air	Compressors for spraying systems, in breweries, pickle lines, plastics, etc.	Oils, odours	no
Gas purification	CO_2 plants	SO_2, H_2S, ethanolamines	yes/no
	Acetylene	Phosphines, hydrochloric acid	yes/no
	Final desulphurisation	H_2S, mercaptans	yes
	Coke oven gas	Naphthalene, organic sulphur compounds	yes/no
Exhaust air purification	Pickle lines	Nitrogen compounds, traces of acids	yes/no
Removal of odours	Mortuaries, bone storage, carcass utilisation plants, fish meal plants	NH_3, amines	yes
	Switch rooms	H_2S, NH_3, acids	yes
	Atomic reactors	Methyl iodide	yes
	Production rooms	SO_2	yes/no
	Pharmaceutical industry, film production	H_2S, aldehydes, butyric acid, nicotine, nitro-benzene, benzaldehyde, camphor, phenols, pyridines	yes/no
	Body and kitchen smells	Palmitic acid, stearic acid, triglycerides, valeric acid, aldehydes, acrolein, caprylic acid, etc.	no
	Coffee-roasting plants	Diacetyl, furfurylalcohol	no
Air conditioning	for offices, hospitals	Traces of inorganic and organic substances	yes/no

*) It must be pointed out, however that activated carbon is suitable only for separating vapour-phase impurities in air or gases; dust and airborne particles must be separated by filters of appropriate design. Mists and aerosols are not retained by adsorption, in this case a combination of cyclones or mechanical filters with activated carbon filters is required.

FIGURE 9.

acid impregnated carbon while hydrocyanide, ammonia or phosgene are converted on activated carbon impregnated with copper or zinc salts. Phosgene or dimethylsulphate can be destroyed or rendered harmless by hydrolysing. Mercaptanes are oxidized into disulphides on alkaline impregnated activated carbon and thus can be removed by adsorption due to the increase of molecular size. Activated carbon treated with bromine converts ethylene - which due to its low molecular weight and low boiling point at normal temperature is extremely difficult to adsorb - into easily adsorbable 1.2 dibromethane. The removal of hydrogen fluoride on activated carbon impregnated with sodium silicate is also technically possible. Chemical reaction (for example with activated carbon impregnated with lead acetate) can in some cases also be used for removal of hydrogen sulphide, preferably by chemical reaction. Iodine - especially radioactive iodine - can be removed by chemical adsorption.

In accordance with the existing problems, combinations of activated carbons treated in different ways can be used in many cases.

Figure 9 contains a selection of applications also indicating possibly required impregnations.

Some gases contain polymerisable compounds such as styrene

and vinylchloride. Phenol and formaldehyde are also frequently
present. These substances are easily adsorbable but they will
polymerize with each other or in the presence of steam and the
products at polymerization cover the pores against further reaction resulting in damage to the activated carbon. In practice,
it is advisable in such cases to previously remove these substances by suitable processes - among others by use of wide-pore activated carbon as prefilter or preliminary layer.

In principle, regeneration of the loaded activated carbon is
possible. In practice, treatment of the activated carbon with
steam - as is usual in solvent recovery - will hardly be discussed.
Depending on the value of the separated substance (for example
iodine, mercury) one or other type of regeneration will be taken
into account. In some cases hot gas desorption or wash with alkaline solutions, acid or water, sometimes also solvent extraction
will be applied. Often the spent activated carbon may be replaced
by new carbon for economic reasons, the so-called throw-away carbon.

Since the use of activated carbon for the removal of radioactive iodine has already been mentioned above (on which more is
to be said in a subsequent lecture) the emission of radioactive
noble gases from nuclear reactors should be briefly discussed.
This emission comes from turbine condensate or from degassing
units of the primary coolant circuit and contains the radioactive
noble gases krypton and xenon from defective fuel elements whose
emission through the stack is not permitted. In principle it
would be possible to allow decay of the radioactivity by storage
in delay tanks. On the other hand it is cheaper, better and less
hazardous to use the gaschromatographic principle with activated
carbon as adsorbent. In this case the noble gases are emitted
after a delay according to the specific retention time of the system Air-Krypton or Air-Xenon respectively.

In connection with the impurities, mention was made earlier
in the lecture of valuable substances whose recovery is profitable.

In many cases their recovery not only improves the profit-
ability of a lot of production processes, but also and not least,
prevents the discharge of toxic substances and is therefore a vital
contribution to the prevention of air pollution. Above all the
recovery process can be used in all industrial operations employing solvents, as is shown in Figure 10.

Figure 11 shows the flow diagram of a recovery plant.

The solvent-laden air is drawn off by fans at the evaporation
points and passed through one or more adsorbers filled with activated carbon. The activated carbon adsorbs the solvent from the

Applications of activated carbon	Air pollutants
Gravure printing	Toluene, benzene, xylene, chlorinated hydrocarbons
Fibre production	Methylene chloride, alcohol, carbon disulphide (hydrogen sulphide), acetone, dimethyl sulphide
Rubber and adhesives industry	Gasoline, toluene, petrolether, benzene, xylene
Coating industry	Ester, alcohols, dimethylformamide, ketones, tetrahydrofurane, chlorinated hydrocarbons, toluene.
Dry cleaning	Gasoline, chlorinated hydrocarbons, fluorinated chlorine hydrocarbons
Plastics industry	alcohols, ester, ketones, ether, chlorinated hydrocarbons

FIGURE 10.

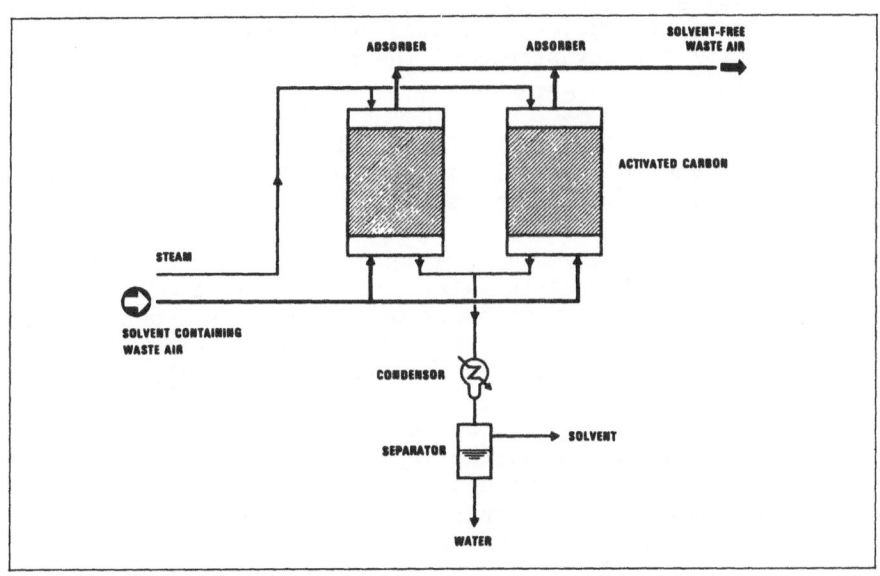

FIGURE 11.   Supersorbon process flow diagram.

air, the purified air leaving the adsorber from the top. Loading of the activated carbon is continued until breakthrough occurs, i.e., traces can be observed in the emerging air or gas. A highly sensitive concentration-measuring instrument indicating this break-through point, the next adsorber can be put into operation auto-matically. This operation can be initiated also by timing the cycles. The saturated adsorber is regenerated (a process step also termed desorption) which is effected by applying superheated steam counter-current to the charging flow. The scouring steam desorbs the organic substances from the activated carbon and the water-solvent-steam mixture is passed to a condenser downstream of the adsorber.

For economic reasons, 100% desorption cannot be carried out and a certain residue remains depending on the organic compound in the activated carbon.

Nowadays, drying and cooling of activated carbon is necessary only in special cases. Generally the temperature of the exhaust air is sufficient to drive out quickly the residual moisture taken up by the activated carbon during the steaming cycle.

Following condensation, water-immiscible solvents may be sepa-rated due to their different specific gravities (e.g. carbon di-sulphide, gasoline, benzene, chlorinated hydrocarbons). This is effected in a separator downstream of the condenser. The organic solvent can immediately be re-used.

For solvents partially or completely miscible in water (e.g. acetone, alcohol, acetic ester) separation can be effected by en-richment condensation or by distillation.

Using the enriching condensation as a regeneration stage as shown in Figure 12, the waste water is largely free of solvent. The water content in the solvent corresponds to the solvent equi-librium. If the water content (mostly between 1 - 3 vol %) is too high, an additional drying stage is required (Figure 13).

In practice, solvent concentrations in the exhaust gas range from 2 to 20 g/m^3. Depending on the explosion limit of the solvent concerned, higher concentrations can be employed (Figure 5). If it is necessary to separate the recovered solvents into the individual compounds, distillation or, better, rectification steps are avail-able.

The design of the adsorption plant depends on the required air throughput, the quantity of solvent to be separated per hour and the required purity of the treated air.

The solvent recovery process mentioned can be used in all

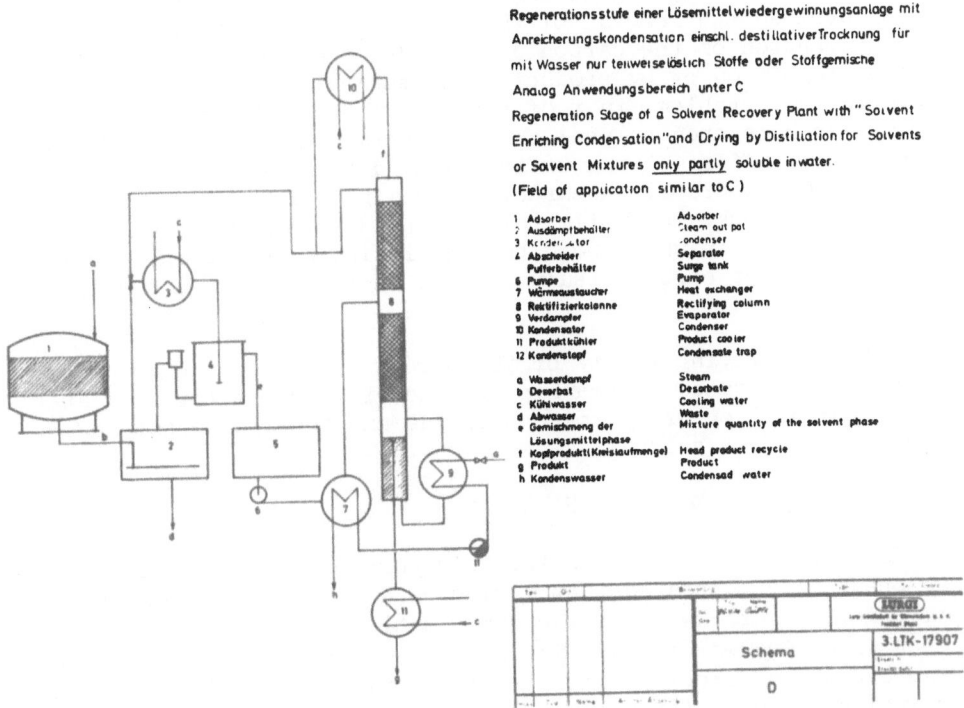

Regenerationsstufe einer Lösemittelwiedergewinnungsanlage
mit Anreicherungskondensation für
mit Wasser teilweise lösliche Stoffe oder Stoffgemische
Regeneration Stage of a Solvent Recovery Plant with "Solvent
Enriching Condensation" for Solvents and Solvent Mixtures partly
soluble in Water

1 Adsorber	Adsorber
2 Ausdämpfbehälter	Steam out pot
3 Kondensator	Condenser
4 Abscheider	Separator

a Wasserdampf	Steam
b Desorbat	Desorbate
c Kühlwasser	Cooling water
d Abwasser	Waste
e Produkt	Product

Anwendungsbereich zum Beispiel für die Wiedergewinnung von:
Applied for the recovery of:

Äthylacetat	Ethyl acetate
Äthylacetat - Alkohol	Ethyl acetate alcohol
Toluol – Alkohol	Toluene - alcohol
Toluol – THF	Toluene - THF
Toluol – MEK	Toluene - MEK

Bemerkung: Das Abwasser ist weitgehend Lösungsmittel frei. Das Produkt enthält
einen Wasseranteil entsprechend dem Lösungsgleichgewicht.
Sollte der Wassergehalt (in den meisten Fällen zwischen 1-3 Gew%)
stören ist eine zusätzliche Trocknungsstufe erforderlich

Remark: The waste water largely free of solvent. The water content in the
correspond to the solvent equilibrium
If the water content (mostly between 1-3 volume%)
is too high, an additional drying stage is required

FIGURE 12. Regeneration stage of a solvent recovery plant.

Regenerationsstufe einer Lösemittelwiedergewinnungsanlage mit
Anreicherungskondensation einschl. destillativer Trocknung für
mit Wasser nur teilweise löslich Stoffe oder Stoffgemische
Analog Anwendungsbereich unter C
Regeneration Stage of a Solvent Recovery Plant with "Solvent
Enriching Condensation" and Drying by Distillation for Solvents
or Solvent Mixtures only partly soluble in water.
(Field of application similar to C)

1 Adsorber	Adsorber
2 Ausdämpfbehälter	Steam out pot
3 Kondensator	Condenser
4 Abscheider	Separator
5 Pufferbehälter	Surge tank
6 Pumpe	Pump
7 Wärmeaustauscher	Heat exchanger
8 Rektifizierkolonne	Rectifying column
9 Verdampfer	Evaporator
10 Kondensator	Condenser
11 Produktkühler	Product cooler
12 Kondenstopf	Condensate trap

a Wasserdampf	Steam
b Desorbat	Desorbate
c Kühlwasser	Cooling water
d Abwasser	Waste
e Gemischmeng der Lösungsmittelphase	Mixture quantity of the solvent phase
f Kopfprodukt (Kreislaufmenge)	Head product recycle
g Produkt	Product
h Kondenswasser	Condensad water

		Schema	3.LTK-17907
			0

FIGURE 13. Regeneration stage of a solvent recovery plant including drying by distillation.

industrial operations employing solvents as shown in Figure 10.

The utilities comprise steam (dependent on the solvent type 2 - 5.5 kg/kg solvent) electrical energy (according to plant conditions up to 0.25 kWh/kg solvent) and cooling water (between 30 and 50 l/kg solvent). The activated carbon consumption is low, ranging between 0.5 and 1 kg per ton of recovered solvent which, in normal operation, is caused by attrition and thermal influence.

The enormous significance of the use of activated carbon, not only as adsorbent but also as a catalyst and as a carrier material, becomes apparent above all in the field of desulphurization of exhaust air from viscose plants.

The previously described recovery process was the origin of a process - named Sulfosorbon process - which now, after several stages of development, has attained a sophisticated technical level and is successfully used in the purification of exhaust air from viscose plants.

The air exhausted from a viscose plant contains both hydrogen sulphide and carbon disulphide. The original Supersorbon process involved the removal of the former by scrubbing followed by the adsorption of the carbon disulphide on activated carbon. Due to the formation of by-products, the scrubbing process resulted in a further problem, that of treating the waste water. In addition, the recovered sulphur could be obtained to the required purity only after melting, extraction, distillation and crystallization. Considerable research, including prolonged laboratory and pilot scale work, led ultimately to the development of the Sulfosorbon process in which both the separation of hydrogen sulphide and recovery of carbon disulphide are carried out in a single-unit operation.

The underlying theory of the process involved the use of a wide-pore activated carbon impregnated with iodine for the catalytic decomposition of hydrogen sulphide and the subsequent adsorption of the free sulphur:

$$H_2S + \tfrac{1}{2} O_2 \rightarrow H_2O + S$$

The activated carbon suited for this purpose had to be newly developed and it appeared that a wide-pore activated carbon was preferable. Figure 14 shows the differential pore distribution of such an activated carbon. The properties of this activated carbon improve the adsorption capacity of a catalyst salt, the high adsorption capacity for sulphur and finally the easy removal of the sulphur on regeneration.

For the physical adsorption of the CS_2, a high-grade fine-

503

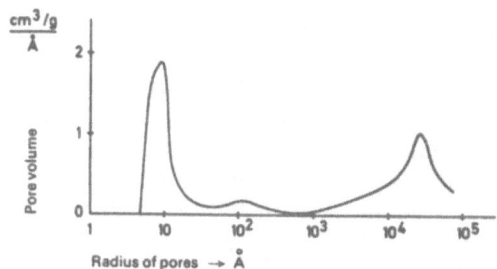

FIGURE 14.   Differential pore distribution of activated carbon
             (open-pored).

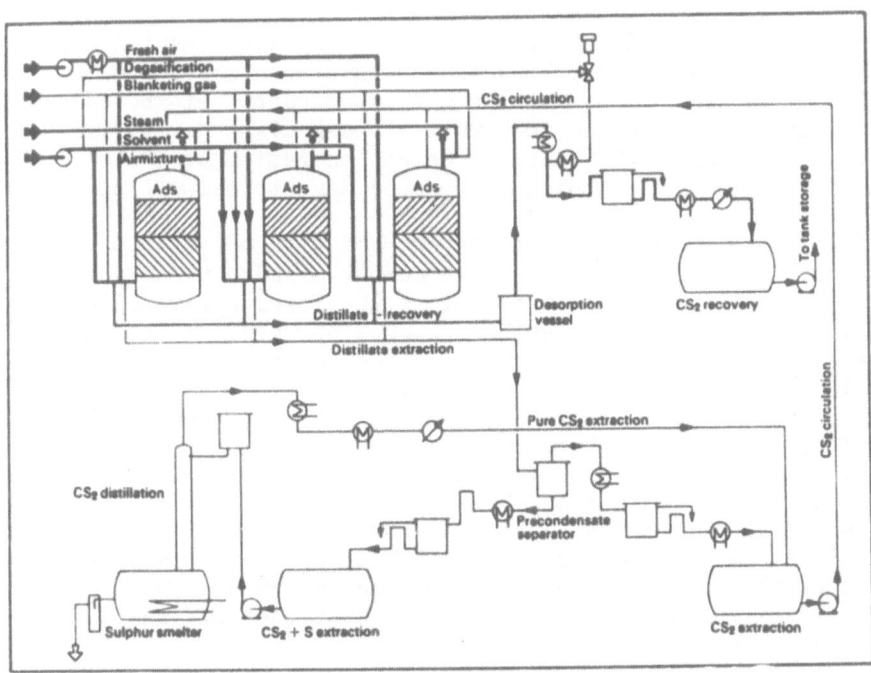

FIGURE 15.   Layout of a Sulfosorbon plant.

pored hydrophobic activated carbon was used.   Both activated carbon
grades are prepared as layers in a single adsorber.   The process
may be explained with reference to a simplified flow diagram (Figure 15).

The exhaust air of the viscose plant (carbon disulphide concentration 1 to 10 g/m^3, hydrogen sulphide 0.5 to 2 g/m^3) is blown through the activated carbon adsorber by a fan. The hydrogen sulphide is converted into elementary sulphur and retained in the lower wide-pored activated carbon layer. The carbon disulphide is, however, adsorbed in the upper fine-pored carbon layer, and desorbed with steam of 110 - 130°C after breakthrough of the first carbon disulphide traces in the adsorber discharge. The steam-CS_2 mixture is passed through a condenser and the CS_2 subsequently recovered from the water in a separator. The recovered carbon disulphide can be re-used in the viscose plant. For safety reasons, the adsorber is filled with inert gas before desorption in order to displace the oxygen. Drying and cooling of the activated carbon with an additional heated clean-air cycle, formerly required after steam desorption, can be avoided in modern plants by application of a hydrophobic activated carbon and by making use of the elevated temperature of the polluted air which is sufficient for drying the activated carbon during charging, thus saving investment and operating costs. Special adsorber designs have been developed to limit the moisture content of the activated carbon at the adsorber wall during desorption and to permit viscose plant air purification down to a residual carbon disulphide concentration of 50 to 100 mg/m^3. The hydrogen sulphide is removed down to below 1 ppm.

Since activated carbon may be charged with up to 100% of its weight of elemental sulphur a large number of adsorption and desorption cycles can be executed before the lower activated carbon layer is saturated. On breakthrough of 1 ppm hydrogen sulphide into the upper layer of activated carbon the adsorber is regenerated. To remove sulphuric acid the adsorber is backwashed with water and the sulphur is extracted with carbon disulphide. After removal of the sulphur the wetted carbon disulphide is desorbed with steam and the moist activated carbon is dried and cooled. All process steps are carried out within the adsorber so that a long useful life of the activated carbon is guaranteed. The carbon disulphide is evaporated from the sulphur-CS_2 mixture and the heat content of the adsorbate vapours from the carbon disulphide recovery cycle may be used as a source of heat energy. The sulphur is melted out at a purity of 99.9%.

For cleaning 100.000 Nm^3/h viscose exhaust air containing on average 10 g CS_2 and 2 g H_2S/m^3 (according to 1 t of recovered carbon disulphide) the following utilities are required:

Steam	10 t
Cooling water	150 m^3
Protective gas	300 Nm^3/h
Electrical energy	700 kWh/h
Activated carbon	2 kg
Catalyst	0.5 kg

Extraction of activated carbon with carbon disulphide for removal of sulphur is possible without difficulty under the prescribed safety conditions. After melting, the sulphur is very pure and requires no additional purification. Catalyst consumption is so low that re-impregnation of the activated carbon is required at the earliest after 10 carbon disulphide extractions.

The combination of the two activated carbon quantities in two layers enables high carbon disulphide charging in the upper layer based on the high sulphur separation in the lower layer so that the required low carbon disulphide content in the exhaust air can be obtained according to the prevailing conditions.

Adsorption processes which are used almost exclusively with activated carbon on account of its special affinity to organic vapours in exhaust air purification, have of course advantages, but also disadvantages. The advantages are the wide range of application regarding the charged activated carbon.

The disadvantages are that adsorption plants, in particular those using activated carbon as adsorbent, should not be used or used only conditionally

- for organic vapours, or compounds which tend to resinify

- for mixtures of organic compounds or individual organic compounds which cannot be re-used

- for organic compounds which after recovery can be further processed only by providing additional equipment (for example distillation, rectification)

- for mixtures of organic compounds with frequently changing concentrations and compositions.

This lecture dealt initially with a process for Claus tail-gas purification using alumina as adsorption catalyst.

That process - jointly developed by SNPA (France) and Lurgi from laboratory scale to commercial scale and known as Sulfreen process - is based on the conversion of SO_2 and H_2S according to the Claus reaction.

According to the flow-sheet (Figure 16), Sulfreen plants consist essentially of at least two reactors containing catalysts, and of the regenerating equipment, such as regenerating gas blower, gas-fired regenerating gas heater and the sulphur condenser, designed as a waste heat boiler.

The tail gas from the Claus plant is passed at the usual temperature of 120 - 140 C through one of the two reactors from

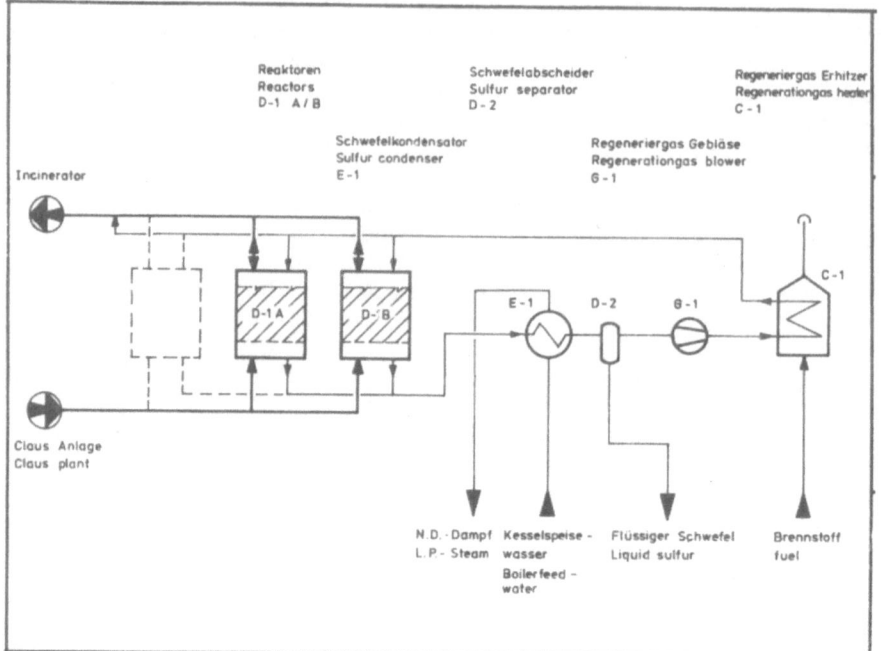

FIGURE 16. Flow-sheet of a Sulfreen plant.

the bottom to the top where by means of the catalyst most of the
sulphur compounds H_2S and SO_2 are converted into elemental sulphur.
This elemental sulphur and the elemental sulphur carried in the
tail gas as vapour and aerosols are adsorbed on the catalyst. The
purified tail gas leaves the reactor and must be further treated
in an incinerator.

During the operating period of the reactor in the tail gas
purification cycle, which has a duration of several hours, the
sulphur-laden catalyst of the other reactor is regenerated within
approx. the same period of time by using a desorption temperature
of approx. 300°C.

The regeneration gas passes from top to bottom through the
reactor to be regenerated, thereby expelling most of the elemental
sulphur as vapour. The major part of the desorbed elemental sulphur is subsequently condensed in the sulphur condenser and the
heat removable from the desorption gas is utilized for the production of low pressure steam.

After sulphur desorption the regeneration gas cycle is purged
with raw gas and the hot catalyst is then cooled by means of

purified gas.

After completion of the cooling period, the regenerated reactor is again ready to be switched to adsorption.

Originally specially impregnated activated carbon was used as adsorption catalyst.

The systematic further development of this process soon revealed that also activated alumina was suitable as adsorbent, the more so as the properties of the alumina could be improved, and by modifying the regeneration process it was possible to keep the activity of the improved catalyst at a high level.

The lower regeneration temperatures made possible the use of carbon steel as material for almost all parts so that the use of alumina also resulted in considerable savings of investment costs.

Under optimum conditions, i.e., if the components H_2S and SO_2 are present in the tailgas at a ratio of 2 : 1, sulphur conversion rates of 85 - 90% can be reached. The conversion rate depends on the concentration of the sulphur compounds in the tail gas of the Claus plant. In practical terms this means that by the combination of the Claus plant with the Sulfreen plant a total yield of above 98% can be reached. This high efficiency also depends on the content of organic sulphur compounds in the tail gas (COS, CS_2) because these compounds are only partially converted under the conditions prevailing in the process. In a large-scale plant which has been in operation since October 1973 a total sulphur conversion of 99.4% related to the Claus furnace + Sulfreen process is still reached. The residual content of H_2S and SO_2 thus is between 1500 and 2000 ppm.

The utilities consumption per ton recovered sulphur is:

Electrical energy	190 kW
Cooling water	1 m^3
Fuel gas	250 Nm^3
Boiler feed water	1 m^3
Catalyst	2 - 4 kg

The process leaves no liquid wastes. The degree of purity of sulphur recovered is equal to that of Claus plants.

PROTECTION AGAINST TOXIC VAPOURS

M. van Zelm and P.C. Stamperius

Chemical Laboratory TNO, Rijswijk, The Netherlands

ABSTRACT. Protection against the inhalation of toxic vapours can be obtained by purification of the breathing air. The filters used for the purification usually contain activated charcoal, having a large capacity for physical adsorption. In addition the charcoal is impregnated with various substances to improve the protection against weakly adsorbed gases. As a result protection may be achieved via different mechanisms. In this paper the attention is focussed upon the effect of water vapour, always present in the air to be purified as regards the various mechanisms.

1. INTRODUCTION

Surrounded by a toxic environment, whether in industry or in case of chemical warfare man has to protect himself. This can be achieved by entering a closed space ventilated with filtered air or by using individual respiratory protective equipment, of which a gas mask is a well-known specimen. The efficiency of the protection provided by such mask is determined to a large extent by the sealing of the facepiece, the proper functioning of valves and the filtration of vapours and aerosols by the filter element. In this paper the filter element will be dealt with, especially its capability of filtrating vapours. The filtration of aerosols and the proper functioning of the mechanical parts of the protective equipment will not be discussed.

Active charcoal is widely used for the filtration of toxic vapours because of its large capacity for physical adsorption. A particular problem is the choice of the grain size of the adsorbent. Especially for use in air-purifying respirators a low pressure drop is required which induces the choice of relatively large pellets. Often cylindrical extruded grains of a diameter of 1 mm and with lengths of two to three times the diameter are used. However, the larger the pellets the less efficient the adsorption process will be and a suitable compromise has to be reached. Anyhow, the pellets need to be of rather uniform size to prevent channeling and they must have sufficient strength to avoid excessive attrition in use.

For some toxic substances physical interaction is too weak to provide sufficient protection. Various impregnations have been applied in the past to enhance the protection afforded against some specific compounds. In the next section some of these impregnations will be discussed.

The physical adsorption as well as the chemical processes taking place on activated charcoal are strongly influenced by water vapour. For a few examples the influence of water in the filtration processes will be discussed. Finally the particular case of protection against carbon monoxide will be dealt with briefly.

2. IMPREGNATION OF CHARCOAL

As already stated the impregnation of charcoal was developed in order to provide sufficient protection against those gases which are only weakly bonded by physical interaction. To arrive at the desired protection various impregnants have been used. Table I lists some of these impregnants (1).

Table I
Examples of impregnations of activated
and corresponding reactions

Impregnation by	effective substances	kind of reaction	stoichiometric reaction
Copper	CuO, Cu_2O	chemical	$Cu_2O + 2HCN \rightarrow H_2O + 2CuCN$
Zinc	ZnO, Na_2ZnO_2	catalytic	$3O_2 + 2AsH_3 \rightarrow As_2O_3 + 3H_2O$
		chemical	$ZnO + 2HCN \rightarrow Zn(CN)_2 + H_2O$
		chemical	$ZnO + COCl_2 \rightarrow ZnCl_2 + CO_2$
Silver	Ag, Ag_2O	catalytic	$3O_2 + 2AsH_3 \rightarrow As_2O_3 + 3H_2O$
Copper	$CuSO_4 \cdot 5aq$	chemical	$CuSO_4 \cdot 5aq + 4NH_3 \rightarrow$ $Cu(NH_3)_4SO_4 + 5H_2O$

By a suitable choice of the impregnants an almost universally effective adsorbent can be prepared. An impregnation commonly used consists of a mixture of copper, chromium and silver salts. An active charcoal impregnated with these substances will provide good protection against physically adsorbed gases as well as hydrogen cyanide, cyanogen chloride and arsine.

Usually the active charcoal is impregnated to provide protection against a limited number of gases only. For industrial use these filters are coded using different colours and a letter code. The codes in the various countries differ considerably but those laid down in German Standard DIN 3181 are well-known in most European countries (see Table II). At present attention is given in the European Committee for Standardization as well as in the International Standards Organisation to achieve international agreement on coding filters for industrial use and, much more important, on the requirements for these filters.

Table II
Code for filters for respiratory protective
devices according to German Standard DIN 3181

letter code	colour	principal application
A	brown	organic vapours, solvents
B	grey	acid gases (e.g. halogens, hydrogen halogenides, hydrocyanic acid, nitrous fumes, arsine)
CO	black band	carbon monoxide
E	yellow	sulfur dioxide
K	green	ammonia

In general impregnation of activated charcoal will decrease the physical adsorption capacity but this capacity is such large that a suitable compromise can usually be found. With such an impregnated activated charcoal several mechanisms may be distinguished with respect to the removal of toxic substances. In addition to physical adsorption chemical reactions with the impregnant or catalytic decomposition induced by the impregnants are possible. A catalytic process is generally to be preferred as a limited amount of catalytic material is capable to convert large quantities of toxic vapours. However, very often the catalyst is poisoned by reaction products or by other constituents from the air. In these cases it would be more appropriate to talk about physical adsorption followed by chemical modification.

In conclusion toxic substances are retained by impregnated active charcoal through one of the following mechanisms:

i. Physical adsorption. Chloropicrin (CCl_3NO_2) or benzene are well-known examples of gases that are adsorbed physically.
ii. Chemical reaction with the impregnants like neutralisation of acid gases, complex formation of ammonia or oxidation of arsine.
iii. Physical adsorption followed by chemical modification. Well-known examples are the filtration of cyanogen chloride which is catalytically decomposed by chromium impregnations on the active charcoal and the hydrolysis of nerve agents.

The hydrolysis mechanism is operative for most chemical warfare agents; like the very toxic nerve agent sarin (isopropyl methylphosphonofluoridate):

$$\text{(CH}_3)_2\text{CH-O-P(=O)(CH}_3)\text{F} + H_2O \rightarrow \text{(CH}_3)_2\text{CH-O-P(=O)(CH}_3)\text{OH} + HF$$

The hydroxysarin formed is much less volatile and harmful and strongly held by the impregnants via its POO group.
The rate of this hydrolysis reaction is rather low.

There could be some discussion whether this last reaction is of the type physical adsorption followed by chemical modification or of the type chemical reaction with the impregnant. The distinction between the two types is that in the first case the compound is adsorbed and a decomposition is induced by the impregnant in which the effectiveness of the impregnant might decrease, whereas in the second case a direct reaction between toxic compound and impregnant provides the protection.

Not much is known about the actual reaction mechanisms and about the kinetics of the reactions occurring. The choice of a certain impregnant is often more a matter of empirism than science.

3. THE EFFECTS OF WATER

Both in collective and individual protection the toxic vapours to be removed are always accompanied by water vapour, profusely present in the atmospheric air. The water vapour generally reduces the protection properties although for the removal of certain toxic gases the presence of some water on the charcoal is essential. In this section some of the effects of water vapour on the protective properties will be discussed using the division in filtration mechanisms given before.

3.1 Water vapour and the physical adsorption capacity.

The adsorption-desorption isotherm of water vapour onto active non-impregnated charcoal generally is of the form as illustrated in Fig. 1.

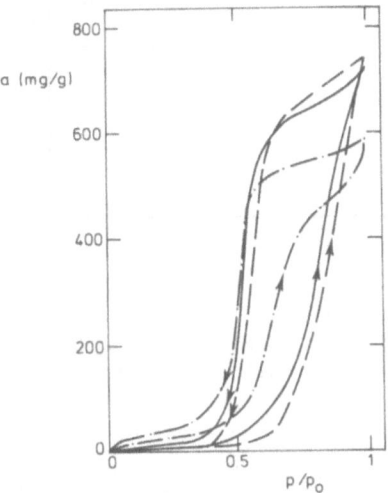

Fig. 1 Water vapour adsorption-desorption isotherms
——— RS, non-impregnated charcoal
–·–·– RSI, impregnated charcoal
– – – RST, RS after removal of inorganic substances and hydrogen treatment at 900°C.

At low relative humidities the water uptake by activated charcoal is relatively small as compared to e.g. silicagel. This can be explained by the hydrophobic nature of the carbon. Above a certain relative pressure of the water vapour the water content increases rapidly due to capillary condensation of water vapour in micropores. The relative pressure at which this process starts depends on the size of the pores and on the number of hydrophylic sites at the surface. In case of impregnated charcoal (RSI) a large amount of hydrophylic material is present and the adsorption branch of the water vapour isotherm shifts to considerably lower relative pressures compared to the non-impregnated adsorbent.
The sharp increase in water content at a given relative pressure is reflected in a sudden decrease in the protection capacity for chloropicrin for example. This is illustrated by the results of the experiments given in table III.

The rate of water uptake under static conditions, i.e. in which the water vapour has to enter a filter by diffusion, is rather small. A gas mask canister attached to the gas mask, being

514

Table III Effect of water on the breakthrough time for chloropicrin adsorbed on active charcoal	
active charcoal in equilibrium with the relative humidities (%)	breakthrough time (min)
20	290
50	275
75	75
90	35

protected at one side by the inlet valve, takes up only a few percent by weight of water when kept in an atmosphere of 60-100% R.H. for a couple of month's time. The water uptake is much more rapid when the filter is used, depending on the relative humidity and the breathing rate.

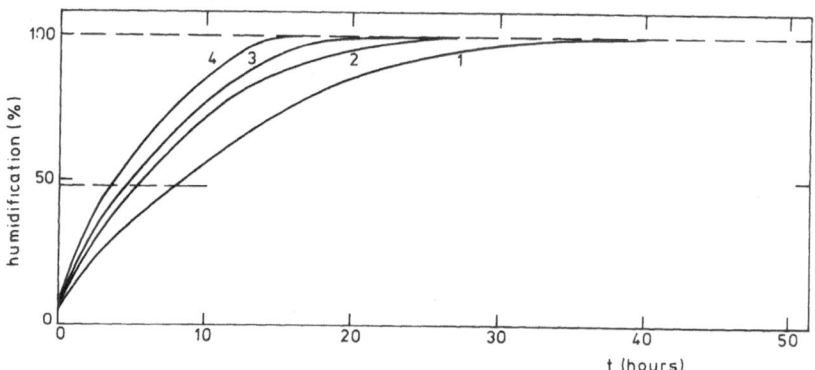

Fig. 2 The rate of water vapour uptake by a gas mask filter worn in an atmosphere of 90% R.H. at different breathing rates

Figure 2 shows for a certain gas mask canister the increase in water content in an atmosphere of 90% R.H. using a breathing machine to simulate human respiration at various breathing volumes and frequencies. It is seen that complete saturation at high relative humidities may take as much as 36 hours. However, it follows from table III that long before that time a considerable decrease in protection capacity has occurred. When the conditions of use are not exactly known it is therefore advisable to test these filters in the laboratory under the most unfavourable condition, i.e. in a highly humid atmosphere.

3.2 Water vapour and the chemical reaction capacity

As an example we take the case of hydrogen cyanide. As was shown
in table I water is one of the reaction products in the reaction
of HCN with the impregnant and one might therefore expect a detrimental influence of an increase of water content on the protection
capacity against HCN. Results of breakthrough tests of HCN on charcoal containing various amounts of water, are shown in figure 3.
Although the data are somewhat scattered no substantial effect of
the water content can be observed. This may be explained by the
fact that the bulk of the water will be in the micro pores and
does not influence the reaction of HCN with the impregnant.

However, we found that a considerable hazard may result from
an additional reaction of the HCN with the impregnant. In fact
this is more an interaction of the group iii type in which a reaction of HCN is induced by the impregnant and water which leads to
the formation of a dangerous compound, that is easily desorbed.
In the HCN tests usually a detection method is used with cupri-
acetate and benzidine-acetate (3). Because of this it will pass
unnoticed that under wet conditions cyanogen ($(CN)_2$) is formed.
Cyanogen is nearly as toxic as hydrogen cyanide (4,5).

The presence of cyanogen in the effluent may be noticed when
the effluent air is led through two washing bottles, the first
containing silver nitrate, the second sodium hydroxide. If HCN is
present then in the first washing bottle silver cyanide will precipitate and in the second bottle CN^--ions will be formed by hydrolysis of $(CN)_2$ and these can be detected and measured by a specific
CN^- electrode.

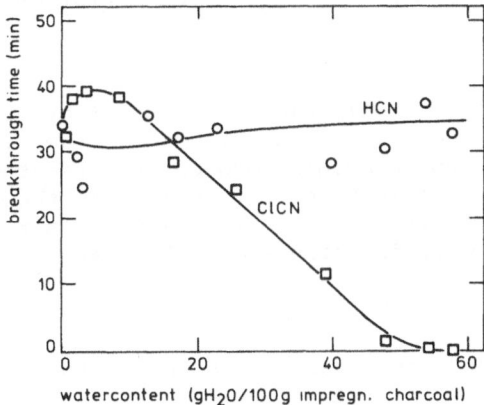

Fig. 3 The effect of water on the breakthrough times of impregnated
charcoal for cyanogen chloride and hydrocyanic acid.

Recently we investigated this for a number of industrial filters designated according to German Standard DIN 3181 with B (protection against acid gases) and BK (ibid, inclusive ammonia). These filters are made to protect against maximum concentrations of 0.2 or 2.0% (vol/vol). In table IV the results are given.

Table IV

The occurrence of cyanogen in the effluent of type B and BK of industrial filters when tested with hydrocyanic acid under wet conditions (flow rate: 30 ℓ/min, relative humidty of test mixture 75%, filters as received)

type of filter	concentration of HCN (vol.%)	breakthrough time for $(CN)_2$ (min)	breakthrough time for HCN (min)
B St 2 %	0.52	34	36
B St 2 %	0,21	> 71	71
B 0.2%	0.21	> 58	58
B 0,2%	0,21	> 44	44
B 0.2%	0.21	> 50	50
BK St 2 %	0.28	14	44
BK St 2 %	0.61	0*	
BK St 2 %	0.21	7	30

*Immediate breakthrough of $(CN)_2$ in concentrations larger than 10 micrograms/ℓ.

With the type B filters only small amounts of cyanogen are formed until the breakthrough of HCN. However, with the type BK filters breakthrough of cyanogen occurs long before breakthrough of HCN. It is obvious that one must be very careful in the choice of impregnants, in particular in combining various impregnations on one and the same charcoal. The effect of water vapour should never be neglected.

3.3 Water vapour and the physical adsorption followed by chemical modification.

Let us consider the effect of water on the filtration of cyanogen chloride. It is assumed that cyanogen chloride is decomposed according to the overall reaction cheme: $ClCN + 2H_2O \rightarrow NH_4Cl + CO_2$.

517

From this reaction scheme it is evident that the presence of some water on the impregnated charcoal is essential. However, the first step has been proven to be a physical adsorption and consequently the presence of large quantities of water is detrimental to the protection.
Figure 3 shows the influence of the water content on the breakthrough times for cyanogen chloride. The figure shows the optimum in breakthrough time as would be expected from the mechanism mentioned before.

Another aspect of the influence of moisture on the cyanogen chloride protection is the loss in protection capacity when the moist charcoal is stored, particularly at high temperatures. The mechanism of this degradation is not known. Figure 4 shows the effect of aging on batches of the same impregnated charcoal, kept dry (less than 2% of water) at room temperature (A), humidified to saturation at 90% R.H. and kept at room temperature (B) and humidified to saturation at 90% R.H. and kept at 50°C.

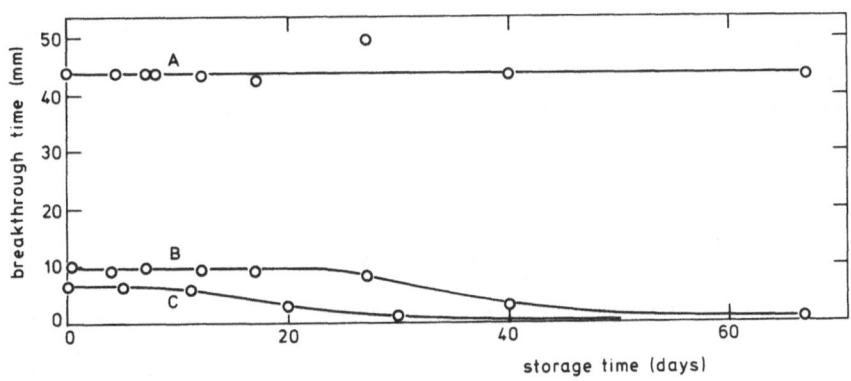

Fig. 4 Influence of ageing under different conditions on the breakthrough time for cyanogen chloride.

Attempts have been made to correlate the gradual decrease of protective capacity with changes in the valency of the chromium ions or with loss of ammonia, used in the impregnation process, but so far no satisfactory explanation has been given.
The addition of pyridines appears to have a favourable effect upon the protective capacity under humid conditions. We have investigated active charcoal that has been impregnated using a solution of $CuCrO_4$ in ammonia and various quantities of pyridine and γ-picoline. After equilibration at 90% R.H. the protective capacity for cyanogen chloride was determined in breakthrough tests. Several batches of impregnated charcoal were prepared. Due to irregularities in the copper and chromium content of the commercially obtained copper

charcoal batch no.	% Cu	% Cr	breakthrough time (min)	addition
			Table V	
			The effect of the addition of pyridine or γ-picoline upon the protective capacity for cyanogen chloride under humid conditions	
45a	3.00	1.00	3	–
45b	3.13	1.00	16½	2% pyridine
45c	2.56	1.00	16	4% pyridine
46a	3.31	1.05	7	–
46b	3.49	1.08	12	0.5% γ-picoline
46c	3.54	1.09	17½	1.0% γ-picoline

chromate a mutual comparison of these batches is not allowed. However, in each series the breakthrough times prove to be increased by adding a few percent of pyridine or γ-picoline (see table V).

Table V shows that the addition of pyridine or γ-picoline in relatively small quantities has a favourable effect on the protective capacity. The effect of the addition of pyridine upon the storage properties under humid conditions has not been investigated as yet.

3.4 Conclusions as to the effect of water vapour

Water vapour strongly affects the physical adsorption capacity of charcoal. The effects are very severe above a given relative humidity when the water vapour isotherm shows a steep rise, due to the filling of the micro pores. The value of the relative humidity at which this steep rise occurs is lower for an impregnated charcoal than for the non-impregnated charcoal.

Chemical reactions between the impregnant and the adsorbate are not influenced to a large extent by water, even if water is one of the reaction products. In the case of physical adsorption followed by chemical modification, as with cyanogen chloride, again a large influence of water vapour is observed. Water vapour may have a detrimental effect upon the quality of the impregnant as is illustrated by the aging of the impregnation for cyanogen chloride.

A striking example of the effect of water is the induction of a chemical reaction producing a desorbing toxic compound as in the case of hydrogen cyanide. In this case apparently well-designed filters offered no protection at all when water vapour is present.

4. PROTECTION AGAINST CARBON MONOXIDE

Activated charcoal itself does not offer any protection against carbon monoxide and no impregnation for activated charcoal is known that will deal with it efficiently. For protection against carbon monoxide a special filter filled with a mixture of desiccant and oxidation catalyst must be used.

So-called hopcalites, mixtures of oxides of various metals, usually cupric oxide and manganese dioxide are used as catalysts. Both carbon monoxide and oxygen are chemisorbed on the surface of the oxides, forming intermediate states. These react with each other in series of steps, resulting in the formation of carbon dioxide which is desorbed.

The hopcalite will give satisfactory protection only as long as it is kept scrupulously dry. Therefore two layers of desiccant are placed at both ends of the canister to protect the hopcalite against water vapour from the ambient. A third layer of drying agent is present between the two layers of hopcalite to protect the second layer against any water that might have been formed in the first layer by catalytic oxidation of hydrogen or hydrocarbons. The filter is exhausted as soon as the layers of drying agents are exhausted, even if no carbon monoxide has ever passed through the filters.

A remedy for the inactivation has been claimed during the second World War (6). In this method divalent or multivalent silver oxide is prepared by reaction between potassium persulphate and silver nitrate. The active compound called argentic oxide AgO, reacts with carbon monoxide to produce metallic silver and carbon dioxide, also in the presence of water vapour, but the rate of reaction is too low to apply the material for protective purposes. Therefore one or more accelerating metals (e.g. oxides of manganese or cobalt) are added at some sacrifice of the stability in the presence of water vapour.

The patents describe a few satisfactory tests in which the material was put onto asbestos as a support. To our knowledge the materials described in the patents have never been applied in industrial carbon monoxide filters, probably because of economical reasons.

REFERENCES

1. Smisek, M., Active Carbon, Elsevier, Amsterdam, 1970.
2. Van Aken, J.G.T., Thesis, Delft, 1969.
3. Bark, L.S., Higson, H.G., Analyst, 88, 751,(1963)
4. Locket, S., Clinical Toxicology, Henry Kimpton, London, 1957.
5. Patty, F.A., Industrial Hygiene and Toxicology, Vol. II, Toxicology, Interscience Publishers John Wiley, New York, 1963.
6. Brit. Pat. 579.809 and 579.817.

TECHNICAL APPLICATION OF SORPTION IN THE NUCLEAR FIELD

J. G. Wilhelm

Radiation Protection and Safety Department/Chemistry
Karlsruhe Nuclear Research Center, Karlsruhe,
Federal Republic of Germany

ABSTRACT. Practical nuclear applications of the sorption of
gases and vapors to solid sorption materials are the removal
of fission product iodine by iodine filters and the delayed
release of noble gases. The use and the design of iodine
filters for nuclear power stations are briefly mentioned. The
removal reactions occurring on the sorption materials are
explained. The requirements to be fulfilled by post accident
recirculating type containment filters and filters for the
head-end removal of iodine from the off-gases of reprocessing
plants are indicated. The paper contains a brief outline of
the use, design and construction of fission gas holdup beds
in nuclear power reactors with data on the source strengths
of activation gases and noble fission gases contained in the
gaseous effluents of boiling water reactors.

1. INTRODUCTION

The ventilation air and the gaseous effluents of reactors and
reprocessing plants contain gaseous fission products especially
of iodine, xenon, krypton and, due to various nuclear reactions,
tritium. Ruthenium can also generate compounds which must be
taken into account because of their high volatility. This
particularly applies to the gaseous effluents generated in
reprocessing spent fuel. In this paper the removal of fission
product iodine and the release of noble gases through fission
gas holdup beds will be dealt with.

The use of impregnated activated charcoal (in the following
the term "charcoal" will always mean activated charcoal)

for trapping gaseous fission product iodine and for the adsorption of noble gases marks the present state of the art of treating ventilation air and gaseous effluents of nuclear facilities. In addition, inorganic iodine sorption materials have been developed for special purposes.

2. REMOVAL OF FISSION PRODUCT IODINE

2.1 Hazards Created by Fission Product Iodine

Today fission product iodine is regarded as the most dangerous fission product with respect to the exposure of the environment by reactors and reprocessing plants. This special position is due to the following properties:

- Some iodine isotopes are produced with a high fission yield.

- Elemental iodine and many iodine compounds are highly volatile.

- Nuclear fission among other products creates iodine isotopes with intermediate and very long halflives. Even long transport times will not sufficiently reduce the activity and activity concentration, respectively, by way of radioactive decay.

- Iodine is largely retained in the human body after having been inhaled or ingested.

- Iodine is enriched in the relatively small thyroid; accordingly, the thyroid may receive a high radiation dose; the probability of thyroid cancer increases with the radiation dose.

- There are enrichment mechanisms, e.g., in the dairy industry the expected enrichment in the milk of iodine precipitated on pastures.

Two iodine isotopes, ^{131}I with a halflife of 8.05 d and ^{129}I with a halflife of 1.7×10^7 a, are particularly important from the safety point of view. ^{131}I is of decisive importance with respect to the environmental exposure in the vicinity of nuclear power stations. The activity level due to ^{131}I in the exhaust air plume may be decisive for the minimum distance between nuclear power stations. Quantities of ^{129}I significant for safety considerations are produced in the gaseous effluents only during reprocessing and the subsequent treatment of radioactive wastes.

The permissible release of ^{131}I with the exhaust air is a
function of the siting criteria of nuclear facilities, but
in general it is extraordinarily low. A maximum release of
^{131}I through the exhaust stack of a nuclear power station
during normal operation on the order of ≤ 0.1 mCi/h requires
a concentration of fission product iodine in the filtered air
of $\leq 10^{-6}$ $\mu g/m^3$, if the volume flow of the gaseous effluent
to be filtered is taken to be 20.000 m^3/h and a ratio of
1 : 20 is assumed to exist between the quantities of ^{131}I and
the other fission product isotopes.

2.2 Iodine Filters in Nuclear Power Stations

Iodine filters are part and parcel of the usual engineered
safeguards of nuclear power stations. Calculations of the
anticipated environmental exposure in general show that
decontamination factors of the gaseous effluent between 10^2
and 10^3 are capable of sufficiently reducing the iodine
release.

Fig. 1 shows a modern iodine filter which is being used in a
number of reactor stations in this or a similar design. It
is remarkable that the type outlined here so far has not
shown any mechanical leak. This finding can be corroborated
by the results of a large number of in-place tests at various
reactor stations. It is impossible to mention at this point
the multitude of filter designs currently in use, but it
should be indicated that the occurrence of mechanical leaks
is not precluded with the required degree of assurance in
most filter designs. This requires expensive in-place tests
and retesting.

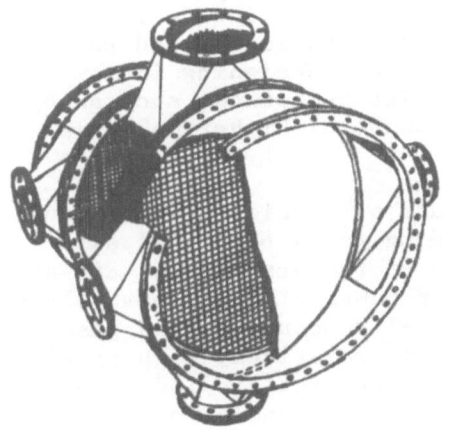

Fig. 1.

Activated charcoal filter
with a plane bed. The
bed arranged between the
perforated plates is
installed vertically. The
activated charcoal can
be filled in from the
upper connection and
removed through the bottom
connection. Air flow is
through the connections
indicated on the side of
the facility in this
design.

The number of iodine filters used in a nuclear power station
are a function of the reactor, containment and ventilation
concepts and the applicable safety rules. Experience has
shown that the number of iodine filters in a nuclear power
plant have increased with the advancing state of the art in
the nuclear field and with the growing of environmental
consciousness. In a modern pressurized water reactor in
Germany iodine filters are installed, e.g., to clean

- effluent air from the equipment area (non accessible
  rooms) of the reactor containment (continuous operation
  for maintenance of a slight negative pressure;

- effluent air from the annular space between the reactor
  containment and the secondary concrete shielding (use
  after incidents);

- recirculated air in the equipment area of the reactor
  containment (use in bypass operation);

- recirculated air in the operating area of the reactor
  containment (use as required).

Similar ventilation concepts are presently being discussed
for boiling water reactors.

2.3 Forms of Fission Product Iodine

Fission product iodine released may be present in various
airborne forms whose fraction of the total amount of fission
product iodine released may vary within broad limits and is
a function of the conditions prevailing during the release
and on the transport path. According to our present knowledge
it is especially elemental iodine, methyl iodide from the
reaction of organic air pollutants with iodine and particulate
iodine which must be borne in mind in designing iodine filters.
Additional inorganic gaseous iodine compounds, such as hydrogen
iodide, and organic iodine compounds of higher molecular
weights must be anticipated.

Under the low mass concentrations of fission product iodine
(on the order of magnitude of $\leq$ pg to mg per m^3 of air)
which must be expected to occur in the ventilation air and
the gaseous effluents of nuclear power stations both elemental
iodine and methyl iodide practically occur only as gases even
at room temperature because of the relatively high partial
vapor pressure. Since in addition particulate iodine must
also be removed, iodine filters always consist of at least
one filter for the removal of aerosols and one gas sorption
filter.

2.4 Reactions of the Adsorbate in Iodine Sorption Materials

Elemental iodine is trapped in charcoal by adsorption and chemisorption. Because of the favorable adsorption behavior of elemental iodine, the porous structure and the extremely large inner surface of the charcoal used (approximately 800 - 1000 m^2/g according to BET), adsorption already constitutes an important mechanism of elemental iodine trapping. The adsorption-desorption equilibria established between iodine in the gas phase and iodine on the surface of charcoal also in humid air result in the removal of elemental iodine with high removal efficiencies.

The adsorption of methyl iodide in charcoal is impaired by the simultaneous adsorption of water vapor from moist air to such an extent that non-impregnated charcoal will no longer perform efficiently at humidities of the air > 30 % R.H. at room temperature. Rising temperatures and humidities of the air will shift the adsorption-desorption equilibrium even more to the unfavorable side.

^{131}I present as methyl iodide is bound to the present type of iodine filter charcoal by isotope exchange or the formation of quarternary ammonia salts $/¯1¯/$. For isotope exchange the activated charcoal is impregnated with major quantities (0.5 - 5 wt. %) of inactive iodine or iodine salts ($^{127}I_2$, $K^{127}I$). The following exchange occurs (which is formulated for the most important iodine isotope for safety considerations, ^{131}I, but likewise applies to other iodine isotopes):

^{127}I (on the charcoal) + CH_3^{131}I (in the gas phase) $\rightleftharpoons$ ^{131}I (on the charcoal) + CH_3^{127}I (in the gas phase).

So radioactive methyl iodide practically enters the impregnated charcoal bed and leaves it again as inactive methyl iodide. Hydrolysis and adsorption are additional removal mechanisms of minor importance. The step determining the rate of isotope exchange is the adsorption and desorption of methyl iodide on the surface of the charcoal. The decontamination factor which can be achieved by isotope exchange is determined by the ratio between inactive ^{127}I and ^{131}I on the surface of the charcoal. Impregnation of the charcoal with large quantities of ^{127}I, in relation to which the quantities of ^{131}I adsorbed are extremely small, shifts the position of the equilibrium in the reaction mentioned above to the far righthand side. Given sufficiently high reaction rates and sufficiently long stay time of the gaseous effluents in the charcoal, high decontamination factors may be achieved.

Another very efficient impregnating agent for the removal of methyl iodide is triethylene diamine:

$$\text{N(CH}_2\text{CH}_2)_3\text{N} + 2\ CH_3 J \longrightarrow [\text{(CH}_3)\text{N}^+(\text{CH}_2\text{CH}_2)_3\text{N}^+(\text{CH}_3)]\ 2\ J^-$$

In this case the removal effect is caused by the bonding to charcoal of methyl iodide in the form of the quaternary ammonia salt constituted by triethylene diamine and methyl iodide. The reactions formulated for methyl iodide to some extent also apply to other organic iodine compounds, such as higher alkyl halides.

Since charcoal cannot be used at high temperature and in gaseous effluents containing NO_x, other sorption materials have been developed on an inorganic basis. Silver or silver-plated copper wire will remove elemental iodine with a high removal efficiency from a fresh metal surface but fail when it comes to iodine in the form of methyl iodide. If silver is built into molecular sieves, such as the Linde-type 13 X molecular sieve, and activated in this way, it is a suitable reaction partner also for removing iodine in the form of methyl iodide /2, 3 7. Molecular sieves containing silver are highly suitable iodine sorption materials also for use at high temperatures. However, the material is very expensive, because high silver contents are necessary to achieve sufficient removal efficiencies and it is mainly the silver contained in an outside layer which will react in the usually short stay times.

Other non-burnable material which can be used for trapping iodine and methyl iodide is produced on the basis of silicic acid or aluminum oxide and impregnated with silver nitrate. With a relatively small amount of silver these materials are capable of achieving high removal efficiencies; in addition, they lend themselves well to the use with the gaseous effluents containing NO_x of reprocessing plants.

Silver built into molecular sieves by ion exchange

reacts with elemental iodine producing silver iodide.
Methyl iodide is also converted into AgI; the reaction
products detected are dimethyl ether and methanol $\underline{/}$ 4 $\underline{/}$.

Silver nitrate (as an impregnation on SiO_2 and Al_2O_3 carrier
materials) probably reacts with elemental iodine according
to the following formulae:

$$AgNO_3 + I_2 \dashrightarrow AgI + INO_3$$

$$2\ INO_3 + AgNO_3 \longrightarrow AgIO_3 + 3\ NO_2 + 1/2\ I_2$$

$$INO_3 \longrightarrow NO_2 + 1/2\ O_2 + 1/2\ I_2$$

Hence, iodine is bound as silver iodide and silver iodate.

Methyl iodide and other alkyl halides react with the silver
nitrate impregnation according to the following formula:

$$AgNO_3 + R - I \longrightarrow (NO_3 \cdot RI \cdot Ag) \longrightarrow RNO_3 + AgI$$

also producing silver iodide.

The end products of the reaction of iodine or methyl iodide
with silver and silver nitrate, respectively, are highly
insoluble. Hence, materials impregnated in this way are
particularly useful for the ultimate storage of ^{129}I which
is produced in large quantities in the gaseous effluents of
reprocessing plants and should be removed from the biocycle.

Out of the multitude of factors influencing the trapping of
iodine and methyl iodide in filters I would like to mention
only some of the most important ones, namely the depth of
the bed and the humidity of the air.

2.5 Dependence of the Decontamination Factor on the Depth
    of the Sorption Bed and the Relative Humidity of the
    Gaseous Effluent

The adsorption of water onto the sorption material above all
influences the removal of radioactive iodine present as
methyl iodide. The removal efficiency of filter beds is
indicated by the decontamination factor (DF). The following
relation applies:

$$DF = \frac{c_o}{c}\ ;$$

c_o = concentration of $CH_3{}^{131}I$ in the unfiltered air;

c = concentration of $CH_3{}^{131}I$ in the filtered air.

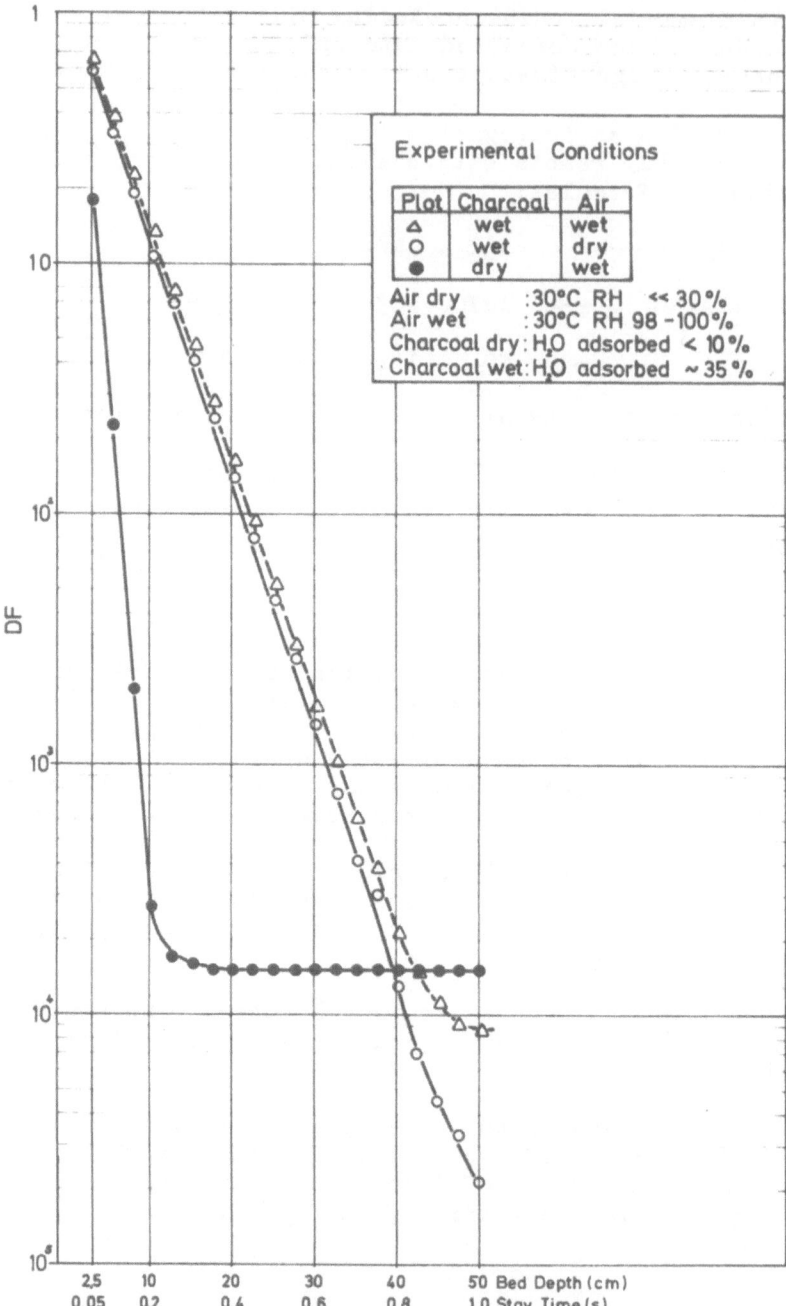

Decontamination Factor of Filter Beds from Impregnated
Activated Charcoal for ^{131}I (loaded in Form of CH_3 ^{131}I)

Fig. 2

Fig. 2 shows the decontamination factor of sorption beds consisting of impregnated activated charcoal for ^{131}I added as methyl iodide, as a function of the depth of the bed. Experiments were performed with previously moistened and dry impregnated charcoal adsorption materials and humid and dry air as the carrier gases for methyl iodide. The curves show that the decontamination effect of impregnated charcoal after preliminary moistening (the water vapor adsorption equilibrium was established at 98 - 100 % R.H. and 30°C) is practically the same with respect to iodine activity in dry ($\ll$ 30 % R.H.) and humid air (98 - 100 % R.H.) up to high decontamination factors.

If dry impregnated charcoal is exposed to methyl iodide in humid air for short periods of time (to prevent larger quantities of water from being adsorbed by the charcoal), much higher decontamination factors result, but the curve shows a bend after the maximum activity has been removed. It must be assumed that there is a reaction of a small fraction of methyl iodide into a compound difficult to trap as a consequence of a reaction with airborne water vapor. The same phenomenon is detected in the removal test conducted at high humidity of the air and impregnated charcoal previously moistened.

Experiments with dry impregnated charcoal and dry air (not shown in the curve) also resulted in high decontamination factors. The curve showed a bend later than in the case of exposure to humid air; accordingly, a lower penetration was achieved through a corresponding depth of the active charcoal bed.

The following statements can be made with respect to the removal performance of iodine filters on the basis of the curves shown in Fig. 2:

- Adsorption of water vapor to impregnated charcoal results in increased penetration.

- A filter previously moistened by adsorbed water will trap radioactive iodine to which it has been exposed as methyl iodide with a reduced removal efficiency also in dry air.

- If a dry impregnated charcoal filter is exposed to radio-active methyl iodide in moist air, a high initial removal efficiency must be anticipated which will decrease with water adsorption onto the surface of the charcoal.

- The most adverse removal conditions for methyl iodide are given by high relative humidity of the air and charcoal previously moistened.

Other removal experiments not described in detail in this paper indicate that

- impregnated charcoal whose macropores had been filled by condensed water (e.g., if the temperature in the charcoal had decreased below the dew point) loses its ability of removing methyl iodide with reasonable removal efficiencies.

In normal operation the humidity of the gaseous effluents of nuclear facilities is on the order of < 70 % R.H., especially when we are talking about the exhaust air of air conditioned rooms which may be entered during operation. Under incident conditions, e.g., due to the failure of seals or lines carrying water or steam, there may be extremely high humidities of the air and it must be anticipated that water vapor condensate is produced. In order to control such incidents iodine filters are equipped with devices to reduce the humidity of the air, such as condensers, humidity separators and preheaters. A reduction of the relative humidity of the air from 98 - 100 % R.H. to 70 % R.H. for instance reduced the penetration of 5 cm of an impregnated charcoal bed (made of KI-impregnated Norit CG 0.8) from 3.2 % to 0.09 % (measured after the adsorption-desorption equilibrium of water vapor on the charcoal had been reached).

The straight sections in the curve in Fig. 2 can be described by a simple exponential function.

The drop in concentration of $CH_3^{131}I$ with the depth of bed (x) follows the following formula:

$$c = c_o \cdot e^{-K''x} ; \qquad (1)$$

It has turned out to be useful to introduce in (1) the respective stay time (t) in the place of the depth of bed

$$c = c_o \cdot e^{-K't} ; \qquad (2)$$

$$\frac{c_o}{c} = e^{K't} = DF; $$

$$\frac{\log DF}{t} = K; \qquad (3)$$

K = constant dependent on the charcoal batch and the removal conditions,

t = stay time; stay time = $\dfrac{\text{bulk volume of charcoal}}{\text{volume flow of gaseous effluent}}$

K is the so-called "index of performance" $\underline{/1/}$, a measure of the removal efficiency of a specific batch of charcoal under fixed removal conditions.

(3) can be used to calculate the decontamination factor achieved at a specific stay time.

The application of (3) to filter designs in which the superficial linear gas velocity greatly deviates from the value under which K had been determined requires some caution. It is also apparent from Fig. 2 that an extrapolation to extremely long stay times is not possible for practical purposes because of the bend in the curve.

2.6 The Effect of the Fission Product Iodine Composition upon the Design of Iodine Filters

The decontamination factors achieved by impregnated charcoal relative to elemental iodine usually are at least 2 orders of magnitude higher than for methyl iodide. For practical purposes this means that the design of an iodine filter is determined by the methyl iodide fraction contained in the fission iodine mixture. This may vary within broad limits and achieve values > 50 % in the gaseous effluent of nuclear facilities.

Elemental iodine (with a vapor pressure comparatively low as against methyl iodide and high chemical reactivity) is partly adsorbed onto surfaces on its transport from the point of release to the stack, partly it also reacts with airborne impurities, especially with organic compounds. The fraction of elemental iodine in the fission product iodine mixture therefore usually decreases with increasing distance from the point of release. It is not possible to set up any binding rules about the composition of the fission iodine mixture in the filter intake air. Accordingly, it can be assumed for a safe filter design that all the iodine present in the air is present as methyl iodide. It should be pointed out that methyl iodide in this connection must be regarded as being representative of a group of iodine compounds difficult to trap (e.g., higher alkyl iodides).

2.7 Aging and Poisoning of Iodine Sorption Materials

While mechanical leaks in iodine filters of suitable filter designs can largely be prevented, aging and poisoning of iodine sorption materials so far have not been brought under statisfactory control. The removal efficiency of iodine filters decreases as the time of operation increases due to the loading with filter poisons which reduce the effective surface of the sorption material by deposits or react with the impregnation and the base material substances, respectively, so as to generate undesirable products. Moreover, and independently, there may be changes in impregnation and in the charcoal which reduce the removal efficiency also of iodine filters not in operation or of the spare batches of sorption material.

The reduction in removal efficiency of the sorption material with time, a summation effect stemming from a multitude of causes (see above), will be summarized here under the general term of "aging".

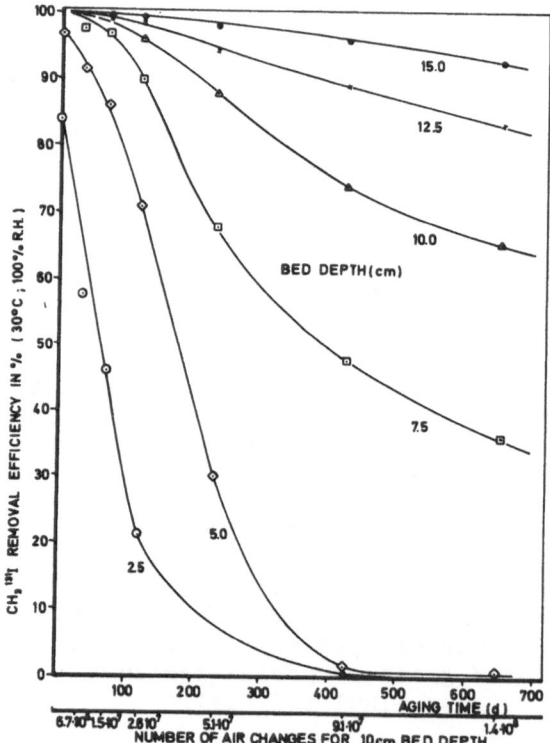

Fig. 3   Aging of impregnated charcoal (NORIT RKI -I) in air from industrial environment

Fig. 3 shows the aging behavior of 6 consecutive filter beds made of KI-impregnated charcoal of 2.5 cm bed depth each. The diagram is a plot of the removal efficiency for ^{131}I, to which the filter was exposed as methyl iodide, as a function of service life and the number of air cycles (referred to 10 cm of bed depth).

During aging the linear air velocity in the charcoal beds was 25 cm/sec. The curve shown in Fig. 3 indicates a relatively fast decrease in removal efficiency at low bed depth, while deeper charcoal beds age much more slowly.

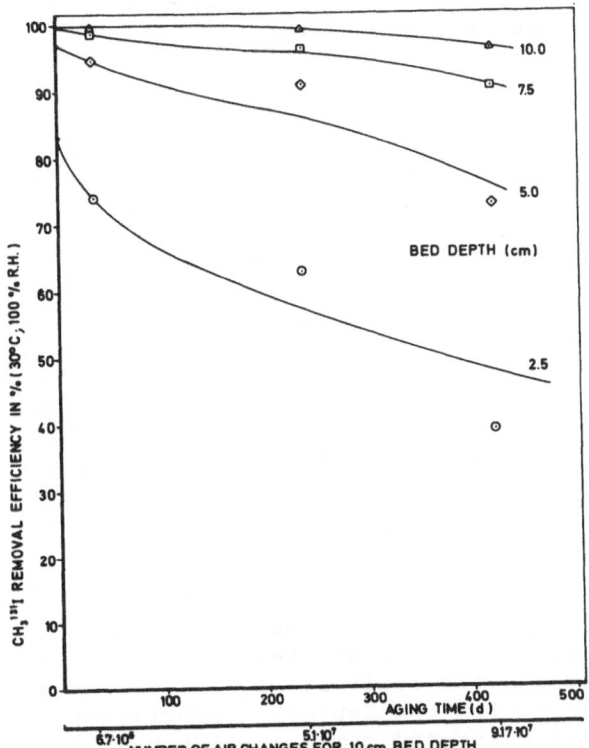

Fig.4 Aging of impregnated charcoal (NORIT RKI-I) in prefiltered air

Fig. 4 shows the removal efficiency for charcoal of the same batch as that shown in Fig. 3; the aging conditions are also alike.

Upstream of the filter beds consisting of KI-impregnated charcoal additional filter beds of non-impregnated activated

charcoal were set up for preliminary removal of filter
pollutants. A comparison between Figs. 3 and 4 shows the
extension of the filter service life achieved in this way.

From our experience it is not possible to describe "aging"
by a mathematical function sufficiently well, since loading
of the impregnated charcoal with filter pollutants, which
decisively contributes to the reduction in the removal
efficiency of impregnated charcoal, varies within broad limits
in terms of place and time and cannot be estimated sufficient]
well.

From extensive aging measurements performed on various iodine
sorption materials it can be stated that the use of flat
filter beds, which are frequently designed without any
consideration for aging, is technically wrong and also
inadmissible from the safety point of view. The financial
expenditure involved in the frequent exchange and, if
necessary, subsequent testing by far exceeds the savings made
in terms of iodine sorption material in one individual filter
unit.

2.8 Designing the Charcoal Layer of Iodine Filters

The charcoal bed of iodine filters in general must be so
dimensioned that the removal efficiencies for radioactively
labeled methyl iodide as the testing agent should be beetween
90 % and 99.9 %, depending upon the use of the system, as
determined by a laboratory test of the charcoal batch in
humid air. A minimum removal efficiency of 90 % is a value
customary for recirculating air-cleanup in the containment
of a reactor, while exhaust air filters require removal
efficiencies of 99 % and more.

The usual bed depths of the charcoal beds are between 2.5
and 50 cm, stay times vary between 0.1 and 2 sec. More
recently, exhaust air filters are characterized by a more
pronounced tendency towards the application of longer stay
times ($\gg$ 0.1 sec) and greater bed depths, respectively. In
addition, provisions for aging and poisoning are made in the
required quantity of charcoal which may give rise to a
pronounced extension of the stay time. Superficial linear
gas velocities between 10 and 50 cm/sec are achieved in the
sorption layer. Loading the impregnated charcoal with
elemental iodine should not exceed 1 mg I_2/g charcoal and
100 μg $CH_3 I$/g charcoal with charcoal impregnated with KI.

2.9 Testing Iodine Filters

Commissioning and follow-up tests of iodine filters are required to prove the removal efficiency relative to fission product iodine. The following tests are usually performed:

(1) Before commissioning samples of the original batch of the activated charcoal used in the iodine filter are tested in the laboratory under standard conditions or under the most adverse operating conditions to be anticipated, mainly with respect to stay time, relative humidity of the air, temperature, pressure and loading. The test agent is radioactively labeled methyl iodide, sometimes elemental iodine.

(2) Prior to commissioning of iodine filter systems used for cleaning the gaseous effluents of a nuclear power station during or after incidents also in-place tests are performed which primarily are conducted to assess mechanical leaks. The testing agent also is radioactively labeled methyl iodide or elemental iodine; in the US also freons are used. These in-place tests are carried out under normal operating conditions of the iodine filter systems. To some extent also filters are checked in place which are used only in normal operation.

(3) At intervals of two months up to several years and after the replacement of charcoal filter cells or an exchange of the charcoal in-place retesting must be performed.

(4) After specified operating periods the removal efficiency of iodine filters is verified by laboratory investigations of charcoal samples from the original filter or from bypass sections to the original filter. These studies are less expensive than in-place tests. The same tests are applied as in the laboratory tests of unused charcoal mentioned initially.

The laboratory tests of newly impregnated charcoals at high relative humidities of air showed marked fluctuations in the removal efficiency.

Under prolonged storage conditions, also in the absence of air, the removal efficiency may greatly decrease. Frequently charcoal in iodine filters shows a marked reduction in removal efficiency, sometimes even within a few months. The adsorption of solvents, light oil /⁻5_7 and other organic compounds has been found as one of the reasons of the decrease in efficiency, but additional effects must also be assumed.

Deep bed filters of a suitable design showed no mechanical leaks in a number of tests in Germany. Iodine filter units equipped with charcoal filter cells (filter elements) showed mechanical leaks due to deficiencies in the contacting device and at the gaskets and the gasket seats. In general it is safe to say that iodine filters equipped with deep bed filters usually show better results in the in-place test than facilities equipped with charcoal filter cells because of their simple design, the absence of clamping devices and of long sealed passages between the contaminated and the clean air sides of the filters and the large bed depths which in most cases can be achieved with a minimum of expense.

2.10 Future Developments of Iodine Filters for Special Applications

Under this heading two aspects should be mentioned above all:

(1) The development of post accident recirculating air-cleanup containment filters.

(2) The development of off gas filters for reprocessing plants.

Ad (1):

Recirculating type filters for post accident conditions are used for the quick reduction of aerosol and iodine concentrations in the containment. Spray systems serving the same purpose can be used only in the containment of water cooled reactors. Their unavoidable disadvantages, such as the generation of corrosion, increased generation of hydrogen, spreading of contamination, production of very large quantities of highly contaminated solutions, the narrow limits of the scrubbing effect by organic iodine compounds and the extensive consequences of a use not justified by the extent of the accident will not occur in the use of recirculating type filter systems during incidents. These filter units must fulfill special requirements for the following reasons:

- high temperature,
- high dose rate,
- high mechanical stress,
- when used in the containment of water cooled reactors, extremely high humidity of the air and condensate production.

Since certain inorganic iodine sorption materials are not flammable and will not desorb iodine which has been removed and are sufficiently resistant to high temperature and ionizing radiation, their use is likely to fulfill the requirements mentioned above.

Ad (2):

Future reprocessing plants are designed to a throughput of up to 1500 t/a. If it is assumed that the fuel to be reprocessed comes from light water reactors and is unloaded at a burnup of 45.000 MWd/t, an annual quantity of 536 kg iodine is produced in the reprocessing plant, consisting of ^{129}I (75 Ci), ^{131}I (21 Ci at 210 d cooling time) and the inactive ^{127}I. The anticipated environmental contamination caused by a reprocessing plant can be avoided only by suitable iodine filters in the off-gas stream. Development work along these lines serves the following purposes:

- Fission product iodine should be removed from the off-gas stream in a form easy to be handled as far upstream in the reprocessing line as possible.

- A sorption material should be developed which will remove iodine from the NO_x-bearing off-gases of reprocessing plants. (Charcoal cannot be used under these conditions because ot the hazard of poisoning and of fire).

- The sorption material must be sufficiently resistant to the filter pollutants occurring in the off-gas of reprocessing plants.

- Iodine should be bound in the sorption material firmly enough to generate a product capable of ultimate storage. The partial vapor pressure of iodine on the sorbent must be negligible from the safety point of view.

- The sorbent must be sufficiently resistant to wear; no major quantities of aerosol should be generated during the replacement.

- The iodine filter should achieve a decontamination factor of 99.9 %.

- The iodine filter must have sufficient capacity for removal of relatively large amounts of iodine from the gaseous effluents in continuous operation.

- The sorbent of the iodine filter must be safely loaded and removed with the respective safety rules being observed.

A highly efficient aerosol filter stage connected upstream should prevent contamination of the sorbens with other fission products and fuel aerosols to such an extent that handling can be carried out with a feasible amount of expenditure.

- The price of the iodine sorption material should be as low as possible because a regeneration process should be avoided.

Extensive laboratory studies $\underline{/}^-$6, 7$\underline{\ }$7 seem to indicate that an iodine filter fulfilling the above requirements can be manufactured using $AgNO_3$ -impregnated iodine sorption materials, such as Bayer Sorbens AC 6120.

3. THE DISCHARGE OF NOBLE GASES THROUGH FISSION GAS HOLDUP BEDS

In order to reduce the environmental burden caused by noble gases most of the effort so far has been concentrated upon the retention of isotopes with relatively short half-lifes. A method which has been applied to a large number of boiling water reactors and more recently also to pressurized water reactors is the use of so-called fission gas holdup beds. In tubes or vessels filled with activated charcoal, through which the gaseous effluent is passed, the noble fission gases xenon and krypton are delayed relative to the carrier gas stream as a consequence of the adsorption-desorption equilibria which will be produced. This is a technical application of gas chromatography. The radioactive xenon and krypton isotopes largely disintegrate within the holdup bed, only ^{85}Kr ($T_{1/2}$ = 10.7 a) is released with a practically undiminished activity.

Besides the noble gases activation gases from the coolant will be present in the off-gas, especially the isotopes of oxygen, nitrogen and argon, which are also passed through the holdup beds.

3.1 Fission Gas Holdup Beds in Boiling Water Reactors (BWR)

According to the present state of the art, the off-gas systems of BWR's are equipped with activated charcoal fission gas holdup beds.

The off-gases from the turbine condenser, which are composed of the activation gases, the radiolytic gases such as hydrogen and oxygen, the noble fission gases from the fuel and air entering through leakages into the condenser, will be removed by vapor jets and passed to a recombiner. In that unit the hydrogen and oxygen diluted by the steam used in the steam jet will be catalytically burned to water. Afterwards the water vapor is precipitated by a cooler and the residual air-gas mixture is passed through a holdup bed in which the short-lived radioisotopes already disintegrate to a large extent. Afterwards the off-gases are passed through a HEPA filter which removes most of the aerosol activity. Prior to introduction into the holdup bed, the gas is dried, e.g. by cooling to approximately -20°C and trapping of the condensed moisture.

The source strength of the radioactive noble gases and the activation gases is given by the specific reactor conditions and the fuel element leakage. The tables outlined below include values which can be assigned to a BWR of 1000 MW$_e$. Some 200 - 300 m^3/h of off-gas are taken in from the turbine condenser. The off-gas can consist of approximately 90 % by volume of radiolysis gases and 10 % by volume of air. In addition, there are the activiation and fission gases whose fraction is less than 0.001 % by volume. Since radiolysis gases are composed in a stochiometric ratio of 2 : 1 (corresponding to H$_2$O), the large initial volume of off-gas of serveral 100 of m^3/h will be reduced to a few 10 of m^3/h after recombination.

The source strengths of the activation gases listed in Table I apply after a decay period of 1 min which is assumed as designed transport time from the outlet of the reactor to the outlet of the steam jets. Only the non-condensable part of the gases is being considered, i.e. the fraction really emerging from the turbine condenser. Because of the very different data provided by various reactor stations and because of the design differences, the source strengths given in Table I should be used as rough outlines only.

It should be indicated that ^{16}N because of its high source strength and its high energy γ-radiation may cause the highest radiation exposure to the off-gas system.

Table I: Activation gases in the off-gas of a BWR

Gas	Half-life	Source strength (μCi/sec)
^3_H	12.3 a	0.3
$^{13}_N$	10 min	6×10^3
$^{16}_N$	7.4 sec	5×10^6
$^{17}_N$	4.2 sec	1
$^{15}_O$	2.1 min	0.2
$^{19}_O$	29 sec	3
$^{41}_{Ar}$	1.8 h	2

Table II shows the noble fission gases whose source strengths must be assumed to exceed 0.1 μCi/sec. Again one minute has been assumed for the transport time from the outlet of the reactor to the outlet of the steam jet. The values shown in Table II for noble gases and iodine may differ by several orders of magnitude in specific cases.

Table II: Fission gases in the gaseous effluents of a BWR

Gas	Half-life	Source strength (μCi/sec)
$^{91}_{Kr}$	8.6 s	5.1×10^3
$^{144}_{Xe}$	9.0 s	1.0×10^1
$^{140}_{Xe}$	13.6 s	2.8×10^4
$^{90}_{Kr}$	32.3 s	1.5×10^5
$^{139}_{Xe}$	40.0 s	2.0×10^5

Gas	Half-life	Source strength (μCi/sec)
^{89}Kr	3.2 min	2.1×10^5
^{137}Xe	3.8 min	2.5×10^5
^{138}Xe	14.2 min	1.8×10^5
$^{135\,m}$Xe	15.7 min	5.2×10^4
^{87}Kr	76 min	4.0×10^4
$^{83\,m}$Kr	1.9 h	6.8×10^3
^{88}Kr	2.8 h	4.0×10^4
$^{85\,m}$Kr	4.4 h	1.2×10^4
^{135}Xe	9.2 h	4.4×10^4
$^{133\,m}$Xe	2.3 d	5.8×10^2
^{133}Xe	5.3 d	1.6×10^4
$^{131\,m}$Xe	12.0 d	3.0×10^1
^{85}Kr	10.7 a	3.0×10^1
^{131}I	8.05 d	6
^{132}I - ^{135}I	$\leqslant$20.8 d	20 - 30

3.2 Fission Gas Holdup Beds in Pressurized Water Reactors (PWR)

During normal operation of a PWR no off-gas must be released from the primary cooling system. Only when air enters the cover gas a continuous release of gas is necessary for volume equalization. This is possibly lower than in a boiling water reactor in which the turbine condenser must be continuously emptied. Apart from this air leaking the

542

quantities of gas to be removed as a consequence of loading
various vessels in a pressurized water reactor must be taken
into account in designing the holdup beds. The volume of
noble fission gases practically plays no role.

The holdup beds at a PWR is fed only a partial stream of
the cover gas circuit, which must be maintained for removal
and combustion of the radiolytically produced hydrogen from
the primary circuit. Usually N_2 is used as cover gas.

The optimum use of the holdup bed with respect to the
environmental burden caused by radioactivity in a PWR depends
on the amount of primary coolant leakage into the room air
of the reactor building. If there is a higher leakage,
continuous degasing should be performed to establish a
relatively low equilibrium activity in the coolant. In the
case of minor leakages it is more advantageous to not
degasify the primary coolant because then less noble gas
activity will be passed on to the stack via the room air than
will be passed through the holdup beds in the case of
degasing.

3.3 Design of an Activated Charcoal Fission Gas Holdup Bed

The amount of charcoal required for a specific holdup time
and an off-gas stream can be calculated by the following
simple relation:

$$t_m = k_d \times \frac{M}{Q} \; ; \qquad (4)$$

where

t_m = mean stay time of the adsorbate (min);

k_d = dynamic adsorption coefficient, characteristic of
the type of activated charcoal, carrier gas and the
adsorbate (cm^3/g);

M = quantity of activated charcoal (g);

Q = volume stream of carrier gas (cm^3/min).

k_d is best determined by direct measurement of the holdup in
a test system using the original charcoal while the
conditions anticipated with respect to volume flow,
temperature, pressure and composition of the off-gas are set.

Since k_d will increase as the temperature decreases, it is
useful to run holdup beds at a reduced temperature. Moreover,
it is possible to raise the total pressure of the off-gas

in the holdup beds. Both measures greatly reduce the required amount of activated charcoal $/\bar{8}\_7$. If air is used as the carrier gas, pressure increase up to 6 ata result in an approximately linear increase in k_d with the pressure. If helium is the carrier gas, there is a comparatively greater increase in k_d with the pressure.

Typical values of k_d for Xe are between 700 and 900 cm^3/g at normal pressure and 25°C; for Kr values between 44 and 55 cm^3/g are quoted under the same conditions. Intercomparison measurements showed higher values for activated charcoal made of coconut shells compared with activated charcoal made of kerosene or coke $/\bar{9}\_7$.

The simultaneous adsorption of water vapor onto the activated charcoal greatly influences the holdup time. In order to achieve good holdup times the charcoal should be kept as dry as possible.

Today holdup beds are designed for holdup times of up to 60 days for xenon and 4 days for krypton. The amount of activated charcoal used for that may be up to 50 t. In this design the release of noble gas activity, which may achieve some 10^6 Ci/a in large boiling reactors, will be reduced to values between less than 1.000 and 20.000 Ci/a.

The advantage of holdup beds above all consists in their simple design and their extremely high availability and operational reliability. More recent studies $/\bar{1}0\_7$ have shown that even in the case of a rupture of a pressurized holdup bed a relatively slight fraction of the adsorbed activity is desorbed. Cooling in the gas and the adsorber material due to the heat of desorption may in addition bring about a delay in the activity release in a rapid depressurization event. One general disadvantage of holdup beds is their relatively large construction volume, especially if they must be operated at atmospheric pressure and room temperature.

REFERENCES

$/\bar{1}\_7$ D.A. Collins, L.R. Taylor, R. Taylor, The Development of Impregnated Charcoals for Trapping Methyl Iodide at High Humidity, TRG 1300(W) (1967).

$/\bar{2}\_7$ W.J. Maeck, D.T. Pence, J.H. Keller, A Highly Efficient Inorganic Adsorber for Airborne Iodine Species, CONF-680821 (1968), p. 185.

$/\bar{3}\_7$ J.G. Wilhelm, Trapping of Fission Product Iodine with Silver-Impregnated Molecular Sieves, Actes du Congrès Saclay, 4. - 6. Nov. (1969), p. 265.

544

/⁻4_7 Ch. Donner, T. Tamberg, Zur Abscheidung von Methyl-jodid und Aethyljodid an Silber-Zeolithen, Atom-wirtschaft 16, 28 (1971).

/⁻5_7 J.G. Wilhelm, K. Gerlach, Ergebnisse und Erfahrungen aus der Untersuchung von Jodfiltern, Proceedings: Seminar on Iodine Filter Testing, Doc. V/559, p. 465 - 479, March 1974.

/⁻6_7 J.G. Wilhelm, H. Schüttelkopf, An Inorganic Adsorber Material for Off-Gas Cleaning in Fuel Reprocessing Plants, CONF-720823 (1972), p. 540.

/⁻7_7 J. Furrer, J.G. Wilhelm, Jodfilterung aus der Abluft von Wiederaufarbeitungsanlagen, Proceedings: Seminar on Iodine Filter Testing, Doc. V/559, p. 185 - 198 (1974).

/⁻8_7 K. Förster, Delaying Radioactive Fission Product Inert Gases in Cover Gas and Off-Gas Streams of Reactors by Means of Activated Charcoal Delay Lines, Kerntechnik, 13, 5, p. 214 (1971).

/⁻9_7 D.P. Siegwarth, C.K. Neulander, R.T. Pao and M. Siegler, Measurement of Dynamic Adsorption Coefficients for Noble Gases on Activated Carbon, CONF-720823, p. 28 - 46 (1972).

/⁻10_7 Dwight W. Underhill, Effect of Rupture in a Fission Holdup Bed, CONF-720823, p. 60 - 69 (1972).

LIST OF PARTICIPANTS

NAME | ADDRESS
(at the time of the meeting)

ABRAHAMSEN, J.   Helly-Hansen A/S, 1501 Moss, Norway

ALBRECHT, G.   Helly-Hansen A/S, 1501 Moss, Norway

BARTHOMEUF, D.   Institut de Recherches sur la Catalyse,
                     69100 Villeurbanne, France

BERG, R.   Institutt for Atomenergi, 2007 Kjeller,
                     Norway

BJERKE, W.   Norsk Hydro A/S, 3901 Porsgrunn, Norway

BJØRSETH, O.   SINTEF, avd. 21, 7034 Trondheim-NTH, Norway

BRENCHLEY, D.L.   Norsk Institutt for Luftforskning, 2007
                     Kjeller, Norway

CENTINCELIK, M.   Nuclear Energy Forum of Turkey, P.O.B. 37,
                     Bakanlikhar, Ankara, Turkey

CARTRAUD, P.   Lab. de Chimie Physique, Université de
                     Poitiers, 86022 Poitiers, France

DAVIES, R.A.   Chemviron, P.O.B. 17, Ixelles 1, 1050
                     Brussels, Belgium

EKEDAHL, E.   Research Institute of National Defence,
                     Box 416, S-172 04 Sundbyberg 1, Sweden

ELMALEH, S.   Lab. de Genie Chimique, Place Eugene
                     Bataillon, 34060 Montpellier, France

FJELLESTAD, K.   OECD Halden Reactor Project, P.O.B. 173,
                     1751 Halden, Norway

GULBRANDSEN, A.   Institutt for Atomenergi, P.O.B. 40,
                     2007 Kjeller, Norway

FREEDMAN, A.M.   Dept. of Chemical Engineering, University
                     of Surrey, Guildford, Surrey, England

HALMØ, T.   SINTEF, 7034 Trondheim-NTH, Norway

HANCKE, I.   SINTEF, avd. 21, 7034 Trondheim-NTH, Norway

HARDWICK, W.H.   Process Technology Division, AERE Harwell,
                     Didcot, Berks. England

HJERMSTAD, H.P.	Norwegian Defence Research Establishment, P.O.B. 25, 2007 Kjeller, Norway
HOLMBERG, K.E.	AB Atomenergi, Fack, 611 01 Nyköping 1, Sweden
HOUGHTON, F.R.	Sutcliffe, Speakman & Co., Ltd., Leigh, Lancs., England
ILSØE-PETTERSEN, O.	The Technical University of Denmark, 2800 Lyngby, Denmark
JOON, K.	Reactor Centrum Nederland, p.t. OECD Halden Reactor Project, P.O.B. 173, 1751 Halden, Norway
JUHOLA, A.J.	MSA Research Corporation, Evans City, Penn. 16033, USA
KABAT, M.J.	Ontario Hydro - Pickering G.S., P.O.B. 175, Pickering, Ontario, Canada
KRILL, H.	Lurgi Apparate-Technik GmbH, Gervinusstr. 17-19, 6000 Frankfurt/M., Germany
LANGE, N.E.K.	The Technical University of Denmark, 2800 Lyngby, Denmark
LANZA, S.	Istituto Impianti Nucleari, Via Diotisalvi 2, 56100 Pisa, Italy
LIESER, K.H.	Technische Hochschule Darmstadt, 61 Darmstadt, Germany
MAGGS, F.A.P.	Chemical Defence Establishment, Porton Down, Salisbury, Wilts. SP4 0JQ, England
MEDEMA, J.	Chemisch Laboratorium TNO, Lange Kleiweg 13, 2100 Rijswijk, The Netherlands
MELTZER, A.	Inst. für Reaktorsicherheit der Technische Überwachungs-Vereine e.V., Glockengasse 2, 5000 Köln 1, Germany
NEEFJES, G.A.	Institutt for Atomenergi, P.O.B. 40, 2007 Kjeller, Norway
NESET, K.M.	Institutt for Atomenergi, P.O.B. 40, 2007 Kjeller, Norway
NILSEN, T.S.	Army Material Command, HFK, Oslo Mil - Oslo 1, Norway
OZIL, P.	Lab. d'Adsorption et Reaction de Gaz sur Solides, St. Martin d'Heres, France
PALLIER, G.	CECA, Carbonisation et Charbons Actifs, B.P. 66, 78140 Velizy-Villacoublay, France

PERSSON, G.   Research Institute for National Defence, Box 416, S-172 04, Sunbyberg 1, Sweden

RANNESTAD, A.   NATO, Scientific Affairs Division, 1110 Brussels, Belgium

RODRIGUES, A.E.   Universidade de Luanda, Dept. de Engenharia Quimica, Luando, Angola

ROUET, J.   CECA, Carbonisation et Charbons Actifs, B.P. 66, 78140 Velizy-Villacoublay, France

SAMMON, D.C.   Membrane Chemistry, AERE, Harwell, Didcot, Berks, England /

SINHUBER, D.   Delbag-Luftfilter GmbH, Schweidnitzer Srr. 11-16, 1 Berlin 31, Germany

SMALL, S.   Noratom-Norcontrol A/S, Holmenveien 20, Oslo 3, Norway

SMITH, T.H.   National Institute of Public Health, Dept. of Environmental Toxicology, Oslo 1, Norway

SOARES, L.D.J.S.   Universidade de Luanda, Dept. de Engenharia Quimica, Luanda, Angola

STAIB, B.O.   Prestfoss Fabrikker A/S, Stortingsgt. 12, Oslo 1, Norway

STOUTE, J.R.D.   Reactor Centrum Nederland, Petten (NH), The Netherlands

SVEEN, S.   Army Material Command, HFK, Oslo Mil - Oslo 1, Norway

TAYLOR, L.R.   UKAEA Windscale Works, Sellafield, Cumberland, U.K.

URBANIC, J.E.   Calgon Corporation, Calgon Center, Box 1346 Pittsburgh, Penn. 15230, USA

de VOOYS, F.   Norit N.V. Research Laboratorium, Textielweg 15, Amersfoort, The Netherlands

WEBER, W.J., Jr.   Water Resources Program, The University of Michigan, Ann Arbor, Michigan 48104, USA

WILHELM, J.G.   Gesellschaft für Kernforschung mbH, Postfach 3640, 7500 Karlsruhe 1, Germany

WINTER, K.   Ceagfilter und Entstaubungstechnik GmbH, Paradiesstr. 35, 46 Dormund-Sölde, Germany

WORLEY, F.L., Jr.   Chemical Engineering Dept., University of Houston, Houston, Texas 7704, USA

van Zelm, M.   Chemisch Laboratorium, TNO, Lange Kleiweg 13, 2100 Rijswijk, The Netherlands

Aas, S.   Institutt for Atomenergi, P.O.B. 40, 2007 Kjeller, Norway